MOLECULAR
BIOLOGY
INTELLIGENCE
UNIT

IgE Receptor (FcεRI) Function in Mast Cells and Basophils

Majed M. Hamawy, Ph.D.
National Institute of Dental Research
National Institutes of Health
Bethesda, Maryland, U.S.A.

Springer-Verlag Berlin Heidelberg GmbH

Molecular Biology Intelligence Unit

IgE RECEPTOR (FcεRI) FUNCTION IN MAST CELLS AND BASOPHILS

Originally published by Springer-Verlag Berlin Heidelberg New York in 1997

Springer

International
ISBN 978-3-662-22024-5

While the authors, editors and publisher believe that drug selection and dosage and the specifications and usage of equipment and devices, as set forth in this book, are in accord with current recommendations and practice at the time of publication, they make no warranty, expressed or implied, with respect to material described in this book. In view of the ongoing research, equipment development, changes in governmental regulations and the rapid accumulation of information relating to the biomedical sciences, the reader is urged to carefully review and evaluate the information provided herein.

Library of Congress Cataloging-in-Publication Data

Hamawy, Majed M., 1958-
IGE receptor (FceRI) function in mast cells and basophils / Majed M. Hamawy
p. cm. — (Molecular biology intelligence unit)
Includes bibliographical references and index.
ISBN 978-3-662-22024-5 ISBN 978-3-662-22022-1 (eBook)
DOI 10.1007/978-3-662-22022-1
1. Immunoglobulin E—Receptors. 2. Fc receptors. I. Title. II. Series.
QR186.8.E2H36 1996
616.07'95—dc20

96-22040
CIP

Publisher's Note

R.G. Landes Company publishes six book series: *Medical Intelligence Unit, Molecular Biology Intelligence Unit, Neuroscience Intelligence Unit, Tissue Engineering Intelligence Unit, Biotechnology Intelligence Unit* and *Environmental Intelligence Unit.* The authors of our books are acknowledged leaders in their fields and the topics are unique. Almost without exception, no other similar books exist on these topics.

Our goal is to publish books in important and rapidly changing areas of bioscience and environment for sophisticated researchers and clinicians. To achieve this goal, we have accelerated our publishing program to conform to the fast pace in which information grows in bioscience. Most of our books are published within 90 to 120 days of receipt of the manuscript. We would like to thank our readers for their continuing interest and welcome any comments or suggestions they may have for future books.

Shyamali Ghosh
Publications Director
R.G. Landes Company

Dedication

To my mother Nawal.

CONTENTS

EDITOR

Majed M. Hamawy, Ph.D.
Laboratory of Immunology
National Institute of Dental Research
National Institutes of Health
Bethesda, Maryland, U.S.A.
Chapter 9, 12

CONTRIBUTORS

Michael A. Beaven, Ph.D.
Laboratory of Molecular Immunology
National Heart, Lung, and Blood Institute
National Institutes of Health
Bethesda, Maryland, U.S.A.
Chapter 4

Marc Benhamou, Ph.D.
Laboratoire d'Immunologie
91 Bd de l'Hopital
Paris, France
Chapter 3

Ross I. Brinkworth, Ph.D.
The Centre for Drug Design and Development
The University of Queensland
Brisbane, Queensland, Australia
Chapter 2

Alasdair M. Gilfillan, Ph.D.
Department of Inflammation and Autoimmune Diseases
Hoffmann La Roche, Inc.
Nutley, New Jersey, U.S.A.
Chapter 10

John R. Gordon, Ph.D.
Department of Veterinary Microbiology
University of Saskatchewan
Saskatoon, Canada
Chapter 11

P. Mark Hogarth, Ph.D.
The Austin Research Institute
The Austin and Repatriation Medical Centre
Heidelberg, Victoria, Australia
Chapter 2

Mark D. Hulett, Ph.D.
The Austin Research Institute
The Austin and Repatriation Medical Centre
Heidelberg, Victoria, Australia
Chapter 2

Tatiana Kassessinoff, Ph.D.
Laboratory of Molecular Immunology
National Heart, Lung, and Blood Institute
National Institutes of Health
Bethesda, Maryland, U.S.A.
Chapter 4

Anna Koffer, Ph.D.
Physiology Department
University College London
London, United Kingdom
Chapter 5

Donald MacGlashan, Jr., M.D., Ph.D.
John Hopkins Asthma and Allergy Center
Baltimore, Maryland, U.S.A.
Chapter 13

Henry Metzger, M.D.
Section on Chemical Immunology, Arthritis and Rheumatism Branch
National Institute of Arthritis and Musculoskeletal and Skin Diseases
National Institutes of Health
Bethesda, Maryland, U.S.A.
Chapter 1

Janet Oliver, Ph.D.
Department of Pathology and Cancer Research and Treatment Center
Division of Cell Pathology
University of New Mexico Health Sciences Center
Albuquerque, New Mexico, U.S.A.
Chapter 8

Janet R. Pfeiffer, Ph.D.
Department of Pathology and Cancer Research and Treatment Center
Division of Cell Pathology
University of New Mexico Health Sciences Center
Albuquerque, New Mexico, U.S.A.
Chapter 8

A.J. (Lin) Rigby, B.Sc.
The Austin Research Institute
The Austin and Repatriation Medical Centre
Heidelberg, Victoria, Australia
Chapter 2

Juan Rivera, Ph.D.
Section on Chemical Immunology
Arthritis and Rheumatism Branch
National Institute of Arthritis and Musculoskeletal and Skin Diseases
National Institutes of Health
Bethesda, Maryland, U.S.A.
Chapter 6

William D. Swaim, Ph.D.
National Institute of Dental Research
National Institutes of Health
Bethesda, Maryland, U.S.A.
Chapter 9

Mark Swieter, Ph.D.
Laboratory of Immunology
National Institute of Dental Research
National Institutes of Health
Bethesda, Maryland, U.S.A.
Chapter 7

Bridget S. Wilson, Ph.D.
Department of Pathology and Cancer Research and Treatment Center
Division of Cell Pathology
University of New Mexico Health Sciences Center
Albuquerque, New Mexico, U.S.A.
Chapter 8

PREFACE

The high affinity IgE receptor (FcεRI) plays a central role in allergic diseases including asthma, allergic rhinitis, atopic dermatitis, and food and drug allergy. Therefore, manipulating the function of this receptor is important for controlling the onset of allergic diseases.

Progress in understanding the structure and function of the FcεRI has been rapid in the past two decades, giving rise to an enormous amount of information. Advances in research on FcεRI make it necessary to have all of the accumulated information in an easily accessible form. The objective of this book is to fulfill that need by providing comprehensive and up-to-date information on all aspects of the FcεRI.

I am honored to be the editor of a book written by an exceptional group of scientists. I am grateful for their cooperation and for their valuable suggestions and assistance in making revisions and reviewing proofs.

Majed M. Hamawy

CHAPTER 1

FcεRI: Perspective

Henry Metzger

The most exciting change in cell biology over recent decades has been the start made in defining the molecular changes that account for cellular phenomena. This progress is providing us with a firmer comprehension of the degree of complexity of living systems—an appreciation that was simply not possible as long as one was limited to studying global phenomena. If I had to pick a single point in time where molecular analysis began in immunology I would select 1939 when Elvin Kabat published the characterization of antibodies.[1] It took another 25 years to identify the principal antibody (IgE) that mediates allergic responses[2] and another 20 years after that until one had at least the outline of the cell surface receptor (FcεRI) through which IgE acts and that serves as the focus of this text.[3]

If different individuals were asked to cite the key events that led to our present state of understanding about the structure and function of FcεRI they would undoubtedly come up with somewhat different lists; here is mine:

I would first of all stress some indirect contributions: studies, such as those of Lefkowitz et al,[4] that led to the modern concept of cell surface receptors in general, of "Fc receptors,"[5] and of the nature of cell membranes.[6] The marvelous generic methods of protein chemistry and genetics must also get their due. More directly, the rat basophilic leukemia cells[7] (now known to be analogues of mucosal mast cells[8]) provided the first source of cells that made feasible the initial biochemical investigations,[9,10] and the development of rat[11] and mouse[12] IgE myelomas markedly enhanced the practicality of investigating this system. As with the other members of the family of "multichain immune recognition receptors" (MIRR),[13] the characterization of the ligand-binding subunit was relatively straightforward.[14-16] However, the analysis and even the recognition of the other subunits proved more difficult because they are held together by the same types of interactions one has to disrupt in order to solubilize the receptor.[17] Nevertheless, the β and γ subunits were ultimately identified.[18,19] Cloning of the genes for the α, β and γ subunits in turn[20-23] led to the current model for FcεRI[23] (Fig. 1.1). Transforming

IgE Receptor (FcεRI) Function in Mast Cells and Basophils, edited by Majed M. Hamawy.

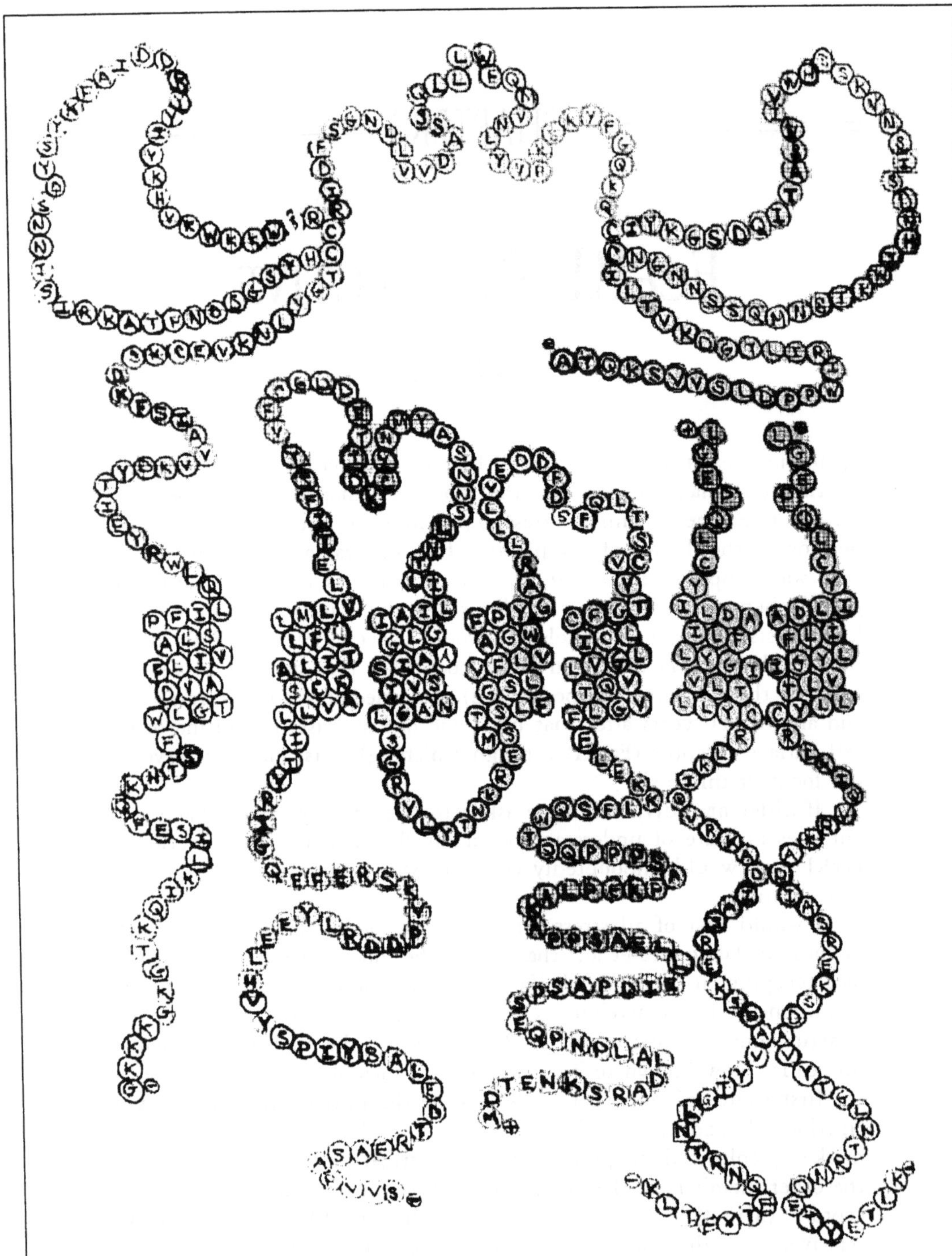

Fig. 1.1. Photograph of original sketch made by the author in preparation for a manuscript[23] that described the cloning of the γ subunit, the expression of the transfected receptors and the proposal for the arrangement of the subunits with respect to each other and to the plasma membrane.

the schematic into a realistic three-dimensional portrait is an eagerly awaited key event.

On the functional side, the studies of Prausnitz and Küstner initiated the modern era.[24] The key experiments of Ishizaka and Campbell[25] (and the elegant ones by Levine in a related system[26]) demonstrated the obligatory role of aggregation. Documentation of the univalence of FcεRI[27] and the use of well-defined oligomers of IgE rigorously substantiated earlier indications that dimers of receptors were the minimal "unit signals."[28] The use of anti-receptor antibodies firmly established that the aggregation of FcεRI rather than the IgE-receptor complex was the necessary and sufficient event to stimulate the cellular response.[29,30] (The central role of FcεRI in anaphylactic responses more generally, was more recently confirmed by gene "knockout" experiments.[31,32]) Numerous studies identified various functional parameters and these were summarized in an important review by Osler et al[33] but the intensive biochemical investigations that characterize the present era were heralded by the studies of Beaven et al.[34,35] Finally, if I had to choose a single paper that initiated the most recent phase, I would select the one by Benhamou et al.[36] Foremost, because rather than just demonstrating an "activity" or an increase in some cellular constituent, it was the first to associate the functional response with a modification in a particular molecule. Second, because it presented the first solid data that in this system like so many others, phosphorylation of tyrosines was a critical initiating event. Earlier studies had suggested that a tyrosine kinase might be associated with the receptor,[37] but it was not until the study of Eiseman and Bolen[38] that the evidence was considerably strengthened and the src-like kinase lyn was identified as the likely culprit. Shortly thereafter Paolini et al made the important observation that the receptor itself becomes phosphorylated in an aggregation-dependent way.[39] Recently, we have provided evidence for a plausible model by which the topological approximation of two or more receptors initiates the cascade of molecular modifications. We propose that by literally bringing substrate to enzyme, aggregation allows constitutively associated lyn kinase to transphosphorylate adjacent receptor(s).[40,41]

The experiments that first implied that FcεRI stimulates cells to do more than degranulate, i.e., it stimulates the synthesis of potent lipidic mediators[42] and turns on genes that lead to the production of cytokines,[43] clearly also broke important new ground. This aspect of the receptor's function may play a significant role in the pathological consequences that follow stimulation of the receptor.

At the time I am writing this introduction I have not seen the chapters that will follow it, but even their titles give an idea of the kinds of questions currently being investigated. What about the future? Identification of the components that contribute to the variety of biochemical cascades stimulated by the receptor is proceeding. As I see it, the major challenge will be to understand how these reactions are integrated with each other and how they are influenced by the various developmental stages of the mast cell. It is a daunting enterprise since even the initial stages of the reaction are difficult to model quantitatively.[44]

It brings me back to the start of this chapter. We are beginning to get an appreciation of just how complicated these systems are. Although the answer is "enormously!," there is no obvious barrier that appears insurmountable. Furthermore, it seems likely that long before we reach the desired end, our partial knowledge will permit us to develop novel therapeutic approaches that will amply repay those who are funding this voyage of discovery.

References

1. Tiselius A, Kabat EA. An electrophoretic study of immune sera and purified antibody preparations. J Exp Med 1939; 69:119-31.
2. Ishizaka K, Ishizaka T, Hornbrook MM. Physico-chemical properties of human reaginic antibody. IV. Presence of a unique

immunoglobulin as a carrier of reaginic activity. J Immunol 1966; 97:75-85.
3. Metzger H, Kinet J-P, Perez-Montfort R, Rivnay B, Wank SA. A tetrameric model for the structure of the mast cell receptor with high affinity for IgE. In: Yamamura Y, Tada T, eds. Progress in Immunology V. Orlando, FL: Academic Press, 1983: 493-501.
4. Lefkowitz RJ, Roth J, Pricer W et al. ACTH receptors in the adrenal: specific binding of ACTH-^{125}I and its relation to adenyl cyclase. Proc Natl Acad Sci USA 1970; 65:745-52.
5. Uhr JW. Passive sensitization of lymphocytes and macrophages by antigen-antibody complexes. Proc Natl Acad Sci USA 1965; 54:1599-606.
6. Singer SJ, Nicolson GL. The fluid mosaic model of the structure of cell membranes. Science 1972; 175:720-31.
7. Eccleston E, Leonard BJ, Lowe JS et al. Basophilic leukaemia in the albino rat and a demonstration of the basopoietin. Nature New Biol 1973; 244:73-6.
8. Seldin DC, Adelman S, Austen KF et al. Homology of the rat basophilic leukemia cell and the rat mucosal mast cell. Proc Natl Acad Sci USA 1985; 82:3871-4.
9. Kulczycki Jr A, Isersky C, Metzger H. The interaction of IgE with rat basophilic leukemia cells. I. Evidence for specific binding of IgE. J Exp Med 1974; 139:600-16.
10. Kulczycki Jr A, Metzger H. The interaction of IgE with rat basophilic leukemia cells. II. Quantitative aspects of the binding reaction. J Exp Med 1974; 140:1676-95.
11. Bazin H, Querinjean P, Beckers A et al. Transplantable immunoglobulin-secreting tumours in rats. IV. Sixty-three IgE-secreting immunocytoma tumours. Immunology 1974; 26:713-23.
12. Liu FT, Bohn JW, Ferry EL et al. Monoclonal dinitrophenyl-specific murine IgE antibody: preparation, isolation, and characterization. J Immunol 1980; 124:2728-37.
13. Keegan AD, Paul WE. Multichain immune recognition receptors: Similarities in structure and signaling pathways. Immunol Today 1992; 13:63-8.
14. Kulczycki Jr A, TA McNearney, Parker CW. The rat basophilic leukemia cell receptor for IgE. I. Characterization as a glycoprotein. J Immunol 1976; 117:661-5.
15. Conrad DH, Froese A. Characterization of the target cell receptor for IgE. II. Polyacrylamide gel analysis of the surface IgE receptor from normal rat mast cells and from rat basophilic leukemia cells. J Immunol 1976; 116:319-26.
16. Newman SA, Rossi G, Metzger H. Molecular weight and valence of the cell-surface receptor for immunoglobulin E. Proc Natl Acad Sci USA 1977; 74:869-72.
17. Rivnay B, Wank SA, Poy G et al. Phospholipids stabilize the interaction between the alpha and beta subunits of the solubilized receptor for immunoglobulin E. Biochemistry 1982; 21:6922-27.
18. Holowka D, Hartmann H, Kanellopoulos J et al. Association of the receptor for immunoglobulin E with an endogenous polypeptide on rat basophilic leukemia cells. J Recept Res 1980; 1:41-68.
19. Perez-Montfort R, Kinet J-P, Metzger H. A previously unrecognized subunit of the receptor for immunoglobulin E. Biochemistry 1983; 22:5722-8.
20. Kinet J-P, Metzger H, Hakimi J et al. A cDNA presumptively coding for the alpha subunit of the receptor with high affinity for immunoglobulin E. Biochemistry 1987; 26:4605-10.
21. Shimizu A, Tepler I, Benfey PN et al. Human and rat mast cell high-affinity immunoglobulin E receptors: characterization of putative alpha-chain gene products. Proc Natl Acad Sci USA 1988; 85:1907-11.
22. Kinet J-P, Blank U, Ra C et al. Isolation and characterization of cDNAs coding for the beta subunit of the high-affinity receptor for immunoglobulin E. Proc Natl Acad Sci USA 1988; 85:6483-7.
23. Blank U, Ra C, Miller L et al. Complete structure and expression in transfected cells of high affinity IgE receptor. Nature 1989; 337:187-9.
24. Prausnitz C, Küstner H. Studien über die Überentfindlichkeit. Zentrabl Bakt 1921; 86:160-9.

25. Ishizaka K, Ishizaka T, Campbell DH. The biological activity of soluble antigen-antibody complexes II. Physical properties of soluble complexes having skin-irritating activity. J Exp Med 1959; 109:127-43.
26. Levine BB, Redmond AP. The nature of the antigen-antibody complexes initiating the specific wheal-and-flare reaction in sensitized man. J Clin Invest 1968; 47:556-67.
27. Mendoza GR, Metzger H. Distribution and valency of receptor for IgE on rodent mast cells and related tumour cells. Nature 1976; 264:548-50.
28. Segal DM, Taurog JD, Metzger H. Dimeric immunoglobulin E serves as a unit signal for mast cell degranulation. Proc Natl Acad Sci USA 1977; 74:2993-7.
29. Ishizaka T, Ishizaka K. Triggering of histamine release from rat mast cells by divalent antibodies against IgE-receptors. J Immunol 1978; 120:800-5.
30. Isersky C, Taurog JD, Poy G et al. Triggering of cultured neoplastic mast cells by antibodies to the receptor for IgE. J Immunol 1978; 121:549-58.
31. Dombrowicz D, Flamand V, Brigman KK et al. Abolition of anaphylaxis by targeted disruption of the high affinity immunoglobulin E receptor α chain gene. Cell 1994; 75:969-76.
32. Takai T, Li M, Sylvestre D et al. FcR gamma chain deletion results in pleiotrophic effector cell defects. Cell 1994; 76:519-29.
33. Osler AG, Lichtenstein LM, Levy DA. In vitro studies of human reaginic allergy. Adv Immunol 1968; 8:183-231.
34. Beaven MA, Rogers J, Moore JP et al. The mechanism of the calcium signal and correlation with histamine release in 2H3 cells. J Biol Chem 1984; 259:7129-36.
35. Beaven MA, Moore JP, Smith GA et al. The calcium signal and phosphatidylinositol breakdown in 2H3 cells. J Biol Chem 1984; 259:7137-42.
36. Benhamou M, Gutkind JS, Robbins KC et al. Tyrosine phosphorylation coupled to IgE receptor-mediated signal transduction and histamine release. Proc Natl Acad Sci USA 1990; 87:5327-30.
37. Quarto R, Metzger H. The receptor for immunoglobulin E: examination for kinase activity and as a substrate for kinases. Mol Immunol 1986; 23:1215-23.
38. Eiseman E, Bolen JB. Engagement of the high-affinity IgE receptor activates *src* protein-related tyrosine kinases. Nature 1992; 355:78-80.
39. Paolini R, Jouvin M-H, Kinet J-P. Phosphorylation and dephosphorylation of the high-affinity receptor for immunoglobulin E immediately after receptor engagement and disengagement. Nature 1991; 353: 855-8.
40. Pribluda VS, Pribluda C, Metzger H. Transphosphorylation as the mechanism by which the high-affinity receptor for IgE is phosphorylated upon aggregation. Proc Natl Acad Sci USA 1994; 91:11246-50.
41. Yamashita T, Mao S-Y, Metzger H. Aggregation of the high-affinity IgE receptor and enhanced activity of $p53/56^{lyn}$ protein-tyrosine kinase. Proc Natl Acad Sci USA 1994; 91:11251-5.
42. Orange RP, Valentine MD, Austen KF. Antigen-induced release of slow reacting substance of anaphylaxis (SRS-A rat) in rats prepared with homologous antibody. J Exp Med 1968; 127:767-82.
43. Plaut M, Pierce JH, Watson CJ et al. Mast cell lines produce lymphokines in response to crosslinkage of FcεRI or to calcium ionophores. Nature 1989; 339:64-7.
44. Wofsy C, Kent UM, Mao S-Y et al. Kinetics of tyrosine phosphorylation when IgE dimers bind to Fcε receptors on rat basophilic leukemia cells. J Biol Chem 1995; 270:20264-72.

CHAPTER 2

The Structural Basis of the Interaction of IgE and FcεRI

Lin Rigby, Mark D. Hulett,
Ross I. Brinkworth and P. Mark Hogarth

INTRODUCTION

FcεRI is capable of inducing one of the most powerful and violent pharmacological responses known. Indeed the association of IgE with FcεRI, and subsequent aggregation is a most important interaction in the induction of human disease, and causes more chronic misery (in the West at least) than the engagement of any other immunological receptor. As approximately one in five people are afflicted with IgE dependent allergies—most notably allergic rhinitis or 'hay fever' and asthma—there has been a large effort made by many groups in studying this receptor, its ligand and the consequences of its activation. The impetus to study this receptor probably stems from its pathological role rather than its physiological one, which is still somewhat undefined, but with evidence pointing to an anti-parasite role.[1]

Yet despite the importance of the interaction of IgE and FcεRI, the widespread interest in this interaction (or its consequences) and its obvious importance to the pharmaceutical industry, we still do not know the structure of these molecules. At the time of writing there is still no published three-dimensional structure of either IgE or its high affinity receptor, or the 'Holy Grail' of all in this field—a structure of FcεRI interacting with its ligand. Nonetheless many groups have contributed to the substantial progress in defining the overall structure of IgE and FcεRI, the regions of these that influence their interaction, and how these structures may sit on the cell surface.

In this chapter we have attempted to bring together the work of many on the structural aspects of FcεRI and IgE. We have drawn on

IgE Receptor (FcεRI) Function in Mast Cells and Basophils,
edited by Majed M. Hamawy.

models of IgE and FcεRI in order to compile and present the data of many groups in defining the nature of this interaction. The FcR nomenclature throughout this review follows that proposed by IUIS/WHO's subcommittee on nomenclature for Fc receptors.[2]

STRUCTURE OF FcεRI AND IgE

FcεRI

Biochemical and molecular structure

The FcεRI has been characterized in three different species, human, rat and mouse. Structurally, it has been defined as a tetrameric surface glycoprotein, comprising an α chain, a β chain and a dimer of disulphide linked γ chains. The α chain has been shown to bind IgE ($>10^{10}M^{-1}$), and to be homologous to the FcγR, while the β and γ chains are involved in cell signaling and surface expression. The characteristics of this receptor from mouse, rat and human are summarized in Table 2.1.

The earliest cell culture model used to demonstrate the interaction of IgE and FcεRI was that of the rat basophilic cell line RBL-2H3[3] which was used for the initial attempts to purify the receptor.[4-6] When a single polypeptide chain with an apparent molecular weight of 50-60 kDa was initially purified (the α chain), the conditions were too stringent to maintain the noncovalent association between the four subunits. This α chain was later found to be associated with other polypeptides, the β chain of 33 kDa,[7-9] and the two disulphide linked γ chains of 7-9 kDa each.[9,10] Purification of the intact complex required less stringent and protective phospholipid, or submicellular concentrations of detergent.[11,12]

Further characterization of the receptor involved monoclonal antibodies (mAbs) raised against the FcεRI complex, and

Table 2.1. Characteristics of FcεRI

Characteristic	Human FcεRI	Mouse FcεRI	Rat FcεRI
Affinity for IgE (Ka)	$10^{10}M^{-1}$ (human)	$10^{10}M^{-1}$ (mouse)	$10^{10}M^{-1}$ (rat)
Specificity	human, mouse or rat IgE	mouse or rat IgE only	rat or mouse IgE only
Associated subunits	α, β, γ	α, β, γ	α, β, γ
Forms of receptor	αβγ2 (Mast cells) αγ2 (Langerhans cells)	αβγ2	αβγ2
Molecular weight (kDa)			
-apparent	45-65 α, 32 β, 7-9 γ	45-65 α, 32 β, 7-9 γ	45-65 α, 32 β, 7-9 γ
-protein backbone	26.4 α, 25.9 β, 7.8 γ	25.8 α, 25.9 β, 7.8 γ	25.2 α, 27 β, 7.8 γ
Chromosome	1q23 α, 11q13 β, 1q23 γ	1 α, 19 β, 1 γ	ND
Cell distribution	Mast cells Basophils Langerhans cells Eosinophils Monocytes (activated)	Mast cells Basophils	Mast cells Basophils

ND = Not Detected

showed the FcεRI-α chain to be highly glycosylated and expressed on the exterior surface of the cell. The β subunit and the disulphide linked γ dimer were not glycosylated.[13,14] Human, rat and mouse cDNAs have been cloned for each of the FcεRI subunits, their molecular structures determined (Table 2.1), and a stylized model of the topology of the FcεRI complex proposed (see Fig. 1.1).[15]

α *Subunit*

FcεRI-α subunit cDNAs have been cloned in the human, mouse and rat,[16-20] and a single transcript has been identified for the human and mouse. In the rat one major full length clone has been identified, as well as three shorter clones comprising varying deletions from the full length FcεRI-α chain gene,[16,18,19] which probably arise from alternative splicing of the genomic DNA.[18,21] The FcεRI-α cDNA encodes a transmembrane glycoprotein comprising an extracellular region of two immunoglobulin(Ig)-like domains, (human 180, mouse 181, rat 181 amino acids), a 21 amino acid transmembrane region, and a cytoplasmic tail of 22-31 amino acids (human 31, mouse 25, rat 22).

A comparison of the amino acid sequences of the individual subunits of FcεRI of different species and also with other receptors is revealing.* The sequence identity between the encoded FcεRI-α proteins of mouse and rat compared to human is approximately 38%, displaying the least homology of the FcεRI subunits (α, β and γ) between species.[20] The FcεRI-α cytoplasmic domains are the least conserved between species (16%) but the FcεRI-α chain transmembrane region is highly conserved (62%), with the eight amino acid motif (LFAVDTGL) present in each species. This sequence is essential for interaction with the FcεRI-γ subunit,[22] and is also conserved in the transmembrane regions of mouse, rat, and human FcγRIII which like FcεRI-α require association with the FcεRI-γ chain for cell surface expression.[23] From the point of view of the interaction of the FcεRI-α subunit with Ig, a comparison with other FcR is interesting.

The FcεRI-α chains exhibit homology with all the FcγR but have the greatest identity with the FcγRIII subclass.[20,24] Comparisons of mouse FcεRI-α and FcγRIII demonstrate this relationship since they exhibit 33% amino acid identity over their entire sequence, with 35% and 48% identity in the extracellular and transmembrane regions, respectively. Of the 95 residues conserved in the mouse, rat and human extracellular region of the FcεRI-α chains, 61 are found in both mouse and human FcγRIII; which suggests that the 34 residues unique to FcεRI-α could be involved in IgE specific binding.[24] Indeed it is clear that all the leukocyte FcR that are Ig-superfamily members have a common evolutionary history.

The rat, mouse and human FcεRI-α chain genes have been cloned and share a common structure of 5 exons; one each encoding the 5'UTR, leader sequence, and each of the two Ig-like domains, and a single exon for the transmembrane and cytoplasmic regions, and 3'UTR.[21,25,26] Indeed the human and mouse FcγRIII genes also exhibit a similar five exon gene structure, suggesting the FcεRI-α and FcγRIII genes arose by gene duplication from a common ancestor.[27,28] The human FcεRI-α chain gene has been mapped to chromosome 1q23 which also contains the low affinity FcγR genes (Table 2.1).[29]

β *Subunit*

cDNA clones of the FcεRI-β subunit have been isolated from the mouse,[20] rat[30] and human,[31] and encode proteins of 243, 235 and 244 amino acids, respectively, which are highly conserved exhibiting 91% amino acid homology and 69% identity. Two mRNA species have been observed in the mouse and rat (1.75 and 2.7 kb) arising from alternate polyadenylation, and two

* The amino acid sequence of the human FcεRI-α chain used in this chapter is from Genbank, accession number X06948, and the numbering system is as shown in Appendix 1.

transcripts of human FcεRI-β have also been described, detected as a doublet around 3.9 kb.[31]

The rat FcεRI-β amino acid sequence, like the mouse and human, has four hydrophobic segments suggestive of transmembrane domains, and no leader peptide. Hydrophobicity plots and studies with monoclonal antibodies suggest that both the N and C termini are positioned in the cytoplasm. Rat FcεRI-β has two linear cytoplasmic domains of 46 and 62 residues[15] and the C-terminal cytoplasmic domain has been shown spectroscopically and by NMR to contain three α-helices possibly important in interaction with the membrane and/or other cytoplasmic domains.[32,33]

The human FcεRI-β gene has been isolated[32] and mapped to chromosome 11q13.[34] It appears to be a single copy gene of seven exons spanning 10 kb. The 5'UTR and part of the N-terminal cytoplasmic domain comprise exon 1, the initial transmembrane region is encoded by exons 2 and 3, the second transmembrane region by exons 3 and 4, the third by exon 5, the fourth by exon 6, and the C-terminal cytoplasmic domain and 3'UTR by exon 7.[31] The mouse and rat FcεRI-β genes have not been isolated, but in the mouse it is believed to be a single gene linked to the *Ly-1* locus which maps to chromosome 19 (Table 2.1).[35,36]

γ *Subunit*

The FcεRI-γ subunit is expressed as a disulphide bonded dimer, with the disulphide bond formed between the cysteine residues located at the N-terminus of the transmembrane region.[22] cDNA clones have been isolated in the rat,[15] mouse,[20] and human[37] and encode highly related proteins that have 86% amino acid identity (Table 2.1).[37]

The FcεRI-γ subunit is an integral membrane protein with an extracellular region of five amino acids, a transmembrane region of 21 amino acids (as does FcεRI-α), and a cytoplasmic region of 36 amino acids.[37] It belongs to a small family of molecules which also contains the ζ and η chains of the TCR complex which are usually homodimeric, but can form heterodimers.[38] Studies using circular dichroism and Fourier-transform infrared spectroscopy (FRET) of the 62 amino acid rat γ subunit[39] agree with the proposed hydrophobicity plot based structure of a five amino acid extracellular domain, an α helix in the transmembrane region and a 36 amino acid cytoplasmic tail.[15] FcεRI-γ has also been shown to associate with rat and mouse FcγRIII, human FcγRIIIa and FcγRII,[40,41] human FcγRI,[42-44] mouse FcγRI (A Gavin, personal communication) and human FcαRI.[45] In addition FcεRI-γ also associates with the TCR/CD3 complex of human and murine T cells.[38,46,47] The human gene has been mapped to chromosome 1q23[29] and that of the mouse to chromosome 1 along with the gene encoding the closely related TCR ζ protein.[36] These regions also contain the respective FcεRI-α and FcγR low affinity loci which could indicate some coordinate regulation of these genes.

Expression of the FcεRI-αβ(γ)$_2$ complex on the cell surface

Cloning of FcεRI-α, β and γ cDNA has been crucial for our understanding of the receptor structure and its expression. Early experimentation on rat FcεRI cDNA indicated that the FcεRI-α subunit expression could not be detected on the surface of COS-7 cells by transfection of FcεRI-α cDNA alone,[16,18] and this is also true for human[17,18] and mouse[20] FcεRI-α. Rat and mouse FcεRI-α chains are not efficiently expressed without cotransfection of the β and γ subunits. However, surface expression can be increased if the β and γ cDNAs of a different species are cotransfected. Optimal expression requires all three subunits to be of the same species. In these species the β subunit has a greater effect on expression than the γ subunit.[20] Surprisingly, the requirements for expression of the human FcεRI-α subunit are different from those of rodent FcεRI-α in that the β chain is not required for cell surface expression. The γ subunit is required and can be of human, rat or mouse origin,[20,37,48] and cotransfection of FcεRI-α and γ subunits with either rat or human β subunit cDNA does not increase expression.

Indeed only 20% of receptors show the FcεRI-αβ(γ)2 phenotype under these conditions, the remaining 80% have been shown to express as FcεRI-α(γ)2.[31] It has been postulated that human FcεRI could exist in vivo in the FcεRI-α(γ)2 form[20,31] without the FcεRI-β subunit, and the existence of this complex has recently been confirmed on human Langerhans cells,[49] where it is capable of signal transduction resulting in calcium mobilization and FcεRI internalization. It has been suggested that the FcεRI-β subunit may be involved in unknown mechanisms related to activation or release of preformed cellular granules found in mast cells, which do not occur in Langerhans cells.[49] The lack of a β subunit in Langerhans cells may not therefore be surprising. However, it cannot be ruled out that Langerhans cells may possess a novel β-like subunit.

Mutageneses of FcεRI-α, β and γ subunits have identified some of the structural requirements for assembly and expression of the FcεRI complex. In transfection studies using COS-7 cells, expression of the rat FcεRI complex has been shown to be unaffected by removal of any single FcεRI-α, β or γ subunit cytoplasmic domain. Indeed, the removal of all the FcεRI-α, β and γ cytoplasmic domains reduces but does not eliminate expression—although it does affect aggregation.[50] Conversely, mutations in the rat FcεRI complex transmembrane regions resulted in either reduced or eliminated expression. These results suggest that the transmembrane regions are critical for rat receptor expression and a model encompassing this has been suggested.[22] Unlike the rat, cotransfection of human FcεRI-α with a truncated rat FcεRI-γ has been shown to ablate expression, indicating that in the human the cytoplasmic domain of FcεRI-γ is necessary for assembly and/or expression and strengthening suggestions that the rat and human receptors assemble in a different manner.

Human FcγRIIIa has been shown to associate with FcεRI-γ and TCR/CD3-ζ subunits.[51] These related homodimer subunits are highly homologous as are the transmembrane regions of the FcεRI-α chains. Thus ζ could be expected to be able to associate with FcεRI-α and substitute for the γ dimer. In the *Xenopus* oocyte expression system rat FcεRI fails to express when FcεRI-α and β chain RNA are injected, although fully functional expression occurs when FcεRI-α, β and γ RNA are coinjected. The receptor function and recognition by mAbs can also be restored by coinjecting human CD3-ζ RNA with the rat FcεRI-α and β RNA. However, this would not be expected to occur in vivo, as CD3-ζ does not occur naturally with FcεRI-α.[52]

Mutation experiments on FcεRI-α have shown that it can be expressed without the FcεRI-γ subunit by fusing the extracellular domains of FcεRI-α to the transmembrane region and cytoplasmic domain of another receptor molecule, for instance p55 IL-2R,[53] FcγRIIa,[54] FcεRI-γ (MH Kershaw, personal communication). These chimeric receptors all express on the cell surface and can bind IgE with an affinity comparable to wild type.

It has also been shown that glycosylation of a soluble form of FcεRI-α is necessary for the proper folding of FcεRI in the endoplasmic recticulum, and also its secretion.[55]

IgE

Like all immunoglobulins, IgE consists of two disulphide bonded heavy chains and two light chains bound to the heavy chains. The ε heavy chain is composed of approximately 550 amino acids that make up a variable region and four constant domains (Cε), unlike IgG, IgD and IgA which have three. The amino acid sequence of the human IgE Fc portion (Cε2 Cε3 Cε4) is shown in Appendix 2. It appears that Cε3 and Cε4 are equivalent domains to the IgG constant domains Cγ2 and Cγ3 which make up the classic IgG-Fc region. The second constant domain Cε2, which is presumed to have the structure of a typical immunoglobulin constant domain, appears to take the place of the hinge region that occurs in IgG, IgD or IgA immunoglobulins.

The heavy chains are attached to each other by two interchain disulphide bonds and their location is unusual in that they occur in distinct regions being either end of a domain. These disulphide bonds occur at Cys241* (between Cε1 and Cε2) and Cys328 (between Cε2 and Cε3) which interact with Cys241 and Cys328, respectively, on the other heavy chain[56] to form parallel disulphide bonds as suggested by Dorrington and Bennich[57] and confirmed experimentally by Helm et al,[56] not the diagonal disulphide bonding that was proposed by Padlan and Davies.[58] It should be noted that the region surrounding the Cys328 is important in binding to FcεRI and in maintaining the conformation of human IgE.[56,59] In rat IgE it is the intrachain disulphide bonds that confer structural stability to the molecule.[60]

Considerable effort has been made to define the structure of IgE; as yet no authentic structure is available but a large body of data has been generated that provides some insight into the probable overall configuration of IgE (Fig. 2.1). Furthermore, with the development of molecular modeling based on authentic homologous structures, several models of IgE have been developed which are supported by experimentally derived data (see below).

It was originally suggested that IgE was a planar molecule,[56,58,61,62] with an extended section between Cε2 and Cε3, that could provide segmental flexibility to this region. The present consensus is that in solution IgE is a compact, fairly rigid and bent Ig (Fig. 2.1),[57,63-66] and FRET measurements indicate that the antigen combining sites lie approximately 7.1 nM from the C-terminus (Cε4) rather than the 17.1 nM expected for a planar Y-shaped molecule like IgG. Furthermore, the apparent compact and rigid nature of IgE is clearly different from IgG1 which is much more flexible (probably because of the hinge) and able to adopt multiple conformations.[67] These studies have been used to refine the model of IgE[56] and indicate that fitting the experimental data to the model requires bending at the junction of Cε2/Cε3 and/or Cε3/Cε4, but most likely at Cε2/Cε3.[68] This also requires a change in the relative orientation of the Cε3 domains by rotating these 10° asymmetrically, and also the rotation of Cε2 by 40-50°.[68] Thus it appears likely that rodent and human IgE is bent in a U-shape, in agreement with the previous experimental data[64,67,69] and recent hydrodynamic studies.[70]

Despite the fact that IgE is more rigid than other Igs, there is some considerable segmental flexibility; this is mostly manifest in the Fab arms where rotation and wagging take place.[69] The asymmetric nature of IgE probably explains the observed 1:1 stoichiometry of IgE:FcεRI interaction, and has several functional implications.[63,64,67,69-73] First, the FcεRI binding site on the concave surface of IgE may be obscured as Cε2 and/or Cε4 are in closer proximity to it, this then allows only the convex surface to bind to FcεRI. Second, the rotation at Cε2/Cε3 may also assist in 'spacing' the antigen combining sites of the Fab arms away from the FcεRI binding site, thereby minimizing any steric hindrance by antigen of IgE/ FcεRI binding. Third, one binding site ensures no receptor aggregation (and therefore activation) in the absence of antigen.

TOPOLOGY OF FcεRI AND IgE COMPLEXES ON THE CELL SURFACE

At present it is believed that the extracellular region of FcεRI lies along the cell membrane rather than being a rigid upright structure. Unlike other related FcR, e.g., FcγRII and FcγRIII, the membrane proximal region has an additional sequence which is likely to provide a rigid

* The amino acid sequence of IgE used in this chapter is from Genbank, accession number L00022. The numbering system is based on that of Bennich,[57] and is shown in Appendix 2.

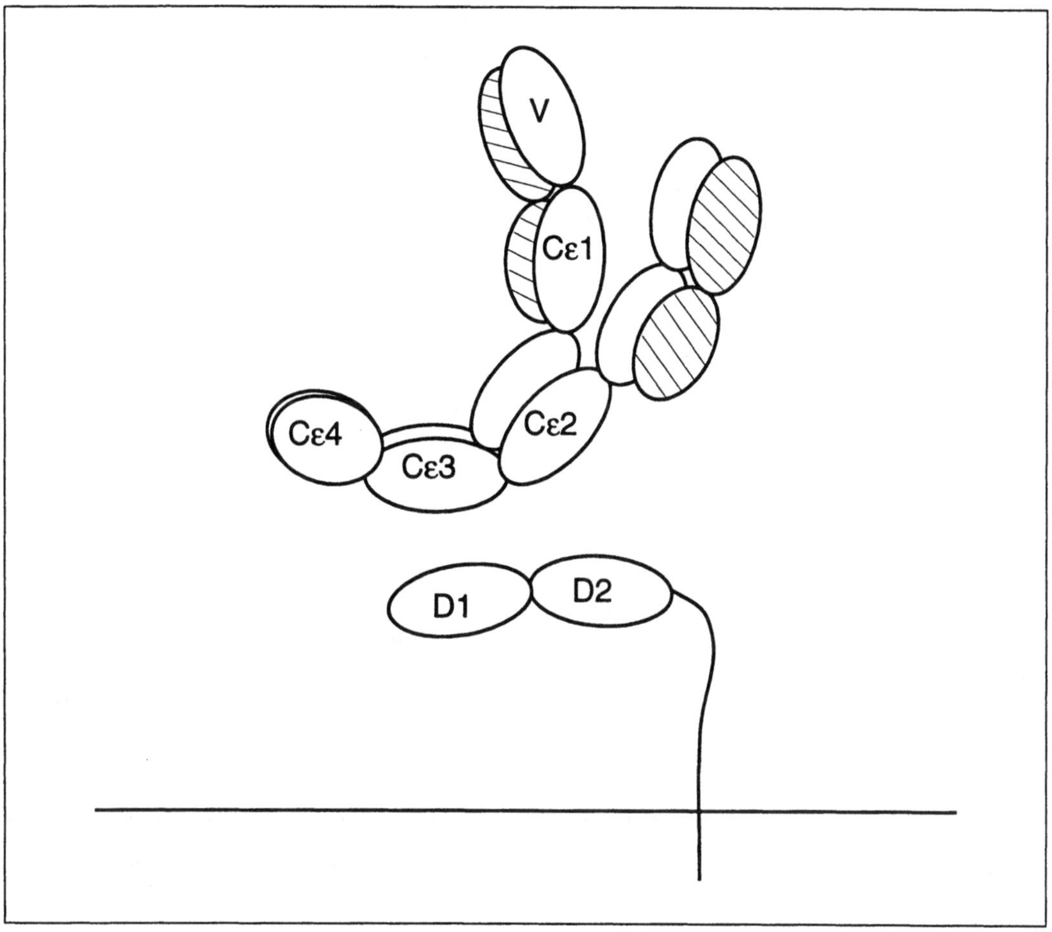

Fig. 2.1. Diagrammatic representation of FcεRI-α on the cell membrane and the bent and rotated IgE. The extracellular domains of FcεRI-α are indicated as D1 and D2; the IgE heavy chain domains as Cε or V and the light chains are shaded. (After Sutton and Gould)

extended stalk that may be required to space the two Ig-like domains 'away' from the membrane.[74]

The bent IgE molecule retains its bent conformation when bound to FcεRI (Fig. 2.1). FRET measurements indicate that when bound to the receptor, the C terminus sits approximately 53Å (5.3 nM) off the membrane with the antigen combining site 69Å (6.9 nM) distant.[64] Overall, the antigen combining sites sit >10 nM from the cell surface[75] and the disulphide bonds of Cε2 are located approximately 4.5 nM from the surface, level with the Cε2/Cε3 junction. FRET measurements also indicate that, in solution, the antigen combining sites lie approximately 7.1 nM from the C-terminus (Cε4) rather than the 17.1 nM expected for a planar Y-shaped molecule like IgG. Moreover the apparent compact and rigid nature of IgE is clearly different from IgG1 which is much more flexible (probably because of the hinge) and able to adopt multiple conformations.[67]

As a consequence of binding to FcεRI, IgE becomes more rigid.[67] However, the magnitude of this change appears to be relatively small.[67,76] Nonetheless it is clear that relative to the segmental motion of FcεRI in solution, movement is reduced but still occurs when receptor bound. This is especially relevant to the Fab arms where most of the segmental flexibility is thought to occur.[69,76]

The measured loss of flexibility is likely to occur at the site of interaction of IgE with FcεRI rather than being a generalized alteration of conformation throughout the IgE molecule.[64,67,76] Thus it seems that the bent IgE molecule, with Fab arms that wag and rotate, binds via its convex surface to FcεRI. At this point a transition from a 'low to high affinity state' may occur[77] when, as a consequence of binding, the structure becomes somewhat more rigid but the Fab arms and their antigen combining sites are orientated away from the cell surface and continue to rotate and move.

Even though most of the studies of the topology of the IgE:FcεRI interaction have been performed in rodent systems, given the similar 'bent' conformation of human IgE[68] it is likely that the same events occur in the interaction of human IgE with human FcεRI.

MOLECULAR BASIS FOR FcεRI:IgE INTERACTIONS

Regions of FcεRI that Influence Binding to IgE

Recent studies by ourselves and others have made significant advances in defining the IgE binding site of FcεRI-α—aspects were well reviewed (Table 2.2, Figs. 2.2., 2.3).[78-82] The FcεRI-α subunit is the IgE binding chain of the FcεRI-$\alpha\beta\gamma_2$ tetrameric receptor complex, as it is capable of binding IgE with high affinity in the absence of associated FcεRI-β or γ subunits. This has been directly demonstrated through the construction of chimeric receptors comprising the extracellular region of the FcεRI-α chain linked to either the transmembrane and cytoplasmic tails of the p55 subunit of the IL-2 receptor,[53] or the transmembrane and cytoplasmic tail of human FcγRII,[54] which retain high affinity

Table 2.2. IgE binding of chimeric FcεRI-α receptors and epitope mapping of anti-FcεRI-α mAb

Chimera[a]		Ig Binding			Inhibitory mAb binding	
Domain 1	Domain 2	hIgE	rIgE	mIgE	anti-hFcεRI-α	anti-rFcεRI-α
H 1	H 2	+	+	+	+	-
R 1	R 2	-	+	+	-	+
II 1	II 2	+[b]	-	-	-	-
III 1	III 2	-	-	-	-	-
II 1	H 2	-[c]	nt	+[c]	+	-
H 1	II 2	-	nt	-	-	-
III 1	H 2	+	+	nt	+	-
H 1	III 2	-	-	nt	-	-
H 1	R 2	-	-	nt	-	-
R 1	H 2	+	+	nt	+	+[d]

a. H = human FcεRI-α domain 1 or 2; R = rat FcεRI-α domain 1 or 2; II = human FcγRII domain 1 or 2; III = human FcγRII domain 1 or 2
b. In a single report mouse FcγRII has been shown to bind mouse IgE with low affinity (Takizawa et al, 1992), however human FcγRII does not bind human or rodent IgE.
c. low affinity binding $<10^7$ M^{-1}
d. binds two inhibitory mAb

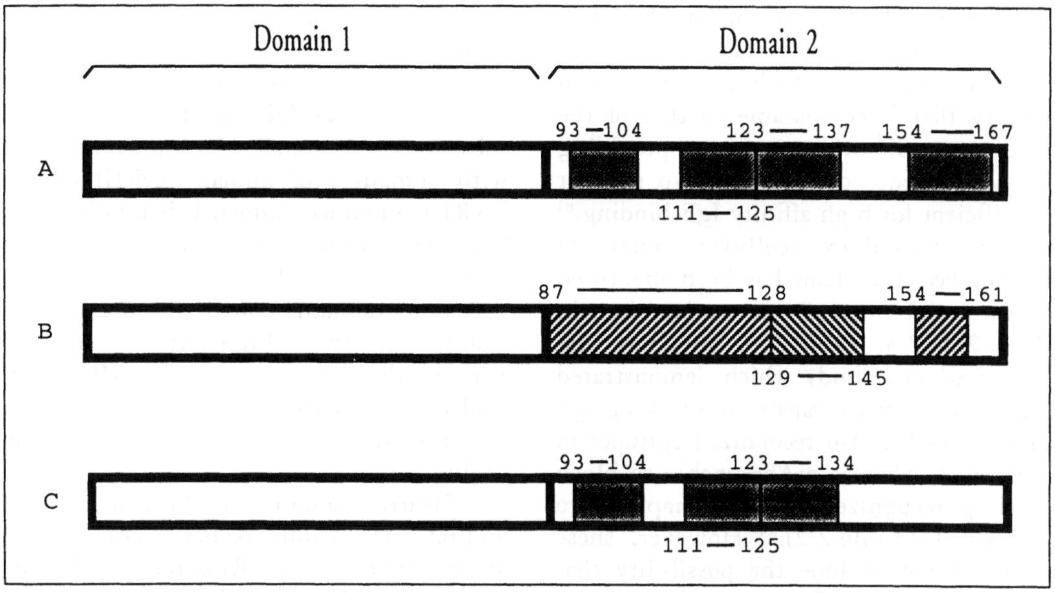

Fig. 2.2. IgE binding regions of human FcεRI-α. Regions of the extracellular domains of human FcεRI-α which influence the binding of (A) human, (B) mouse and (C) rat IgE are shaded, and flanking residues numbered. See text for details.

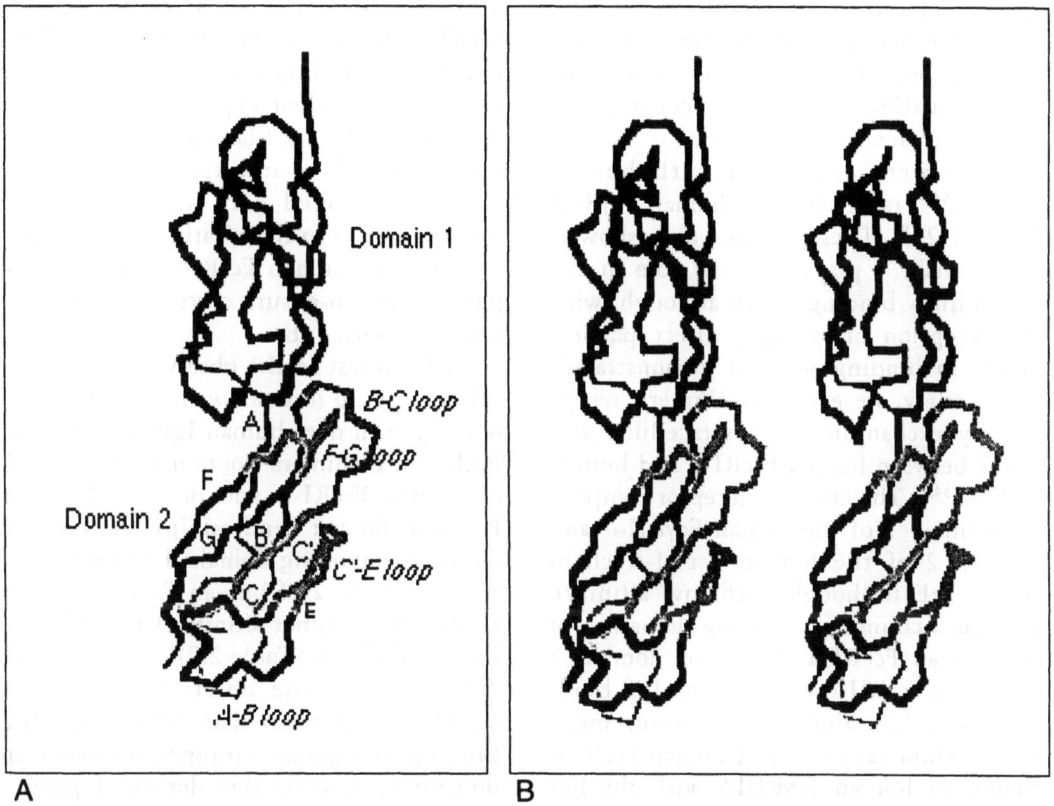

Fig. 2.3. Postulated alpha carbon backbone of the extracellular domains of human FcεRI-α. (A) Domains 1 and 2 are shown, the β-strands of Domain 2 are labeled. Consensus regions known to be involved in the interaction with IgE as defined in chimeric receptor studies are shown in gray, loops between β-strands in this region are labeled in italics. (B) Stereo view of FcεRI-α model labeled in (A).

IgE binding.[53,54] Soluble human FcεRI-α, consisting of only the extracellular region, has also been shown to bind IgE with an affinity that is comparable to that of the wild-type cell-surface receptor, providing further evidence that the FcεRI-α subunit is sufficient for high affinity IgE binding.[83]

The second extracellular domain of human FcεRI-α chain has been identified as the principle IgE interactive domain (Figs. 2.2, 2.3, Table 2.2). This was first suggested in a study which demonstrated that FcεRI-α mAbs capable of blocking IgE binding to FcεRI-α recognized epitopes in domain 2, whereas mAbs unable to block binding recognized epitopes mapping to domain 1 (Table 2.2).[84] However, these data did not exclude the possibility that the blocking of IgE binding by the domain 2 mAbs was a result of steric hindrance from a site distant to the actual binding site, or due to a conformational change induced in the binding site.

Direct evidence for the role of domain 2 as the binding domain has come from studies utilizing chimeric receptors in which extracellular domains of human FcεRI-α were exchanged with human FcγRIIa,[54,85] rat FcεRI-α or human FcγRIII (Table 2.2).[86] These studies also showed that domain 1 plays a crucial role in the high affinity binding of IgE although, with the exception of rat IgE, direct participation in binding was not demonstrated. In our study, we generated chimeric receptors by exchanging the extracellular domains between human FcεRI-α and human FcγRIIa.[54,85] The chimeric receptor comprising domain 1 of the human FcγRII-α and domain 2 of the human FcεRI-α chain bound IgE (although with low affinity), whereas a chimera containing domain 1 of the human FcεRI-α chain and domain 2 of human FcγRII-α exhibited no IgE binding (Table 2.2). Similar studies using interspecies chimeras of rat and human FcεRI-α chains, or human FcγRIIIA with the human FcεRI-α chain, have also demonstrated that domain 2 of the human FcεRI-α chain directs the binding of IgE. In these studies, the substitution of domain 2 of human FcεRI-α with domain 2 of human FcγRIII or rat FcεRI-α (neither of which bind human IgE) was found to result in the loss of human IgE binding, whereas the substitution of domain 1 of human FcεRI-α with domain 1 of human FcγRIII or rat FcεRI maintained human IgE binding.[73,86] The major difference in our studies and those of Mallamaci[86] was that our substitution of domain 1 with human FcγRII resulted in loss of high affinity binding, whereas the use of human FcγRIII or rat FcεRI-α as described by Mallamaci et al[86] had no effect on high affinity binding (Table 2.2).

Clearly domain 1 of human FcγRIII and rat FcεRI-α must be better able to substitute for human FcεRI-α domain 1 than human FcγRII domain 1 to maintain correct receptor conformation. This is perhaps not surprising as both human FcγRIII and rat FcεRI-α domain 1 exhibit significantly higher amino acid identity to human FcεRI-α domain 1 than human FcγRII-α. Indeed, the finding that domain 2 of human FcεRI-α when expressed as a single domain in either a filamentous phage display system[73,87] or in a transient COS cell system[86] exhibited only weak or no IgE binding, respectively, clearly indicates that domain 1 of human FcεRI-α plays an important role to ensure correct receptor interaction with IgE.

Of interest is the observation that rat IgE appears to interact with rat FcεRI differently than does human IgE with human FcεRI.[86] A chimera containing domain 1 of human FcεRI-α and domain 2 of rat FcεRI-α did not bind rat IgE; however, a chimera containing domain 1 of rat FcεRI-α and domain 2 of human FcεRI-α bound rat IgE with higher affinity than wild-type human FcεRI-α (Table 2.2). This finding, together with the observation that rat FcεRI-α chain mAb that inhibits rat IgE binding recognizes epitopes localized in domain 1, suggests that domain 1 plays a crucial role in the interaction of rat IgE with rat FcεRI-α. These data suggest that the focus of the interaction of human and rat IgE with their respective receptors is different, which may not be surprising given that human and rat FcεRI-α share only 38% amino acid sequence identity. Thus it is possible that distinct IgE binding sites do exist in human and rat FcεRI-α, with domain 2 of human FcεRI-α containing a binding site for human IgE and domain 1 of rat FcεRI-α a binding site for rat IgE. It is interesting to note that rat IgE was found to bind to a chimera comprising domain 1 of human FcγRIII and domain 2 of human FcεRI-α, suggesting that human FcεRI-α interacts with rat IgE principally through domain 2, in contrast to rat FcεRI-α which appears to bind rat IgE through domain 1. Another possible explanation for the apparent difference in the binding of rat and human IgE to FcεRI-α is that certain domains are not able to fully substitute for others in the context of specific chimeric receptors to ensure conservation of correct receptor structure. Therefore, while it is apparent from the chimeric receptor studies that specific domains interact with IgE, the precise role of other regions or domains still remains unclear, e.g., domain 2 of rat FcεRI-α.

The IgE binding regions within domain 2 of human FcεRI-α have been further mapped to subregions using homologous scanning mutagenesis. Two separate studies have both identified multiple regions of domain 2 as crucial in the interaction with IgE[54,86] (Figs. 2.2, 2.3). We used human FcγRII-α as a scaffold to display regions of human FcεRI-α by replacing segments of human FcγRII-α domain 2 with the corresponding regions of human FcεRI-α domain 2.[54] Using this 'gain of function' approach (i.e., FcγR binding IgE) we have identified at least three independent regions of human FcεRI-α domain 2 capable of *directly* binding IgE. The human FcεRI-α domain 2 regions encompassed residues Trp87 to Lys128, Tyr129 to Asp145 and Lys154 to Glu161 which when inserted into human FcγRII-α were each independently found to *impart* mouse IgE binding to human FcγRII-α (Fig. 2.2A). These human FcεRI-α/FcγRII-α chimeric receptors bound IgE only in the form of immune complexes, implying that all three regions (together with domain 1 as discussed above) contribute to the formation of a high affinity IgE binding site. A similar approach utilizing human FcεRI-α/ human FcγRIII chimeras, employing a loss of function strategy, has also identified multiple IgE binding sites of human FcεRI-α domain 2.[86] This approach involved the insertion of homologous regions of human FcγRIIIa into human FcεRI-α, and the *loss* of IgE binding function was determined.

The substitution of three regions of human FcεRI-α with the equivalent regions of human FcγRIII resulted in the complete loss of human IgE binding; Ser93 to Phe104, Arg111 to Glu125 and Asp123 to Ser137 (Fig. 2.2B). These regions correlate with two of the three IgE binding regions defined in our study, as both the Ser93 to Phe104 and Arg111 to Glu125 regions identified by Mallimaci et al[86] are situated in our Trp87 to Lys128 region, and the Asp123 to Ser137 region of Mallimaci et al[86] overlaps our Tyr129 to Asp145 region. Substitution of a fourth region of hFcεRIα (Lys154 to Ile167) resulted in a partial loss of IgE binding and corresponds with the third of the direct IgE binding regions defined in our study, i.e., Lys154 to Glu161. As the readout for the Mallimaci experiments was loss of IgE binding, these data can also be interpreted as the replacement of these regions of FcεRI-α simply resulting in a conformational change in the receptor, thus altering IgE binding to a distant binding site. With the exception of the Lys154 to Ile167 substitution, none of the domain 2 chimeric receptors were recognized by any of the inhibitory mAb, thus the possibility of an alteration in conformation being responsible for loss of IgE binding cannot be excluded.

The binding of rat IgE to the human FcεRI-α/human FcγRIII chimeras was also examined, and found to exhibit the same pattern of binding as human IgE with the exception that residues Lys154 to Ile167 of human FcεRI-α did not influence rat IgE binding in this system. These data suggest that rat IgE interacts with human FcεRI-α differently to human IgE. However, rat and human IgE do share common binding regions, i.e., Ser93 to Phe104 and Arg111 to Glu125 and Asp123 to Ser137 (Fig. 2.2C). The rat IgE binding region contained in the last of these regions can be further located to residues Asp123 to His134 based on overlap with the His134 to Glu163 region identified as not required for binding.[86]

While the role of domain 2 has been well characterized, the role of domain 1 is still unclear. The substitution of two regions in domain 1 of human FcεRI-α, namely residues 35-46 and 80-92, was also found to result in a loss of human IgE binding, suggesting these regions may also play a role in human IgE binding.[86] However, both of these chimeras, although expressed on the cell surface, were not recognized by any of the FcεRI-α mAb tested. Thus, they are likely to be incorrectly folded making conclusions about the role of these regions in IgE binding difficult. Therefore in summary, the results of these two studies clearly demonstrate that domain 2 of human FcεRI-α is an IgE interactive domain of human FcεRI, and identify at least four regions contributing to the binding of IgE, Ser93 to Phe104, Arg111 to Glu125, Tyr129 to Ile137,* and Lys154 to Glu161.

In the absence of a genuine 3-dimensional structure, a number of molecular models of human FcεRI-α have been proposed.[80,85,88] We have generated a model of human FcεRI-α domain 2 based on the structure of domain 2 of CD4.[85] The model is a truncated C2-SET domain comprising seven β-strands forming two antiparallel β-sheets, linked by a disulphide bridge between strands B and F (Fig. 2.3). Similar models have also been proposed based on CD2 domain 2,[80,89] or antibody domains.[88] The latter group has proposed two models which differ in the orientation of the two extracellular domains relative to one another. One model has the two domains positioned end-to-end allowing only longitudinal interaction between the domains, with the second proposing a bent conformation promoting more lateral interaction between the two domains. Based on all of these models, the identified IgE binding regions of human FcεRI domain 2 are situated predominantly in loop regions juxtaposed at the interface with domain 1, specifically the F-G, C'-E and B-C loops, with contributions also from the B and C strands (Fig. 2.3).

The localization of the domain 2 IgE interactive sites to this region of domain 2, together with the finding that domain 1 also plays a key role, suggests it is this interdomain interface between domains 1 and 2 that comprises the IgE binding site of human FcεRI-α. In support of this model, the mAb 15A5, which recognizes an epitope encompassed by residues 100-115 of human FcεRI-α (corresponding to the B-C loop and B strand), can completely block the binding of IgE to FcεRI, suggesting the multiple IgE binding sites are indeed juxtaposed.[84]

Based on the studies described herein for FcεRI-α and those described elsewhere for the FcγR, i.e., FcγRI,[90] FcγRII[91,92] and FcγRIII,[93] it is becoming apparent that there are a number of structural similarities in the molecular basis of how these receptors interact with their respective ligands. The two Ig-like domain structure of the extracellular regions of the leukocyte FcRs clearly represents a conserved binding motif of this receptor family.[90] The second extracellular domain of all these receptors is responsible for the direct binding of Ig, with the first domain playing a crucial role in maintaining optimal binding.[54,73,86,93] The mapping of the Ig binding regions in domain 2 of human FcεRI and human FcγRII has indicated that the interaction of these receptors with their Ig ligands involves similar regions. Three homologous regions of both human FcεRI-α and human FcγRIIa domain 2 play crucial

* Overlapping region of the 123-137 region identified by Mallimaci et al,[86] and the 129-145 region identified by Hulett et al.[54]

roles in the binding of IgE and IgG, respectively.[54,86,92]

Based on model domain structures, the Ig binding regions of both receptors are located in similar regions of domain 2 juxtaposed at the domain 1 interface, namely the B-C, C'-E and F-G loops (Fig. 2.3).[80,85,88,92] The influence of domain 1 on the Ig binding by domain 2 in both human FcγRII and human FcεRI is consistent with the location of the binding regions of these receptors in close proximity with domain 1, i.e., the interface of domains 1 and 2. Thus, the finding that these loop regions are involved in the binding of Ig by two functionally distinct FcRs, in conjunction with the conserved nature of the two domain extracellular binding 'unit' of the leukocyte FcRs, strongly suggests that this region will also comprise the key interactive site of all members of this family. Based on this observation, it can be postulated that the structurally conserved 'Ig folds' of the second extracellular domains of the leukocyte FcRs are providing the 'scaffolding' to display 'variable' loop regions which contain determinants directing the specificity of these receptors for their Ig ligands.

Regions of IgE that Influence Binding to FcεRI

Intense interest in how IgE binds to FcεRI has involved many groups over a considerable time. These data are summarized in Figure 2.4 and in the model structures of Figure 2.5. In these the essential binding regions are indicated. In the model (Fig. 2.5) the relationships of different regions of IgE, shown by different investigators to influence binding to FcεRI, are presented. It is noteworthy that despite obvious differences in the extent of the mutations of IgE, there are clearly overlapping regions that influence binding; these are shown in stereo in Figure 2.5E.

Early studies using papain digestion of IgE to produce Fc fragments capable of blocking IgE binding to mast cell FcεRI, localized binding to Cε2, Cε3 and Cε4 (Fig. 2.4). However, attempts to further localize this IgE binding region by continued fragmentation of IgE were unsuccessful. Indeed, Fab'2 fragments or isolated Cε2 failed to block IgE binding. Thus these experiments suggested that the tertiary and quaternary structure of IgE-Fc was important for receptor binding.[60,94,95] A separate study observed that the junction region of rat Cε2/Cε3 was partially protected from tryptic digestion when cell bound, and proposed this region as the FcεRI binding site.[96] Circular dichroism with thermal inactivation of human myeloma IgE indicated that Cε3 and Cε4, but not Cε2, were likely to be implicated in binding.[57]

Synthetic peptides have been used in separate approaches to identify the FcεRI binding site of IgE. These include use as competitive inhibitors of IgE binding to FcεRI and in epitope mapping studies using monoclonal anti-peptide antibodies or anti-IgE antibodies. An early study reported that a synthetic pentapeptide of amino acids residues Asp330-Gly335 (320-324)* in the human Cε2/Cε3 junction region was able to partially inhibit the Prausnitz-Küstner reaction,[97] although the experiment could not be reproduced.[98]

More recently the residues Asp330-Arg334 in this junction region again have been implicated in FcεRI-α binding,[80] and it has been suggested that mutagenesis of Pro333 alters binding affinity of human IgE-Fc to FcεRI-α and thus could impart high affinity to this interaction.[66] The Pro333 residue is located in the equivalent position to the Leu235 residue in IgG1 that, with its associated residues Leu234-Gly237, is crucial for binding to the high affinity gamma receptor (FcγRI).[99,100] Mutagenesis of residues Asp330(361), Asn332(363) and Arg334(365) to alanine has also been shown to reduce binding by one-third.[101]

* Amino acid numbers in the sequence of IgE used in this chapter are based on those of Dorrington and Bennich 1978 (Appendix 2). It should be noted that residue numbers used by Hamburger[97] and Presta et al[101] are shown in parenthesis.

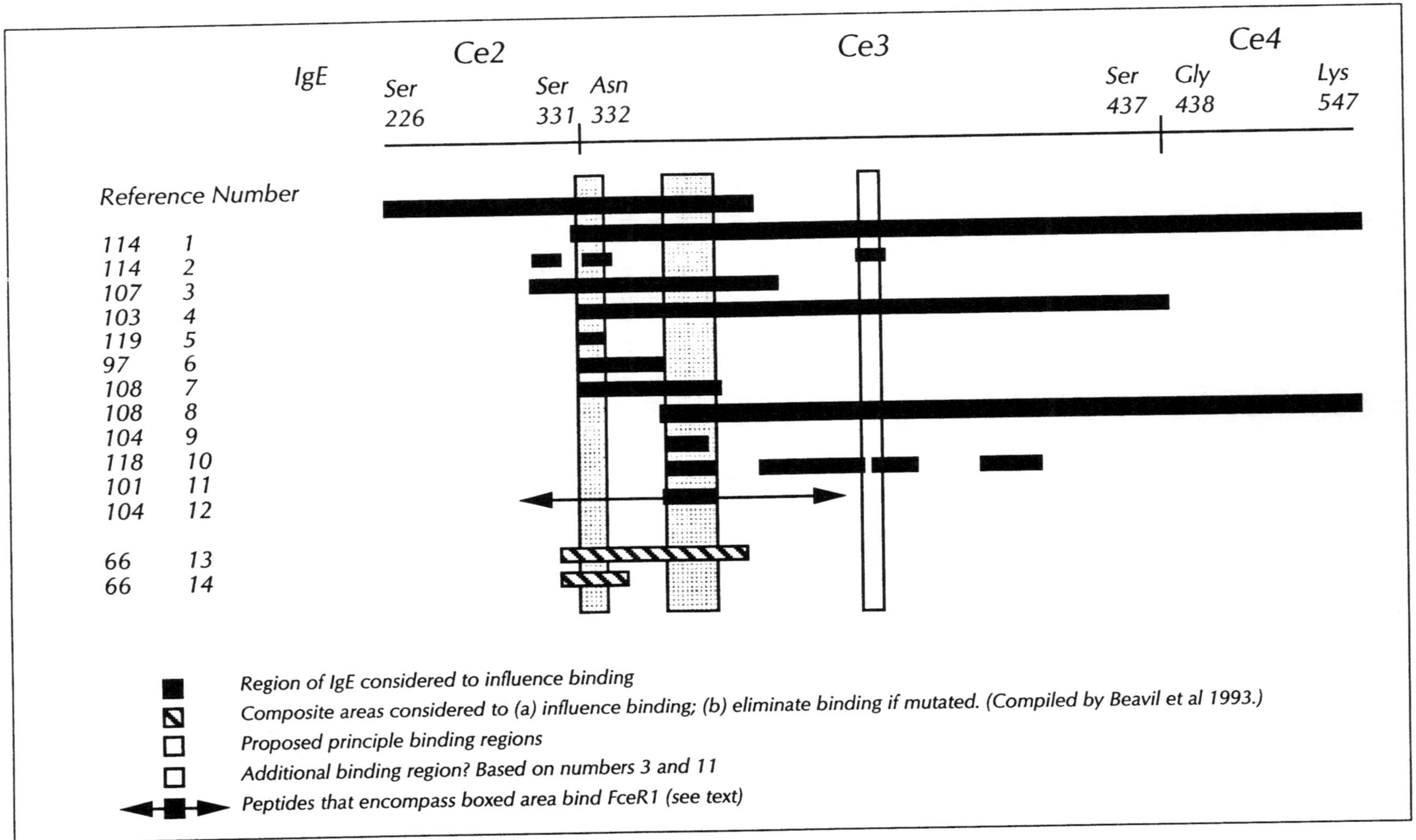

Fig. 2.4. Compilation of data from a number of studies which identify regions of IgE influencing binding to FcεRI. Sources of the data are from indicated references. (After Beavil)

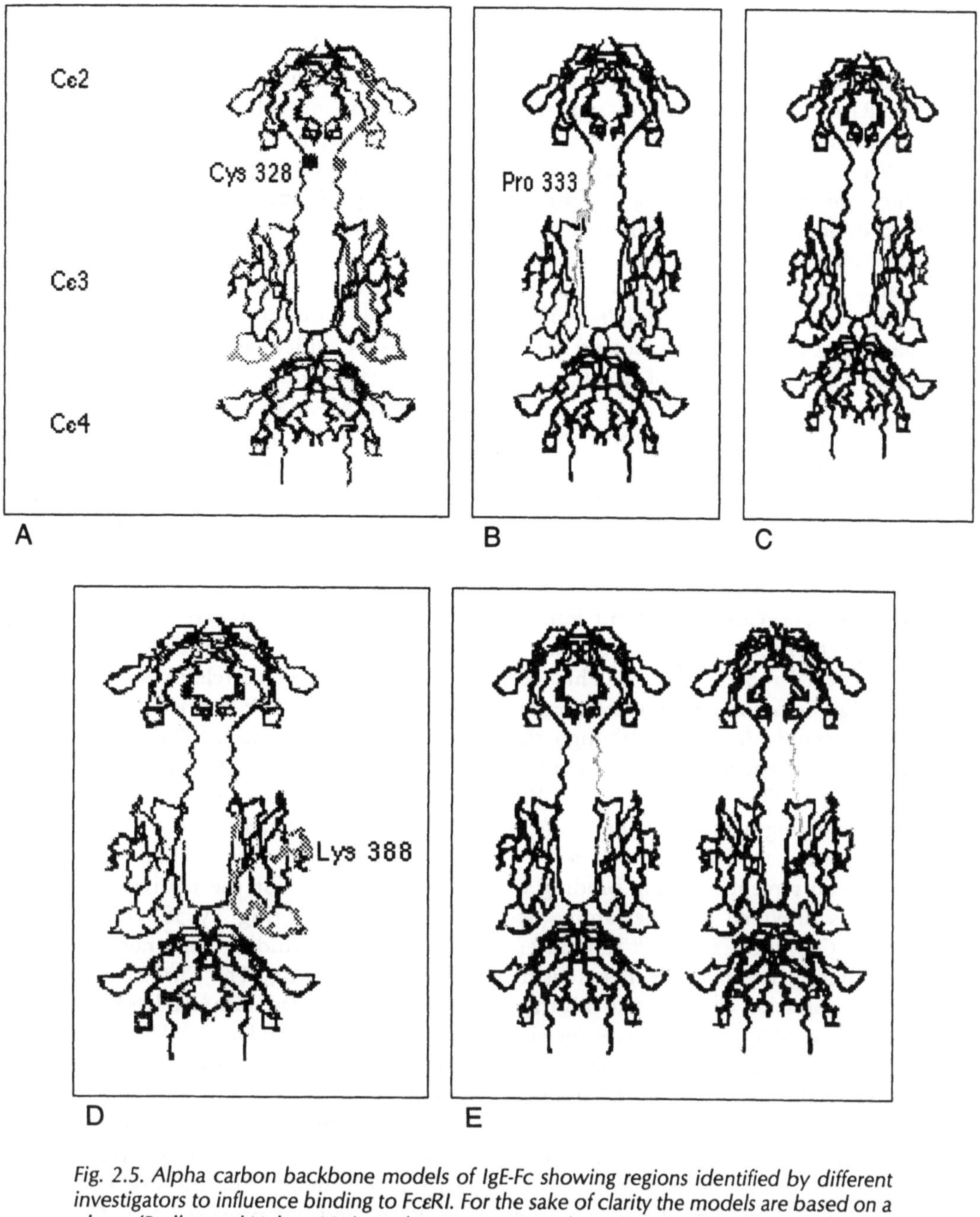

Fig. 2.5. Alpha carbon backbone models of IgE-Fc showing regions identified by different investigators to influence binding to FcεRI. For the sake of clarity the models are based on a planar (Padlan and Helm 1991), not bent structure and are designed merely to indicate the extent and location of various residues involved in the IgE:FcεRI interaction. Data from (A) Helm, darker gray indicates the region equivalent to the original peptide; lighter gray, the region which large peptides need to encompass in order to block binding.(1996) (B) Beavil (1993) darker gray, region of residues necessary for binding; lighter gray, regions in which mutations destroy binding. (C) Takemoto(1994) (D) Presta(1994) (E) Stereo view of regions in Cε3 common to more than one of the studies above.

An *E. coli*-derived peptide of amino acid residues Gln301-Lys376 was the first *E. coli*-derived peptide demonstrated to bind FcεRI-α,[102] and it encompasses this often implicated Asp330-Gly335 residue block (Fig. 2.5A). However, attempts to reproduce these findings using a similar, active, *E. coli*-derived IgE-Fc fragment were unsuccessful.[81,103] The explanation for this discrepancy may be in the experience of Liu et al[95] who also used active *E. coli*-derived fragments, but suggest that incorrect folding reduces the proportion of active molecules.

Notwithstanding the potential problems in the native folding of the *E. coli* produced IgE, recent studies have shown that *E. coli*-derived, glutathione-S-transferase (GST) ε heavy chain fusion proteins, containing approximately 120 amino acids from Cε2,Cε3 and Cε4 will inhibit human IgE binding to FcεRI-α if they *encompass* a Cε3 sequence Pro343-Ser353. This sequence of Cε3 is positioned in the loop that joins the β-strands A and B, although the 18 amino acid sequence Leu340-Thr357 fused directly to GST did not bind.[104] For example, large fragments need the Pro343-Ser353 sequence to bind, but there is a structural requirement enabling binding to occur that is provided by other amino acids in Cε2, Cε3 and Cε4.[56,59,60] It is interesting to note that in experiments using eukaryotic expression of truncated IgE fragments fused to p55 (the IL-2 receptor signal peptide) the smallest functional fragment produced was composed of residues Ala329-Lys547; deletions from either the N or C termini rendered the fragments inactive.[103]

A recent review compiling data from several studies[43,102,105,106] using *E. coli* fragments and chimeric IgE antibodies has indicated that residues Cys328-Val361 (encompassing the Cε2/Cε3 junction and the AB loop) appear to be sufficient for binding/inhibition activity, while deletions in residues Cys328-Tyr339 destroy activity,[66] which agrees with the findings of Helm et al (Fig. 2.5A,B). Epitope mapping using monoclonal antibodies to IgE or IgE-derived peptides is an alternative approach to define the FcεRI-α binding site on IgE. The action of such antibodies that block the binding of IgE to FcεRI-α has also been studied, and while introducing the problem of steric hinderance of receptor binding by the antibody, it also shows that Cε3 contains the major binding site(s).

Two related approaches were used; in the first, Fab fragments of mouse anti-human IgE monoclonal antibodies—the epitopes of which were localized to residues Ser306-Gln313, Thr311-Thr320 of Cε2, Ser331-Ala338 or Val382-Lys391 of Cε3—could inhibit IgE binding to FcεRI-α, though not completely. However, some combinations of these were more potent which implies that either multiple contact sites (defined by these epitopes) are involved in binding, or that these epitopes are close to, but not part of, the binding region (Fig. 2.5C).[107] However, the location of these epitopes correlates with FcεRI-α binding sites in IgE defined by use of recombinant IgE proteins. Thus the Thr311-Ala338 encompasses the Cε2/Cε3 junction region, and residues Val382-Lys391 coincide with the β-strand D where Presta[101] (Fig. 2.5D) also found that exchanging Lys388-Arg393 for the equivalent section of IgG or changing Lys388(423) to Pro could remove binding, but not Ala could obliterate binding. The Thr311-Ala338 region is within the Gln301-Lys376 critical binding region of Helm,[61,102] and Ser331-Ala338 overlaps the Cys328-Val361 segment of Beavil,[66] but not the Pro343-Ser353 segment recently defined.[104] More recent work[108] has confirmed that human Cε3 is the main binding site for human FcεRI-α, and has proposed that Cε2 is required for conformation and stability and that FcεRI-α binding is probably not restricted to any one site on IgE.

It is not surprising that these data, accumulated in various expression systems, with different assay methods, differ in their conclusions. The Cε2/Cε3 junction region has been implicated many times, as has the Cε3 AB loop; other regions showing binding interaction are those on the Cε3 EF loop and CD loop[101] facing the same Cε2/Cε3 cleft as the Cε3 AB loop, also the Cε3 FG loop which impinges into the spatial area of the Asp330-Gly335 (Cε2/Cε3) junction site. The sensitivity of IgE binding to mutation of the Cε3 D strand implies the segment may impart structural stability to Cε3.[101] These site directed mutagenesis experiments, have led Presta[101] to propose that the exposed face of the human Cε3 domain binds human FcεRI-α by electrostatic interaction. However, the radical amino acid replacements that were employed in this study could have, by their size or charge difference, caused major structural alterations.[104]

Rat and mouse IgE have been shown to bind to rodent FcεRI-α[3] with a stoichiometry of 1:1[109] and also to human FcεRI-α[110] with lower affinity but the same stoichiometry. However, human IgE does not bind to rodent FcεRI-α.[104,110] The differences in specificity have been exploited to localize binding sites by building chimeric immunoglobulins from rodent or human IgE with IgG. Human IgG1/mouse IgE chimeras have shown that both Cε2 and Cε3 are necessary for rodent (mouse) IgE to bind rodent (rat) FcεRI,[111] but human IgE/ mouse IgE chimeras suggest that only Cε3 is necessary.[106] This could imply that γ at C2 obstructs binding, and human or mouse Cε2 is required for conformational stability.[106] Only Cε3 and Cε4 are required for high affinity binding in human IgE,[106] although Cε2 has been implicated.[107] Rodent (mouse) Cε4 is not required for binding rodent (rat) FcεRI.[111,112] Further human IgE/mouse IgE chimeras of Cε3 have shown that exchanges to mouse at residues Ser300-Phe346 in the Cε2/Cε3 junction region of human IgE cannot impart the ability to bind rodent FcεRI-α, nor can this exchange with mouse Cε2 replacing human Cε2.[108] Both chimeras, however, bind human FcεRI-α.

In human IgE the Cε2/Cε3 junction region and the Cε3 AB loop have been implicated in high affinity binding. In rodent FcεRI binding, the AB loop exchanged for rat in a human IgE/rat IgE chimera, does not confer rodent FcεRI-α binding to the human IgE, and the chimera continues to bind human FcεRI-α.[104] Chimeras containing both these regions (Ser300-Ile356) as rodent, with either mouse or human Cε2, do not bind rodent FcεRI-α, but retain the ability to bind human FcεRI-α. An additional Cε3 chimera containing more Cε3-derived sequence (Thr357-Val399) did not bind rat FcεRI-α and had reduced binding to human FcεRI-α, confirming the need for a complete rodent Cε3 for rodent FcεRI binding.[106,108,112,113] Cε4 is not required for binding, but is important in conformation.[113]

The binding region of IgE would appear to involve several sites, with the major sites in the Cε2/Cε3 junction region[80] and the Cε3 AB loop (residues Pro343-Ser353).[114] The former region has homology with the region of IgG1 that imparts high affinity binding to FcγRI and is conserved in human and rodent IgE. The latter region projects into the cleft between Cε3 and Cε4, which has homology to a site on rat IgG1 that contains many histidine residues, and has been shown by crystallography to bind the side of the neonatal rat receptor FcRn at the junction of domain 1 and domain 2.[115] However, this IgG receptor is unrelated to FcεRI-α or any other Fc receptor thus the significance of this interaction is difficult to assess.

In summary, human IgE can bind human FcεRI, but not rodent FcεRI; the main regions of binding appear to be at the Cε2/Cε3 junction and the Cε3 AB loop with, perhaps, the D β-strand and/or a face of Cε2 fronting the Cε2/Cε3 cleft assisting. The Cε2 and Cε4 domains have been implicated more for their donation to structure than high affinity binding. The rodent IgE molecule appears to be more sensitive than human IgE, and requires the Cε3 domain to be intact and totally rodent in order to bind to rodent FcεRI, but is capable of binding human FcεRI with Cε3 in intact rodent or chimeric form. As with the human, the major function of the Cε2 and Cε4 domains appears to involve supplying a stable supporting structure. Noting the small but significant differences between human and rodent FcεRI binding regions on their IgE, it is interesting to return to the suggestion that *domain 1* of rodent FcεRI-α is important in *rat IgE* binding, whereas human FcεRI-α *domain 2* is the principle binding region for both human and rat IgE.[86]

CONCLUDING REMARKS

How Does the IgE:FcεRI Interaction Take Place?

We have come a long way in attempting to define the molecular basis of the IgE:FcεRI interaction but the details of how IgE and FcεRI react are, at present, difficult to ascertain. The prodigious effort in attempting to define how this interaction takes place has improved our knowledge of the likely portions of IgE and FcεRI that govern the interaction. Structure:function studies, especially those of the last decade, have provided us with a better understanding of key regions of IgE and FcεRI that influence their interaction. Together with biophysical experiments, these have given us a general picture of a bent IgE with several areas of its convex face interacting with a receptor whose two domains also contain multiple regions that enable it to interact with its ligand.

There are still many issues to be resolved, and none more sought after than the elucidation of the precise interatomic associations that collectively define this high affinity interaction. It is almost certain that such information will come from solving the structure of FcεRI:IgE cocrystals. Such information has been elusive, but as protein expression systems and crystallization technologies improve, the day will come when such information will become available to us—even if only through the patent literature! At present, however, we can develop plausible hypotheses on how this interaction takes place.

It has been suggested that when bound IgE lies along the receptor on its 2-fold axis of symmetry,[66,80,116] but it has not been determined whether the binding site(s) are all on one, or on both IgE heavy chains. Indeed there is ample evidence that the latter may be the case: (1) the IgE:FcεRI stoichiometry is 1:1; (2) the Cε2/Cε3 junction of IgE is protected from trypsinization when bound;[96] (3) this protection is bilaterally symmetrical[96] and IgE monomer does not bind;[117] (4) whole IgE will not bind if its critical interchain (human, intrachain-rat) disulphide bonds are not intact;[60] and (5) it has been suggested that Cε2 is rotated.[68] This could occur if the Cε2 (a) chain was rotated to align with protruding portions of the Cε3 (b) chain. It is also possible that the IgE molecule could bind at an angle across the domain 1/domain 2 junction of the receptor (and not along the long axis of the receptor), with different segments of Cε2/Cε3 interacting with the receptor. Do the Cε2/Cε3 junction and the Cε3 AB loop interact with different segments of FcεRI?

Thus, the challenge of the moment is to refine the models on the basis of solid data, and accept or reject these. At the end of the day resolution of cocrystal structures will be informative, and in combination with the already assembled data will provide the impetus for the development of effective strategies for the treatment of IgE induced inflammation.

References

1. Gounni AS, Lamkhioued B, Ochiai K et al. High-affinity IgE receptor on eosinophils is involved in defence against parasites. Nature 1994; 367:183-86.
2. Conrad D, Cooper M, Fridman WH et al. IUIS/WHO Subcommittee on nomenclature of Fc receptors. Nomenclature of Fc receptors. Bulletin of the World Health Organization 1994; 72(5):809.
3. Kulczycki Jr A, Metzger H. The interaction of IgE with rat basophilic leukemia cells. II. Quantitative aspects of the binding reaction. J Exp Med 1974; 140:1676-95.
4. Conrad DH, Froese A. Characterization of the target cell receptor for IgE. II. Polyacrylamide gel analysis of the surface IgE

Appendix 2.1. Amino acid sequence of human FcεRI-α. Genbank accession # X06948

1									10										20					
Val	Pro	Gln	Lys	Pro	Lys	Val	Ser	Leu	Asn	Pro	Pro	Trp	Asn	Arg	Ile	Phe	Lys	Gly	Glu	Asn	Val	Thr	Leu	Thr
26									35										45					
Cys	Asn	Gly	Asn	Asn	Phe	Phe	Glu	Val	Ser	Ser	Thr	Lys	Trp	Phe	His	Asn	Gly	Ser	Leu	Ser	Glu	Glu	Thr	Asn
51									60										70					
Ser	Ser	Leu	Asn	Ile	Val	Asn	Ala	Lys	Phe	Glu	Asp	Ser	Gly	Glu	Tyr	Lys	Cys	Gln	His	Gln	Gln	Val	Asn	Glu
76									85										95					
Ser	Glu	Pro	Val	Tyr	Leu	Glu	Val	Phe	Ser	Asp	Trp	Leu	Leu	Leu	Gln	Ala	Ser	Ala	Glu	Val	Val	Met	Glu	Gly
101									110										120					
Gln	Pro	Leu	Phe	Leu	Arg	Cys	His	Gly	Trp	Arg	Asn	Trp	Asp	Val	Tyr	Lys	Val	Ile	Tyr	Tyr	Lys	Asp	Gly	Glu
126									135										145					
Ala	Leu	Lys	Tyr	Trp	Tyr	Glu	Asn	His	Asn	Ile	Ser	Ile	Thr	Asn	Ala	Thr	Val	Glu	Asp	Ser	Gly	Thr	Tyr	Tyr
151									160										170					
Cys	Thr	Gly	Lys	Val	Trp	Gln	Leu	Asp	Tyr	Glu	Ser	Glu	Pro	Leu	Asn	Ile	Thr	Val	Ile	Lys	Ala	Pro	Arg	Glu
176									185										195					
Lys	Tyr	Trp	Leu	Gln	Phe	Phe	Ile	Pro	Leu	Leu	Val	Val	Ile	Leu	Phe	Ala	Val	Asp	Thr	Gly	Leu	Phe	Ile	Ser
201									210										220					
Thr	Gln	Gln	Gln	Val	Thr	Phe	Leu	Leu	Lys	Ile	Lys	Arg	Thr	Arg	Lys	Gly	Phe	Arg	Leu	Leu	Asn	Pro	His	Pro
226				230																				
Lys	Pro	Asn	Pro	Lys	AsnAsn	-																		

Appendix 2.2. Amino acid sequence of IgE. Genbank accession #L00022.

220 Ce2 230 240
Lys Thr|| Phe Ser Val Cys Ser Arg Asp Phe Thr Pro Pro Thr Val Lys Ile Leu Gln Ser Ser **Cys** Asp Gly Gly Gly His Phe Pro Pro

250 260 270
Thr Ile Gln (Leu) Leu Cys Leu Val Ser Gly Tyr Thr Pro Gly Thr Ile Asn Ile Thr Trp Leu Glu Asp Gly Gln Val Met Asp Val Asp

280 290 300
Leu Ser Thr Ala Ser Thr Thr Gln Glu Gly Glu Leu Ala Ser Thr Gln Ser Glu Leu Thr Leu Ser Gln Lys His Trp Leu Ser Asp Arg

310 320 330 Ce2 Ce3
Thr Tyr Thr Cys Gln Val Thr Tyr Gln Gly His Thr Phe Glu Asp Ser Thr Lys Lys **Cys** Ala Asp Ser|| Asn Pro Arg Gly Val Ser Ala

340 350 360
Tyr Leu Ser Arg Pro Ser Pro Phe Asp Leu Phe Ile Arg Lys Ser Pro Thr Ile Thr Cys Leu Val Val Asp Leu Ala Pro Ser Lys Gly

370 380 390
Thr Val Asn Leu Thr Trp Ser Arg Ala Ser Gly Lys Pro Val Asn His Ser Thr Arg Lys Glu Glu Lys Gln Arg Asn Gly Thr Leu Thr

400 410 420
Val Thr Ser Thr Leu Pro Val Gly Thr Arg Asp Trp Ile Glu Gly Glu Thr Tyr Gln Cys Arg Val Thr His Pro His Leu Pro Arg Ala

430 Ce3 Ce4 440 450
Leu Met Arg Ser Thr Thr Lys Thr Ser|| Gly Pro Arg Ala Ala Pro Glu Val Tyr Ala Phe Ala Thr Pro Glu Trp Pro Gly Ser Arg Asp

460 470 480
Lys Arg Thr Leu Ala Cys Leu Ile Gln Asn Phe Met Pro Glu Asp Ile Ser Val Gln Trp Leu His Asn Glu Val Gln Leu Pro Asp Ala

490 500 510
Arg His Ser Thr Thr Gln Pro Arg Lys Thr Lys Gly Ser Gly Phe Phe Val Phe Ser Arg Leu Glu Val Thr Arg Ala Glu Trp Glu Gln

520 530 540 Ce4.
Lys Asp Glu Phe Ile Cys Arg Ala Val His Glu Ala Ala Ser Pro Ser Gln Thr Val Gln Arg Ala Val Ser Val Asn Pro Gly Lys

Interchain disulphide bond Cys residues are underlined and in bold.

receptor from normal rat mast cells and from rat basophilic leukemia cells. J Immunol 1976; 116:319-26.

5. Kulczycki Jr A, McNearney TA, Parker CW. The rat basophilic leukemia cell for IgE. I. Characterization as a glycoprotein. J Immunol 1976; 117:661-65.
6. Kanellopolous J, Rossi G, Metzger H. Preparative isolation of the cell receptor for immunoglobulin E. J Biol Chem 1979; 254(16):7691-97.
7. Holowka D, Hartmann H, Kanellopolous J et al. Association of the receptor for immunoglobulin E with an endogenous polypeptide on rat basophilic leukemia cells. J Recept Res 1980; 1:41-68.
8. Holowka D, Metzger H. Further characterization of the β-component of the receptor for immunoglobulin E. Mol Immunol 1982; 19(2):219-27.
9. Perez-Montfort, R Kinet J-P, Metzger H. A previously unrecognised subunit of the receptor for immunoglobulin E. Biochemistry 1983; 22:5722-28.
10. Alcaraz G, Kinet J-P, Kumar N et al. Phase separation of the receptor for immunoglobulin E and its subunits in Triton X-114 J Biol Chem 1984; 259(23):14922-27.
11. Rivnay B, Wank SA, Poy G et al. Phospholipids stabilize the interaction between the α and β subunits of the solubilized receptor for immunoglobulin E. Biochemistry 1982; 21:6922-27.
12. Kinet J-P, Alcaraz G, Leonard A et al. Dissociation of the receptor for immunoglobulin E in mild detergents Biochemistry 1985; 24:4117-24.
13. Metzger H, Kinet J-P, Perez-Montfort R et al. A tetrameric model for the structure of the mast cell receptor for immunoglobulin E. Prog Immunol 1983; 5:493-501.
14. Metzger H, Alcaraz G, Hohman R et al. The receptor with high affinity for immunoglobulin E. Ann Rev Immunol 1986; 4:419-70.
15. Blank U, Ra C, Miller L et al. Complete structure and expression in transfected cells of high affinity IgE receptor. Nature 1989; 337:187-89.
16. Kinet J-P, Metzger H, Hakimi J et al. A cDNA presumptively coding for the α subunit of the receptor with high affinity for immunoglobulin E. Biochemistry 1987; 26:4605-10.
17. Kochan J, Pettine LF, Hakimi J et al. Isolation of the gene coding for the α subunit of the human high affinity IgE receptor. Nuc Acid Research 1988; 16:3584.
18. Shimizu A, Tepler I, Benfey PN et al. Human and rat mast cell high-affinity immunoglobulin E receptor: characterization of putative α chain gene products. Proc Natl Acad Sci USA 1988; 85:1907-11.
19. Liu F-T, Albrandt K, Robertson MW. cDNA heterogeneity suggests structural variants related to the high-affinity IgE receptor. Proc Natl Acad Sci USA 1988; 85:5639-43.
20. Ra C, Jouvin M-HE, Kinet J-P. Complete structure of the mouse mast cell receptor for IgE (FcεRI) and surface expression of chimeric receptors (rat-mouse-human) on transfected cells. J Biol Chem 1989b; 264(26):15323-27.
21. Tepler I, Shimizu A, Leder P. The gene for the rat mast cell high affinity IgE receptor α chain. Structure and alternative mRNA splicing patterns. J Biol Chem 1989; 264(10):5912-15.
22. Varin-Blank N, Metzger H. Surface expression of mutated subunits of the high affinity mast cell receptor for IgE. J Biol Chem 1990; 265(26):15685-94.
23. Farber DL, Sears DW. Rat CD16 is defined by a family of class III Fcγ receptors requiring coexpression of heteroprotein subunits. J Immunol 1991; 146(12):4352-61.
24. Ravetch JV, Kinet J-P. Fc Receptors. Annu Rev Immunol 1991; 9:457-92.
25. Ye Z-S, Kinet J-P, Paul WE. Structure of the gene for the α chain of the mouse high affinity receptor for IgE FcεRI. J Immunol 1992; 149(3):897-900.
26. Pang J, Taylor GR, Munroe DG et al. Characterization of the gene for the human high affinity IgE receptor (FcεRI) α-chain. J Immunol 1993; 151:6166-74.
27. Ravetch JV, Perussia B. Alternative membrane forms of FcγRIII (CD16) on human NK cells and neutrophils. Cell-type spe-

cific expression of two genes which differ in single nucleotide substitutions. J Exp Med 1989; 170:481-91.

28. Qui WQ, De Bruin D, Brownstein BH et al. Organization of the human and mouse low-affinity FcγR genes: duplication and recombination. Science 1990; 248:732-35.
29. Le Conait M, Kinet J-P, Berger R. The human genes for the α and γ subunits of the mast cell receptor for immunoglobulin E are located on human chromosome band 1q23. Immunogenetics 1990; 32:183-86.
30. Kinet J-P, Blank U, Ra C et al. Isolation and characterization of cDNAs coding for the β subunit of the high-affinity receptor for immunoglobulin E. Proc Natl Acad Sci USA 1988; 85:6483-87.
31. Küster H, Zhang L, Brini AT et al. The gene and cDNA for the human high affinity immunoglobulin E receptor β chain and expression of the complete human receptor. J Biol Chem 1992; 267(18):12782-87.
32. Zloh M, Anderson G, Clark-Lewis I et al. N.m.r. studies of the cytoplasmic C-terminal β-subunit domain of the high-affinity IgE receptor. Biochem Soc Trans 1994a; 22:1027-29.
33. Zloh M, Anderson G, Clark-Lewis I et al. Spectroscopic and conformational studies of the C-terminal cytoplasmic β subunit 46-peptide of the high affinity IgE receptor. Biochem Soc Trans 1994b; 22:450S.
34. Sandford AJ, Shirakawa T, Moffat MF et al. Localisation of atopy and β subunit of high-affinity IgE receptor (FcεRI) on chromosome 11q. Lancet 1993; 341:332-34.
35. Hogarth PM, Houlden BA, Latham SE et al. Ly-12.1, a lymphocyte specific alloantigen encoded by a gene linked to Ly-1. Immunogenetics 1988; 27:383-87.
36. Huppi K, Siwarski D, Mock BA et al. Gene mapping of the three subunits of the high affinity FcR for IgE to mouse chromosomes 1 and 19. J Immunol 1989; 143:3787-91.
37. Küster H, Thompson H, Kinet J-P. Characterization and expression of the gene for the human Fc receptor γ subunit: Definition of a new gene family. J Biol Chem 1990; 265(11):6448-52.
38. Orloff DG, Ra C, Frank SJ et al. Family of disulphide-linked dimers containing the ζ and η chains of the T cell receptor and the γ chain of Fc receptors. 1990 Nature 347:189-91.
39. Anderson GJ, Haris PI, Chapman D et al. Synthesis and spectroscopy of membrane receptor proteins. The γ subunit of the IgE receptor. Eur J Biochem 1992; 207:51-54.
40. Masuda M, Roos D. Association of all three types of FcγR (CD64, CD32, CD16) with a γ chain homodimer in cultured human monocytes. J Immunol 1993; 151:7188-95.
41. Ra C, Jouvin M-H, Blank U et al. A macrophage Fcγ receptor and the mast cell receptor for immunglobulin E share an identical subunit. Nature 1989a; 341:752-54.
42. Ernst LK, Duchemin AM, Anderson CL. Association of the high affinity receptor for IgG (FcγRI) with the γ subunit of the IgE receptor. Proc Natl Acad Sci USA 1993; 90:6023-27.
43. Geha RS, Helm BA, Gould H. Inhibition of the Prausnitz-Küstner reaction by an immunoglobulin ε-chain fragment synthesised in *E. coli.* Nature 1985; 315:577-78.
44. Scholl PR, Geha RS. Physical association between the high-affinity IgG receptor (FcγRI) and the γ subunit of the high-affinity IgE receptor. Proc Natl Acad Sci USA 1993; 90:8847-50.
45. Morton HC, Van den Herikoudijk IE, Vossebeld P et al. Functional association between the human myeloid immunoglobulin A Fc receptor (CD89) and FcR γ chain—molecular basis for CD89/FcR γ chain association. J Biol Chem 1995; 270(50):29781-87.
46. Viver E, Rochet N, Kochan J et al. Structural similarity between Fc receptors and T cell receptors. Expression of the γ-subunit of FcεRI in human T cells, natural killer cells and thymocytes. J Immunol 1991; 147(12):4263-70.
47. Qian D, Sperling AI, Lancki DW et al. The γ chain of the high-affinity receptor for IgE is a major functional subunit of the T cell antigen receptor complex in γδ T lymphocytes. Proc Natl Acad Sci USA 1993; 90:11875-79.
48. Miller L, Blank U, Metzger H et al. Expression of high-affinity binding of human immunoglobulin E by transfected cells.

Science 1989; 244:334-37.

49. Jürgens M, Wollenberg A, Hanau D et al. Activation of human epidermal Langerhans cells by engagement of the high affinity receptor for IgE, FcεRI. J Immunol 1995; 155:5184-89.
50. Mao S-Y, Varin-Blank N, Edidin M et al. Immobilization and internalization of mutated IgE receptors in transfected cells. J Immunol 1991; 146(3):958-66.
51. Anderson P, Caligiuri M, O'Brien C et al. Fcγ receptor type III (CD16) is included in the ζ NK receptor complex expressed by human natural killer cells. Proc Natl Acad Sci USA 1990; 87:2274-78.
52. Howard FD, Rodewald H-R, Kinet J-P et al. CD3ζ subunit can substitute for the γ subunit of FcεRI receptor type I in assembly and functional expression of the high-affinity IgE receptor: evidence for inter-receptor complementation. Proc Natl Acad Sci USA 1990; 87:7015-19.
53. Hakimi JC, Seals C, Kondas JA et al. The α subunit of the human IgE receptor (FcεRI) is sufficient for high affinity IgE binding. J Biol Chem 1990; 265(36):22079-81.
54. Hulett MD, McKenzie IFC, Hogarth PM. Chimeric Fc receptors identify immunoglobulin-binding regions in human FcγRII and FcεRI. Eur J Immunol 1993; 23:640-45.
55. Letourneur O, Sechi S, Willette-Brown J et al. Glycosylation of human truncated FcεRI α chain is necessary for efficient folding in the endoplasmic recticulum. J Biol Chem 1995; 270(14):8249-56.
56. Helm BA, Ling Y, Teale C et al. The nature and importance of the inter-ε chain disulphide bonds in human IgE. Eur J Immunol 1991; 21:1543-48.
57. Dorrington KJ, Bennich HH. Structure-function relationship in human immunoglobulin E. Imunol Rev 1978; 41:3-25.
58. Padlan EA, Davies DR. A model of the Fc of immunoglobulin E. Mol Immunol 1986; 23(10):1063-75.
59. Takatsu K, Ishizaka T, Ishizaka K. Biologic significance of disulphide bonds in human IgE molecules. J Immunol 1975; 114: 1838-45.
60. Rousseaux-Prevost R, Rousseaux J, Bazin H et al. Differential reduction of the inter-chain disulphide bonds of rat immunoglobulin E: relation to biological activity. Mol Immunol 1984; 21:233-41.
61. Helm BA, Marsh P, Vercelli D et al. The mast cell binding site on human immunoglobulin E. Nature 1988; 331:180-83.
62. Padlan EA, Helm BA. A modeling study of the α-subunit of human high-affinity receptor for immunoglobulin-E. Receptor 1992; 2:129-44.
63. Holowka D, Wensel T, Baird B. A nanosecond fluorescence depolarization study on the segmental flexibility of receptor-bound immunoglobulin E. Biochemistry 1990; 29:4607-12.
64. Zheng Y, Shopes B, Holowka D et al. Conformations of IgE bound to its receptor FcεRI and in solution. Biochemistry 1991; 30(38):9125-32.
65. Davis KG, Glennie M, Harding SE et al. A model for the solution conformation of rat IgE. Biochem Soc Trans 1990; 18:935-36.
66. Beavil AJ, Beavil RL, Chan CMW et al. Structural basis of the IgE-FcεRI interaction. Biochem Soc Trans 1993; 21(4): 968-72.
67. Zheng Y, Shopes B, Holowka D et al. Dynamic conformations compared for IgE and IgG1 in solution and bound to receptors. Biochemistry. 1992; 31(33):7446-56.
68. Beavil AJ, Young RJ, Sutton BJ et al. Bent domain structure of recombinant human IgE-Fc in solution by X-ray and neutron scattering in conjunction with an automated curve fitting procedure. Biochemistry 1995; 34:14449-61.
69. Holowka D, Conrad DH, Baird B. Structural mapping of membrane-bound immunoglobulin E-receptor complexes: Use of monoclonal anti-IgE antibodies to probe the conformation of receptor-bound IgE. Biochemistry 1985; 24:6260-67.
70. Keown M, Ghirlando R, Young RJ et al. Hydrodynamic studies of a complex between the Fc fragment of human IgE and a soluble fragment of the FcεRI α chain. Proc Natl Acad Sci 1995; 92:1841-54.
71. Kanellopolous J, Liu TY, Poy G et al. Composition and subunit structure of the cell receptor for immunoglobulin E. J Biol Chem 1980; 255(19):9060-66.

72. Ra C, Kuromitsu S, Hirose T et al. Soluble human high-affinity receptor for IgE abrogates the IgE-mediated allergic reaction. International Immunology 1993; 5(1): 47-54.
73. Robertson MW. Phage and Escherichia coli expression of the human high affinity immunoglobulin E receptor α-subunit ectodomain. J Biol Chem 1993; 268(17): 12736-43.
74. Kershaw MH, Darcy PK, Hulett MD et al. Redirected cytotoxic effector function: requirements of chimeric single chain high affinity IgE receptors. J Biol Chem 1996 (in press).
75. Baird B, Holowka D. Structural mapping of Fc receptor bound immunoglobulin E: proximity to the membrane surface of the antibody combining site and another site in the Fab segments. Biochemistry 1985; 24:6252-59.
76. Slattery J, Holowka D, Baird B. Segmental flexibility of receptor-bound immunoglobulin E. Biochemistry 1985; 24:7810-20.
77. Ortega E, Licht A, Biener Y et al. A glycolipid-specific monoclonal antibody modulates Fcε receptor stimulation of mast cells. Molec Immunol 1990; 27(12):1269-77.
78. Metzger H. Molecular aspects of receptors and binding factors for IgE. Adv Immunol 1988; 43:277-12.
79. Burton DR, Woof JM. Human antibody effector function. Adv Immunol 1992; 51:1-84.
80. Sutton BJ, Gould HJ. The human IgE network. Nature 1993; 366:421-28.
81. Kochan JP, Mallamaci M, Gilfillan A et al. Characterization of the human IgE Fc-FcεRIα interaction. In: M.Z. Atassi (ed). Immunobiology of Proteins and Peptides VII. NY: Plenum Press 1994; 31-38.
82. Hulett MD, Hogarth PM. Molecular basis of Fc receptor function. Adv Immunol 1994; 57:1-27.
83. Blank U, Ra C, Kinet J-P. Characterization of truncated α chain products from human, rat and mouse high affinity receptor for immunoglobulin E. J Biol Chem 1991; 266(4):2639-46.
84. Riske F, Hakimi J, Mallamaci M et al. High affinity human IgE receptor (FcεRI). Analysis of functional domains of the α-subunit with monoclonal antibodies. J Biol Chem 1991; 266(17):11245-51.
85. Hogarth PM, Hulett MD, Ierino FL et al. Identification of the immunoglobulin binding regions (IBR) of FcγRII and FcεRI. Immunological Reviews 1992; 125:21-35.
86. Mallamaci MA, Chizzonite R, Griffin M et al. Identification of sites on the human FcεRIα subunit which are involved in binding human and rat IgE. J Biol Chem 1993; 268(29):22076-83.
87. Scarselli E, Esposito G, Traboni C. Receptor Phage. Display of functional domains of the human high affinity IgE receptor on the M13 phage surface. FEBS Let 1993; 329:223-26.
88. Padlan EA, Helm BA. Modelling study of IgE/receptor interactions. Biochem Soc Trans 1993; 21(4):963-67.
89. McDonnell JM, Beavil AJ, Mackay et al. Structure based design and characterisation of peptides that inhibit IgE binding to its high affinity receptor. Nature Struct Biol 1996; 3(5):419-426.
90. Hulett MD, Osman N, McKenzie IFC et al. Chimeric Fc receptors identify functional domains of the murine FcγRI high affinity receptor for IgG. J Immunol 1991; 147: 1863-68.
91. Hulett MD, Witort E, Brinkworth RI et al. Identification of the IgG binding site of the low affinity receptor for IgG FcγRII: enhancement and ablation of binding site directed mutagenesis. J Biol Chem 1994; 269:15287:93.
92. Hulett MD, Witort E, Brinkworth RI et al. Multiple regions of human FcγRII (CD32) contribute to the binding of IgG. J Biol Chem 1995; 270(36):21188-94.
93. Hibbs ML, Tolvanen M, Carpen O. Membrane-proximal Ig-like domain of FcγRIII (CD16) contains residues critical for ligand binding. J Immunol 1994; 152:4466-74.
94. Ishizaka T, Ishizaka K. Biology of immunoglobulin E. Prog. Allergy 1975; 19:60-21.
95. Liu F-T, Albrandt K, Bry CG et al. Expression of a biologically active fragment of human IgE ε chain in *Escherichia coli.* Proc Natl Acad Sci USA 1984; 81:5369-73.
96. Perez-Montfort R, Metzger H. Proteolysis

of soluble IgE-receptor complexes: Localization of sites on IgE which interact with the Fc receptor. Mol Immunol 1982; 19: 1113-25.

97. Hamburger R. Peptide inhibition of the Prausnitz-Küstner reaction. Science 1975; 189:389-90.

98. Bennich H, Ragnarsson U, Johansson SGO et al. Failure of the putative IgE pentapeptide to compete with IgE for receptors on basophils and mast cells. Int Archs Allergy appl Immun 1977; 53:459-68.

99. Duncan AR, Woof JM, Partridge LJ et al. Localization of the binding site for the human high-affinity Fc receptor on IgG. Nature 1988; 332:563-64.

100. Canfield SM, Morrison SL. The binding affinity of human IgG for its high affinity Fc receptor is determined by multiple amino acids in the CH2 domain and is modulated by the hinge region. J Exp Med 1991; 173:1483-91.

101. Presta L, Shields R, O'Connell L et al. The binding site on human immunoglobulin E for its high affinity receptor. J Biol Chem 1994; 269(42):26368-73.

102. Helm BA, Kebo D, Vercelli D et al. Blocking of passive sensitization of human mast cells and basophil granulocytes with IgE antibodies by a recombinant human-ε chain fragment of 76 amino acids. Proc Natl Acad Sci USA 1989; 86:9465-69.

103. Basu M, Hakimi J, Dharm E et al. Purification and characterization of human recombinant IgE-Fc fragments that bind to the human high affinity IgE receptor. J Biol Chem 1993; 268(18): 13118-27.

104. Helm BA, Sayers I, Higginbottom A et al. Identification of the high-affinity receptor binding region in immunoglobulin E. J Biol Chem 1996; 271:7494-7500.

105. Stanworth DR, Humphrey JH, Bennich H et al. Inhibition of Prausnitz-Küstner reaction by proteolytic-cleavage fragments of a human myeloma protein of immunoglobulin class E. Lancet 1968; 11:17-18.

106. Nissim A, Jouvin M-H, Eshhar Z. Mapping of the high affinity Fcε receptor binding site to the third constant region domain of IgE. EMBO J 1991; 10:101-107.

107. Takemoto H, Nishimura S, Kosada Y et al. Anti-human IgE monoclonal antibodies recognizing epitopes related to the binding sites of high and low affinity IgE receptors. Microbiol Immunol 1994; 38(1):63-71.

108. Nissim A, Schwartzbaum S, Siraganian R et al. Fine specificity of the IgE interaction with the low and high affinity Fc receptor. J Immunol 1993; 150(4):1365-74.

109. Baniyash M, Eshhar Z, Rivnay B. Relationships between epitopes on IgE recognized by defined monoclonal antibodies and by the Fcε receptor on basophils. J Immunol 1986; 136:588-93.

110. Conrad DH, Wingard JR, Ishizaka T. The interaction of human and rodent IgE with the human basophil IgE receptor. J Immunol 1983; 130(1):327-33.

111. Weetall M, Shopes B, Holowka D et al. Mapping the site of interaction between murine IgE and its high affinity receptor with chimeric Ig. J Immunol 1990; 145(11):3849-54.

112. Keegan AD, Fratazzi C, Shopes B et al. Characterisation of new rat anti-mouse IgE monoclonals and their use along with chimeric IgE to further define the site that interacts with FcεRII and FcεRI. Molec Immunol 1991; 28:1149-54.

113. Schwarzbaum S, Nissim A, Alkalay I et al. Mapping of murine IgE epitopes involved in IgE-Fcε receptor interactions. Eur J Immunol 1989; 19:1015-23.

114. Helm BA. Is there a link between the nature of agents that trigger mast cells and the induction of immunoglobulin (Ig)E synthesis? In: Atassi MZ, ed. Immunobiology of Proteins and Peptides VII. New York: Plenum Press, 1994:1-10.

115. Burmeister WP, Huber AH, Bjorkman PJ. Crystal structure of the complex of rat neonatal Fc receptor with Fc. Nature 1994; 372:379-83.

116. Burt DS, Hastings GZ, Healy J et al. Analysis of the interaction between rat immunoglobulin E and rat mast cells using antipeptide antibodies. Mol Immunol 1987; 24(4):379-89.

117. Coleman JW, Helm BA, Stanworth DR et al. Inhibition of mast cell sensitization in vitro by a human immunoglobulin ε-chain fragment synthesized in Escherichia coli. Eur

J Immunol 1985; 15:966-69.
118. Nio N, Seguro K, Ariyoshi Y et al. Inhibition of histamine release by synthetic human IgE peptide fragments: Structure-activity studies. In: Yanaihara N, ed. Peptide Chemistry 1989. Osaka: Protein Res Found, 1990:204-208.
119. Nissim A, Eshhar Z. The human mast cell receptor binding site maps to the third constant domain of immunoglobulin E. Molec Immunol 1992; 29(9):1065-72.
120. Takizawa F, Adamczewski M, Kinet J-P. Identification of the low affinity receptor for immunoglobulin E on mouse mast cells and macrphages as FcγRII and FcγRIII. J Exp Med 1992; 176:469-75.

CHAPTER 3

FcεRI-Induced Protein Tyrosine Phosphorylation

Marc Benhamou

INTRODUCTION

Phosphorylation of proteins on tyrosine residues modulates the biological activity of a number of polypeptides and controls many protein-protein interactions thereby regulating the subcellular location of these proteins.[1-4] For instance, p60*src*-related protein-tyrosine kinases (PTKs) are regulated by the phosphorylation state of a critical carboxy-terminal tyrosine residue.[1] Also, SH2 domains that are present in many proteins such as phospholipase C-γ1 bind to phosphotyrosine-containing polypeptides thereby promoting association between proteins subsequently to tyrosine phosphorylation.[2] Thus, tyrosine phosphorylation emerges as a major regulatory mechanism for cell signaling.

Receptors such as growth factor receptors are transmembrane PTKs. Others, such as FcεRI, are associated with genuine PTKs and, upon aggregation, lead to tyrosine phosphorylation of several proteins, including the receptor itself. FcεRI tyrosine phosphorylation is exclusively on cytoplasmic domain of FcεRI-β and FcεRI-γ chains.[5-7] These domains contain one Immunoreceptor Tyrosine-based Activation Motif (ITAM)[8] that is the site for tyrosine phosphorylation[5-7] and that is critical for signal transduction in many receptor systems. This ITAM ($D/EX_7D/EX_2YX_2L/IX_7YX_2L/I$), first described by Reth,[9] is shared by receptors that form the Multichain Immune Recognition Receptor (MIRR) family.[10] These receptors include T cell and B cell receptors for antigen, FcεRI, FcγRI, FcγRIIα, FcγRIIIα and FcαR. FcεRI-β-ITAM (βITAM) and FcεRI-γ–ITAM (γITAM) are not equivalent since βITAM tyrosine phosphorylation does not depend on γITAM tyrosine residues,[5] whereas γITAM tyrosine phosphorylation greatly depends on βITAM tyrosine residues.[6] In addition,

IgE Receptor (FcεRI) Function in Mast Cells and Basophils,
edited by Majed M. Hamawy.

these ITAMs differ from each other by their signaling capabilities (see below). Whatever these differences, tyrosine phosphorylation of FcεRI is critical for cell activation since association with the receptor of a large signaling complex that includes tyrosine phosphorylated proteins and PTKs depends on it.[11]

Dependency of FcεRI-mediated secretory signal on these phosphorylations was evidenced by the strong inhibitory effect of all PTK inhibitors tested to date.[12-19] The extent of these phosphorylations results from the balance between PTK activity and phosphotyrosine phosphatase (PTPase) activity that removes the phosphate group from phosphotyrosine residue.[20-23] This is best evidenced in experiments where tyrosine phosphorylation induced after FcεRI aggregation by multivalent antigen disappears upon addition of excess monovalent hapten.[11,24-26] In this review I shall concentrate on the "PTK side" of these phosphorylations. Regulation of FcεRI-dependent tyrosine phosphorylations by PTPases is presented elsewhere (see chapter 7 in this book).

THE FcεRI SIGNALING UNIT

Phosphorylation of FcεRI-β and -γ chains on tyrosine is the first molecular alteration observed after receptor aggregation.[25,27] Since the receptor lacks any intrinsic tyrosine kinase activity,[28] this phosphorylation results from activation of associated PTKs and/or inactivation of associated PTPases. Whereas the existence of FcεRI-associated PTK activity has been demonstrated in 1986 by Quarto and Metzger,[28] it is only recently that a PTPase activity was detected that is physically associated with the receptor in a constitutive way.[23] This PTPase dephosphorylates in vitro the phosphorylated FcεRI chains but not other receptor-associated PTK substrates. It was proposed that this PTPase is responsible for the down-regulation of FcεRI phosphorylation and signaling.[23] This aspect of FcεRI signaling is discussed in chapter 7.

MECHANISMS OF FcεRI-INDUCED TYROSINE PHOSPHORYLATION

Tyrosine Phosphorylation of FcεRI Subunits

Tyrosine phosphorylation of FcεRI-β and γ subunits is detected within 5 second stimulation with multivalent antigen and is fully reversed upon addition of excess monovalent hapten.[25,26] In 1987, Seagrave et al reported that addition of excess hapten to antigen-stimulated mast cells resulted in immediate cessation of activation signal but not in immediate disaggregation of aggregated receptors.[29] These data suggested that hapten acted by preventing recruitment of new receptors rather than by disrupting aggregates of receptors. It seemed therefore that tyrosine phosphorylation of aggregated receptors resulted from the aggregation process rather than from the aggregation state. To clarify this question several recent studies were carefully conducted.

To allow for mathematical analysis of data the stimulation system was simplified and dimeric or trimeric IgE were used instead of multivalent antigen. Addition of excess monomeric IgE after 2 minute stimulation with IgE oligomers blocked any further recruitment of new receptors but did not induce disaggregation of aggregated receptors.[30] Under these conditions, the level of FcεRI tyrosine phosphorylation that was reached after 2 minute stimulation stopped increasing. Yet, it declined only very slowly and not abruptly as in the antigen/hapten system.[31] Furthermore, the aggregated FcεRI were still active inasmuch as they were still able to induce continued β-hexosaminidase release. Similar results were obtained with permeabilized mast cells. In addition, under the latter conditions, chelating Mg^{2+} (required for PTKs activity), 6 minutes after addition of IgE monomer, resulted in immediate disappearence of oligomer-induced tyrosine phosphorylation of the receptor.[31] Thus, when recruitment of new receptors was blocked by IgE monomers, aggregated FcεRI were still actively phosphorylated by virtue of a balance between PTK and PTPase activities that was in favor of PTKs. Additional studies demonstrated that during the first 20 minute stimulation with dimeric IgE the kinetics of tyrosine phosphorylation of the receptor paralleled the *total number* of receptor aggregates and not the *rate* of aggregation.[30] To account for the apparent discrepancy between these results and those obtained in the antigen/hapten system, Wofsy et al have postulated that within receptor aggregates, immobilized/desensitized (i.e., irreversibly inactivated) receptors coexist in increasing proportion together with mobile/active receptors.[30] This model takes into account previous observations that activation leads to time-dependent stable interaction of FcεRI with cytoskeleton[32,33] resulting in immobilization of the former. This is accompanied by a time-dependent increase in ligand rebinding to IgE-sensitized FcεRI at the cell surface[34] and in receptor desensitization.[35] Interaction of immobilized receptors with antigen would be predictably more stable than interaction of mobile receptors. Addition of hapten would release the mobile/active receptors from aggregates but not the immobilized/inactive ones, resulting in immediate cessation of the signal without dissociation of antigen from the cell surface.[30] By contrast, IgE monomers would be unable to displace oligomeric IgE from FcεRI because of the extremely low dissociation rate between oligomeric IgE and FcεRI. Thus, it is not the aggregation process but the aggregation state that is responsible for tyrosine phosphorylation of FcεRI and activation signal.

In membrane preparations, aggregation of FcεRI leads to tyrosine phosphorylation of FcεRI-β and -γ that is halted upon addition of excess hapten.[26] Moreover, in contrast to the whole cell system, FcεRI tyrosine phosphorylation in membrane preparations is almost exclusive of other substrates. These data indicate that the *minimal* machinery responsible for FcεRI aggregation-dependent tyrosine phosphorylation does not require cytosolic specimens. This machinery includes an FcεRI-associated PTK activity[28] that can phosphorylate exogenous substrates as well as FcεRI and is increased after FcεRI aggregation by antigen.[36] After receptor aggregation and solubilization, phosphorylation of exogenous substrate by FcεRI-associated PTK is linear with time whereas that of FcεRI rapidly reaches a plateau. After disaggregation of these solubilized receptors by 1 hour incubation with excess hapten, this PTK activity remains elevated toward exogenous substrate but not toward the disaggregated receptor itself.[36] These observations are consistent with transphosphorylation as the mechanism by which FcεRI is phosphorylated and provide an explanation as to why FcεRI aggregation is needed for activation.

Role of p53/56^{lyn}

Several PTKs could account for this FcεRI-associated PTK activity. These include the p60^{src}-related p53/56^{lyn},[37] the ZAP70-related p72^{syk}/PTK72[38,39] and the *tec*-related p77^{btk}.[40] Yet, the latter has not been detected in anti-FcεRI immunoprecipitates[41] and p72^{syk} is coprecipitated with FcεRI only after stimulation.[39] Thus, the only membrane-associated PTK interacting with FcεRI before (and after) stimulation that has been reported to date is p53/56^{lyn}.[37] This makes p53/56^{lyn} the prime candidate for the PTK responsible for FcεRI tyrosine phosphorylation. Consistent with this comment is the observation that p53/56^{lyn} is activated shortly after FcεRI engagement[37] and that p53/56^{lyn} depletion from cell lysates before and after stimulation removes the FcεRI-associated PTK activity.[36,42]

P60^{src}-related PTKs share a common general organization with an amino-terminal myristilation site that anchors them to the cytoplasmic aspect of the plasma membrane, followed by one SH2 and one SH3 domain and the catalytic domain.[1] Thus, these PTKs are perfectly localized to interact with trans-membrane receptors and with effector proteins by association through SH domains. Regarding FcεRI, experiments with cleavable crosslinkers led to an estimate of 3-20% receptors that are associated with $p53/56^{lyn}$ under resting conditions.[42]

Receptor aggregation leads to $p53/56^{lyn}$ activation and to recruitment of new $p53/56^{lyn}$ molecules to the tyrosine phosphorylated receptor.[42] Estimates show a 3- to 10-fold increase in the $p53/56^{lyn}$ association with FcεRI together with an up to 12-fold increase in FcεRI-associated $p53/56^{lyn}$-autophosphorylating activity per $p53/56^{lyn}$ molecule.[42] This could account for the up to 100-fold increase in FcεRI-associated PTK activity toward FcεRI-β and γ that is observed after receptor ligation.[36,42] This good correlation between FcεRI-associated $p53/56^{lyn}$ kinase activity and the phosphorylating activity toward FcεRI-β and γ that is associated with the receptor strongly suggests that $p53/56^{lyn}$ is the PTK responsible for FcεRI tyrosine phosphorylation.

In a recent study where vaccinia virus was used as a vector to express receptors and PTKs in "null" cell lines, aggregation-dependent tyrosine phosphorylation of FcεRI required coexpression of $p53/56^{lyn}$.[7] The mechanism for further recruitment of $p53/56^{lyn}$ to activated receptors seems to involve SH2/phosphotyrosine interaction since it is dependent on kinase activity[42] and since the increase in associated PTK activity correlates with the extent of tyrosine phosphorylation of the receptor.[36] Thus, the current model involves transphosphorylation of adjacent FcεRI by receptor-associated, aggregation-activated $p53/56^{lyn}$ and further recruitment and activation of $p53/56^{lyn}$ through interaction of lynSH2 with tyrosine phosphorylated FcεRI. This mechanism would propagate receptor phosphorylation within a receptor aggregate thereby amplifying the activation signal.

Since FcεRI is composed of four subunits ($\alpha\beta(\gamma)_2$) the question arises as to which is (are) the subunit(s) that physically interacts with $p53/56^{lyn}$. To answer that question investigators have designed chimeric molecules with the cytoplasmic C-terminal domain of FcεRI-β or the cytoplasmic domain of FcεRI-γ fused to irrelevant Tac transmembrane and extracellular molecules. These chimeras, named TTβ and TTγ, respectively, were expressed into RBL-2H3 rat mast cells and their association with $p53/56^{lyn}$ was assessed.[6,43]

Coprecipitation experiments demonstrated that $p53/56^{lyn}$ associates with unstimulated TTβ but not with TTγ,[6] suggesting that FcεRI-β is the receptor subunit that interacts with $p53/56^{lyn}$ under a resting state. Yet, aggregation of TTβ did not result in $p53/56^{lyn}$ activation, TTβ tyrosine phosphorylation or cell activation,[6,43] i.e., the first and last events following receptor clustering were not reproduced in this artificial system. The capacity of $p53/56^{lyn}$ to interact with tyrosine phosphorylated receptors was also investigated with the SH2 domain of $p53/56^{lyn}$ fused to GST.[44,45] This construct bound tyrosine phosphorylated FcεRI-β (but not FcεRI-γ) immobilized on nitrocellulose membranes and could be used to precipitate the tyrosine phosphorylated receptor.[44,45] This precipitation was inhibited in the presence of tyrosine phosphorylated βITAM peptide but not phospho-γITAM peptide.[45] Thus $p53/56^{lyn}$ associates preferentially, if not exclusively, with the FcεRI-β chain before and after stimulation.

Interestingly, P815 mouse mast cell line is negative for FcεRI-α and β but positive for FcεRI-γ. Cotransfection of the missing subunits allowed reconstitution of functional FcεRI expression.[46,47] By contrast, no signal was generated upon aggregation of FcεRI when the cotransfected FcεRI-β had the C-terminal cytoplasmic domain deleted.[47] When the cotransfected FcεRI-β chain had all the ITAM tyrosines replaced by phenylalanines, aggregation-induced tyrosine phosphorylation of the FcεRI-β chain was abolished and that of the FcεRI-γ chain was dramatically reduced.[6] FcεRI-associated PTK activity was absent in Y-to-F mutated FcεRI-β transfectants and did not increase after stimulation. This correlated with no detectable constitutive association of p53/56^{lyn} with the mutated receptor.[6] Thus, whereas p53/56^{lyn} association with FcεRI-β is dependent on the tyrosine residues within βITAM (Y-to-F mutation abolishes this interaction),[6] it does not depend on phosphorylation of these residues (TTβ and FcεRI-β associate with p53/56^{lyn} under resting conditions where tyrosine phosphorylation is either absent in the case of TTβ, or minimal in the case of FcεRI-β).[6] Altogether, these data are consistent with p53/56^{lyn} phosphorylating FcεRI and associating with FcεRI-β before and after stimulation through different types of interaction. The constitutive interaction would be independent of tyrosine-phosphorylated βITAM, but dependent of βITAM, whereas recruitment would be through tyrosine phosphorylated βITAM/lynSH2 interaction.

Role of p72^{syk}

P72^{syk} (also named PTK72) is a cytosolic PTK that contains two SH2 domains in tandem.[39,48,49] Tyrosine phosphorylation of FcεRI promotes rapid activation of p72^{syk} that is independent of calcium influx or PKC activation.[38,39,50] Yet, whereas we reproducibly observed association of p72^{syk} with tyrosine phosphorylated receptor only,[39] others reported constitutive association of p72^{syk} with FcεRI.[38] These contradictory data echo those obtained with other members of the MIRR family: some authors observed constitutive association of p72^{syk} with T cell and B cell receptor[51-54] whereas others reported such association only after receptor engagement.[55,56] Such discrepancy could be explained by many parameters such as the difference in the anti-p72^{syk} antibodies used in these studies, the detergent used for cell lysis or the degree of basal activation of the cells before stimulation. Thus, the question of the constitutive association of p72^{syk} with the receptor needs further clarification.

Syk-SH2/GST fusion proteins were used to analyze p72^{syk} interaction with FcεRI subunits. The syk-SH2/GST fusion proteins bound to tyrosine phosphorylated FcεRI-β and FcεRI-γ subunits: although no binding was detected with amino-terminal Syk-SH2 domain, poor binding was observed with carboxy-terminal Syk-SH2 domain and strong binding when both domains were fused in tandem to GST.[44,45] The latter fusion protein (Syk-SH2$_{NC}$/GST) inhibited association of cellular p72^{syk} with the tyrosine phosphorylated receptor.[44] Conversely, phospho-γITAM and phospho-βITAM peptides inhibited interaction of Syk-SH2$_{NC}$/GST fusion protein with tyrosine phosphorylated FcεRI.[45] Yet, more phospho-βITAM than phospho-γITAM peptides were necessary to inhibit this interaction and more endogenous phospho-FcεRI-γ than phospho-FcεRI-β were precipitated by sykSH2$_{NC}$-GST fusion protein.[44,45] As well, more cellular p72^{syk} was precipitated with phospho-γITAM than with phospho-βITAM peptides, and no p72^{syk} was precipitated with the nontyrosine phosphorylated peptides.[45] Thus, p72^{syk} preferentially interacts with tyrosine phosphorylated FcεRI-γ compared to tyrosine phosphorylated FcεRI-β. Interestingly, aggregation of TTγ, but not TTβ, on rat mast cells transfected with these constructs resulted in p72^{syk} activation[43] and phosphorylation.[6,43] Therefore FcεRI-γ is the chain that governs p72^{syk} mobilization after FcεRI ligation.

The question of p72^{syk} activation was addressed in several reports. A B cell line expressing wild-type p53/56^{lyn} predominantly over other *src*-related PTKs showed clear activation of p72^{syk} after B cell receptor ligation. In *lyn*-negative mutants obtained by disruption of *lyn* gene, no receptor-dependent activation of p72^{syk} was observed anymore.[57] In another study, tyrosine phosphorylation and activation of p72^{syk} after FcεRI aggregation was dramatically increased when p53/56^{lyn} was coexpressed with the receptor in a system using "null" cell lines infected with vaccinia virus as a vector to express FcεRI and PTKs.[7] Thus, p53/56^{lyn} appears pivotal to p72^{syk} activation.

In an artificial system consisting of COS cells transfected with *lyn* and *syk* cDNA separately or in combination, with or without a point mutation abrogating their kinase activity, there was substantial increase in p72^{syk} kinase activity when p53/56^{lyn} was coexpressed with it.[57] In addition, p72^{syk} tyrosine phosphorylation was dependent on p53/56^{lyn} coexpression.[57] Interestingly, p72^{syk} tyrosine phosphorylation was dramatically less intense when the kinase activity of p72^{syk} was abrogated demonstrating both hetero- and auto-phosphorylation of p72^{syk}.[57]

In the "null" cell model described above, p53/56^{lyn} was capable of phosphorylating a kinase domain-truncated p72^{syk} (SykT) after FcεRI clustering, confirming p53/56^{lyn}-dependent hetero-phosphorylation of p72^{syk} in a FcεRI system.[7] Therefore, a possible mechanism for FcεRI-dependent p72^{syk} activation involves tyrosine phosphorylation of p72^{syk} by p53/56^{lyn}. Yet, another mechanism for p72^{syk} activation that indirectly involves p53/56^{lyn} was recently suggested. Incubation of RBL cell lysates with peptides corresponding to γITAM or βITAM followed by p72^{syk} immunoprecipitation revealed that doubly tyrosine phosphorylated γITAM peptides induced increased p72^{syk} tyrosine phosphorylation.[58] Doubly tyrosine phosphorylated βITAM peptides as well as γITAM peptides phosphorylated on one tyrosine residue only induced lower phosphorylation of p72^{syk}.[58] Importantly, incubation of immunoprecipitated p72^{syk} with doubly phosphorylated γITAM peptides greatly increased the in vitro activity of p72^{syk}.[58] It is a reasonable assumption that this activation resulted from direct binding of sykSH2 tandem domains to phospho-γITAM peptides. Thus, there are two possible mechanisms for p72^{syk} activation that are not mutually exclusive and that involve binding of p72^{syk} to ITAM motif phosphorylated on tyrosine by p53/56^{lyn} (such as γITAM), and tyrosine phosphorylation of p72^{syk} by p53/56^{lyn}, respectively.

Interestingly, p72^{syk} activation per se does not seem to be sufficient for cell activation. Indeed, p50^{csk}-negative B cells showed high constitutive activity of p53/56^{lyn} and p72^{syk}.[59] Yet, aggregation of the B cell receptor (BCR) was still required for calcium mobilization,[59] an event that is dependent on p72^{syk} activity in BCR-stimulated B cells.[60] In addition, a p72^{syk} chimera made of p72^{syk} fused to irrelevant transmembrane and extracellular domains was constitutively tyrosine phosphorylated in vivo in mast cells, probably as a result of membrane association that renders the p72^{syk} portion of the chimera more accessible to membrane-anchored *src*-related PTKs (p53/56^{lyn}?).[61] Yet, no mast cell activation was observed unless the p72^{syk} chimera was aggregated, an event that was accompanied by increased tyrosine phosphorylation of the latter.[61] Altogether, the data are consistent with aggregation-dependent trans-autophosphorylation of activated p72^{syk} on specific tyrosine residues that would allow p72^{syk} to achieve its biological function that is to recruit and phosphorylate specific effector molecules. This hypothesis would be in agreement with a recent study where association of the p72^{syk}-related PTK ZAP-70 with tyrosine phosphorylated CD3ζ and η chains enhanced autocatalytic activity of ZAP-70 thereby generating specific binding sites for SH2-containing signaling proteins.[62] Yet, one should keep in mind that alternate, phosphorylation-independent, activation of p72^{syk} cannot be ruled out since this has been also reported although its mechanism is still unknown.[63]

The importance of p72^{syk} in FcεRI-generated activation signal has been investigated in recent studies. A chimeric PTK was designed with human p72^{syk} fused to irrelevant transmembrane and extracellular domains.[61] The resulting p72^{syk} chimera was expressed in RBL-2H3 rat mast cell line. Although the p72^{syk} portion of the chimera was associated with the inner aspect of the plasma membrane there was no evident spontaneous cell activation. Clustering of this p72^{syk} chimera induced tyrosine phosphorylation of the chimera itself but neither of FcεRI subunits nor of endogenous p72^{syk}.[61] Yet, ligation of the chimera induced a profile of tyrosine phosphorylated substrates that was identical to that resulting from receptor engagement.[61] In addition, upon aggregation of the chimera complete cell activation was clearly achieved and to an extent comparable to that resulting from FcεRI engagement: there was release of preformed granular markers such as β-hexosaminidase, generation of newly formed lipid mediators such as leukotrienes and activation of gene transcription such as those encoding c-fos and IL-3.[61] Therefore, the full range of cell activation induced by FcεRI aggregation was mimicked by $p72^{syk}$ chimera ligation.

In other studies, the effect of inhibition of $p72^{syk}$ on cell activation was analyzed. In the first one, a PTK inhibitor, piceatannol, was used that preferentially inhibits $p72^{syk}$ over $p53/56^{lyn}$ (in some instances $p53/56^{lyn}$ activity was even increased after piceatannol treatment).[64] Although piceatannol treatment resulted in inhibition of $p72^{syk}$ activity, there was no decrease in antigen-induced FcεRI-β and γ tyrosine phosphorylation confirming that $p72^{syk}$ is not the PTK responsible for phosphorylation of the receptor. Yet, inhibition of $p72^{syk}$ was accompanied by inhibition of inositol trisphosphate generation and of serotonin release.[64] However, piceatannol inhibited to a significant extent serotonin release induced by a calcium ionophore[64] —a condition where $p72^{syk}$ is not activated in mast cells.[39] Thus, it is possible that part of piceatannol action is independent of inhibition of $p72^{syk}$.

In another study, overexpression of SykT (a kinase domain-truncated $p72^{syk}$) in the rat mast cell line RBL-2H3 resulted in increased tyrosine phosphorylation of aggregated FcεRI, a result of phosphotyrosine protection by SykT-SH2 domains.[7] Yet, there was decreased tyrosine phosphorylation and activation of endogenous $p72^{syk}$ as a result of competition between SH2 domains of $p72^{syk}$ and of SykT for recruitment to the phosphorylated receptor. This correlated with decreased tyrosine phosphorylation of all other substrates as well as decreased calcium mobilization.[7] In a comparable study, recombinant Syk-SH2 domains were introduced into permeabilized RBL-2H3 rat mast cells.[65] This resulted in inhibition of FcεRI-induced association of $p72^{syk}$ with the tyrosine phosphorylated receptor. There was also inhibition of FcεRI-induced tyrosine phosphorylation of proteins (among which $p72^{syk}$ and PLC-γ1), and of β-hexosaminidase release and leukotriene generation.[65]

A summary of all the cellular events induced by FcεRI ligation that are under the control of $p72^{syk}$ activation are listed in Table 3.1. So far, all data converge to suggest that the role of FcεRI tyrosine phosphorylation subsequently to FcεRI aggregation would be solely to lead to $p72^{syk}$ mobilization. Although additional investigations will be required to confirm this provocative hypothesis, $p72^{syk}$ already appears as a key component of the signaling pathway activated by FcεRI aggregation.

I conclude from all these studies that FcεRI with the associated $p53/56^{lyn}$ and PTPase that collectively modulate $p72^{syk}$ activity, together with $p72^{syk}$, form the FcεRI signaling unit.

Unfit Data and Unanswered Questions

Thus, the current model[6,7] is that aggregation-induced trans-phosphorylation of FcεRI by FcεRI-β chain-associated $p53/56^{lyn}$ leads to further recruitment (and activation) of $p53/56^{lyn}$ through lynSH2/phospho-βITAM interaction with resulting amplification. As a consequence, $p72^{syk}$ is recruited through Syk-SH2/phospho-γITAM interaction, and stimulated with resulting cell activation (see Table 3.2). Although this model successfully accounts for most collected data, a careful review of the literature indicates that the model needs to be refined. Indeed, this model attributes a pivotal role to the FcεRI-β chain. Yet, TTγ chimeric molecule (that is not associated with FcεRI-β) can activate mast cells after aggregation, i.e., it induces tyrosine phosphorylation of proteins, inositol trisphosphate production, calcium mobilization, cytoskeleton reorganization and secretion.[6,43,66,67]

Transfection of human FcεRI-α chain in P815 mouse mast cells results in expression of a human FcεRI-α-mouse FcεRI-γ chimeric receptor. This FcεRI-β-deficient receptor is capable of transducing activation signal including tyrosine phosphorylation and calcium mobilization.[47] These data have been confirmed recently in human FcεRI⁺ (but FcεRI-β chain-negative) Langerhans cells[68] (see below). Therefore, FcεRI-β appears to be at least partially dispensable for FcεRI signaling. The reason for such dispensability may be linked to the ability of TTγ to activate p53/56^{lyn} as well as p72^{syk}, yet through a mechanism that is still not understood.[43] Thus, the exact role of FcεRI-β needs further clarification. Whatever the role of FcεRI-β, it

Table 3.1. FcεRI-induced cellular events and their control by p72^{syk} activation

	Cellular Event	Assessed Through	Reference
Under the control of p72^{syk} activation	Tyrosine phoshorylation	direct observation	7, 61, 65
	calcium mobilization	Plc-γ1 tyrosine phoshorylation	61,65
		inositol trisphosphate generation	64
		increase in free internal calcium	7
	degranulation	β-hexosaminidase release	61, 65
		serotonin release	64
	arachidonic acid metabolism	arachidonic acid generation	102
		leukotriene generation	61, 65
	cytokine generation	IL-3 gene transcription	61
	gene transcription	IL-3 gene transcription	61
		c-fos gene transcription	61
	cytoskeleton reorganization	morhpological change	64
NOT under the control of p72^{syk} activation	p53/56^{lyn} mobilization	p53/56^{lyn} tyrosine phoshorylation	64
		p53/56^{lyn} activation	64
	FcεRI tyrosine phoshorylation	direct observation	7, 61,64,65

Table 3.2. Current model for the initial steps of FcεRI-induced cell activation

Resting state		*FcεRI is phosphorylated on serine and threonine, NOT on tyrosine* *A subpopulation of FcεRI is associated with p53/56^{lyn}*
Activated state	*Step 1*	*Following aggregation, FcεRIβ-associated p53/56^{lyn} phosphorylates tyrosine residues within βITAM and γITAM domains of receptors that are in trans-position within an aggregate*
	Step 2	*p72^{syk} is recruited to engaged receptors through interaction of sykSH2 tandem domains with phospho-gITAM*
	Step 3	*Recruited p72^{syk} is activated by both its interaction with phospho-γITAM and by FcεRI-associated p53/56^{lyn}, with subsuquent trans-(auto)-phosphorylation of p72^{syk}*
	Step 4	*Auto-(trans)-phosphorylated p72^{syk} recruits SH2 domains-containing effector molecules*

This activated state is amplified throughout activation by recruitment of new p53/56^{lyn} molecules to phospho-βITAM through interaction with lynSH2, with the subsequent tyrosine phosphorylation of new receptors within an aggragate, followed by recruitment and activation of additional p72^{syk} and mobilization of additional effector molecules includes quantitative and qualitative modulation of FcεRI-γ function. In a recent study, FcεRI-γ-containing FcεRI and FcεRI-γ-containing TCR/CD3 complex (in the absence of FcεRI-γ-related CD3ζ) were coexpressed in the same cell line.[5] Independent aggregation of either type of receptor generated profiles of tyrosine phosphorylated substrates that were overlapping but not identical and that were abolished by Y-to-F mutation of FcεRI-γ.[5] Thus, FcεRI-γ function seems to vary according to the receptor context to which it is associated. Interestingly, coaggregation of TTβ chimera with FcεRI in TTβ-transfected rat mast cells resulted in decreased serotonin release, calcium mobilization, tyrosine phosphorylation and $p53/56^{lyn}$ and $p72^{syk}$ activation.[43]

The model also confers a critical role to FcεRI-γ tyrosine phosphorylation for $p72^{syk}$ mobilization. Yet, although aggregation of TTγ activates $p72^{syk}$,[6,43] no tyrosine phosphorylation of aggregated TTγ could be detected.[6,43] In addition, attempts to coprecipitate kinase activity with these chimeras have been unsuccessful so far.[67] Thus, the way in which these chimeric molecules activate $p72^{syk}$ is still a matter of speculation.

There are several important questions that are not addressed by the model. One of these is the mechanism for $p53/56^{lyn}$ constitutive association with FcεRI in resting cells. Similar to what has been reported for the interaction of the $p60^{src}$-related $p56^{lck}$ with CD4 and CD8, $p53/56^{lyn}$ interaction with cytoplasmic domain of FcεRI-β could be mediated through the "unique" juxtamembrane N-terminal region of the kinase.[69,70] Alternatively, similar to what has been demonstrated for several SH2 domains in a different system, binding of lynSH2 to serine-phosphorylated FcεRI-β in resting cells could be envisaged.[71,72] The latter hypothesis would have the advantage of providing a role for the reported constitutive serine phosphorylation of FcεRI-β in resting cells[73] as well as an explanation as to why not more than 20% FcεRI are associated with $p53/56^{lyn}$ under resting conditions.[42] Yet, it has not been confirmed so far in binding experiments using GST-lynSH2 fusion protein and filter-bound FcεRI-β.[44] Another hypothesis has been recently suggested: GD_{1b} ganglioside derivatives that associate with a minority of FcεRI and independently with $p53/56^{lyn}$ could provide a link between the receptor and the PTK.[74,75] In this case, FcεRI association with the ganglioside would have to be through the FcεRI-β chain. This question has not been solved yet.

$P60^{src}$-related PTKs such as $p53/56^{lyn}$ are inactivated and activated by phosphorylation and dephosphorylation, respectively, of a critical carboxy-terminal tyrosine residue.[1] The phosphorylation state of this residue is under the control of the PTK $p50^{csk}$ and of the transmembrane PTPase CD45 in several cell systems.[1] A role for CD45 in FcεRI-mediated activation has been demonstrated recently.[20,22] Yet, no $p50^{csk}$/CD45 involvement in FcεRI-dependent $p53/56^{lyn}$ activation has been reported to date.

CONNECTING TYROSINE PHOSPHORYLATION WITH OTHER SIGNALING MECHANISMS

PTKs and tyrosine phosphorylations are intertwined within a complex network of signaling mechanisms where each component influences and is influenced by another. An excellent example of integration between different regulatory mechanisms for cell signaling is provided by the Ras/Raf/MEK/ERK pathway where PTKs (MEK) cooperate with G proteins ($p21^{ras}$) and serine/threonine kinases (Raf-1, ERK1/2) to achieve activation signal (see below). Here, I shall concentrate on the different bridges by which cross-talks between tyrosine phosphorylation and other signaling mechanisms proceed.

CONNECTIONS WITH CALCIUM MOBILIZATION: PLC-G1/G2

Inhibitors of PTKs inhibit FcεRI-dependent calcium influx into RBL-2H3 rat mast cell line (M Benhamou and RP

Siraganian, unpublished observations), whereas calcium influx induced by FcεRI ligation or by calcium ionophores results in tyrosine phosphorylation of 105-130 kDa substrates.[14,76,77] Thus, PTKs are connected with calcium mobilization both upstream and downstream of it.

Aggregation of FcεRI induces phosphatidylinositol (PI) hydrolysis with generation of diacylglycerol (DAG) and inositol phosphates (IP). Inositol-1,4,5 trisphosphate induces calcium release from intracellular stores into the cytosol with the subsequent influx of calcium from extracellular milieu (see chapter 4). PI breakdown is under the control of PI-specific phospholipase (PL) C. Thus, PLC activation is central to calcium mobilization in FcεRI-controlled systems. Several isotypes of PLC have been described, some of which are regulated through tyrosine phosphorylation (see chapter 4). Inhibitors of PTKs were found to inhibit IP generation by FcεRI ligation,[12,13,15-17] suggesting that tyrosine phosphorylation was required for PLC activation. Yet, contradictory results were reported.[18,19] In addition, of the biphasic DAG production after FcεRI aggregation, only the second PLD-dependent phase was inhibited by PTK inhibitors whereas the first phase, that should be related to PLC activation, was not.[78] The reason for such discrepancies remains so far unclarified.

PTK-regulated PLC-γ1 and PLC-γ2 isotypes are translocated to the plasma membrane as a result of receptor aggregation.[79,80] In addition, a PTK activity that is apparently associated with a 44 kDa protein coprecipitated with PLC-γ1. This PTK activity increased shortly and transiently after FcεRI engagement.[81] This correlated with rapid and transient tyrosine phosphorylation of PLC-γ1 after receptor ligation.[13,15,17,27,81] Yet, there was discrepancy between the transient tyrosine phosphorylation of PLC-γ1 and the sustained production of IP.[13,15] Thus, whereas PLC-γ1 is involved in FcεRI-dependent PI breakdown, other PLC (such as PLC-γ2 or non-PTK-regulated PLCs) may participate in it. Interestingly, PLC-γ1 contains two SH2 and one SH3 domain.[2] These domains could promote interaction of PLC-γ1 with upstream or downstream effector molecules.[81] No association of PLC-γ1 with FcεRI has been detected,[81] but activation of PLC-γ1 was observed after selective activation of $p72^{syk}$.[61] Thus, $p72^{syk}$ could be the PTK responsible for PLC-γ1 tyrosine phosphorylation. Since the PTPase inhibitor phenylarsine oxide (PAO) completely abolishes tyrosine phosphorylation of PLC-γ1 but not that of FcεRI subunits,[82] it would be interesting to know whether PAO blocks $p72^{syk}$ activation and tyrosine phosphorylation. If not, this would suggest that additional components take place between $p72^{syk}$ and PLC-γ1.

Connections with Serine/Threonine Kinase

Connections with protein kinase C (PKC) activation: PLC-g1, PLD, PKC-d, $p77^{btk}$, $p72^{emt}$

As noted for calcium signal, there are both upstream and downstream connections between PTK and PKC activity since PKCs can be targets for PTKs, and PTKs can be targets for PKCs. Stimulation of PKC by FcεRI aggregation or by phorbol esters leads to tyrosine phosphorylation of the 105-130 kDa proteins described above.[76] Conversely, PTKs are indirectly involved in regulation of PKC activity since DAGs produced after PI hydrolysis by PLC activate PKCs.

Interestingly, DAG is also produced as a result of PLD activity following a two-step process: PLD hydrolyzes phosphatidylcholine (PC) into phosphatidic acid which is in turn converted into DAG by phosphatidate hydrolase activity. FcεRI-dependent activation of PLD is inhibited by cell pretreatment with PTK inhibitors.[18,78] Moreover, PLD activity was recovered in increased amount in antiphosphotyrosine precipitates after FcεRI aggregation, suggesting that PLD is either tyrosine phosphorylated or tightly associated with a phosphoprotein.[78] Thus, FcεRI-dependent

DAG generation follows a two-phase kinetics, the first one attributed to PLC activity and the second phase resulting from PLD activity.[78] The DAGs generated as a result of both these activities are different since their phospholipid source is PI in the case of PLC and PC in the case of PLD. Since several PKC isoforms (α, β, δ, ε, ζ) are expressed in rat mast cells that are differently involved in cell responses to antigen,[83-86] it would be interesting to know the relative sensitivity of each isoform to PI- and PC-derived DAGs, respectively. These could provide a timed, sequential control of PKC isozymes mobilization.

Modulation of PKC activity by PTKs is not only indirect (through control of PLC and PLD activity) but also direct through tyrosine phosphorylation of PKC. Indeed, PKC-δ is a target for PTK activity.[87] This PKC isozyme is found associated with FcεRI-β and is responsible for FcεRI-γ threonine phosphorylation.[86] Following FcεRI aggregation, PKC-δ becomes rapidly membrane-associated, and membrane-associated PKC-δ is rapidly and transiently tyrosine phosphorylated.[87] As a consequence, its catalytic activity towards FcεRI-γ in vitro is significantly diminished. Yet, although activated FcεRI is still associated with PKC-δ, *tyrosine phosphorylated* PKC-δ is not associated with FcεRI.[87] It is not known whether tyrosine phosphorylation of PKC-δ induces its dissociation from FcεRI or whether FcεRI-associated PKC-δ is not a target for tyrosine phosphorylation. However, tyrosine phosphorylation clearly modulates the specific activity of this PKC isozyme toward a given substrate.

Another connection between PTKs and PKCs is provided by the PTK $p77^{btk}$. This kinase is the product of the *btk* gene whose mutations are responsible for X-linked agammaglobulinemia.[88,89] In murine mast cells it was first described as a PTK named *emb*.[90] $P77^{btk}$ contains one SH2 and one SH3 domain like $p60^{src}$-related PTKs, but has neither a C-terminal regulatory tyrosine residue nor an N-terminal myristylation site.[88-90] It forms a family together with the PTKs' $p72^{emt}$ (or *itk*)[90] and *tec*.[89] Aggregation of FcεRI leads to rapid and transient tyrosine phosphorylation of $p77^{btk}$.[41] This correlated with increased kinase activity toward acid-treated enolase as an exogenous substrate.[41] A small proportion (5%) of $p77^{btk}$ was found associated with the membrane fraction under resting conditions. FcεRI ligation induced a rapid and sustained translocation of $p77^{btk}$ to the plasma membrane that correlated with translocation of PKC.[41] Actually, it was recently found that $p77^{btk}$ was constitutively associated with PKC-βI.[91] This association is through the so-called pleckstrin homology (PH) domain of $p77^{btk}$, a domain originally recognized in the PKC substrate pleckstrin.[91] PKC association with $p77^{btk}$ did not increase after FcεRI-mediated cell activation but the PKC activity was found to downregulate both in vivo tyrosine phosphorylation and in vitro autocatalytic activity of $p77^{btk}$.[91] However, PKC-$p77^{btk}$ association did not result in detectable decrease (or increase) in in vitro kinase activity of PKC-β, indicating that $p77^{btk}$ is a target for PKC but that the reverse is not true.

The recent observation that a point mutation in $p77^{btk}$ PH domain results in activation of this PTK confirmed the importance of the PH domain in regulation of the catalytic activity of $p77^{btk}$.[92] A suggested substrate for $p77^{btk}$ is the γ chain of FcεRI.[40] Yet, as noted earlier, $p77^{btk}$ has not been found associated with FcεRI before or after stimulation.[41] Thus, the exact role of this PTK in FcεRI-mediated signaling remains elusive. A potential target for $p77^{btk}$ is the protooncogene $p120^{cbl}$. This 120 kDa protein is found associated with the SH3 domain of $p53/56^{lyn}$ and of $p77^{btk}$ in HL60 and B cells, respectively.[93,94] $P120^{cbl}$ is phosphorylated on tyrosine after aggregation of T cell and B cell receptors as well as FcγRs.[93-95] In mast cells, FcεRI aggregation results in tyrosine phosphorylation of 105-130 kDa substrates that have not all been characterized.[14,76,77,96] It is therefore tempting to speculate that one of them is $p120^{cbl}$.

Another *tec*-related PTK, $p72^{emt}$, is tyrosine phosphorylated and activated after FcεRI aggregation.[96a] Like $p77^{btk}$, $p72^{emt}$ interacts with PKC in mouse mast cells through its PH domain.[96a] This interaction is not modulated by FcεRI aggregation. However, FcεRI-dependent $p72^{emt}$ activation is dependent on PKC activity as it is abrogated by pretreatment of the cells with the PKC inhibitor calphostin, or by depletion of cellular PKC upon long-term treatment with high concentration of PMA.[96a] Although both calcium-dependent (α, βI, βII) and -independent (ε, ζ, θ) PKC isoforms associate with $p72^{emt}$ in mouse mast cells, only the calcium-dependent isoforms seem to regulate $p72^{emt}$ activity as observed in PKC overexpression experiments.[96a] The targets for $p72^{emt}$ have not been identified yet.

Connections with MAP kinases ERK1/2 activation

Mitogen-activated protein kinases (MAPKs) form a family of serine/threonine kinases that are activated by multiple extracellular signals and that include JNK, ERK1 and ERK2, and HOG (see ref. 97 for a comprehensive review on MAPK). Downstream effectors of MAPK can be membrane-bound, cytosolic, cytoskeleton-associated as well as nuclear. MAPKs thus represent a major crossroads in cell signaling. Aggregation of FcεRI induces MAPK activation that results from tyrosine (but also serine and threonine) phosphorylation of ERK1 and ERK2.[98-103] This ERK1/2 activation is under the control of FcεRI-γ but not of FcεRI-β, since TTγ, but not TTβ, aggregation results in ERK1/2 tyrosine phosphorylation and activation.[102] Consistent with this observation, FcεRI-dependent ERK1/2 mobilization was potentiated by overexpression of $p72^{syk}$ and inhibited by overexpression of SykT, the kinase domain-truncated $p72^{syk}$.[102] In addition, selective activation of $p72^{syk}$ leads to tyrosine phosphorylation of ERK1/2.[61] Although this phosphorylation is rapid, pharmacological approaches have suggested that it depends on calcium influx and on activity of PKC and of GTP-binding proteins.[101,103]

One ERK1/2 activation pathway includes a cascade of activations sequentially involving $p95^{vav}$ (see below), the small GTP-binding protein $p21^{ras}$, the serine kinase Raf-1 and the serine/threonine/tyrosine kinase MEK1, the latter being directly responsible for the phosphorylation/activation of ERK1/2.[97,104] Evidence for this ERK1/2 activation pathway in FcεRI-activated rat mast cells has been accumulated[102,103] and suggest that $p72^{syk}$-activated $p95^{vav}$ could directly connect FcεRI aggregation to ERK1/2 activation pathway (see below). Potential targets for ERK1/2 include the serine kinase $p90^{rsk}$ that can phosphorylate ribosomal protein S6 (ref. 98) and PLA_2 that releases arachidonic acid, the precursor of prostaglandins and leukotrienes, from membrane phospholipids.[102,103] These data and the fact that secretion closely correlates with ERK1/2 activity[98-103] underscore the importance of these enzymes in FcεRI-mediated cell activation (see chapters 6 and 10 for more detailed analysis).

Potential connections with other serine/threonine kinases: Nck

Although they have no enzymatic activity, adaptor molecules play an important role in signal transduction by connecting effector molecules with one another. Adaptors are constituted almost exclusively of SH2 and SH3 domains through which connections are achieved. Recruitment to upstream effectors is frequently through interaction between effector phosphotyrosine and adaptor SH2 domain, whereas recruitment of downstream effectors results from interaction between adaptor SH3 domain and proline-rich regions of effector molecules. Thus, adaptors represent indispensable relays at nodal points of the signaling network. One such molecule, Nck, contains one SH2 and three SH3 domains. It is phosphorylated on tyrosine residue shortly after FcεRI aggregation.[105] Although no Nck downstream effector has been reported to date in the FcεRI system, recent evidence demonstrates that a serine/threonine kinase named NAK is complexed with Nck through interaction with the second of Nck SH3 domains in a variety of cell types including NIH3T3, PC12, A431 and HeLa cells.[106] Thus, Nck represents a potential connection with serine/threonine kinases.

Connections with G Proteins Activity: p95^{vav}, Shc, Grb2, So

Again, FcεRI-dependent tyrosine phosphorylations are regulated by as well as regulate G protein activity. Indeed, FcεRI induction of tyrosine phosphorylation of 105-130 kDa substrates is mimicked by G protein activators.[19,76,103] However, the steps leading from G proteins to these tyrosine phosphorylations have not been characterized yet. In addition, tyrosine phosphorylation of at least ERK1/2 is inhibited by GDPβS, an inactive competitor of GTP.[103] It is thought that this is due to the involvement of the small GTP-binding protein p21^{ras} upstream ERK1/2 activation (see above). Of note, the potential importance of this pathway in FcεRI-mediated cell activation is underscored by experiments in which microinjection of p21^{ras} into rat mast cells stimulated degranulation.[107]

G proteins are inactive when associated with GDP and in an active conformation when associated with GTP. Cleavage of GTP to GDP through GTPase activity of G proteins returns the latter to their resting state. Small G proteins activation is regulated upstream by guanine nucleotide exchange factors (GEFs) that increase the rate of exchange of bound GDP for GTP, and downstream by GTPase-activating proteins that accelerate the degradation by small G proteins of bound GTP into GDP (see chapter 3). The protooncogene p95^{vav} is one such GEF.[108] P95^{vav} contains one SH3 and two SH2 domains.[108] Its GEF activity is induced either by direct tyrosine phosphorylation[109] or by phorbol esters/diglycerides[110] by virtue of a phorbol ester/diacylglycerol binding domain.[108] FcεRI aggregation leads to tyrosine phosphorylation of p95^{vav}.[102,111] In growth-factor receptor systems, p95^{vav} upstream effector for tyrosine phosphorylation is the tyrosine phosphorylated receptor itself that associates with p95^{vav} through p95^{vav} SH2 domain.[112] In a non-growth-factor receptor system such as B cells activated by ligation of antigen receptors, p95^{vav} rapidly associates with a 70 kDa tyrosine phosphorylated protein (Vap-1), an interaction mediated by the SH2 domain of p95^{vav}.[113] The identity of Vap-1 remains unknown. Interestingly, FcεRI-induced tyrosine phosphorylation of p95^{vav} is potentiated by overexpression of p72^{syk} and inhibited by overexpression of SykT.[102] Direct activation/phosphorylation of p95^{vav} by p72^{syk} is therefore an attractive hypothesis that could explain why selective activation of p72^{syk} results in ERK1/2 phosphorylation.[61]

Whereas it seems clear that one important target of p95^{vav} could be the small G protein p21^{ras},[109,110] another p21^{ras} activation pathway has been characterized. It involves association between the adaptor molecule Grb2 and the GEF Son of Sevenless (Sos) through their SH3 domains and proline-rich region, respectively.[114] Stimulation through surface receptors leads to association of Grb2-SH2 domain with either tyrosine phosphorylated receptor or tyrosine phosphorylated adaptor Shc, resulting in increased local concentration of Sos near plasma membrane-anchored p21^{ras} with subsequent activation of the latter.[114] In RBL-2H3 rat mast cells, a constitutive phospho-Shc-Grb2-Sos complex is observed with no increased tyrosine phosphorylation of Shc after FcεRI aggregation.[115] However, after FcεRI stimulation Grb2-Sos complex associates with a 33 kDa tyrosine phosphorylated protein (that is *not* FcεRI-β) through interaction with Grb2-SH2 domain, suggesting a Shc-independent recruitment of Grb2-Sos.[115]

Other Substrates: p75^{HS1}/SPY75

P75^{HS1} is a protein with a helix-turn-helix motif and an SH3 domain that is a major substrate for PTK in FcεRI as well as BCR signaling.[50,116,117] FcεRI-induced tyrosine phosphorylation of $p75^{HS1}$ is rapid and transient. In addition, $p75^{HS1}$ coprecipitates with 150 and 95 kDa tyrosine-phosphorylated proteins after FcεRI aggregation.[116] Yet, no association between $p75^{HS1}$ and PLC-γ1/2, $p95^{vav}$ or FcεRI-β or γ has been detected.[116] In addition, although $p75^{HS1}$ was reported to be associated with $p53/56^{lyn}$ after BCR ligation,[117] no such association and no association with a PTK activity has been observed after FcεRI aggregation.[50,116] Therefore, both the PTK responsible for its phosphorylation and its target(s) remain unknown allowing only to speculate on the role of $p75^{HS1}$.

REGULATION OF LATE EVENTS: CYTOSKELETON REORGANIZATION

Increased adhesion of mast cells to extracellular matrix occurs as a response to FcεRI aggregation (see chapter 8). This is linked to cytoskeleton reorganization, notably to assembly of F-actin plaques at sites of focal adhesion.[118] A recent study demonstrated that PTKs are involved in this phenomenon since PTK inhibitors block actin polymerization and since tyrosine phosphorylated proteins accumulate at these sites.[118] Cytoskeleton reorganization is a late event in FcεRI-mediated signaling. Therefore, tyrosine phosphorylated substrates that are likely to participate in it should be phosphorylated late after FcεRI aggregation.

Late substrates have been reported to be phosphorylated on tyrosine downstream calcium influx, PKC and G protein activation.[14,76] They include 105-130 as well as 66-71 kDa proteins.[14,76] Since secretion was consistently accompanied by tyrosine phosphorylation of these substrates it was proposed that these were involved in modulation of degranulation process.[76] Independent approaches demonstrated that FcεRI-induced tyrosine phosphorylation of the high molecular weight late substrates was dependent on cell adhesion to extracellular matrix.[77] One of these high molecular weight substrates has been identified as the cytosolic PTK $pp125^{FAK}$.[96] This PTK accumulates at sites of focal adhesion and is a substrate for $p60^{src}$.[119] Interestingly, FcεRI-dependent activation of $p60^{src}$ has been reported,[37] yet seems too early and too transient to be directly responsible for $pp125^{FAK}$ phosphorylation. The targets for $pp125^{FAK}$ are likely to be focal adhesion proteins. One such potential target is the 68 kDa protein paxillin.[120-123] Tyrosine phosphorylation of paxillin after FcεRI aggregation occurs downstream calcium influx and PKC activation and is dramatically potentiated by cell adhesion.[124] Tyrosine phosphorylated paxillin is then able to interact with $p53/56^{lyn}$ in vitro through Lyn-SH2 domain.[125]

Another target recently reported is a 77 kDa substrate (FAP) that associates with nontyrosine phosphorylated $pp125^{FAK}$ in mast cells.[126] FcεRI aggregation of adherent mast cells induces tyrosine phosphorylation of both $pp125^{FAK}$ and FAP resulting in dissociation of FAP from $pp125^{FAK}$.[126]

PTKs AND FcεRI SIGNALING IN HUMANS

Understanding FcεRI signaling in humans is a critical issue, given both the important occurrence of allergies within the world population and the central role played by IgE and FcεRI in these reactions. In addition, the differences between species bar any automatic extrapolation from murine systems to human systems. For instance, expression of human FcεRI on the cell surface minimally requires FcεRI-α and γ chains but not FcεRI-β, whereas expression of murine FcεRI requires *all* the chains of the receptor.[127] These differences could have pronounced consequences on the signal generated upon FcεRI aggregation.

In humans, FcεRI is not only expressed at the surface of mast cells and basophils but also on Langerhans cells,[128,129] monocytes from atopic patients[130] and eosinophils from hypereosinophilic patients.[131] Recently, evidence started to accumulate on the involvement of PTKs in human FcεRI signaling in mast cells, basophils and Langerhans cells. In basophils, antibodies against the transmembrane PTPase CD45 and inhibitors of PTKs inhibited IgE-mediated histamine release.[132-134]

Using normal basophils in vitro-derived from bone marrow precursor cells as well as tumoral basophils, we recently demonstrated that FcεRI aggregation resulted in very rapid, extracellular calcium-independent tyrosine phosphorylation of a number of substrates.[135] There was correlation between histamine release and tyrosine phosphorylation in antigen dose-responses as well as in dose-responses of the PTK inhibitor genistein.[135] Furthermore, FcεRI ligation on human basophils resulted in rapid and calcium-independent tyrosine phosphorylation and activation of $p72^{syk}$.[135] The importance of $p72^{syk}$ activation is underscored by the strong inhibitory effect of the $p72^{syk}$ inhibitor piceatannol on the secretory response to FcεRI aggregation on human basophils.[134] Yet, it is not clear which of the two potential $p72^{syk}$ isoforms expressed in human basophils is involved in this signal.[136]

In mast cells, only pharmacological evidence with PTK inhibitors[133] and piceatannol[134] so far indicates an involvement of PTKs in FcεRI signaling. In Langerhans cells, FcεRI aggregation also induces tyrosine phosphorylation of many proteins.[68] Yet, no FcεRI-β chain is detected in these cells demonstrating that PTK activation is not dependent on FcεRI-β expression in this system.[68] Moreover, FcεRI aggregation on Langerhans cells from atopic dermatitis patients, *but not from normal individuals*, results in calcium mobilization in the absence of detectable difference in the pattern of tyrosine phosphorylation between the two populations.[68]

As observed in human basophils, anti-CD45 antibodies inhibited calcium mobilization in FcεRI-stimulated Langerhans cells from atopic dermatitis patients.[137] Thus, human cells provide promising tools for investigation of the role of FcεRI-β chain, of CD45, as well as of the relationship between PTK activation and other signaling mechanisms.

CONCLUDING REMARKS

Much progress has been made in only half a decade since the first report that rat FcεRI aggregation initiates protein-tyrosine phosphorylations that are coupled to activation signal.[24] Although additional studies are required to refine the current model, we have now an almost complete picture of the mechanism leading to FcεRI tyrosine phosphorylation and of its role in activation of the signal. Yet, this progress represents only the very beginning of our understanding of how PTKs and their substrates fit into FcεRI-mediated signaling.

Indeed, our knowledge of events involving PTK activity downstream $p72^{syk}$ mobilization still resembles a scattered puzzle whose pieces are being more or less randomly collected. Many substrates and PTKs remain to be characterized.[50,76,99,138,139] Although the importance of CD45 is inferred from many studies,[20,22,132,137] its precise site of intervention has not been identified yet, and many more PTPases await identification (see chapter 7). In addition, phosphotyrosine and SH2 domains provide an important means of molecular interactions that help integrate many pathways into a cohesive activation network. For instance, the same adaptor can be used by cells for different independent connections between molecules involved in FcεRI signaling: the SH2 domain of Grb2 can associate with phospho-Shc but also with a 33 kDa phospho-protein, whereas its SH3 domains can associate with Sos but also with a 75 kDa protein.[115] Many such connections await identification and analysis. This is enough to keep an army of investigators busy for years to come...

ACKNOWLEDGMENTS

I gratefully acknowledge Thomas Bieber, Toshiaki Kawakami, Jean-Pierre Kinet, Henri Metzger, Janet Oliver and Juan Rivera for communicating data before publication. I am indebted to Reuben P. Siraganian for his continuous support and encouragements.

References

1. Cooper J, Howell B. The when and how of Src regulation. Cell 1993; 73:1051-54.
2. Koch CA, Anderson D, Moran MF et al. SH2 and SH3 domains: elements that control interactions of cytoplasmic signaling proteins. Science 1991; 252:668-74.
3. Bork P, Margolis B. A phosphotyrosine interaction domain. Cell 1995; 80:693-94.
4. Kavanaugh WM, Williams LT. An alternative to SH2 domains for binding tyrosine-phosphorylated proteins. Science 1994; 266:1862-65.
5. Paolini R, Renard V, Vivier E et al. Different roles for the FcεRI γ chain as a function of the receptor context. J Exp Med 1995; 181:247-55.
6. Jouvin M-H, Adamczewski M, Numerof R et al. Differential control of the tyrosine kinases Lyn and Syk by the two signaling chains of the high affinity immunoglobulin E receptor. J Biol Chem 1994; 269:5918-25.
7. Scharenberg AM, Lin S, Cuenod B et al. Reconstitution of interactions between tyrosine kinases and the high affinity IgE receptor which are controlled by receptor clustering. EMBO J 1995; 14:3385-94.
8. Cambier J, Daëron M, Fridman W et al. New nomenclature for the Reth motif (or ARH1/TAM/ARAM/YXXL). Immunol Today 1995; 16:110.
9. Reth M. Antigen receptor tail clue. Nature 1989; 338:383-84.
10. Keegan A, Paul WE. Multichain immune recognition receptors: similarities in structure and signaling pathways. Immunol Today 1992; 13:63-68.
11. Paolini R, Numerof R, Kinet J-P. Phosphorylation/dephosphorylation of high-affinity IgE receptors: a mechanism for coupling/uncoupling a large signaling complex. Proc Natl Acad Sci USA 1992; 89:10733-37.
12. Deanin GG, Martinez AM, Pfeiffer JR et al. Tyrosine kinase-dependent phosphatidylinositol turnover and functional responses in the FcεRI signalling pathway. Biochem Biophys Res Comm 1991; 179:551-57.
13. Park DJ, Min HK, Rhee SG. IgE-induced tyrosine phosphorylation of phospholipase C-γ1 in rat basophilic leukemia cells. J Biol Chem 1991; 266:24237-40.
14. Yu K-T, Lyall R, Jariwala N et al. Antigen- and ionophore-induced signal transduction in rat basophilic leukemia cells involves protein tyrosine phosphorylation. J Biol Chem 1991; 266:22564-68.
15. Schneider H, Cohen-Dayag A, Pecht I. Tyrosine phosphorylation of phospholipase Cγ1 couples the Fcε receptor mediated signal to mast cells secretion. Int Immunol 1992; 4:447-53.
16. Kawakami T, Inagaki N, Takei M et al. Tyrosine phosphorylation is required for mast cell activation by FcεRI crosslinking. J Immunol 1992; 148:3513-19.
17. Yamada K, Jelsema CL, Beaven MA. Certain inhibitors of protein serine/threonine kinases also inhibit tyrosine phosphorylation of phospholipase Cγ1 and other proteins and reveal distinct roles for tyrosine kinase(s) and protein kinase C in stimulated, rat basophilic RBL-2H3 cells. J Immunol 1992; 149:1031-37.
18. Kumada T, Miyata H, Nozawa Y. Involvement of tyrosine phosphorylation in IgE receptor-mediated phospholipase D activation in rat basophilic leukemia (RBL-2H3) cells. Biochem Biophys Res Comm 1993; 191:1363-68.
19. Stephan V, Benhamou M, Gutkind JS et al. FcεRI-induced protein tyrosine phosphorylation of pp72 in rat basophilic leukemia cells (RBL-2H3). Evidence for a novel signal transduction pathway unrelated to G protein activation and phosphatidylinositol breakdown. J Biol Chem 1992; 267: 5434-41.
20. Berger SA, Mak TW, Paige CJ. Leukocyte common antigen (CD45) is required for immunoglobulin E-mediated degranulation of mast cells. J Exp Med 1994; 180:471-76.
21. Hampe CS, Pecht I. Protein tyrosine phosphatase activity enhancement is induced upon Fc_{ε} receptor activation of mast cells. FEBS Lett 1994; 346:194-98.
22. Adamczewski M, Numerof RP, Koretsky GA et al. Regulation by CD45 of the tyrosine phosphorylation of high affinity IgE receptor β- and γ-chains. J Immunol 1995; 154:3047-55.
23. Swieter M, Berenstein EH, Siraganian RP. Protein tyrosine phosphatase coprecipitates

with the high affinity IgE receptor and dephosphorylates the receptor subunits but not Lyn or Syk. J Immunol 1995; 155: 5330-36.
24. Benhamou M, Gutkind JS, Robbins KC et al. Tyrosine phosphorylation coupled to IgE receptor-mediated signal transduction and histamine release. Proc Natl Acad Sci USA 1990; 87:5327-30.
25. Paolini R, Jouvin M-H, Kinet J-P. Phosphorylation and dephosphorylation of the high-affinity receptor for immunoglobulin E immediately after receptor engagement and disengagement. Nature 1991; 353: 855-58.
26. Pribluda VS, Metzger H. Transmembrane signaling by the high-affinity IgE receptor on membrane preparations. Proc Natl Acad Sci USA 1992; 89:11446-50.
27. Li W, Deanin GG, Margolis B et al. FcεRI-mediated tyrosine phosphorylation of pultiple proteins, including phospholipase Cγ1 and the receptor $\beta\gamma_2$ complex, in RBL-2H3 rat basophilic leukemia cells. Mol Cell Biol 1992; 12:3176-82.
28. Quarto R, Metzger H. The receptor for immunoglobulin E: examination for kinase activity and as a substrate for kinases. Mol Immunol 1986; 23:1215-23.
29. Seagrave JC, Deanin GG, Martin JC et al. DNP-phycobiliproteins, fluorescent antigens to study dynamic properties of antigen-IgE-receptor complexes on RBL-2H3 rat mast cells. Cytometry 1987; 8:287-95.
30. Wofsy C, Kent UM, Mao S-Y et al. Kinetics of tyrosine phosphorylation when IgE dimers bind to Fcε receptors on rat basophilic leukemia cells. J Biol Chem 1995; 270:20264-72.
31. Kent UM, Mao S-Y, Wofsy C et al. Dynamics of signal transduction after aggregation of cell-surface receptor: Studies on the type I receptor for IgE. Proc Natl Acad Sci USA 1994; 91:3087-91.
32. Robertson D, Holowka D, Baird B. Crosslinking of immunoglobulin E-receptor complexes induces their interaction with the cytoskeleton of rat basophilic leukemia cells. J Immunol 1986; 136:4565-72.
33. Apgar JR. Antigen-induced crosslinking of the IgE receptor leads to an association with the detergent-insoluble membrane skeleton of rat basophilic leukemia (RBL-2H3) cells. J Immunol 1990; 145:3814-22.
34. Erickson JW, Posner RG, Goldstein B et al. Bivalent ligand dissociation kinetics from receptor-bound immunoglobulin E: evidence for a time-dependent increase in ligand rebinding at the cell surface. Biochemistry 1991; 30:2357-63.
35. Weetall M, Holowka D, Baird B. Heterologous desensitization of the high affinity receptor for IgE (FcεRI) on RBL cells. J Immunol 1993; 150:4072-83.
36. Pribluda VS, Pribluda C, Metzger H. Transphosphorylation as the mechanism by which the high-affinity receptor for IgE is phosphorylated upon aggregation. Proc Natl Acad Sci USA 1994; 91:11246-50.
37. Eiseman E, Bolen JB. Engagement of the high-affinity IgE receptor activates *src* protein-related tyrosine kinases. Nature 1992; 355:78-80.
38. Hutchcroft JE, Geahlen RL, Deanin GG et al. FcεRI-mediated tyrosine phosphorylation and activation of the 72 kDa protein-tyrosine kinase, PTK72, in RBL-2H3 rat tumor mast cells. Proc Natl Acad Sci USA 1992; 89:9107-11.
39. Benhamou M, Ryba NJP, Kihara H et al. Protein-tyrosine kinase $p72^{syk}$ in high affinity IgE receptor signaling. Identification as a component of pp72 and association with the receptor γ chain after receptor aggregation. J Biol Chem 1993; 268:23318-24.
40. Price DJ, Kawakami Y, Kawakami T et al. Purification of a major tyrosine kinase from RBL-2H3 cells phosphorylating $Fc_{\varepsilon}RI$ γ-cytoplasmic domain and identification as the Btk tyrosine kinase. Biochim Biophys Acta 1995; 1265:133-42.
41. Kawakami Y, Yao L, Miura T et al. Tyrosine phosphorylation and activation of Bruton tyrosine kinase upon FcεRI cross-linking. Mol Cell Biol 1994; 14:5108-13.
42. Yamashita T, Mao S.-Y., Metzger H. Aggregation of the high-affinity IgE receptor and enhanced activity of $p53/56^{lyn}$ protein-tyrosine kinase. Proc Natl Acad Sci USA 1994; 91:11251-55.
43. Wilson BS, Kapp N, Lee RJ et al. Distinct functions of the FcεRI γ and β subunits in

the control of FcεRI-mediated tyrosine kinase activation and signaling responses in RBL-2H3 mast cells. J Biol Chem 1995; 270:4013-22.

44. Kihara H, Siraganian RP. Src homology 2 domains of Syk and Lyn bind to tyrosine-phosphorylated subunits of the high affinity IgE receptor. J Biol Chem 1994; 269:22427-32.
45. Shiue L, Green J, Green OM et al. Interaction of $p72^{syk}$ with the γ and β subunits of the high-affinity receptor for immunoglobulin E, FcεRI. Mol Cell Biol 1995; 15:272-81.
46. Miller L, Alber G, Varin-Blank N et al. Transmembrane signaling in P815 mastocytoma cells by transfected IgE receptors. J Biol Chem 1990; 265:12444-53.
47. Alber G, Miller L, Jelsema CL et al. Structure-function relationships in the mast cell high affinity receptor for IgE. Role of the cytoplasmic domains and of the β subunit. J Biol Chem 1991; 266:22613-20.
48. Taniguchi T, Kobayashi T, Kondo J et al. Molecular cloning of a porcine gene *syk* that encodes a 72 kDa protein-tyrosine kinase showing high suceptibility to proteolysis. J Biol Chem 1991; 266:15790-96.
49. Rowley RB, Bolen JB, Fargnoli J. Molecular cloning of rodent $p72^{Syk}$. Evidence of alternative mRNA splicing. J Biol Chem 1995; 270:12659-64.
50. Minoguchi K, Benhamou M, Swaim WD et al. Activation of protein tyrosine kinase $p72^{syk}$ by FcεRI aggregation in rat basophilic leukemia cells. $p72^{syk}$ is a minor component but the major protein tyrosine kinase of pp72. J Biol Chem 1994; 269:16902-08.
51. Couture C, Baier G, Altman A et al. $p56^{lck}$-independent activation and tyrosine phosphorylation of $p72^{syk}$ by T cell antigen receptor/CD3 stimulation. Proc Natl Acad Sci USA 1994; 91:5301-05.
52. Couture C, Baier G, Oetken C et al. Activation of $p56^{lck}$ by $p72^{syk}$ through physical association and N-terminal tyrosine phosphorylation. Mol Cell Biol 1994; 14: 5249-58.
53. Hutchcroft JE, Harrison ML, Geahlen RL. Association of the 72 kDa protein-tyrosine kinase PTK72 with the B cell antigen receptor. J Biol Chem 1992; 267:8613-19.
54. Yamada T, Taniguchi T, Yang C et al. Association with B cell-antigen receptor with protein-tyrosine kinase $p72^{syk}$ and activation by engagement of membrane IgM. Eur J Biochem 1993; 213:455-59.
55. Burkhardt AL, Stealey B, Rowley RB et al. Temporal regulation of nontransmembrane protein tyrosine kinase enzyme activity following T cell antigen receptor engagement. J Biol Chem 1994; 269:23642-47.
56. Saouaf S, Mahajan S, Rowley RB et al. Temporal differences in the activation of three classes of nontransmembrane protein tyrosine kinases following B cell antigen receptor surface engagement. Proc Natl Acad Sci USA 1994; 91:9524-28.
57. Kurosaki T, Takata M, Yamanashi Y et al. Syk activation by the Src-family tyrosine kinase in the B cell receptor signaling. J Exp Med 1994; 179:1725-29.
58. Shiue L, Zoller MJ, Brugge JS. Syk is activated by phosphotyrosine-containing peptides representing the tyrosine-based activation motifs of the high affinity receptor for IgE. J Biol Chem 1995; 270:10498-502.
59. Hata A, Sabe H, Kurosaki T et al. Functional analysis of Csk in signal transduction through the B cell antigen receptor. Mol Cell Biol 1994; 14:7306-13.
60. Takata M, Sabe H, Hata A et al. Tyrosine kinases Lyn and Syk regulate B cell receptor-coupled Ca^{2+} mobilization through distinct pathways. EMBO J 1994; 13:1341-49.
61. Rivera V, Brugge JS. Clustering of Syk is sufficient to induce tyrosine phosphorylation and release of allergic mediators from rat basophilic leukemia cells. Mol Cell Biol 1995; 15:1582-90.
62. Neumeister EN, Zhu Y, Richard S et al. Binding of ZAP-70 to phosphorylated T cell receptor ζ and η enhances its autophosphorylation and generates specific binding sites for SH2 domain-containing proteins. Mol Cell Biol 1995; 15:3171-78.
63. Sada K, Yanagi S, Yamamura H. Activation of p72syk by thrombin in a cell-free system. Biochem Biophys Res Comm 1994; 200:1-7.

64. Oliver JM, Burg DL, Wilson BS et al. Inhibition of mast cell FcεRI-mediated signaling and effector function by the Syk-selective inhibitor, Piceatannol. J Biol Chem 1994; 269:29697-703.
65. Taylor JA, Karas JL, Ram MO et al. Activation of the high-affinity immunoglobulin E receptor FcεRI in RBL-2H3 cells is inihibited by Syk SH2 domains. Mol Cell Biol 1995; 15:4149-4157.
66. Letourneur F, Klausner RD. T cell and basophil activation through the cytoplasmic tail of T cell-receptor ζ family proteins. Proc Natl Acad Sci USA 1991; 88:8905-09.
67. Eiseman E, Bolen JB. Signal transduction by the cytoplasmic domains of FcεRI-γ and TCR-ζ in rat basophilic leukemia cells. J Biol Chem 1992; 267:21027-32.
68. Jürgens M, Wollenberg A, Hanau D et al. Activation of human epidermal Langerhans cells by engagement of the high affinity receptor for IgE, FcεRI. J Immunol 1995; 155:5184-89.
69. Shaw AS, Amrein KE, Hammond C et al. The *lck* tyrosine protein kinase interacts with the cytoplasmic tail of the CD4 glycoprotein through its unique amino-terminal domain. Cell 1989; 59:627-36.
70. Turner JM, Brodsky MH, Irving BA et al. Interaction of the unique N-terminal region of tyrosine kinase $p56^{lck}$ with cytoplasmic domains of CD4 and CD8 is mediated by cystein motifs. Cell 1990; 60:755-65.
71. Pendergast AM, Muller AJ, Havlik MH et al. BCR sequences essential for transformation by the *BCR-ABL* oncogene bind to the ABL SH2 regulatory domain in a nonphosphotyrosine-dependent manner. Cell 1991; 66:161-71.
72. Muller AJ, Pendergast AM, Havlik MH et al. A limited set of SH2 domains binds BCR through a high-affinity phosphotyrosine-independent interaction. Mol Cell Biol 1992; 12:5087-93.
73. Fewtrell C, Goetze A, Metzger H. Phosphorylation of the receptor for immunoglobulin E. Biochemistry 1982; 21:2004-10.
74. Minoguchi K, Swaim WD, Berenstein EH et al. Src tyrosine kinase $p53/56^{lyn}$, a serine kinase and FcεRI associate with α-Galactosyl derivatives of ganglioside GD_{1b} in rat basophilic leukemia RBL-2H3 cells. J Biol Chem 1994; 269:5249-54.
75. Swaim WD, Minoguchi K, Oliver C et al. The anti-ganglioside monoclonal antibody AA4 induces protein tyrosine phosphorylations, but not degranulation, in rat basophilic leukemia cells. J Biol Chem 1994; 269:19466-73.
76. Benhamou M, Stephan V, Robbins KC et al. High-affinity IgE receptor-mediated stimulation of rat basophilic leukemia (RBL-2H3) cells induces early and late protein-tyrosine phosphorylations. J Biol Chem 1992; 267:7310-14.
77. Hamawy MM, Mergenhagen SE, Siraganian RP. Cell adherence to fibronectin and the aggregation of the high affinity immunoglobulin E receptor synergistically regulate tyrosine phosphorylation of 105-115 kDa proteins. J Biol Chem 1993; 268:5227-33.
78. Lin P, Fung W-J, Li S et al. Temporal regulation of the IgE-dependent 1,2-diacylglycerol production by tyrosine kinase activation in a rat (RBL 2H3) mast-cell line. Biochem J 1994; 299:109-14.
79. Atkinson TP, Kaliner MA, Hohman RJ. Phospholipase C-γ1 is translocated to the membrane of rat basophilic leukemia cells in response to aggregation of IgE receptors. J Immunol 1992; 148:2194-2200.
80. Atkinson TP, Lee C-W, Rhee SG et al. Orthovanadate induces translocation of phospholipase C-γ1 and -γ2 in permeabilized mast cells. J Immunol 1993; 151:1448-55.
81. Fukamachi H, Kawakami Y, Takei M et al. Asociation of protein-tyrosine kinase with phospholipase C-γ1 in bone marrow-derived mouse mast cells. Proc Natl Acad Sci USA 1992; 89:9524-28.
82. Adamczewski M, Paolini R, Kinet J-P. Evidence for two distinct phosphorylation pathways activated by high affinity immunoglobulin E receptors. J Biol Chem 1992; 267:18126-32.
83. Ozawa K, Yamada K, Kazanietz MG et al. Different isozymes of protein kinase C mediate feedback inhibition of phospholipase C and stimulatory signals for exocytosis in rat RBL-2H3 cells. J Biol Chem 1993; 268:2280-83.

84. Ozawa K, Szallasi Z, Kazanietz MG et al. Ca^{2+}-dependent and Ca^{2+}-independent isozymes of protein kinase C mediate exocytosis in antigen-stimulated rat basophilic RBL-2H3 cells. Reconstitution of secretory responses with Ca^{2+} and purified isozymes in washed permeabilized cells. J Biol Chem 1993; 268:1749-56.
85. Razin E, Szallasi Z, Kazanietz MG et al. Protein kinases C-β and C-ε link the mast cell high-affinity receptor for IgE to the expression of c-fos and c-jun. Proc Natl Acad Sci USA 1994; 91:7722-26.
86. Germano P, Gomez J, Kazanietz MG et al. Phosphorylation of the γ chain of the high affinity receptor for immunoglobulin E by receptor-associated protein kinase C-δ. J Biol Chem 1994; 269:23102-07.
87. Smith H, Chang E-Y, Szallasi Z et al. Tyrosine phosphorylation of protein kinase C-δ in response to the activation of the high affinity receptor for immunoglobulin E modifies its substrate recognition. Proc Natl Acad Sci USA 1995; 92:9112-16.
88. Tsukada S, Saffran DC, Rawlings DJ et al. Deficient expression of a B cell cytoplasmic tyrosine kinase in human X-linked agammaglobulinemia. Cell 1993; 72:279-90.
89. Vetrie D, Vorechovsky I, Sideras P et al. The gene involved in X-linked agammaglobulinemia is a member of the *src* family of protein-tyrosine kinases. Nature 1993; 361:226-33.
90. Yamada N, Kawakami Y, Kimura H et al. Structure and expression of novel protein-tyrosine kinases, Emb and Emt, in hematopoietic cells. Biochem Biophys Res Comm 1993; 192:231-40.
91. Yao L, Kawakami Y, Kawakami T. The pleckstrin homology domain of Bruton tyrosine kinase interacts with protein kinase C. Proc Natl Acad Sci USA 1994; 91: 9175-79.
92. Li T, Tsukada S, Satterthwaite A et al. Activation of Bruton's tyrosine kinase (BTK) by a point mutation in its Pleckstrin homology (PH) domain. Immunity 1995; 2:451-60.
93. Marcilla A, Rivero-Lezcano OM, Agarwal A et al. Identification of the major tyrosine kinase substrate in signaling complexes formed after engagement of Fcγ receptors. J Biol Chem 1995; 270:9115-20.
94. Cory GOC, Lovering RC, Hinshelwood S et al. The protein product of the c-*cbl* protooncogene is phosphorylated after B cell receptor stimulation and binds the SH3 domain of Bruton's tyrosine kinase. J Exp Med 1995; 182:611-15.
95. Donovan JA, Wange RL, Langdon WY et al. The protein product of the c-*cbl* protooncogene is the 120 kDa tyrosine-phosphorylated protein in Jurkat cells activated via the T cell antigen receptor. J Biol Chem 1994; 269:22921-24.
96. Hamawy MM, Mergenhagen SE, Siraganian RP. Tyrosine phosphorylation of $pp125^{FAK}$ by the aggregation of high affinity immunoglobulin E receptors requires cell adherence. J Biol Chem 1993; 268:6851-54.
96a. Kawakami Y, Yao L, Tashiro M, Gibson S, Mills GB, Kawakami T. Activation and interaction with protein kinase C of a cytoplasmic tyrosine kinase, Itk/Tsk/Emt, on FcεRI crosslinking on mast cells. J Immunol 1995; 155:3556-62.
97. Seger R, Krebs EG. The MAPK signaling cascade. FASEB J 1995; 9:726-35.
98. Tsai M, Chen R-H, Tam S-Y et al. Activation of MAP kinases, $pp90^{rsk}$ and pp70-S6 kinases in mouse mast cells by signaling through the c-*kit* receptor tyrosine kinase or $Fc_{\varepsilon}RI$: rapamycin inhibits activation of pp70-S6 kinase and proliferation in mouse mast cells. Eur J Immunol 1993; 23: 3286-91.
99. Fukamachi H, Takei M, Kawakami T. Activation of multiple protein kinases including a MAP kinase upon FcεRI crosslinking. Int Arch Allergy Immunol 1993; 102: 15-25.
100. Santini F, Beaven MA. Tyrosine phosphorylation of a mitogen-activated protein kinase-like protein occurs at a late step in exocytosis. Studies with tyrosine phosphatase inhibitors and various secretagogues in rat RBL-2H3 cells. J Biol Chem 1993; 268:22716-22.
101. Offermans S, Jones SVP, Bombien E et al. Stimulation of mitogen-activated protein

kinase activity by different secretory stimuli in rat basophilic leukemia cells. J Immunol 1994; 152:250-61.

102. Hirasawa N, Scharenberg A, Yamamura H, Beaven MA, Kinet J-P. A requirement for Syk in the activation of the microtubule-associated protein kinase/phospholipase A_2 pathway by FcεRI is not shared by a G protein-coupled receptor. J Biol Chem 1995; 270:10960-67.
103. Hirasawa N, Santini F, Beaven MA. Activation of the mitogen-activated protein kinase/cytosolic phospholipase A_2 pathway in a rat mast cell line. Indications of different pathways for release of arachidonic acid and secretory granules. J Immunol 1995; 154:5391-402.
104. Moodie SA, Willumsen BM, Weber MJ et al. Complexes of Ras.GTP with Raf-1 and mitogen-activated protein kinase. Science 1993; 260:1658-61.
105. Li W, Hu P, Skolnik EY et al. The SH2 and SH3 domain-containing Nck protein is oncogenic and a common target for phosphorylation by different surface receptors. Mol Cell Biol 1992; 12:5824-33.
106. Chou MM, Hanafusa H. A novel ligand for SH3 domains. The Nck adaptor protein binds to a serine/threonine kinase via an SH3 domain. J Biol Chem 1995; 270: 7359-64.
107. Bar-Sagi D, Gomperts B. Stimulation of exocytotic degranulation by microinjection of the *ras* oncogene protein into rat mast cells. Oncogene 1988; 3:463-69.
108. Boguski MS, Bairoch A, Attwood TK et al. Proto-*vav* and gene expression. Nature 1992; 358:113.
109. Gulbins E, Coggeshall KM, Baier G et al. Tyrosine kinase-stimulated guanine nucleotide exchange activity of Vav in T cell activation. Science 1993; 260:822-25.
110. Gulbins E, Coggeshall KM, Baier G et al. Direct stimulation of Vav guanine nucleotide exchange activity for Ras by phorbol esters and diglycerides. Mol Cell Biol 1994; 14:4749-58.
111. Margolis B, Hu P, Katzav S et al. Tyrosine phosphorylation of *vav* proto-oncogene product containing SH2 domain and transcription factor motifs. Nature 1992; 356:71-74.
112. Bustelo XR, Ledbetter JA, Barbacid M. Product of *vav* proto-oncogene defines a new class of tyrosine protein kinase substrates. Nature 1992; 356:68-71.
113. Bustelo XR, Barbacid M. Tyrosine phosphorylation of the *vav* proto-oncogene product in activated B cells. Science 1992; 256:1196-99.
114. MacCormick F. How receptors turn Ras on. Nature 1993; 363:15-16.
115. Turner H, Reif K, Rivera J et al. Regulation of the adapter molecule Grb2 by the FcεR1 in the mast cell line RBL2H3. J Biol Chem 1995; 270:9500-06.
116. Fukamachi H, Yamada N, Miura T et al. Identification of a protein, SPY75, with repetitive helix-turn-helix motifs and an SH3 domain as a major substrate for protein tyrosine kinase(s) activated by FcεRI crosslinking. J Immunol 1994; 152:642-52.
117. Yamanashi Y, Okada M, Semba T et al. Identification of HS1 protein as a major substrate of protein-tyrosine kinase(s) upon B cell antigen receptor-mediated signaling. Proc Natl Acad Sci USA 1993; 90:3631-35.
118. Pfeiffer JR, Oliver JM. Tyrosine kinase-dependent assembly of actin plaques linking FcεR1 crosslinking to increased cell substrate adhesion in RBL-2H3 tumor mast cells. J Immunol 1994; 152:270-79.
119. Guan JL, Shalloway D. Regulation of $pp125^{FAK}$ both by cellular adhesion and by oncogenic transformation. Nature 1992; 358:690-92.
120. Turner CE, Glenney JR, Burridge K. Paxillin: a new vinculin-binding protein present in focal adhesions. J Cell Biol 1990; 111:1059-68.
121. Turner CE, Miller JT. Primary sequence of paxillin contains putative SH2 and SH3 domain binding motifs and multiple LIM domains: identification of a vinculin and $pp125^{FAK}$-binding region. J Cell Sci 1994; 107:1583-91.
122. Hildebrand JD, Schaller MD, Parsons JT. Paxillin, a tyrosine phosphorylated focal adhesion-associated protein binds to the carboxyl terminal domain of focal adhesion kinase. Mol Biol Cell 1995; 6:637-47.
123. Tachibana K, Sato T, D'Avirro N, Morimoto C. Direct association of $pp125^{FAK}$ with

paxillin, the focal adhesion-targeting mechanism of pp125^{FAK}. J Exp Med 1995; 182:1089-100.

124. Hamawy MM, Swaim WD, Minoguchi K et al. The aggregation of the high affinity IgE receptor induces tyrosine phosphorylation of paxillin, a focal adhesion protein. J Immunol 1994; 153:4655-62.

125. Minoguchi K, Kihara H, Nishikata H et al. Src family tyrosine kinase Lyn binds several proteins including paxillin in rat basophilic leukemia cells. Mol Immunol 1994; 31:519-29.

126. Hamawy MM, Minoguchi K, Swaim WD et al. A 77 kDa protein associates with pp125^{FAK} in mast cells and becomes tyrosine-phosphorylated by high affinity IgE receptor aggregation. J Biol Chem 1995; 270:12305-09.

127. Miller L, Blank U, Metzger H et al. Expression of high-affinity binding of human immunoglobulin E by transfected cells. Science 1989; 244:334-37.

128. Bieber T, de la Salle H, Wollenberg A et al. Human epidermal Langerhans cells express the high affinity receptor for immunoglobulin E (FcεRI). J Exp Med 1992; 175:1285-90.

129. Wang B, Tieger A, Kilgus O et al. Epidermal Langerhans cells from normal human skin bind monomeric IgE via FcεRI. J Exp Med 1992; 175:1353-65.

130. Maurer D, Fiebiger E, Reininger B et al. Expression of functional high affinity immunoglobulin E receptors (FcεRI) on monocytes of atopic individuals. J Exp Med 1994; 179:745-50.

131. Gounni AS, Lamkhioued B, Ochiai K et al. High-affinity IgE receptor on eosinophils is involved in defence against parasites. Nature 1994; 367:183-86.

132. Hook WA, Berenstein EH, Zinsser FU et al. Monoclonal antibodies to the leukocyte common antigen (CD45) inhibit IgE-mediated histamine release from human basophils. J Immunol 1991; 147:2670-76.

133. Lavens SE, Peachell PT, Warner JA. Role of tyrosine kinases in IgE-mediated signal transduction in human lung mast cells and basophils. Am J Respir Cell Mol Biol 1992; 7:637-44.

134. Lavens SE, Warner JA. The role of the tyrosine kinase, Syk, in IgE-mediated signal transduction in human lung mast cells (HLMC) and basophils. FASEB J 1995; 9:A781.

135. Benhamou M, Feuillard J, Lortholary O et al. Protein tyrosine kinases in activation signal of human basophils through the immunoglobulin E receptor type I. J Leukoc Biol 1996; 59:461-70.

136. Yagi S, Suzuki K, Hasegawa A et al. Cloning of the cDNA for the deleted Syk kinase homologous to ZAP-70 from human basophilic leukemia cell line (KU812). Biochem Biophys Res Comm 1994; 200:28-34.

137. Bieber T, Jürgens M, Wollenberg A et al. Characterization of the protein tyrosine phosphatase CD45 on human epidermal Langerhans cells. Eur J Immunol 1995; 25:317-21.

138. Connelly PA, Farrell CA, Merenda JM et al. Tyrosine phosphorylation is an early signaling event common to Fc receptor crosslinking in human neutrophils and rat basophilic leukemia cells (RBL-2H3). Biochem Biophys Res Comm 1991; 177: 192-201.

139. Keegan AD, Pierce JH, Artrip J et al. Ligand stimulation of transfected and endogenous growth factor receptors enhances cytokine production by mast cells. EMBO J 1991; 10:3675-82.

CHAPTER 4

Role of Phospholipases, Protein Kinases and Calcium in FcεRI-Induced Secretion

Michael A. Beaven and Tatiana Kassessinoff

INTRODUCTION

Tissue mast cells and their cognate blood basophils exhibit several distinct responses to IgE-directed antigens including release of preformed inflammatory mediators from secretory granules (chapter 9) and synthesis of arachidonic acid-derived metabolites (chapter 10) and various cytokines (chapter 11). The mechanisms for the generation of cytokines and arachidonic acid are described in chapters 10 and 11. Here we focus on the role of phospholipases, protein kinase C (PKC) and calcium in the FcεRI-mediated release of secretory granules and on where the mechanisms of secretion diverge from the signaling pathways for synthesis of arachidonic acid and cytokines. These mechanisms are best understood from studies in the rat RBL-2H3 cell line and its mutated sublines that have been made to express G protein-coupled receptors.

RBL-2H3 cells respond to antigen via FcεRI through sequential activation of the tyrosine kinases, Lyn and Syk (see chapter 3), and to adenosine analogs via the adenosine A_3 receptor.[1,2] Stimulation of cells via FcεRI and the adenosine A_3 receptor leads to activation of phospholipases C and D, with the resultant generation of inositol phosphates and diglycerides, increased concentration of cytosolic Ca^{2+} ($[Ca^{2+}]_i$), activation of PKC (see also chapter 6) and calcium-activated myosin light chain kinase. A second cascade of signals is generated through Ras to induce phosphorylation and activation of components of the mitogen

IgE Receptor (FcεRI) Function in Mast Cells and Basophils,
edited by Majed M. Hamawy.

activated protein (MAP) kinase/phospholipase A_2 pathway (see also chapter 10). Potential links between FcεRI and phospholipase C and between FcεRI and Ras have been identified, but those between FcεRI and phospholipase D have not. With the exception of the phosphorylation of phospholipase A_2 by MAP kinase or those noted in chapter 6, little is known about the identity or relevance of the proteins that are phosphorylated by PKC or MAP kinase in RBL-2H3 cells. In general, the identification of distal signals that are generated through these two kinases and, as yet unidentified, calcium-activated kinases remain the challenge for future research.

Other receptors that have been introduced into RBL-2H3 cells through gene transfection include the G protein-coupled muscarinic m1[3] and m3[4] receptors (referred to as RBL-2H3-m1 or -m3 cells, respectively), as well as epitope-tagged receptors for the chemotactic peptide, fMLP,[5] C5a[6,7] platelet activating factor,[8] IL-8[9] and thrombin.[10] Studies with the epitope tagged receptors indicate that heterologous desensitization is associated with phosphorylation of receptors by PKC and homologous desensitization with phosphorylation of receptors by G protein-coupled receptor kinases (GRKs).[9,11] All transfected sublines respond to the appropriate agonists with the resultant stimulation of the phospholipase C-mediated cascade of signals and the release of secretory granules. Thus, RBL-2H3 cells possess coupling components for a wide variety of G protein- and tyrosine kinase-dependent receptors. Except for the RBL-2H3-m1 cells, these sublines will not be discussed further.

Variations in the extent and timecourse of responses to various receptor-ligands and pharmacologic probes as well as reconstitution studies with washed permeabilized cells suggest the following sequence of events. The activation of phospholipase C and D generates diglycerides for the activation of PKC and the activation of phospholipase C provides a mechanism for increasing $[Ca^{2+}]_i$. The activation of certain isoforms of PKC along with a modest but sustained elevation of $[Ca^{2+}]_i$ provide the necessary signals for secretion. The secretory machinery in RBL-2H3 cells responds promptly to these two stimulatory signals to indicate that earlier events are rate limiting for secretion. With respect to distal events, secretion is closely correlated with a stoichiometric phosphorylation of myosin light and heavy chains by calcium-activated myosin light chain kinase and PKC but whether phosphorylation and secretion are causally related has not been established. There is evidence also that G proteins participate at late steps in the secretory process. MAP kinase does not appear to regulate secretion but it may regulate, by phosphorylation of cytosolic phospholipase A_2, the release of arachidonic acid. In the remainder of this chapter, we will discuss each of these mechanisms in turn and their relationship to secretion.

PROCEDURES FOR STIMULATION OF RBL-2H3 CELLS

Stimulation of RBL-2H3 Cells via FcεRI

Of the rat immunocytomas that secrete IgE,[12] one (HI-DNP-ε-26.82) has been widely used[13] as a source of DNP(dinitrophenyl)-specific mouse IgE for sensitizing RBL-2H3 cells to the antigen dinitrophenylated bovine serum albumin (DNP-BSA) or horse serum albumin (DNP-HSA). Typically, RBL-2H3 cells are incubated with the monoclonal DNP-specific IgE overnight in complete growth medium. Radiolabeled inositol, serotonin, arachidonic acid or other radiolabeled metabolites may be included to label metabolic pools. The next day, this medium is replaced with a simple buffered salt-glucose medium. The cultures are then stimulated by addition of DNP-BSA, or other receptor ligand, for measurement of release of granule constituents (e.g., histamine, [^{3}H]serotonin or hexosaminidase)[14,15] or radiolabeled metabolites such as inositol phosphates,[16] arachidonic acid,[17] phosphatidic acid and various phospholipids,[18] as well as unlabeled products such as the diglycerides by enzymatic assay procedures.[15,18]

Chemical Stimulants and Inhibitors

Stimulants that bypass early receptor-mediated stimulatory events include the Ca^{2+}-ionophores, ionomycin and A23187, and activators of PKC such as phorbol 12-myristate 13-acetate. Low concentrations of these reagents induce a substantial elevation in $[Ca^{2+}]_I$ and activation of PKC, respectively. By themselves, they do not induce release of secretory granules but, in combination, they are able to stimulate secretion.[19,20] Thapsigargin, which elevates $[Ca^{2+}]_I$ by blocking uptake of Ca^{2+} into inositol 1,4,5-trisphosphate-sensitive stores, also stimulates secretion but only at concentrations (>100 nM) far in excess of those (10 nM) required for elevation of $[Ca^{2+}]_I$.[21]

In addition to the above agents, inhibitors of calcium-influx and PKC such as EGTA, La^{3+}, staurosporine and Ro31-7549 have been widely used to unmask the roles of calcium and PKC in secretion. While such agents are useful pharmacologic probes, they should be used prudently as will become evident later.

MECHANISMS OF ACTIVATION OF PHOSPHOLIPASES VIA FcεRI AND OTHER RECEPTORS

Phospholipase C

At least fourteen isoforms of phospholipase C (designated as β, γ and δ and by Arabic numerals to indicate subspecies) exist in mammalian tissues, all of which show significant similarity in two regions (X and Y regions) and little similarity elsewhere.[22] Another category, formerly designated phospholipase C-α, lacks the X and Y regions and the putative cDNA was shown to encode an unrelated protein[23] that is devoid of phospholipase C activity in RBL-2H3 cells.[24] The remaining isozymes catalyze the hydrolysis of the phosphatidylinositols with phosphatidylinositol 4,5-bisphosphate as the preferred substrate, and exhibit minor differences in Ca^{2+} requirements. Reconstitution studies with purified proteins demonstrate that phospholipase C-β1 can be activated through the pertussis toxin-resistant G_q subfamily of G proteins. Multimeric immune receptors, including FcεRI, utilize the tyrosine phosphorylation of phospholipase C-γ1 and 2 as an alternate mechanism, a process that probably requires the interaction of tyrosine kinase with a *src*-homology domain (SH2), specifically present in this isoform. The presence of another *src*-homology domain (SH3) is thought to promote binding of the isoform to the membrane-associated cytoskeleton.[22] The phosphorylated and unphosphorylated forms of phospholipase C-γ exhibit the same activities in vitro except in the presence of the actin/phosphatidylinositol 4,5-bisphosphate-binding protein, profilin, when only the phosphorylated form remains active.[22] Thus, the availability of phosphatidyl 4,5-bisphosphate may further regulate activity of the phospholipase.

RBL-2H3 cells possess the β2, β3, γ1 and γ2 forms of phospholipase C (N Hirasawa and MA Beaven, unpublished observations). The recruitment of the tyrosine kinase, Syk ($p72^{syk}$), by the γ chain of FcεRI results in the tyrosine phosphorylation of phospholipase C-γ1[25] and γ2 (N Hirasawa and MA Beaven, unpublished observations), among other proteins, the apparent activation of these isozymes[25] and the sustained hydrolysis of phosphatidylinositols in RBL-2H3 cells.[26] It has not been formally established that Syk catalyzes the phosphorylation of phospholipase C-γ, but it has been shown that the phospholipase C-mediated cascade of signals, including granule secretion, is dependent on Syk. These events are blocked in RBL-2H3 cells by the introduction of a dominant negative gene for a truncated Syk lacking the kinase domain.[27]

Stimulation of RBL-2H3 cells via adenosine A_3 receptors, in contrast to stimulation through FcεRI, results in transient activation of phospholipase C[1] by a process that is inhibited by cholera and pertussis toxins[28] and markedly enhanced by dexamethasone-treatment, such that cells now secrete in response to adenosine analogs.[29,30] As the toxins induce the expected ADP-ribosylation of $G_{\alpha s}$ and $G_{\alpha i\text{-}2 \text{ and } 3}$ respectively in RBL-2H3 cells[31] and dexamethasone increases the expression of the A_3 receptor as well as the α and β subunits of trimeric G proteins,[32] it has been suggested that both the α and βγ subunits of G proteins contribute to the activation of phospholipase C.[2]

The expressed muscarinic m1 receptors in RBL-2H3-m1 cells are coupled to phospholipase C-β3 through $G_{\alpha q/11}$ as indicated by selective downregulation of these proteins with prolonged exposure to carbachol but not to antigen. Also, unlike antigen, carbachol does induce tyrosine phosphorylation of phospholipase C-γ1 and γ2 (OH Choi, K Yamada and MA Beaven, unpublished observations). Both stimulants, nevertheless, elicit sustained phospholipase C-mediated responses.[3,33]

Phospholipase D

The mechanisms of activation of phospholipase D are still undetermined, but the recent cloning of a gene that encodes one form of phospholipase D should facilitate studies of these mechanisms.[34] Measurement of phospholipase D in cell free extracts is problematic because of low recovery of enzyme activity.[35] Roles for G proteins and protein phosphorylation have been proposed for the activation of this enzyme (reviewed in ref. 36). Recent reports indicate activation of membrane-associated phospholipase D by small G proteins such as the ADP-ribosylation factor, ARF,[37,38] Rho[39,40] or both[41] and a requirement for activation by phosphatidylinositol 4,5-bisphosphate.[42] So far, two types of membrane-associated enzyme activities have been characterized in cell free systems. One is dependent on oleic acid,[43-45] the other on small G proteins as noted above.[37-41] The recombinant protein (PLD1) from the cloned-gene appears to fit the latter type in that it is stimulated by ARF-1 and phosphatidylinositol 4,5-bisphosphate and inhibited by oleate. The protein is selective for phosphatidylcholine and may be but one member of a unique phospholipase D gene family.[34]

Prior work had shown that receptor-mediated activation of phospholipase D results in the production of phosphatidic acid which is then rapidly converted to diglycerides through the action of phosphatidate hydrolase.[46,47] Both lipid metabolites are thought to serve as intracellular messengers for mitogenesis and the activation of PKC and phosphatyldinositol 4-kinase (reviewed in refs. 48-50). Phosphatidylcholine is the major substrate for the enzyme in mast cells[51] and RBL-2H3 cells[52] where, as in other types of cells, the enzyme appears to be activated indirectly through a rise in $[Ca^{2+}]_i$ and the action of PKC[52] to produce sustained increases in phosphatidic acid and diglycerides. Consistent with studies in other types of cells,[48] phospholipase D appears to be responsible for the bulk of the diglycerides that are generated during mast-cell stimulation, especially at late stages of the cell response.

Patterns of Activation of Phospholipases C and D Through FcεRI and Other Receptors

Aggregating FcεRI on isolated rat peritoneal mast cells and RBL-2H3 cells results in substantial and sustained increases in inositol phosphates, phosphatidic acid and diglycerides through the actions of phospholipase C and D, respectively.[52-55] Phospholipase D, however, is the major contributor of diglycerides[56] and mass analysis of another product of phospholipase D-catalyzed reactions, choline, reveals 5- to 10-fold increases in this metabolite after stimulation.[51] In RBL-2H3 cells, diglycerides exhibit a rapid transient rise followed by an increase which is sustained for at least 30 minutes. The sustained increase has been attributed to activation of phospholipase D by a tyrosine kinase.[57]

Stimulation via the A_3 receptors causes a sustained activation of phospholipase D and PKC[2] in addition to the transient activation of phospholipase C mentioned previously. These processes are also sensitive to inhibition by cholera and pertussis toxins[28] and enhancement by dexamethasone,[29] again inferring phospholipase regulation by both α and βγ subunits of G proteins. The sustained activation of phospholipase D and PKC probably accounts for the ability of adenosine analogs to synergize secretory responses to antigen and other secretagogues in RBL-2H3 cells.[2] The pattern of response to adenosine analogs strongly supports the notion that the activation of phospholipase D reinforces and sustains primary signals transduced via phospholipase C.

In many types of cells, activation of phospholipase C, but not of phospholipase D, can be suppressed by activation of PKC with phorbol esters.[58] This is also true for RBL-2H3 cells. Hydrolysis of phospholipids and rise in $[Ca^{2+}]_I$ in response to carbachol (OH Choi, K Yamada and MA Beaven, unpublished observations) or FcεRI aggregation[59] are suppressed by phorbol 12-myristate 13-acetate and enhanced 2- to 3-fold by the PKC inhibitor, Ro31-7549. This would indicate that stimulation of phospholipase C by G protein- or tyrosine kinase-dependent mechanisms is subject to substantial feedback inhibition by PKC. Phospholipase D-mediated reactions are stimulated rather than suppressed by phorbol ester, thus indicating a positive regulatory-role for PKC in RBL-2H3 cells.[55]

MOBILIZATION OF CALCIUM AND ACTIVATION OF SERINE/ THREONINE KINASES

MOBILIZATION OF INTRACELLULAR AND EXTRACELLULAR CALCIUM IONS

Early studies established that degranulation of rat peritoneal mast cells is dependent on external Ca^{2+} and is associated with influx of Ca^{2+} ($^{45}Ca^{2+}$) and other divalent cations[60] (also see citations in ref. 61). Blockade of influx by addition of calcium chelators or cations that were known to block calcium uptake substantially or totally blocked secretion from basophils and various types of mast cells.[62] Various mechanisms were proposed for the regulation of Ca^{2+}-influx,[63,64] but most recent studies favor the view that Ca^{2+}-influx is associated with an emptying of intracellular calcium stores through the generation of second messengers. As has been observed in other types of electrically nonexcitable cells, Ca^{2+}-influx occurs in two phases: an initial increase in $[Ca^{2+}]_I$ through a rapid release of Ca^{2+} from inositol 1,4,5-trisphosphate-sensitive Ca^{2+}-stores, followed by an influx of extracellular Ca^{2+}.[65,66]

In the RBL-2H3 cell, FcεRI aggregation causes a transient increase in $[Ca^{2+}]_I$ in the absence of external Ca^{2+}.[3,67] These increases occur abruptly after a delay of 50-100 seconds,[68] which has been attributed to release of Ca^{2+} from intracellular stores by inositol 1,4,5-trisphosphate for the following reasons: First, the increase in $[Ca^{2+}]_I$ coincides with an early appearance of this inositol phosphate.[69,70] Second, Ca^{2+} is released from permeabilized RBL-2H3 cells by submicromolar amounts of inositol 1,4,5-trisphosphate in a cooperative fashion.[71] Third, the increase is not observed when the intracellular stores are first depleted by carbachol in RBL-2H3-m1 cells.[3,33]

The provision of external calcium, after the FcεRI-induced transient increase in $[Ca^{2+}]_I$, results in a prompt and sustained increase in $[Ca^{2+}]_I$. This sustained elevation in $[Ca^{2+}]_I$, as determined by Ca^{2+}-sensitive fluorescent probes, is totally dependent on influx of external Ca^{2+}. Blockade of Ca^{2+}-influx by use of EGTA or La^{3+} causes a rapid decline in $[Ca^{2+}]_I$ to basal levels and subsequent neutralization of these reagents rapidly restores $[Ca^{2+}]_I$ to its former elevated levels.[72] The increase in $[Ca^{2+}]_I$ is associated with an increase in total intracellular Ca^{2+} through uptake of Ca^{2+} into mitochondrial stores when $[Ca^{2+}]_I$ is elevated above basal levels.[21,73] Influx of Ca^{2+} is also evident from an accelerated uptake of $^{45}Ca^{2+}$ by RBL-2H3 cells for several minutes after the addition of antigen.[61,74,75] Other cations impede Ca^{2+}-uptake, either by blocking entry of Ca^{2+} at the cell surface, e.g., La^{3+} and Zn^{2+}, or by competing for Ca^{2+}-entry into the cell, e.g., Sr^{2+}, Ba^{2+} and Mn^{2+}.[61] The cation influx pathway is apparently not voltage-activated[74,76] nor blocked by organic blockers of voltage-activated Ca^{2+}-channels.[77] Indeed, entry of Ca^{2+} is suppressed when cells are depolarized with high concentrations of external K^{+}[74,76] and a repolarizing current may be required to maintain influx.[76,78-82] The same influx mechanism appears to operate when RBL-2H3 cells are stimulated via m1 receptors[33] or with thapsigargin which depletes inositol 1,4,5-trisphosphate-sensitive Ca^{2+} stores by blocking uptake into these stores.[21]

When external calcium is present throughout the stimulatory period, the initial transient response may fuse with the sustained response which may be oscillatory or sinusoidal in nature. There is, however, considerable variation in the responses of individual cells for reasons that are still unclear (for a detailed review see ref. 83).

In other types of electrically nonexcitable cells, including rat peritoneal mast cells,[60] influx of Ca^{2+} is closely associated with the emptying of inositol 1,4,5-trisphosphate-sensitive Ca^{2+}-stores[65,66] and the generation of a diffusible messenger molecule, called CIF for "calcium influx factor", from intracellular organelles.[84-88] CIF has been characterized as a mixture of several active components of low molecular weight (<1000) which are inactivated by treatment with alkaline phosphatase.[86,87] CIF is released into the cytosol when intracellular stores are depleted by stimulation of cells with receptor ligands or thapsigargin. CIF activity has been demonstrated by introduction of the cytosolic extracts, or partially purified preparations therefrom, into patch-clamped unstimulated cells.

Influx of Ca^{2+} is thought to occur through a low conductance current, designated I_{crac} for "calcium release-activated calcium current", which has been characterized in mast cells[89] and RBL-2H3 cells.[90] This current appears to be highly selective for Ca^{2+} ions[89] and its activation is dependent on GTP, possibly acting via a low molecular weight G protein.[90,91] To date, Ca^{2+}-conductance, including I_{crac}, has been examined in mast cells by the use of reagents such as GTP and the chemical secretagogue, compound 48/80,[89,90,92,93] but there is no reason to suspect that I_{crac} would not be activated in antigen stimulated mast cells. An additional and novel sphingolipid-gated Ca^{2+}-gated efflux channel (from Ca^{2+}-storage organelles) was recently described in RBL-2H3 cells[94] but its physiological relevance remains undetermined.

Recruitment of PKC

PKC is a family of serine/threonine kinases that are rapidly activated in response to elevated $[Ca^{2+}]_i$ and the generation of diglycerides via phospholipases C and D. Phorbol esters can substitute for diacylglycerol in activating PKC and have been widely used in unmasking PKC-mediated phosphorylations and actions in vivo. Based on their diverse properties and their historical sequence of discovery, the isoforms have been categorized as conventional or calcium-dependent (α, the alternatively spliced variants βI and βII, and γ), novel or calcium-independent (δ, ε, η, θ and μ) and atypical (ζ and λ). The atypical isoforms are the least understood category but they fail to respond to phorbol esters.[36] The structural differences among these categories account for their diverse properties and possibly permit the individual isoforms to perform different functions within the cell.[95,96] Indeed, studies in RBL-2H3 cells strongly support this notion. As this topic is discussed in detail in chapter 6, our comments are restricted to a discussion of the role of PKC in FcεRI-mediated secretion.

RBL-2H3 cells contain the Ca^{2+}-dependent α, βI and βII isoforms and the Ca^{2+}-independent δ, ε and ζ isoforms of PKC.[97-99] Following FcεRI aggregation, all isoforms, except δ, rapidly associate with the membrane fraction. In the absence of external Ca^{2+}, only the Ca^{2+}-independent δ and ε isoforms do so.[97] As noted below, studies with washed permeabilized cells demonstrate that this association is necessary for transduction of signals for secretion.

All isozymes readily diffuse from washed permeabilized RBL-2H3 cells and as a consequence, the cells fail to secrete in response to FcεRI crosslinking,[97] but stimulatory signals such as hydrolysis of inositol phospholipids are enhanced by as much as 2- to 3-fold.[59] The secretory response to FcεRI aggregation can be restored by provision of recombinant isozymes of PKC. Such studies suggested that FcεRI-induced secretion was mediated primarily by PKC-β and δ[97] and feedback-inhibition of phospholipase C was mediated primarily by PKC-α and ε.[59] The caveat is that provision of exogenous PKC may allow access of the isozymes to sites that they do not normally occupy.[95] The studies, nevertheless, demonstrate that release of secretory granules is dependent on PKC and suggest that signals are transduced by specific isoforms.

The requirement for PKC is also evident from studies with phorbol 12-myristate 13-acetate, both as an activator or downregulator of PKC, and inhibitors of the kinase. The isoforms of PKC exhibit different rates of degradation when RBL-2H3 cells are continuously exposed to phorbol 12-myristate 13-acetate. PKC-β is degraded within minutes, and PKC-α within hours, whereas the Ca^{2+}-independent isozymes (δ, ε and ζ) resist degradation.[97,100,101] After prolonged exposure to the phorbol ester, the secretory response to antigen is reduced by about 80%. Therefore, Ca^{2+}-dependent isozymes probably contribute a substantial stimulatory signal for secretion.[97,100]

FcεRI-mediated secretion is also inhibited by various inhibitors of PKC. This is true for mast cells, basophils and cultured mast cell lines for a variety of secretagogues. It should be noted, however, that kinase inhibitors rarely have the selectivity intended when used in vivo. In RBL-2H3 cells for example, the PKC inhibitors, staurosporine and Go-6976, as well as the myosin light chain kinase inhibitor, KT5926, all inhibit FcεRI-induced tyrosine phosphorylation of phospholipase C-γ and generation of inositol phosphates.[102] Also, the myosin light chain kinase inhibitor, ML-7, inhibits activation of PKC[103] and the tyrosine kinase inhibitor, herbimycin, effectively blocks the phosphoinositide cycle, whether this cycle is activated via a G protein, as in the case with carbachol stimulation, or tyrosine kinase, as in the case with antigen stimulation.[104] Of the many drugs that we have tested, only the Ro series of PKC inhibitors[105] have proved to be selective when tested against the systems that we employ routinely in our studies. Stimulatory systems not inhibited by the Ro compounds in RBL-2H3 cells include the activation of tyrosine kinases, myosin light chain kinase, MAP kinase and phospholipases C and A_2. As will be noted later, Ro31-7549, nonetheless, suppresses release of secretory granules and TNF while release of arachidonic acid remains untouched.[102,103,106,107]

The Activation of MAP Kinase

Tyrosine kinase-dependent receptors, such as the EGF receptor and the multimeric immune receptors, utilize the SH2-containing protein, Shc, the adaptor protein, Grb2, and the guanine nucleotide exchange factor, Sos, to convert Ras (p21*ras*) to its active GTP-bound state[108-111] which, in turn, activates the MAP kinase pathway via Raf1.[111] Activation is accomplished through Ras-mediated translocation of the kinase Raf1, which, when phosphorylated by an unidentified kinase, phosphorylates and activates a unique tyrosine/threonine kinase, MEK, which then phosphorylates and activates MAP kinase. In RBL-2H3 cells other signal transducers, for example Vav, may operate (see also chapter 6).

The phosphorylation of $p42^{mapk}$ is associated with a shift in the electrophoretic migration of $p42^{mapk}$ and an increase in MAP kinase activity.[106,112-114] Expression of porcine Syk, or a dominant-negative truncated Syk (Syk-T) that lacks the kinase domain,[114,115] via a vaccinia expression system into RBL-2H3-m1 cells, has indicated that Syk is essential for the activation of MAP kinase and release of arachidonic acid through phosphorylation of phospholipase A_2 upon antigen, but not carbachol, stimulation.[114] This and other studies indicate that the MAP kinase/phospholipase A_2 pathway may be activated by other stimulants through alternate pathways.[106] Current studies indeed suggest that elevated $[Ca^{2+}]_I$ and PKC provide alternate, albeit transient, signals for activation whereas FcεRI is capable of producing a prolonged signal for stimulating the Ras/MAP kinase pathway (C Zhang, N Hirasawa and MA Beaven, unpublished observations). The studies in total strongly suggest that MAP kinase plays little or no role in FcεRI-mediated secretion but has an essential role in the activation of phospholipase A_2 (unpublished observations; see also chapter 10).[106,114]

POTENTIAL TARGETS FOR CALCIUM AND PKC THAT MAY MEDIATE FcεRI-INDUCED SECRETION

Myosin Light and Heavy Chains; Substrates for Phosphorylation by PKC and Ca^{2+}-Activated Myosin Light Chain Kinase

Studies of myosin phosphorylation have been undertaken with the idea that the interaction of myosin and actin might underlie the exocytotic process in mast cells. Stimulated RBL-2H3 cells exhibit rapid morphological changes (see chapter 8), formation of f-actin[116,117] and the redistribution of actin[118] and myosin.[119] Studies with various types of cells indicate that the light chains of myosin are phosphorylated by PKC and Ca^{2+}-activated myosin light chain kinase and the heavy chains of myosin are phosphorylated by PKC and other kinases. These phosphorylations promote myosin filament assembly and disassembly and phosphorylation of the light chains by myosin light chain kinase increases the actin-activated Mg/ATPase activity of myosin. The physiological relevance of these phosphorylations, however, is unclear.[120]

In antigen-stimulated RBL-2H3 cells, myosin light chains and heavy chains are phosphorylated stoichiometrically by PKC.[121] The myosin light chains are constitutively phosphorylated by myosin light chain kinase predominantly at serine-19 in unstimulated cells, but stimulation with antigen or other stimulants causes phosphorylation at additional sites by myosin light chain kinase and as well as by PKC at serine 1 or 2.[103,122] All these additional phosphorylations show close correlation with the rate and extent of degranulation.[103,121] Other observations indicate a correlation with these phosphorylations and release of secretory granules. Secretion occurs only when phosphorylation by both enzymes is apparent, for example with high, but not low, concentrations of the Ca^{2+}-ionophore, A23187 or with low concentrations of A23187 in combination with phorbol 12-myristate 13-acetate. In contrast, phorbol ester alone induces phosphorylation of the light chains by PKC without inducing secretion. Also, secretion is not observed after FcεRI aggregation when phosphorylation by either PKC or myosin light chain kinase is suppressed with, respectively, Ro31-7549 or KT5926.[103]

While there appears to be a remarkably close association between degranulation and the coincident phosphorylation of the myosin chains by PKC and by myosin light chain kinase, it remains to be established whether these events are epiphenomena or causally related. RBL-2H3 cells are unusual in that, of the two known isoforms of myosin (i.e., A and B), they express only myosin-A mRNA and protein.[123] Thus, studies with antibodies against myosin-A in permeabilized cells should be instructive.

G Proteins as Mediators of Events Distal to Calcium

Another candidate for mediating a late step in FcεRI-mediated exocytosis is the hypothetical G protein, G_e, where "e" stands for exocytosis.[124,125] A role for G_e

was deduced from the observation of Gomperts and co-workers that Ca^{2+} and nonhydrolyzable GTP analogs together were sufficient to induce maximal secretory response in permeabilized[126] and patch-clamped mast cells.[127] Because elevation in $[Ca^{2+}]_I$ eliminated the requirement for the GTP analog, and vice versa, the sequence: $Ca^{2+} \rightarrow Ca^{2+}$-binding protein (the hypothetical receptor, C_e) $\rightarrow G_e$ was proposed for the secretory pathway.[124] Variants of the model have been proposed.[128,129] The essential core of the hypothesis was that a Ca^{2+}-receptor required a G protein for further transduction of signals. Since the work of Gomperts and co-workers, interest has expanded to include low molecular weight as well as the heterotrimeric G proteins and their role in vesicle trafficking and membrane fusion (see chapter 5 for further details).

With respect to G_e specifically, there is evidence that G_e is a pertussis toxin-sensitive, trimeric G protein[128-130] and that chemical stimulants such as compound 48/80 and mastoparan[129] induce secretion by direct activation of G_e. Compound 48/80 partially penetrates the plasma membrane of mast cells on account of its hydrophobic core[131] and could presumably interact directly with membrane bound G_e. A potential candidate is G_{i3}, because this G protein is present in the plasma membrane of rat peritoneal mast cells. In addition, the secretory actions of 48/80 can be blocked by the introduction of either a synthetic peptide that corresponds to the carboxy-terminus of $G_{\alpha i\text{-}3}$ or an antibody to this peptide into permeabilized rat mast cells.[132]

COORDINATION OF SIGNALING EVENTS FOR SECRETION AND OTHER FUNCTIONAL RESPONSES

Facts and Speculations

In the remaining sections we will attempt to tie together the discussions in the preceding sections and indicate our assessment of what has been established and what has not. In an earlier review,[62] one of us outlined a number of uncertainties, some of which are still pertinent today. One of these was whether or not we had a complete understanding of the mechanisms of calcium mobilization via FcεRI. The kinetics of production of the various inositol phosphates in response to antigen-stimulation of RBL-2H3 cells is atypical and the actual increase in levels of inositol 1,4,5-trisphosphate are relatively small.[33,69,133,134] The small increase is especially evident when compared to increases obtained in response to carbachol in RBL-2H3-m1 cells.[33] The uncertainties with respect to the mechanisms of activation of phospholipase C, phospholipase A_2 and PKC have largely been eliminated because of the details about these enzymes, as well as the discovery of the MAP kinase pathway, that have emerged since the earlier review. We can now state with some conviction that increase in $[Ca^{2+}]_I$ and activation of PKC by themselves will induce release of secretory granules (see next section). Apart from the observations and speculations regarding myosin and G proteins as discussed above, the scenario of events that lead to secretion, beyond calcium and PKC, is obscure. Our final note of caution is that because of the similarities in the effects of pharmacologic probes on basophils, isolated mast cells, RBL-2H3 cells, and other types of cultured mast cells, we cannot assume all these types of cells utilize the same mechanisms until such cells have been subject to the same experimental rigor as RBL-2H3 cells.

Release of Secretory Granules; Indications of a Role for Calcium and PKC

Except for the caveats noted earlier, the studies with pharmacologic inhibitor agents suggest that elevation of $[Ca^{2+}]_I$ and activation of PKC are necessary signals for secretion and the studies with pharmacologic activators suggest that these two are sufficient to initiate secretion in mast cells and basophils. Secretion can be induced in RBL-2H3 cells, as in normal mast cells and basophils, by low, nonsecretory concentrations of Ca^{2+}-ionophore and phorbol 12-myristate 13-acetate.[19] Secretion that is induced by high concentrations of Ca^{2+}-ionophore alone is associated with hydrolysis of inositol phospholipids[20] and activation of PKC.[103] Such findings, however, do not prove that elevation of $[Ca^{2+}]_I$ and activation of PKC are the physiologic signals for secretion of granules in antigen-simulated RBL-2H3 cells,[62] although the pattern of phosphorylation of myosin chains, as discussed above, implies that these two signals are generated following FcεRI aggregation.

The difficulty in interpreting studies with pharmacologic activators is this: the signals that are generated by the combination of Ca^{2+}-ionophore and phorbol 12-myristate 13-acetate are more intense than those generated by optimal concentrations of antigen. Moreover, antigen is a much more potent stimulant of secretion than the drug combination when doses are matched to give the same increase in $[Ca^{2+}]_I$. The possibility exists, therefore, that additional synergistic signals may be generated by antigen.[19] Our most recent measurements in individual fura2-loaded RBL-2H3 cells have indicated a basal $[Ca^{2+}]_I$ of 70 nM to 85 nM with increases to between 500 nM to 1200 nM in cells stimulated with concentrations of antigen (20 ng/ml DNP-BSA) that are optimal for secretion (40-50% release of secretory granules). Substantial secretion (15-20% release of granules) is observed, however, at 200 nM $[Ca^{2+}]_I$ when cells are stimulated with sub-optimal concentrations of antigen (1-3 ng/ml DNP-BSA). Therefore, relatively modest increases in $[Ca^{2+}]_I$ are sufficient to promote substantial secretion by antigen. These increases are much smaller than those elicited by Ca^{2+}-ionophores or thapsigargin even at concentrations that can induce secretion in combination with phorbol ester.[20,21]

The most definitive information on the role of calcium and PKC has come from reconstitution studies with permeabilized cells. As previously mentioned, washed permeabilized cells lose all isozymes of PKC and fail to secrete in response to antigen. A full secretory response to antigen is restored by subsequent addition of nM concentrations of either the Ca^{2+}-dependent PKC-β or the Ca^{2+}-independent PKC-δ (other isozymes were much less effective), but only in the presence of 1 μM free Ca^{2+} to indicate separate roles for Ca^{2+} and PKC in exocytosis.[97]

The above data did not exclude the existence of another necessary pathway as, for example, the activation of tyrosine kinases should this pathway remain intact in permeabilized cells. Additional studies have since indicated that elevation of $[Ca^{2+}]_I$ and activation of PKC alone were sufficient signals for secretion.[135] Provision of the PKC agonist, 1-oleoyl 2-acetyl-sn-glycerol (OAG), in the presence of the calcium-independent PKC-δ and physiologically relevant concentrations of calcium were sufficient to elicit a maximal secretory response. The elevation of $[Ca^{2+}]_I$ and provision of PKC-δ alone did not stimulate secretion. The data indicated a half maximal response at 200 nM $[Ca^{2+}]_I$, and near maximal response at 700-1000 nM $[Ca^{2+}]_I$, or responses that were comparable to those observed in intact, antigen-stimulated cells. When $[Ca^{2+}]_I$ was raised to nonphysiological concentrations of 10 μM, a modest release of inositol phosphates, diglycerides and secretory granules was apparent, presumably because of the activation of PKC by lipid metabolites. The studies also indicated that, with respect to the calcium-dependent PKC-β, relatively small increases in $[Ca^{2+}]_I$ (>250 nM) were required for full activation of this isoform. Thus, in the intact cell, a 2- to 3-fold increase in $[Ca^{2+}]_I$ is sufficient to allow recruitment of PKC-β and an additional increase is necessary for secretion in combination with the activation of PKC-β and δ. The signaling pathways for secretion are depicted in Figure 4.1.

An intriguing phenomenon is that FcεRI-induced secretion in washed permeabilized cells exhibits a quantal response when the concentration of PKC-β or δ is less than optimal.[135] This observation could

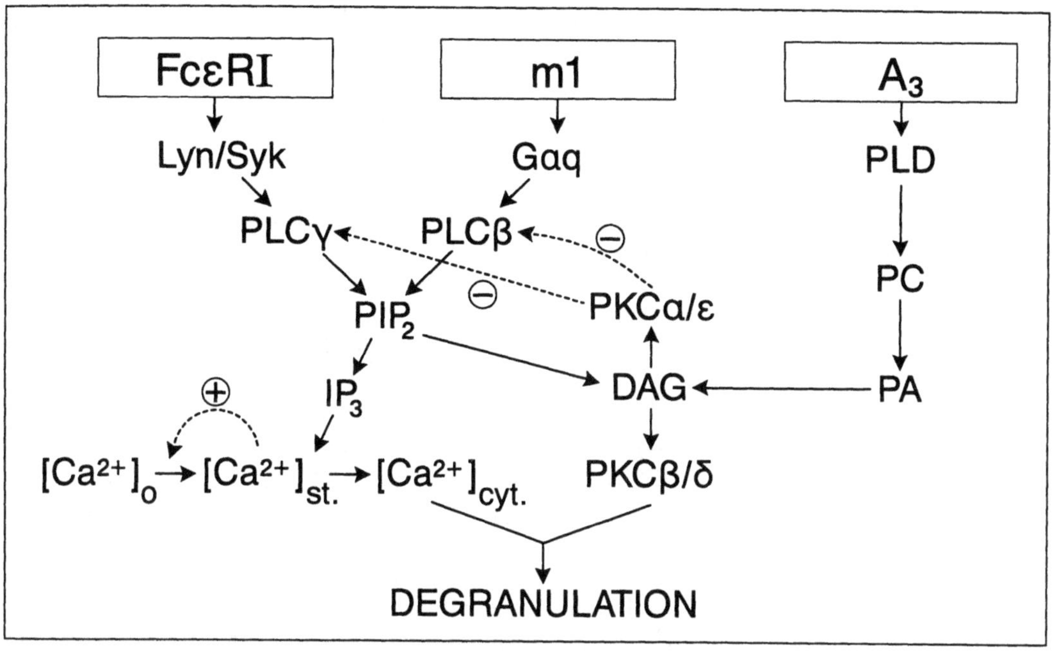

Fig. 4.1. Signaling pathways for release of secretory granules in RBL-2H3 cells. FcεRI-mediated signaling pathways and, for comparison, those from other receptors are depicted. FcεRI activates phospholipase C (PLC)γ through the tyrosine kinases, Lyn and Syk, and the transfected muscarinic m1 receptor activates phospholipase C-β via the G protein $G_{\alpha q/11}$. These activations result in the generation of inositol 1,4,5-trisphosphate (IP_3) and diacylglycerol (DAG) from phosphatidylinositol 4,5-bisphosphate (PIP_2) with the resultant increase in concentration of cytosolic calcium ions ($[Ca^{2+}]_{cyt}$) through release from intracellular stores ($[Ca^{2+}]_{st}$) and influx of external calcium ions ($[Ca^{2+}]_o$) as well as the activation of protein kinase C (PKC) by diacylglycerol. Influx of calcium may be mediated through generation of a diffusible messenger molecule, CIF, from depleted intracellular stores (broken arrow). The β and δ isoforms of protein kinase C and the increase in cytosolic calcium, provide the necessary signals for degranulation while the α and ε isoforms of protein kinase C transduce negative signals for feedback inhibition of phospholipase C (broken arrows). Stimulation of adenosine A_3 receptors synergizes antigen- (via FcεRI) or carbachol- (via m1 receptors) induced secretion through activation of phospholipase D (PLD) with the resultant generation of phosphatidic acid (PA), and therefrom diglycerides, from phosphatidylcholine (PC). Phospholipase D is activated by the other two receptors as well (not shown) and probably provides an additional signal for secretion by enlarging the pool of diglycerides. MAP kinase does not appear to play a role in degranulation but has an essential role in the activation of phospholipase A_2 (see chapter 10).

imply that each molecule of membrane-associated PKC had access to limiting amounts of substrate that was critical for secretion. The identity of this substrate, or substrates, is currently unknown.

Signaling Pathways for Generation of Arachidonic Acid and Cytokines

These topics are covered in chapters 10 and 11 but several relevant points are noted here. With respect to release of arachidonic acid, the deduced amino acid sequence of the cDNA for cytosolic phospholipase A_2 shows high homology between the human and murine forms of the enzyme with a common Ca^{2+}-dependent phospholipid binding domain and a single serine-containing consensus site for phosphorylation by MAP kinase.[136] These and other findings suggest that enzyme activity is regulated primarily by calcium and MAP kinase[137,138] as appears to be true for RBL-2H3 cells.[106,114] In RBL-2H3-m1 cells, stimulation of MAP kinase/phospholipase A_2 with antigen, carbachol, A23187 or thapsigargin is blocked by relatively low concentrations of quercetin but is not blocked by Ro31-7549. Therefore, the pathway appears to be dependent on a quercetin-sensitive protein kinase that is not PKC.[106] Overexpression of truncated Syk also suppresses activation of the pathway when cells are stimulated by FcεRI aggregation but not when stimulated by carbachol[114] indicating that this pathway, like that for secretion, is dependent on Syk.[114] These studies demonstrate that release of arachidonic acid is attributable exclusively to the regulation of cytosolic phospholipase A_2 by MAP kinase (for activation of the enzyme) and Ca^{2+} (for association of $cPLA_2$ with the membrane). Secretion of granules, in contrast, is regulated by Ca^{2+} and PKC but not by MAP kinase as was discussed earlier (see Fig. 4.1).

The induction of synthesis of cytokines is discussed elsewhere (chapter 11), but we should note that, in RBL-2H3 cells, the cytokine, TNF, is not constitutively expressed nor incorporated into secretory granules but is generated de novo upon cell stimulation.[107] It is then released by a process analogous to constitutive secretion in that brefeldin A, an agent known to disrupt Golgi membranes in these cells, inhibits this release without inhibiting release of secretory granules.[107] Unlike constitutive secretion, however, the secretion of TNF is highly regulated by Ca^{2+} and PKC. Studies with various stimulants and inhibitors indicate that simultaneous mobilization of Ca^{2+} and activation of PKC are sufficient signals for secretion although optimal production of TNF may be dependent on additional synergistic signals. Inhibition of either Ca^{2+} mobilization or PKC alone abrogates release of TNF, and the TNF retained within the cell is degraded.[107] The release and downstream effects of this cytokine should, therefore, be sensitive to therapeutic blockade.

CONCLUDING REMARKS

In conclusion, it would appear that FcεRI-mediated processes for release of secretory granules, arachidonic acid (and its lipid metabolites) and inflammatory cytokines should be amenable to therapeutic intervention. Inhibitors of PKC, such as Ro31-7549 and its cognates,[139] appear to be effective inhibitors of secretion of both secretory granules and TNF-containing Golgi-derived secretory vesicles. Quercetin and possibly selective inhibitors of the MAP kinase pathway, if such can be developed, might be effective suppressants of release of arachidonic acid-derived lipid metabolites. Finally, the differences in the regulation of PKC-dependent trafficking of secretory granules and cytokine-containing vesicles should be an interesting challenge for future research.

The recent cloning of the intracellular sphingolipid-gated Ca^{2+} channel (see reference 94, page 60)[140] and the finding that sphingosine 1-phosphate is responsible for substantial release of Ca^{2+} ions from the intracellular stores in RBL-2H3 cells[140,141,142] indicate that calcium homostasis in antigen-stimulated RBL-2H3 cells is regulated by both phosphoinositide and sphingolipid metabolites. The findings explain certain anomalies (discussed on page 63) arising from previous studies.[142]

REFERENCES

1. Ramkumar V, Stiles GL, Beaven MA, Ali H. The A_3R is the unique adenosine receptor which facilitates release of allergic mediators in mast cells. J Biol Chem 1993; 268:16887-90.
2. Ali H, Choi OH, Fraundorfer PF, Yamada K, Gonzager HMS, Beaven MA. Sustained activation of phospholipase D via adenosine A_3 receptors is associated with enhancement of antigen- and Ca^{2+}-ionophore-induced secretion in a rat mast cell line cell. J Pharm Exp Ther 1996; 276:837-45.
3. Jones SVP, Choi OH, Beaven MA. Carbachol induces secretion in a mast cell line (RBL-2H3) transfected with the m1 muscarinic receptor gene. FEBS Lett 1991; 289:47-50.
4. Stephan V, Benhamou M, Gutkind JS, Robbins KC, Siraganian RP. Fc epsilon RI-induced protein tyrosine phosphorylation of pp72 in rat basophilic leukemia cells (RBL-2H3). Evidence for a novel signal transduction pathway unrelated to G protein activation and phosphatidylinositol hydrolysis. J Biol Chem 1992; 267:5434-41.

5. Ali H, Richardson RM, Tomhave ED, Didsbury JR, Snyderman R. Differences in phosphorylation of formylpeptide and C5a chemoattractant receptors correlate with differences in desensitization. J Biol Chem 1993; 268:24247-54.
6. Tomhave ED, Richardson RM, Didsbury JR, Menard L, Snyderman R, Ali H. Cross-desensitization of receptors for peptide chemoattractants. Characterization of a new form of leukocyte regulation. J Immunol 1994; 153:3267-75.
7. Ali H, Richardson RM, Tomhave ED, Didsbury JR, Snyderman R. Differences in phosphorylation of formylpeptide and C5a chemoattractant receptors correlate with differences in desensitization. J Biol Chem 1993; 268:24247-54.
8. Ali H, Richardson RM, Tomhave ED, DuBose RA, Haribabu B, Snyderman R. Regulation of stably transfected platelet activating factor receptor in RBL-2H3 cells. Role of multiple G proteins and receptor phosphorylation. J Biol Chem 1994; 259:24557-83.
9. Richardson RM, DuBose RA, Ali H, Tomhave ED, Haribabu B, Snyderman R. Regulation of human interleukin-8 receptor A: Identification of a phosphorylation site involved in modulating receptor functions. Biochemistry 1995; 34:14193-201.
10. Ali H, Tomhave ED, Richardson RM, Haribabu B, Snyderman R. Thrombin primes responsiveness of selective chemoattractant receptors at a site distal to G protein activation. J Biol Chem 1996; 271:3200-6.
11. Richardson RM, Ali H, Tomhave ED, Haribabu B, Snyderman R. Crossdesensitization of chemoattractant receptors occurs at multiple levels: Evidence for a role for inhibition of phospholipase C activity. J Biol Chem 1995; 270:27829-33.
12. Bazin H, Querinjean P, Beckers A, Heremans JF, Dessy F. Transplantable immunoglobulin-secreting tumours in rats. IV. Sixty-three IgE-secreting immunocytoma tumours. Immunology 1974; 26:713-23.
13. Liu FT, Bohn JW, Ferry EL, Yamamoto H, Molinaro CA, Sherman LA, Klinman NR, Katz DH. Monoclonal dinitrophenyl-specific murine IgE antibody: Preparation, isolation and characterization. J Immunol 1980; 124:2728-37.
14. Beaven MA, Ludowyke RI. Stimulatory signals for secretion in mast cells and basophils. In: Mond JJ, Cambier JC, Weiss A, eds. Advances in Regulation of Cell Growth, Volume 1: Regulation of Cell Growth and Activation. New York: Raven Press Ltd., 1989:245-85.
15. Baumgartner RA, Beaven MA. Mediator release by mast cells and basophils. In: Herzenberg LA, Herzenberg L, eds. Handbook of Experimental Immunology. 5th ed. Cambridge: Blackwell Science, Inc., 1995 (in press).
16. Dean NM, Beaven MA. Review: Methods for the analysis of inositol phosphates. Anal Biochem 1989; 183:199-209.
17. Garcia-Gil M, Siraganian RP. Phospholipase A_2 stimulation during cell secretion in rat basophilic leukemia cells. J Immunol 1986; 136:259-63.
18. Kennerly DA. Quantitative analysis of water-soluble products of cell-associated phospholipase C- and phospholipase D-catalyzed hydrolysis of phosphatidylcholine. Methods Enzymol 1991; 197:191-07.
19. Beaven MA, Guthrie DF, Moore JP, Smith GA, Hesketh TR, Metcalfe JC. Synergistic signals in the mechanism of antigen-induced exocytosis in 2H3 cells: Evidence for an unidentified signal required for histamine release. J Cell Biol 1987; 105:1129-36.
20. Lo TN, Saul W, Beaven MA. The actions of Ca^{2+} ionophores on rat basophilic (2H3) cells are dependent on cellular ATP and hydrolysis of inositol phospholipids. A comparison with antigen stimulation. J Biol Chem 1987; 262:4141-45.
21. Ali H, Maeyama K, Sagi-Eisenberg R, Beaven MA. Antigen and thapsigargin promote influx of Ca^{2+} in rat basophilic RBL-2H3 cells by ostensibly similar mechanisms that allow filling of inositol 1,4,5-trisphosphate-sensitive and mitochondrial Ca^{2+} stores. Biochem J 1994; 304:431-40.
22. Rhee SG, Choi KD. Regulation of inositol phospholipid-specific phospholipase C isozymes. J Biol Chem 1992; 267:12393-96.

23. Crooke ST, Bennett CF. Mammalian phosphoinositide-specific phospholipase C isozymes. Cell Calcium 1989; 10:309-23.
24. Martin JL, Pumford NR, LaRosa AC, Martin BM, Gonzaga HMS, Beaven MA, Pohl LR. A metabolite of halothane covalently binds to an endoplasmic reticulum protein that is highly homologous to phosphatidylinositol-specific phospholipase C-α but has no activity. Biochem Biophys Res Comm 1991; 178:679-85.
25. Park DJ, Min HK, Rhee SG. IgE-induced tyrosine phosphorylation of phospholipase C-γ1 in rat basophilic leukemia cells. J Biol Chem 1991; 266:24237-40.
26. Beaven MA, Moore JP, Smith GA, Hesketh TR, Metcalfe JC. The calcium signal and phosphatidylinositol breakdown in 2H3 cells. J Biol Chem 1984; 259:7137-42.
27. Jouvin M-H E, Adamczewski M, Numerof R, Letourneur O, Valle A, Kinet J-P. Differential control of the tyrosine kinases lyn and syk by the two signaling chains of the high affinity immunoglobulin E receptor. J Biol Chem 1994; 269:5918-25.
28. Ali H, Cunha-Melo JR, Saul WF, Beaven MA. The activation of phospholipase C via adenosine receptors provides synergistic signals for secretion in antigen stimulated RBL-2H3 cells: Evidence for a novel adenosine receptor. J Biol Chem 1990; 265:745-53.
29. Collado-Escobar D, Cunha-Melo JR, Beaven MA. Treatment with dexamethasone downregulates IgE-receptor mediated signals and upregulates adenosine-receptor mediated signals in a rat mast cell (RBL-2H3) line. J Immunol 1990; 144:244-50.
30. Linden J. Cloned adenosine A_3 receptors: pharmacological properties, species differences and receptor functions. Trends Pharmacol Sci 1994; 15:298-306.
31. Hide M, Ali H, Price SR, Moss J, Beaven MA. The GTP-binding protein, $G_{\alpha z}$: Its down-regulation by dexamethasone and its credentials as a mediator of antigen-induced responses in RBL-2H3 cells. Mol Pharm 1991; 40:473-79.
32. Ramkumar V, Wilson M, Dhanraj DN, Gettys TW, Ali H. Dexamethasone upregulates A_3 adenosine receptors in rat basophilic leukemia (RBL-2H3) cells. J Immunol 1995; 154:5436-43.
33. Choi OH, Lee JH, Kassessinoff T, Cunha-Melo JR, Jones SVP, Beaven MA. Carbachol and antigen mobilize calcium by similar mechanisms in a transfected mast cell line (RBL-2H3 cells) that expresses m1 muscarinic receptors. J Immunol 1993; 151: 5586-95.
34. Hammond SM, Altshuller YM, Sung T, Rudge SA, Rose K, Engebrecht J, Morris AJ, Frohman MA. Human ADP-ribosylation factor-activated phosphatidylcholine-specific phospholipase D defines a new and highly conserved gene family. J Biol Chem 1995; 270:29640-43.
35. Nakamura S-I, Shimooku K, Akisue T, Jinnai H, Hitomi T, Kiyohara Y, Ogino C, Yoshida K, Nishizuka Y. Mammalian phospholipase D: Activation by ammonium sulfate and nucleotides. Proc Natl Acad Sci USA 1995; 92:12319-22.
36. Nishizuka Y. Protein kinase C and lipid signaling for sustained cellular responses. FASEB J 1995; 9:484-96.
37. Brown AH, Gutowski S, Moomaw CR, Slaughter C, Sternweiss PC. ADP-Ribosylation factor, a small GTP-dependent regulatory protein, stimulates phospholipase D activity. Cell 1993; 75:1137-44.
38. Cockcroft S, Thomas GM, Fensome A, Geny B, Cunningham E, Gout I, Hiles I, Totty NF, Truong O, Hsuan JJ. Phospholipase D: a downstream effector of ARF in granulocytes. Science 1994; 263:523-26.
39. Bowman EP, Uhlinger DJ, Lambeth JD. Neutrophil phospholipase D is activated by a membrane-associated Rho family of small molecular weight GTP-binding protein. J Biol Chem 1993; 268:21509-12.
40. Malcolm KC, Ross AH, Qui R-G, Symons M, Exton JH. Activation of rat liver phospholipase D by the small GTP-binding protein RhoA. J Biol Chem 1994; 269: 25951-54.
41. Singer WD, Brown HA, Bokoch GM, Sternweis PC. Resolved phospholipase D activity is modulated by cytosolic factors other than ARF. J Biol Chem 1995; 270:14944-50.

42. Pertile P, Liscovitch M, Chalifa V, Cantley LC. Phosphatidylinositol 4,5-bisphosphate synthesis is required for activation of phospholipase D in U937 cells. J Biol Chem 1995; 270:5130-35.
43. Chalifa V, Mohn H, Liscovitch M. A neutral phospholipase D activity from rat brain synaptic plasma membranes. Identification and partial characterization. J Biol Chem 1990; 265:17512-19.
44. Massenburg D, Han JS, Liyanage M, Patton WA, Rhee SG, Moss J, Vaughan M. Activation of rat brain phospholipase D by ADP-ribosylation factors 1,5 and 6: separation of ADP-ribosylation factor-dependent and oleate-dependent enzymes. Proc Natl Acad Sci USA 1994; 91:11718-22.
45. Okamura S, Yamashita S. Purification and characterization of phosphatidylcholine phospholipase D from pig lung. J Biol Chem 1994; 269:31207-13.
46. Billah MM, Anthes JC. The regulation and cellular functions of phosphatidylcholine hydrolysis. Biochem J 1990; 269:281-91.
47. Exton JH. Signaling through phosphatidylcholine breakdown. J Biol Chem 1990; 265:1-4.
48. Liscovitch M, Cantley LC. Lipid second messengers. Cell 1994; 77:329-34.
49. Nakamura S, Nishizuka Y. Lipid mediators and protein kinase C activation for the intracellular signaling network. J Biochem (Tokyo) 1994; 115:1029-34.
50. Boarder MR. A role for phospholipase D in control of mitogenesis. Trends Pharmacol Sci 1994; 15:57-62.
51. Dinh TT, Kennerly DA. Assessment of receptor-dependent activation of phosphatidylcholine hydrolysis by both phospholipase D and phospholipase C. Cell Regul 1991; 2:299-309.
52. Lin P, Wiggan GA, Gilfillan AM. Activation of phospholipase D in a rat mast (RBL-2H3) cell line: A possible unifying mechanism for IgE-dependent degranulation and arachidonic acid metabolite release. J Immunol 1991; 146:1609-16.
53. Beaven MA, Metzger H. Signal transduction by Fc receptors: the Fcε case. Immunol Today 1993; 14:222-26.
54. Kennerly DA. Diacylglycerol metabolism in mast cells. Analysis of lipid metabolic pathways using molecular species analysis of intermediates. J Biol Chem 1987; 262: 16305-13.
55. Lin P, Gilfillan AM. The role of calcium and protein kinase C in the IgE-dependent activation of phosphatidylcholine-specific phospholipase D in a rat mast (RBL-2H3) cell line. Eur J Biochem 1992; 207:163-68.
56. Kennerly DA. Phosphatidylcholine is a quantitatively more important source of increased 1,2-diacylglycerol than is phosphatidylinositol in mast cells. J Immunol 1990; 144:3912-19.
57. Lin P, Fung SJ, Li S, Chen T, Repetto B, Huang KS, Gilfillan AM. Temporal regulation of the IgE-dependent 1,2-diacylglycerol production by tyrosine kinase activation in a rat (RBL 2H3) mast-cell line. Biochem J 1994; 299:109-14.
58. Exton JH. Phosphatidylcholine breakdown and signal transduction. Biochim Biophys Acta 1994; 1212:26-42.
59. Ozawa K, Yamada K, Kazanietz MG, Blumberg P, Beaven MA. Different isozymes of protein kinase C mediate feedback inhibition of phospholipase C and stimulatory signals for exocytosis in rat RBL-2H3 cells. J Biol Chem 1993; 268:2280-83.
60. Foreman J, Hallet M, Mongar J. The relationship between histamine secretion and [^{45}Ca]uptake by mast cells. J Physiol 1977; 271:193-214.
61. Hide M, Beaven MA. Calcium influx in a rat mast cell (RBL-2H3) line: Use of multivalent metal ions to define its characteristics and role in exocytosis. J Biol Chem 1991; 266:15221-29.
62. Beaven MA, Cunha-Melo JR. Membrane phosphoinositide-activated signals in mast cells and basophils. Progr Allergy 1988; 42:123-84.
63. Pearce FL. Calcium and histamine secretion from mast cells. Prog Med Chem 1982; 19:59-109.
64. Metzger H, Alcaraz G, Hohman R, Kinet JP, Pribluda V, Quarto R. The receptor with high affinity for immunoglobulin E. Annu Rev Immunol 1986; 4:419-70.

65. Putney JW, Jr. Capacitative calcium entry revisited. Cell Calcium 1990; 11:611-24.
66. Putney JW. Excitement about calcium signalling in inexcitable cells. Science 1993; 262:676-78.
67. Millard PJ, Ryan T, Webb WW, Fewtrell C. Immunoglobulin E receptor crosslinking induces oscillations in intracellular free ionized calcium in individual tumor mast cells. J Biol Chem 1989; 264:19730-39.
68. Millard PJ, Gross D, Webb WW, Fewtrell C. Imaging asynchronous changes in intracellular Ca^{2+} in individual stimulated tumor mast cells. Proc Natl Acad Sci USA 1988; 85:1854-58.
69. Cunha-Melo JR, Dean NM, Moyer JD, Maeyama K, Beaven MA. The kinetics of phosphoinositide hydrolysis in rat basophilic leukemia (RBL-2H3) cells varies with the type of IgE receptor crosslinking agent used. J Biol Chem 1987; 262:11455-63.
70. Pribluda VS, Metzger H. Calcium-independent phosphoinositide breakdown in rat basophilic leukemia cells. Evidence for an early rise in inositol 1,4,5-trisphosphate which precedes the rise in other inositol phosphates and in cytoplasmic calcium. J Biol Chem 1987; 262:11449-54.
71. Meyer T, Holowka D, Stryer L. Highly cooperative opening of calcium channels by inositol 1,4,5-trisphosphate. Science 1988; 240:653-56.
72. Beaven MA, Rogers J, Moore JP, Hesketh TR, Smith GA, Metcalfe JC. The mechanism of the calcium signal and correlation with histamine release in 2H3 cells. J Biol Chem 1984; 259:7129-36.
73. Mohr FC, Fewtrell C. The effect of mitochondrial inhibitors on calcium homeostasis in tumor mast cells. Am J Physiol 1990; 258:C217-C226.
74. Kanner BI, Metzger H. Initial characterization of the calcium channel activated by the crosslinking of the receptors for immunoglobulin E. J Biol Chem 1984; 259: 10188-93.
75. Fewtrell C, Sherman E. IgE receptor-activated calcium permeability pathway in rat basophilic leukemia cells: Measurement of the unidirectional influx of calcium using quin2-buffered cells. Biochemistry 1987; 26:6995-7003.
76. Mohr FC, Fewtrell C. Depolarization of rat basophilic leukemia cells inhibits calcium uptake and exocytosis. J Cell Biol 1987; 104:783-92.
77. Middleton Jr E, Drzewiecki G, Triggle D. Effects of smooth muscle calcium antagonists on human basophil histamine release. Biochem Pharmacol 1981; 30:2867-69.
78. Mohr FC, Fewtrell C. The relative contributions of extracellular and intracellular calcium to secretion from tumor mast cells. Multiple effects of the proton ionophore carbonyl cyanide m-chlorophenylhydrazone. J Biol Chem 1987; 262:10638-43.
79. McCloskey MA, Quian Y-X. Selective expression of potassium channels during mast cell differentiation. J Biol Chem 1994; 269:14813-19.
80. Labrecque GF, Holowka D, Baird B. Antigen-triggered membrane potential changes in IgE-sensitized rat basophilic leukemia cells: evidence for a repolarizing response that is important in the stimulation of cellular degranulation. J Immunol 1989; 142:236-43.
81. Penner R, Matthews G, Neher E. Regulation of calcium influx by second messengers in rat mast cells. Nature 1988; 334:499-504.
82. Romanin C, Reinsprecht M, Pecht I, Schindler H. Immunologically activated chloride channels involved in degranulation of rat mucosal mast cells. EMBO J 1991; 10:3603-08.
83. Fewtrell C. Ca^{2+} oscillations in nonexcitable cells. Annu Rev Physiol 1993; 55:427-54.
84. Randriamampita C, Tsien RY. Emptying of intracellular Ca^{2+} stores releases a novel small messenger that stimulates Ca^{2+} influx. Nature 1993; 364:809-14.
85. Parekh AP, Terlau H, Stumer W. Depletion of $InsP_3$ stores activates a Ca^{2+} and K^+ current by means of a phosphatase and a diffusible messenger. Nature 1993; 364: 814-18.
86. Kim HY, Thomas D, Hanley MR. Chromatographic resolution of an intracellular calcium influx factor from thapsigargin-activated Jurkat cells. J Biol Chem 1995; 270:9706-08.
87. Thomas D, Hanley MR. Evaluation of calcium influx factors from stimulated Jurkat

T-lymphocytes by microinjection into *Xenopus* oocytes. J Biol Chem 1995; 270:6429-32.
88. Bird GSJ, Bian X, Putney JW. Calcium entry signal? Nature 1995; 373:481-82.
89. Hoth M, Penner R. Depletion of intracellular calcium stores activates a calcium current in mast cells. Nature 1992; 355: 353-56.
90. Fasolato C, Hoth M, Penner R. A GTP-dependent step in the activation mechanism of capacitative calcium influx. J Biol Chem 1993; 268:20737-40.
91. Bird GSJ, Putney JW. Inhibition of thapsigargin-induced calcium entry by microinjected guanine nucleotide analogs. Evidence for the involvement of a small G-protein in capacitative calcium entry. J Biol Chem 1993; 268:21486-88.
92. Matthews G, Neher E, Penner R. Second messenger-activated calcium influx in rat peritoneal mast cells. J Physiol (Lond) 1989; 418:105-30.
93. Kuno M, Okada T, Shibata T. A patch-clamp study: secretagogue-induced currents in rat peritoneal mast cells. Am J Physiol 1989; 256:C560-68.
94. Kindman LA, Kim S, McDonald TV, Gardner P. Characterization of a novel intracellular sphingolipid-gated Ca^{2+}-permeable channel from rat basophilic leukemia cells. J Biol Chem 1994; 269:13088-91.
95. Dekker LV, Parker PJ. Protein kinase C-a question of specificity. Trends Biochem Sci 1994; 19:73-77.
96. Newton AC. Protein kinase C: structure, function and regulation. J Biol Chem 1995; 270:28495-98.
97. Ozawa K, Szallasi Z, Kazanietz MG, Blumberg PM, Mischak H, Mushinski JF, Beaven MA. Ca^{2+}-Dependent and Ca^{2+}-independent isozymes of protein kinase C mediate exocytosis in antigen-stimulated rat basophilic RBL-2H3 cells: Reconstitution of secretory responses with Ca^{2+} and purified isozymes in washed permeabilized cells. J Biol Chem 1993; 268:1749-56.
98. Germano P, Gomez J, Kazanietz MG, Blumberg PM, Rivera J. Phosphorylation of the the γ chain of the high affinity receptor for immunoglobulin E by receptor-associated protein kinase C-δ. J Biol Chem 1994; 269:23102-7.
99. Nakamura Y, Nakashima S, Ojio K, Banno Y, Miyata H, Nozawa Y. Ceramide inhibits IgE-mediated activation of phospholipase D, but not of phospholipase C, in rat basophilic leukemia (RBL-2H3) cells. J Immunol 1996; 156:256-62.
100. Cunha-Melo JR, Gonzaga HMS, Ali H, Huang FL, Huang K-P, Beaven MA. Studies of protein kinase C in the rat basophilic leukemia (RBL-2H3) cell reveal that antigen-induced signals are not mimicked by the actions of phorbol myristate acetate and Ca^{2+} ionophore. J Immunol 1989; 143: 2617-25.
101. Huang FL, Yoshida Y, Cunha-Melo JR, Beaven MA, Huang K-P. Differential downregulation of protein kinase C isozymes. J Biol Chem 1989; 264:4238-43.
102. Yamada K, Jelsema CL, Beaven MA. Certain inhibitors of protein serine/threonine kinases also inhibit tyrosine-phosphorylation of phospholipase Cγ1 and other proteins and reveal distinct roles for tyrosine kinase(s) and protein kinase C in stimulated, rat basophilic RBL-2H3 cells. J Immunol 1992; 149:1031-37.
103. Choi OH, Adelstein RS, Beaven MA. Secretion from rat basophilic RBL-2H3 cells is associated with phosphorylation of myosin light chains by myosin light chain kinase as well as phosphorylation by protein kinase C. J Biol Chem 1994; 269:536-41.
104. Santini F, Beaven MA. Nonselective actions of herbimycin A. Immunology Today 1993; 14:369-70.
105. Wilkinson SE, Parker PJ, Nixon JS. Isoenzyme specificity of bisindolylmaleimides, selective inhibitors of protein kinase C. Biochem J 1993; 294:335-37.
106. Hirasawa N, Santini F, Beaven MA. Activation of the mitogen-activated protein kinase/cytosolic phospholipase A_2 pathway in a rat mast cell line-Indications of different pathways for release of arachidonic acid and secretory granules. J Immunol 1995; 154:5391-402.
107. Baumgartner RA, Yamada K, Deramo VA, Beaven MA. Secretion of tumor necrosis factor (TNF) from a rat mast cell line is a

brefeldin A-sensitive and a calcium/protein kinase C-regulated process. J Immunol 1994; 153:2609-17.
108. Pelicci G, Lanfrancone L, Grignani F, McGlade J, Cavallo F, Forni G, Nicoletti I, Grignani F, Pawson T, Pelicci PG. A novel transforming protein (SHC) with an SH2 domain is implicated in mitogenic signal transduction. Cell 1992; 70:93-104.
109. Skolnik EY, Batzer A, Li N, Lee CH, Lowenstein E, Mohammadi M, Margolis B, Schlessinger J. The function of GRB2 in linking the insulin receptor to Ras signaling pathways. Science 1993; 260:1953-55.
110. Skolnik EY, Lee CH, Batzer A, Vicentini LM, Zhou M, Daly R, Myers MJ, Jr, Backer JM, Ullrich A, White MF et al. The SH2/SH3 domain-containing protein GRB2 interacts with tyrosine-phosphorylated IRS1 and Shc: implications for insulin control of ras signalling. EMBO J 1993; 12:1929-36.
111. Smit L, de Vries-Smits AMM, Bos JL, Borst J. B cell antigen receptor stimulation induces formation of a Shc-Grb2 complex containing multiple tyrosine-phosphorylated proteins. J Biol Chem 1994; 269:20209-12.
112. Santini F, Beaven MA. Tyrosine phosphylation of a Mitogen Activated Protein (MAP) kinase-like protein occurs at a late step in exocytosis: Studies with tyrosine phosphatase inhibitors and various secretagogues in RBL-2H3 cells. J Biol Chem 1993; 268:22716-22.
113. Offermanns S, Bombien E, Schultz G. Stimulation of tyrosine phosphorylation and mitogen-activated-protein (MAP) kinase activity in human SH-SY5Y neuroblastoma cells by carbachol. Biochem J 1993; 294:545-50.
114. Hirasawa N, Scharenberg A, Yamamura H, Beaven MA, Kinet J-P. A requirement for Syk in the activation of the MAP kinase/phospholipase A_2 pathway by FcεRI is not shared by a G protein-coupled receptor. J Biol Chem 1995; 270:10960-67.
115. Scharenberg AM, Lin S, Cuenod B, Yamamura H, Kinet JP. Reconstitution of interactions between tyrosine kinases and the high affinity IgE receptor which are controlled by receptor clustering. EMBO J 1995; 14:3385-94.
116. Apgar JR. Regulation of the antigen-induced F-actin response in rat basophilic leukemia cells by protein kinase C. J Cell Biol 1991; 112:1157-63.
117. Apgar JR. Polymerization of actin in RBL-2H3 cells can be triggered through either the IgE receptor or the adenosine receptor but different signaling pathways are used. Mol Biol Cell 1994; 5:313-22.
118. Pfeiffer JR, Seagrave JC, Davis BH, Deanin GG, Oliver JM. Membrane and cytoskeletal changes associated with IgE-mediated serotonin release from rat basophilic leukemia cells. J Cell Biol 1985; 101:2145-55.
119. Spudich A. Myosin reorganization in activated RBL cells correlates temporally with stimulated secretion. Cell Motil Cytoskeleton 1994; 29:345-53.
120. Sellers JR, Goodson HV. Motor proteins 2:myosin. Protein Profile 1995; 2:1342-442.
121. Ludowyke RI, Peleg I, Beaven MA, Adelstein RS. Antigen-induced secretion of histamine and the phosphorylation of myosin by protein kinase C in rat basophilic leukemia cells. J Biol Chem 1989; 264:12492-501.
122. Teshima R, Suzuki K, Ikebuchi H, Terao T. Enhancement of the phosphorylation of membrane bound myosin light chain by antigen stimulation in rat basophilic leukemia cells. Mol Immunol 1989; 26:641-48.
123. Choi OH, Park C-S, Itoh K, Adelstein RS, Beaven MA. Cloning of the cDNA encoding rat myosin heavy chain-A and evidence for the absence of myosin heavy chain-B in cultured rat mast (RBL-2H3) cells. J Muscle Res Cell Motility 1996; 17:69-77.
124. Gomperts BD. G_e: A GTP-binding protein mediating exocytosis. Annu Rev Physiol 1990; 52:591-606.
125. Gomperts BD, Barrowman MM, Cockcroft S. Dual role for guanine nucleotides in stimulus-secretion coupling. Fed Proc 1986; 45:2156-61.
126. Howell TW, Cockcroft S, Gomperts BD. Essential synergy between Ca^{2+} and guanine nucleotides in exocytotic secretion from permeabilized rat mast cells. J Cell Biol 1987; 105:191-97.
127. Fernandez JM, Neher E, Gomperts BD. Capacitance measurements reveal stepwise

fusion events in degranulating mast cells. Nature 1984; 312:453-55.

128. Lillie THW, Gomperts BD. Nucleotides and divalent cations as effectors and modulators of exocytosis in permeabilized rat mast cells. Phil Trans R Soc Lond B 1992; 336:25-34.

129. Aridor M, Traub LM, Sagi-Eisenberg R. Exocytosis in mast cells by basic secretagogues: Evidence for direct activation of GTP-binding proteins. J Cell Biol 1990; 111:909-17.

130. Aridor M, Sagi-Eisenberg R. Neomycin is a potent secretagogue of mast cells that directly activates a GTP-binding protein involved in exocytosis. J Cell Biol 1990; 111:2885-91.

131. Ortner MJ, Chignell CF. Spectroscopic studies of rat mast cells, mouse mastocytoma cells and compound 48/80-II. The synthesis and some binding properties of spin-labeled 48/80. Biochem Pharmacol 1981; 30:283-88.

132. Aridor M, Rajmilevich G, Beaven MA, Sagi-Eisenberg R. Activation of exocytosis by the heterotrimeric G-protein G_{i3}. Science 1993; 262:1569-72.

133. Cunha-Melo JR, Dean NM, Ali H, Beaven MA. Formation of inositol 1,4,5-trisphosphate and inositol 1,3,4-trisphosphate from inositol 1,3,4,5-tetrakisphosphate and their pathways of degradation in RBL-2H3 cells. J Biol Chem 1988; 263:14245-50.

134. Ali H, Cunha-Melo J, Beaven M. Receptor-mediated release of inositol 1,4,5-trisphosphate and inositol 1,4-bisphosphate in rat basophilic leukemia RBL-2H3 cells permeabilized with streptolysin 0. Biochim Biophys Acta 1989; 1010:88-99.

135. Beaven MA, Ozawa K. Role of calcium, protein kinase C and MAP kinase in the activation of mast cells. Allergol Internat 1996; 45:(in press).

136. Clark JD, Lin L, Kriz RW, Ramesha CS, Sultzman LA, Lin AY, Milona N, Knopf JL. A novel arachidonic acid-selective cytosolic PLA_2 contains a Ca^{2+}-dependent translocation domain with homology to PKC and GAP. Cell 1991; 65:1043-51.

137. Lin LL, Lin AY, Knopf JL. Cytosolic phospholipase A_2 is coupled to hormonally regulated release of arachidonic acid. Proc Natl Acad Sci USA 1992; 89:6147-51.

138. Kramer RM. Structure, function and regulation of mammalian phospholipases A_2. In: Brown BL, Dobson PRM, eds. Advances in Second Messenger and Phosphoprotein Research. Vol. 28. New York: Raven Press, Ltd., 1993:81-89.

139. Birchall AM, Bishop J, Bradshaw D et al. Ro 32-0432, a selective and orally active inhibitor of protein kinase C prevents T cell activation. J Pharm Exptl Therap 1994; 268:922-29.

140. Mao C, Kim SH, Almenoff JS et al. Molecular cloning and characterization of SCaMPER, a sphingosine Ca^{2+} release-mediating protein from endoplasmic reticulum. Proc Natl Acad Sci USA. 1996; 93:1993-1996.

141. Choi OH, Kim J-H, Kinet J-P. Calcium mobilization via the sphingosine kinase in signalling by the FcεRI antigen receptor. Nature. 1996; 380:634-6.

142. Beaven MA. Calcium signalling: Sphingosine kinase versus phospholipase C. Curr Biol. 1996; 6:798-801.

CHAPTER 5

Role of GTP-Binding Proteins in FcεRI Signaling

Anna Koffer

INTRODUCTION

Direct evidence for the involvement of specific GTPases in the response of mast cells or the related rat basophilic leukemia (RBL-2H3) cells to FcεRI activation is limited. However, several lines of circumstantial evidence point to the crucial role of GTP-binding proteins in the response of mast or RBL-2H3 cells to FcεRI activation. These include the effects of GTP depletion, of inhibition of the isoprenoid pathway and of various inhibitors and stimulators of GTP-binding proteins. Such experiments will be described in the first section of this chapter. Due to my interest in the mechanism of exocytosis, this and the last section are biased towards the secretory responses to activation of FcεRI and of GTP-binding proteins.

Tyrosine phosphorylation of various proteins following FcεRI crosslinking activates a number of signaling pathways. Activation of phospholipase C-γ (PLC-γ) was shown to be essential for mediating responses to FcεRI, initiating a set of well-documented responses: hydrolysis of PIP_2, production of IP_3 and DAG, Ca^{2+} mobilization and activation of protein kinase C (PKC).[1-3] PLC-γ activation constitutes a major signaling pathway from the FcεRI receptor. In addition to PLC-γ activation, another intensively studied pathway leads from the activated tyrosine kinases to activation of a small GTPase, Ras.[4] Various changes in the components of the Ras signaling pathway following receptor crosslinking have been demonstrated in mast cells. These will be described in the second part of this chapter with a particular emphasis on the similarities between the mechanisms of Ras activation in mast cells and lymphocytes.

Finally, part three will deal with several named GTP-binding proteins that have been implicated in regulation of secretion from mast cells. The emphasis will be on the small GTPases, Rac and Rho, that, as we

IgE Receptor (FcεRI) Function in Mast Cells and Basophils,
edited by Majed M. Hamawy. © 1997 R.G. Landes Company.

have shown, directly control both secretory and cytoskeletal responses of mast cells. Although their relevance to FcεRI activation is not yet established, our evidence indicates that they are crucial to both secretory and cytoskeletal responses of permeabilized cells and therefore very probably also involved in those induced by FcεRI activation.

INHIBITORS AND ACTIVATORS OF GTP-BINDING PROTEINS

This section deals with effects of mostly nonspecific inhibitors and activators of GTP-binding proteins on various mast cell functions, but particularly on secretion. Such experiments were crucial for implying the role of GTPases in cellular signaling. Since both inhibitory and stimulatory GTP-binding proteins may be present, interpretation of the results of such nonselective stimulation or inhibition of GTPases is limited. Further, more focused investigations are required to identify the processes and the GTPases involved.

Mycophenolic Acid (MPA): GTP-Depletion

Depletion (about 60%) of guanine nucleotides with mycophenolic acid, an inhibitor of de novo GTP synthesis, was found to suppress FcεRI-mediated degranulation of RBL-2H3 cells. The rate of Ca^{2+} influx was also substantially inhibited, while the Ag-stimulated IP3 production, the membrane ruffling and pinocytosis were largely unaffected.[5] These results implicated GTP-binding proteins in the control of the FcεRI-dependent signal transduction, particularly that of the Ca^{2+} influx. A proportion of GTPases with high affinity for GTP may be present in an active state, resistant to GTP depletion and therefore unaffected by this treatment.

Lovastatin: Inhibition of the Isoprenoid Pathway

GTP-binding proteins undergo a variety of post-translational lipid modifications, which affect their ability to interact with membranes and other proteins and consequently their biological activity. Isoprenylation (farnesylation or geranyl geranylation at the C-terminal) is the most common form of processing of small GTPases Ras, Rho, Rac and Rab and also occurs on γ subunits of G proteins.[6] The production of mevalonic acid, an intermediate of the isoprenoid biosynthetic pathway, is inhibited by lovastatin. In RBL-2H3 cells, this drug inhibited FcεRI-induced Ca^{2+} mobilization, 1,4,5-inositol trisphosphate production, morphological responses and secretion.[7] These results suggest that nonsterol isoprenoid synthesis is required for FcεRI-activated signal transduction. Lovastatin was also found to inhibit the level of tyrosine phosphorylation in FcεRI-stimulated RBL-2H3 cells.[8] This may also contribute to its inhibitory effects on the secretory response and indicates that some tyrosine kinases may be controlled by GTPases.

Pertussis and Cholera Toxins: ADP-Ribosylation

Pertussis toxin (Ptx)

The use of toxins, that are specifically directed at various GTP-binding proteins, has provided valuable information about the involvement of GTPases in FcεRI signaling. It is well documented that Ptx ADP-ribosylates the α subunit of the heterotrimeric G protein G_i preventing its interaction with a receptor and the consequent dissociation into α and βγ subunits. Regulation of adenylate cyclase and other possible effectors is thus impaired. Ptx does not inhibit FcεRI-induced secretion from mast cells or basophils,[9-11] suggesting that it is independent of a G_i-like protein. Ptx, however, does inhibit the response of mast cells to polycationic secretagogues such as compound 48/80, indicating that different signaling pathways are involved in this response.[9] In addition, Ptx treatment inhibits thrombin-induced secretion from bone marrow derived mast cells, F-Met-Leu-Phe-induced secretion from basophils,[9,10] and a response of RBL-2H3 cells to activation of the adenosine receptor.[11] Thus, a G_i-like protein, although not directly involved in the FcεRI signaling pathway, may exert a modulatory/synergistic effect on the response to FcεRI following activation of another type of receptor.

Cholera toxin (CT)

CT ADP ribosylates the α subunit of another G protein, G_s. This modification inactivates the GTPase activity of Gs and thus results in its constitutive activation. In RBL-2H3 cells, CT was found to potentiate Ca^{2+} influx, production of inositol phosphates and secretion which are induced by FcεRI crosslinking.[11-13] This effect was independent of any increase in cAMP level suggesting that adenylate cyclase is not involved. In contrast, CT inhibited the response of RBL-2H3 cells to activation of the adenosine receptor.[11] Again, these results indicate that different G proteins are associated with different receptors: a CT-sensitive GTP-binding protein, activated by a receptor other than FcεRI, amplifies the FcεRI signaling pathway.

Presence of Gα subunits

Substrates for ADP-ribosylation by cholera and pertussis toxin have been identified in RBL-2H3 cells: $G\alpha_s$, $G\alpha_{i2}$, $G\alpha_{i3}$.[14] The results were confirmed using mRNA hybridization with Gα-specific oligonucleotides and specific antibodies to the proteins. In addition, another G protein, $G\alpha_z$, that is not a substrate for CT or Ptx, was found. This protein is a good candidate for a G protein mediating the FcεRI response as evidenced from studies with RBL-2H3 cells. Dexamethasone treatment of these cells suppresses their responsiveness to FcεRI aggregation and, at the same time, markedly decreases the amount of $G\alpha_z$ mRNA and its membrane-associated protein product, $G\alpha_z$. In mast cells, two substrates for Ptx have also been identified: $G\alpha_{i2}$ and $G\alpha_{i3}$.[15]

ATP_4^- and Streptolysin-O: Cell Permeabilization

The GTP-binding protein G_e

A pivotal role of GTP-binding proteins in the control of regulated exocytosis has been shown by studies of permeabilized mast cells. Initial experiments were made with cells permeabilized with ATP (in its ATP_4^- form) and with whole-cell patch-pipette technique.[16,17] Later studies of Streptolysin-O permeabilized cells confirmed the involvement of GTPases and led to the postulation of a guanine nucleotide binding protein termed G_e. It operates downstream of phospholipase C and its effects are synergistic with those of a Ca^{2+} binding protein.[18,19] It is postulated that G_e controls the late stages of the exocytotic process and therefore is likely to be the ultimate target of any signaling pathway leading from a receptor to degranulation (Fig. 5.1).

Permeabilized cells and GTP-γ-S

Other studies of permeabilized mast or RBL-2H3 cells confirmed the involvement of GTP-binding proteins in regulation of secretion,[20-23] breakdown of phosphoinositides,[18,24] phosphatidic acid formation,[25] arachidonate release,[26-29] Ca^{2+} channels,[30] K^+ channels,[31] and cytoskeleton organization.[32,33] Some conflicting results, obtained in these studies, may be due to different methods of permeabilization,[34,35] different ionic conditions[36] and varying pH.[20] Factors that both stimulate and inhibit secretion are present in the cytosol of mast cells and leak out at different rates, which greatly affects the responsiveness of permeabilized cells.[35] The ability of extensively washed permeabilized mast cells to respond, even in the absence of Ca^{2+}, to GTP-γ-S indicates that the GTPases and their regulatory proteins (e.g., guanine nucleotide exchange factor [GEF]) are tightly associated with the membranes and/or the cytoskeleton.[37]

Permeabilized cells and Ca^{2+}

In the presence of ATP, permeabilized mast cells are able to secrete in response to Ca^{2+} independently of guanine nucleotides.[36-38] The response to Ca^{2+}/ATP is relatively resistant to various inactivators of GTP-binding proteins (GDP-β-S, GDP, deoxyGDP), but a GTP-dependent component is clearly present.[21,37,39,40] This indicates an involvement of a GTPase that has either a very high affinity for GTP or

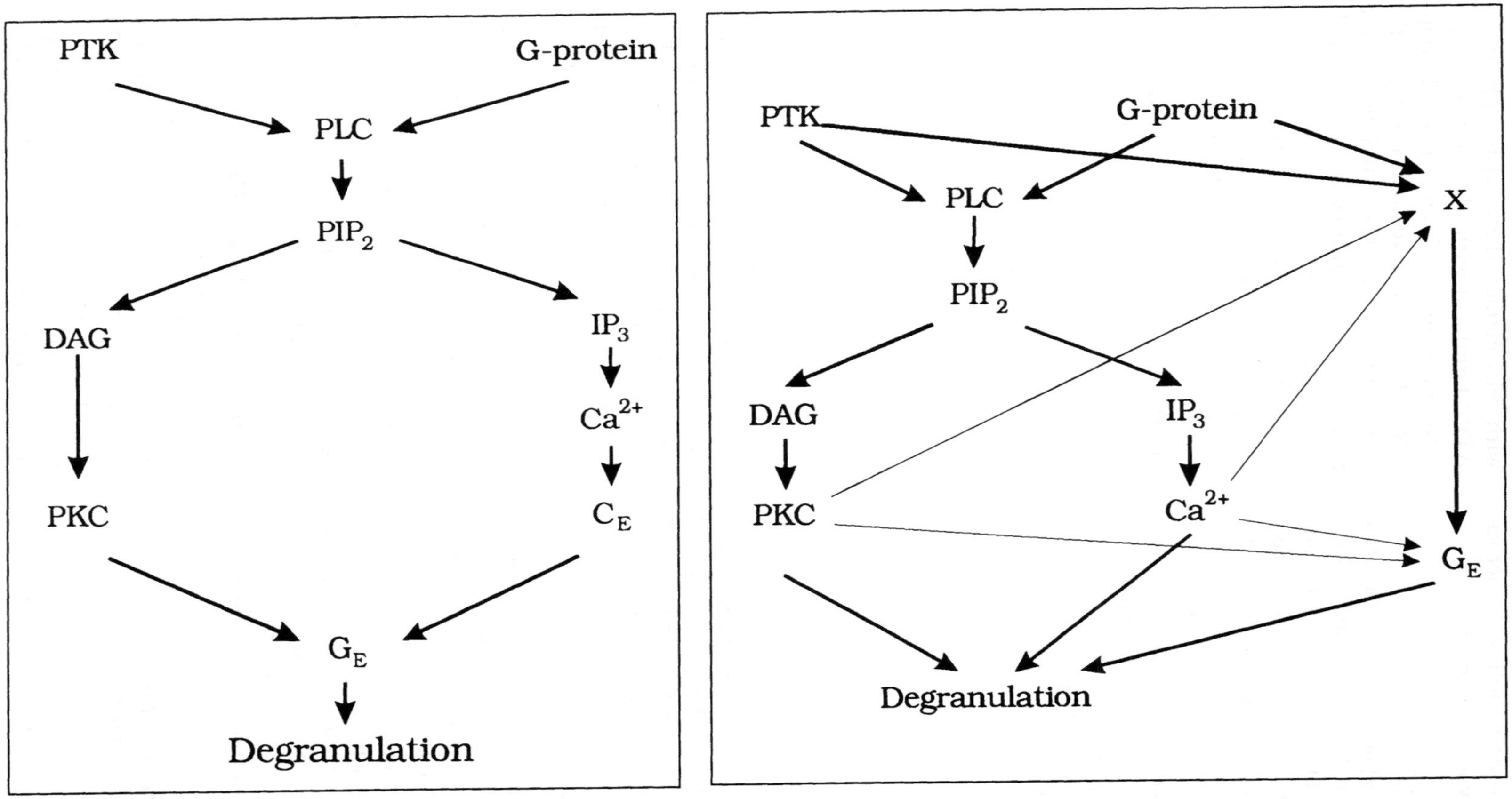

Fig. 5.1. Signaling pathways to secretion: in series or parallel? (A) PLC stimulation, following activation of either the FcεRI and the associated PTK or a G protein-coupled receptor, leads (via PLC) to calcium mobilization and activation of PKC. Both enhance the activity of a putative GTP-binding protein, G_e, positioned downstream of PLC and controlling a late step in the exocytotic process. A calcium binding protein is postulated to act as a guanine nucleotide exchange factor. The scheme is based on ref. 39. (B) Receptor activation leads to stimulation of PLC (via PTK or a G protein) and of an unknown factor X, that in turn activates a putative GTP binding protein, G_e. Calcium and PKC can elicit secretion independently of G_e, and, vice versa, activation of G_e leads to secretion independently of calcium, but the two pathways act in synergy. Calcium enhances activities of the components of the G_e pathway and a cross-talk between the components of the two pathways may occur at many levels.

that may be activated by a Ca^{2+}-dependent transphosphorylation reaction (from ATP to GDP). In washed permeabilized cells, the transphosphorylation would have to be catalyzed by a membrane-bound nucleoside diphosphate kinase. The rate and the extent of the response to Ca^{2+}/ATP is markedly increased by addition of GTP-γ-S.[37-39] It was suggested that the synergistic effects of Ca^{2+} and GTP-γ-S are due to a presence of a calcium-binding protein, C_e, that acts as a GEF for G_e. In this model, G_e acts in series with C_e (Fig. 5.1A).[39] However, the possibility that the calcium- and the GTP-dependent pathways are independent but act in synergy cannot be excluded and a scheme for this is shown in Figure 5.1B.

Relevance to the FcεRI Receptor?

Whether the responses to GTP-γ-S are relevant to events which follow crosslinking of the FcεRI receptor is difficult to establish by approaches described above since alternative signaling pathways seem to exist for most cellular functions. For example, with respect to PLC activation, it now appears that two pathways exist; the protein tyrosine kinase (PTK)-dependent activation of PLC-γ and the PTK-independent activation of PLC-γ and PLC-β. The PTK-induced activation of PLC-γ is independent of Ras.[41] In mast cells, PLC-γ may also be activated by the βγ subunits of G proteins (e.g., G_i and G_s), acting in synergy with the PTK pathway. PLC-β can be activated by both βγ subunits, which bind to the N-terminal region of PLC-β, as well as by $G_{q\alpha}$, which binds to the C-terminal.[42] Thus, inhibition of the GTP-binding protein pathway will still allow the PTK dependent activation of PLC. In addition to PLC activation, resulting in an increase of $Ca^{2+}{}_i$ and activation of PKC, another major signaling pathway leads from FcεRI activated tyrosine kinases to a small GTPase, Ras. Similarly to PLC, activation of the Ras signaling pathway can be initiated by multiple alternative mechanisms.

Ras SIGNALING PATHWAY

Ras signaling pathway from PTK receptors is well established. Although the multichain immune system receptors do not possess PTK activity, their mechanism for Ras activation parallels that of receptor tyrosine kinases in many aspects. The Ras pathway in B and T-cells has been studied more extensively than that in mast cells but it appears that the two systems are very similar.

Ras and its Regulatory Proteins

The Ras belongs to a superfamily of small (20-30 kDa) GTP-binding proteins that, at present, comprises over 50 members.[43,44] These proteins are divided into four main groups based on their sequential, structural and functional homology: Ras-, Rab-, Arf- and Rho-families.[45] Additionally, a large number of Ras-regulatory proteins have been identified, which control the transition between active (GTP-bound) and inactive (GDP-bound) states. These include guanine nucleotide exchange factors (GEFs), guanine nucleotide dissociation inhibitors (GDIs) and GTPase activating proteins (GAPs) (Fig. 5.2). These regulatory proteins often contain several recognizable sequences, such as the Src homology (SH) domains SH2 and SH3, and the pleckstrin homology (PH) domains, indicating that they may serve multiple regulatory functions.[46] The SH2 domains bind to phosphotyrosine residues, while SH3 domains bind to proline-rich regions that are often present on cytoskeletal proteins.[47] The PH domains, the major PKC substrate in platelets, bind to both PIP_2 and to βγ subunits of G proteins and may serve for docking of the PH-containing proteins to the plasma membrane.[48] These multiple domains offer the potential for regulation of a diverse range of signaling molecules and for initiating cross-talk between different signaling pathways. Indeed, many Ras-regulatory proteins act on more than one member of the Ras family.

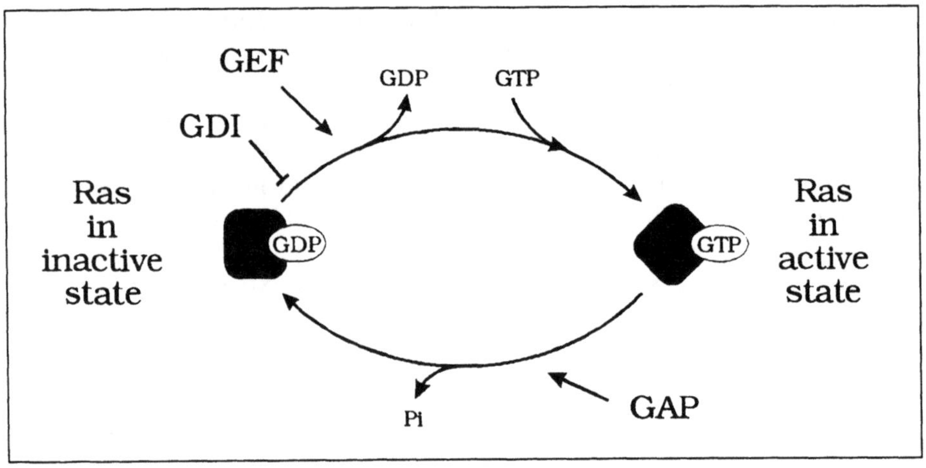

Fig. 5.2. Cycling of Ras between active and inactive state. The inactive GDP-Ras is converted to an active form by an exchange of GDP for GTP, which enables Ras interaction with its target proteins. Exchange activity is promoted by guanine nucleotide exchange factors (GEFs) and inhibited by guanine nucleotide dissociation inhibitor (GDI). Hydrolysis of bound GTP to GDP, promoted by GTPase-activating proteins GAPs), returns Ras to its inactive form.

Activation of Ras by PTK Receptors

In the case of PTK receptors, such as the growth factor receptors, ligand binding was found to stimulate activity of Ras guanine nucleotide exchange factor (Ras-GEF), leading to an increased level of Ras-GTP. A proline-rich protein Sos was identified as a mammalian Ras-GEF[49] and recent evidence indicates that the link between tyrosine kinase receptors and the Ras-GEF, Sos, is mediated via an intermediary adapter protein Grb2.[50] Grb2 contains a single SH2 domain and two flanking SH3 domains. Through its SH2 domain, Grb2 binds to the tyrosine-phosphorylated receptor either directly or indirectly, via another phosphotyrosine-containing protein, i.e., adapter protein p52 Shc. The SH3 domains of Grb2, on the other hand, are bound to the proline-rich carboxy-terminal tail of Sos. Thus, stimulation of the receptor induces binding of the Grb2-Sos complex to the phosphorylated receptor, leading to the activation of Sos which in turn causes recruitment of Ras to the plasma membrane and consequently increase in its activity. Another SH2 domain containing adapter protein, p52 Shc, is also an early substrate for the receptor activated PTKs, but its role is not yet clear. It may modulate the activity of Sos and/or extend the range of signaling pathways that are activated by the receptor.[49,51-53]

Activation of Ras in Lymphocytes and Mast Cells

In B lymphocytes, regulation of Grb2 complexes by the B cell antigen receptor (BCR) is indicated by the following observations: tyrosine phosphorylation of the p52 Shc protein and formation of Shc-Grb2-Sos complexes after BCR crosslinking (Fig. 5.3).[54,55] In T cells, no formation of Shc-Grb2-Sos complexes was observed in response to TCR crosslinking, although Shc was found to be tyrosine phosphorylated. Instead, activation of Ras in T cells may be controlled by a mechanism involving formation of a trimeric complex of Grb2, Sos and a novel substrate for TCR-activated PTKs, p36 (Fig. 5.3). This membrane-bound tyrosine phosphoprotein binds to the SH2 domain of Grb2.[56,57] In addition, the p36 protein also forms trimeric complexes with Grb2 and phospholipase Cγ.[58] Another substrate for TCR-activated PTKs is p75,

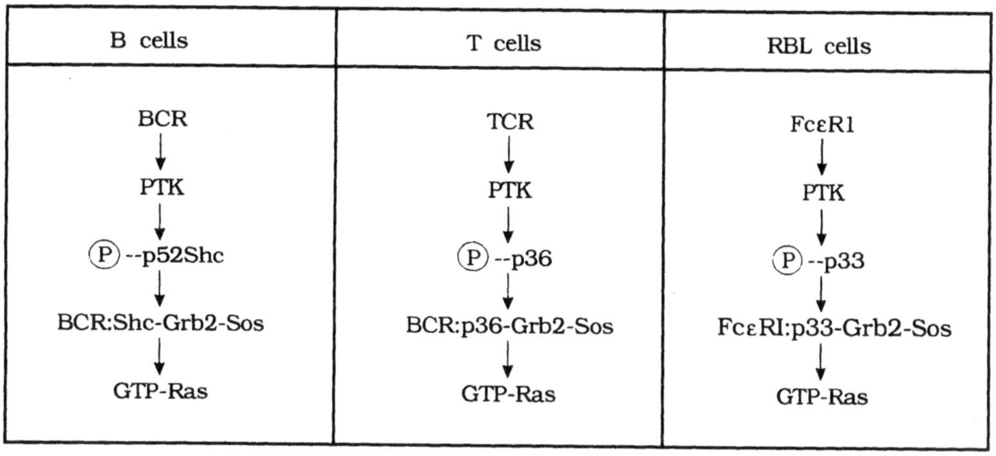

Fig. 5.3. Ras activation in B cells, T cells and RBL 2H3 cells. Activation of PTKs after receptor crosslinking leads to phosphorylation of adapter proteins Shc, p36 and p33. These proteins bind to another adapter protein Grb2 that is complexed to a Ras-GEF, Sos. The trimeric complexes associate with the crosslinked receptors, recruiting Sos to the plasma membrane and allowing it to act on Ras, producing its active GTP-bound form.

a hematopoietic cell-specific protein that is also complexed with Grb2 via its SH3 domains. Thus Grb2, via its two SH3 domains, can couple to several effectors, which have the potential to initiate multiple signaling pathways. (Fig. 5.4).[59] It should be noted however, that the tyrosine kinase route is not the only one to Ras activation. In T cells, PKC, activated by antigen receptor stimulation via the PLC (and/or PLD) pathway, also induces an increase in Ras activity and the suggested mechanism for this is a PKC-induced inhibition of a Ras GAP.[60,61]

More recent studies of RBL-2H3 cells have shown that the PTK dependent mechanism of Ras activation may be similar to that of T cells.[62] Again, a formation of a trimeric complex, in this case Sos-Grb2-p33, was observed after FcεRI crosslinking. The p33 protein, which is a substrate for FcεRI-activated PTKs, was found associated with the SH2 domain of Grb2, while Sos was bound to the SH3 domain of Grb2 (Fig. 5.3). The authors suggested that p36 in T cells and p33 in RBL-2H3 cells may have an Shc-like role, linking signals from FcεRI to the Grb2 pathway. A role for Ras in the exocytotic response of mast cells is suggested by microinjection experiments: microinjected *ras* oncogene protein induced exocytotic degranulation while the proto-oncogenic protein had no effect. However, this response was preceded by long delays, indicating that the effect is indirect.[63]

The Proto-Oncogene *Vav*

Another potential Ras-GEF is the protein Vav. Vav is the product of a protooncogene *vav*, which is expressed specifically in hematopoietic cells, and is another target protein for the tyrosine phosphorylation induced by antigen receptor crosslinking.[64] Vav shares sequence homology, designated Dbl homology (DH) domain, with the products of both the human *dbl* gene and the yeast *cdc24* gene. The protein products possess GEF activity also for Rho-related GTPases[65] suggesting that Vav may serve a dual function. In addition, Vav contains other sequence motifs including one SH2 and two SH3 domains, pleckstrin homology domain, as well as sequences similar to those found in some transcription factors: helix-loop-helix, leucine zipper and zinc finger. Therefore, Vav may function in hematopoietic cells as a transducer of

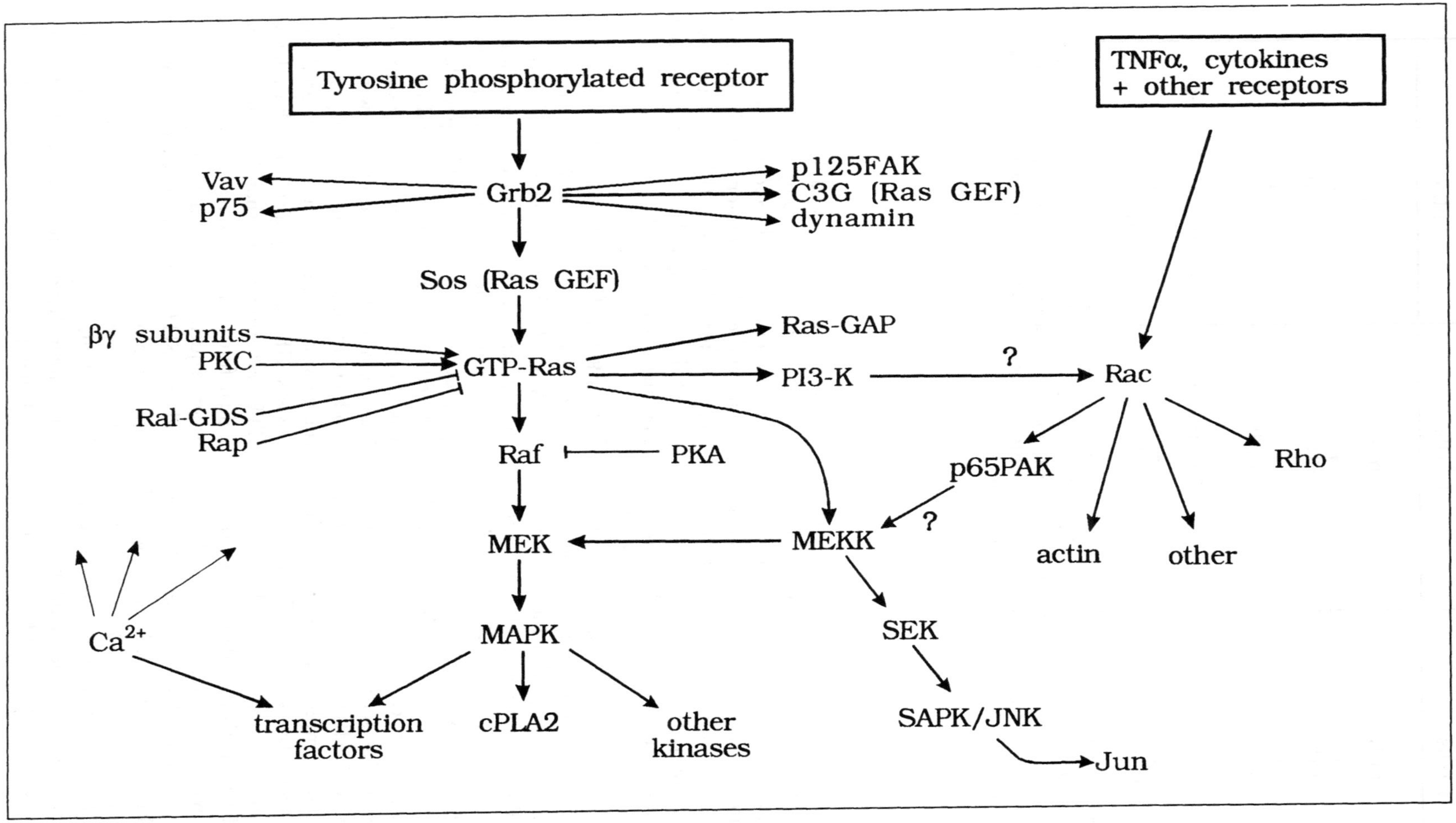

Fig. 5.4. MAPK pathway as a part of an extensive signaling network. Receptor activation initiates multiple signaling pathways that branch off the central MAPK cascade. Calcium contributes to the regulation of many components of the network. The adapter protein Grb2 interacts with multiple signaling molecules. Ras controls several effector molecules including PI3-K that, in turn, may control Rac. In addition to PTKs, Ras is controlled by G proteins and PKC. A divergent cascade of kinases is initiated by MEKK, a kinase that seems to be regulated by both Ras and Rac/Cdc42.

tyrosine phosphorylation signals to Ras and/or Rho signaling pathways as well as to transcriptional events.[49,64,66-68] In RBL-2H3 cells, tyrosine phosphorylation of Vav following the crosslinking of the FcεRI receptor is dependent on the activation of Syk kinase.[69]

The function of Vav, however, is still controversial. In both T and B cells, Vav was reported to act as a GEF for Ras and its nucleotide exchange activity was found to be stimulated by receptor induced tyrosine phosphorylation.[70,71] Moreover, binding of phorbol esters or 1,2-diacyl glycerol to Vav, presumably to the cysteine-rich zinc finger-like sequence, also induced an increase of its Ras-GEF activity, but this activation was tyrosine kinase independent.[72] On the other hand, Vav did not exhibit any Ras-GEF activity in standard nucleotide exchange assays. Furthermore, the morphology of Vav-transformed cells was found to be different from that induced by Ras transformation but similar to that found in RhoA-transformed cells.[73,74] The direct regulation of Ras-GEF activity of Vav by diacylglycerides is also controversial since Vav was found unable to bind phorbol esters in an in vitro assay.[75] Consequently, the nature of the GTPase that is regulated by Vav remains to be established. Since Vav is a multidomain protein, its nucleotide exchange activity towards Ras may be indirect. Vav binds to Grb2[76] and may interact with the proline-rich domain of a Ras-GEF such as Sos. This interaction enables Vav to coordinately regulate the activities of both Ras and Rho (Fig. 5.4).

Targets for Ras

Raf

Raf binds to GTP-Ras

Interaction of activated Ras (Ras in a GTP-bound form) with its effector molecules is expected to lead to various cellular responses. The best characterized of these is the activation of the cytoplasmic serine/threonine kinase, Raf. The N-terminal, regulatory, domain of Raf directly interacts with the effector region of GTP-Ras. This interaction leads to a translocation of Raf to the plasma membrane where it is activated. This initiates a phosphorylation cascade (Ras/Raf/mitogen-activated tyrosine kinase kinase (MEK)/mitogen-activated tyrosine kinase (MAPK)), leading to the stimulation of MAPK1 and 2 (also known as ERKs for extracellular signal-regulated kinases). The activation of MAPK leads in turn to the activation of other kinases, lipases (PLA2) and various transcription factors that control cell proliferation or differentiation (Fig. 5.4).[77,78]

MAPK pathway

Activation of the Ras/Raf/MEK/MAPK pathway following crosslinking of the antigen receptor has been demonstrated in both B and T-cells.[79,80] In T-cells, the MAPK cascade was shown to synergize with calcineurin, a calcium-calmodulin dependent serine-threonine phosphatase, to induce translocation of the transcription factor NFAT (the nuclear factor of activated T cells) into the nucleus, leading to the expression of interleukin-2.[81] In RBL-2H3 cells, MAPK is tyrosine phosphorylated[82] and activated[83] following crosslinking of the FcεRI. Calcineurin activity in RBL-2H3 cells is required for the FcεRI-induced transcription of cytokines but not for secretion.[84]

The activation of MAPK in RBL-2H3 cells is required for serine phosphorylation of cytosolic phospholipase 2 (cPLA2) which results in the activation of this enzyme and subsequent release of arachidonic acid. In contrast to the secretory response, activation of MAPK and cPLA2 was independent of PKC but dependent on a GTP-binding protein, probably Ras. Indeed, the MAPK/cPLA2 pathway was shown to be correlated with the FcεRI-induced tyrosine phosphorylation of Vav, the presumed GEF for Ras and/or Rho. The Vav phosphorylation was dependent on the activation of the tyrosine kinase Syk. Thus it appears that in RBL-2H3 cells, FcεRI/Syk/Ras/Raf/MEK/MAPK pathway leads to the activation of cPLA2, while another PKC-dependent pathway leads to secretion;[29,69] both pathways may act in synergy.

Multiple pathways for activating MAPK

Raf activation can also be achieved by PTK-independent mechanisms that may, in T cells, involve PKC.[85,86] Whether this pathway bypasses Ras is not clear, but regulation of Raf and of the MAPK cascade can obviously occur at many different levels (Fig. 5.4), involving heterotrimeric G proteins, second messengers (calcium, cAMP, diacylglycerides), PKC and PKA.[87] For example, MAPK could be stimulated independently of the tyrosine kinase Syk or of the tyrosine phosphorylated Vav in RBL-2H3 cells, that were transfected with the muscarinic m1 receptor and activated by carbachol.[69] The muscarinic receptor m1 is a seven-membrane-spanning domain receptor coupled to a heterotrimeric G protein, Gq. Some heterotrimeric G proteins, including Gq, have been reported to activate Ras (and MAPK) via their βγ subunits binding to the PH region of a Ras regulatory protein, i.e., Sos, p120GAP or Vav.[88] Thus, G protein coupled receptors can modulate the MAPK cascade through both Ras-dependent and -independent pathways, using either βγ or α subunits. The type of regulation is different for various cells as well as receptors.[87]

The receptor/Ras/Raf/MEK/MAPK pathway forms a central channel of a diverse signaling network (Fig. 5.4). Positive or negative stimuli feed into the network at each step of the cascade, and multiple signaling pathways branch off the central route via specific effectors that respond to the activation of the individual components.[89] For example, the binding of the adapter Grb2 to the phosphorylated FcεRI receptor may cause a recruitment of various proline-rich signaling molecules to the SH3 domains of Grb2: in addition to Sos, other proteins such as p75,[59] C3G (another GEF for Ras),[90] dynamin,[91] $pp125^{FAK}$,[92] or Vav[76] may be recruited. The FcεRI-dependent activation of an exchange factor for Ras may lead to a coordinated regulation of Ras and other GTPases, such as Rac and Rho. This is expected because Ras-GEFs often possess domains that are known to function as Rho/Rac exchange factors.[46,49] In addition to the Ras-induced activation of Raf, there are other known Ras targets, i.e., candidates for Ras effectors, such as Ras-GAP, phosphatidylinositol 3-kinase (PI3-K), the GTPases Ral and Rap and the mitogen-activated protein kinase kinase kinase (MEKK or MAPKKK) (Fig. 5.4).

Ras-GAP

Similarly to Raf, Ras-GAP (p120-GAP, Ras GTPase activating protein) binds to the effector region of the GTP-bound Ras. Ras-GAP acts as a negative feedback regulator of Ras because it stimulates its intrinsically low GTPase activity and thus downregulates it.[46] Activation of the PTK receptor EGFR leads to the association of the SH2-containing Ras-GAP with the receptor. This results in its phosphorylation and inhibition of its activity and consequently in an increased level of Ras-GTP.

Receptor stimulation also leads to an increased association of Ras-GAP with a protein p190.[93,94] p190 is a phosphoprotein which shows a variable degree of GAP activity on all members of the Rho family of GTPases.[93,95,96] The formation of multimolecular complexes that contain Ras-GAP, Ras, p190 and Rho may thus provide another opportunity for a coordinated regulation of both Ras and Rho mediated effects. In accord with the reported effects of the Rho-like GTPases on the cytoskeleton (see below), expression of the p120-GAP N-terminus in fibroblasts resulted in an enhanced p190-GAP activity and induced the dissolution of actin stress fibers, an effect which is consistent with downregulation of Rho.[97] Crosslinking of the BCR receptors of B lymphocytes also leads to tyrosine phosphorylation of Ras-GAP and its increased association with p190.[98-100] It can be expected that a similar mechanism is initiated after the crosslinking of FcεRI receptors.

Phosphatidylinositol 3-kinase (PI3-K)

p110 binds to GTP-Ras

Generation of 3'phosphorylated inositol phospholipids is promoted by PI3-K, an enzyme consisting of a 110 kDa catalytic subunit and an 85 kDa regulatory subunit. GTP-Ras, again through its effector region, interacts directly with the catalytic subunit of PI3-K.[101] In neutrophils, an increase in 3'phosphorylated inositol phospholipids is induced by receptor stimulation via GTP-binding proteins.[102] In PC12 cells, growth factor induced PI3-K activation is inhibited by a dominant negative Ras mutant, N17Ras.[101] Thus, PI3-K is another possible effector for Ras, providing another diverging pathway downstream of Ras. In both B and T-cells, PI3-K is activated following receptor crosslinking.[103,104] The importance of this pathway for mast cell function is shown by the inhibitory effects of wortmannin, a PI3-K inhibitor, on histamine secretion from RBL-2H3 cells.[105] Moreover, wortmannin also inhibited some other responses of RBL-2H3 cells to FcεRI activation, such as IP_3 synthesis and membrane ruffling.[106]

p85 subunit

The control of PI3-K, however, is very complex and does not occur only via GTP-binding proteins. The regulatory p85 subunit is tyrosine and/or threonine phosphorylated in response to receptor stimulation and binds to activated PTK receptors via its SH2 domains.[107-109] This may provide an alternative control mechanism. Further complexity arises from the existence of different isoforms. Activation of T cells resulted in a differential phosphorylation of α and β isoforms of the p85 subunits, indicating that they may serve different functions.[109] The p85 subunit also contains a proline-rich region that has been implicated in binding to the SH3 domain of the Src-related kinases such as Lyn or Fyn. The binding of p85 to SH3-containing proteins would provide an opportunity for an assembly of a signaling complex involving both PTK- and non-PTK-receptors.[4]

Multiple regulation of PI3-K

In B cells, PI3-K has been found associated with Vav,[110] suggesting that this tyrosine phosphorylated protein may also transduce signals to PI3-K. Vav, with its Dbl homology domain, may coordinate activities of Ras and Rho-like proteins. The latter have also been shown to act both upstream and downstream of PI3-K.[111-115] Yet another level of regulation is that by the βγ subunits of some heterotrimeric GTP-binding proteins, observed in neutrophils and endothelial cells. The isoforms, that were activated by βγ, were different from those that responded to PTK receptor activation.[115,116] This multilayered control of PI3-K points to its central role in cell physiology although there are still many unresolved questions.

Ral-GDS, MEKK, Rap

Other possible candidates which can act as Ras effectors/modulators will be mentioned briefly.

Ral-GDS

Ral-GDS is a guanine nucleotide dissociation stimulator (or GEF) for the small Ras-related GTPase, Ral. This protein has been identified as a potential Ras effector due to its GTP-dependent interaction with the Ras effector region. It also has the ability to compete with Raf for binding to Ras.[117,118] Thus, Ral-GDS may inhibit interactions between Ras p21 and its effectors.

MEKK

In a manner similar to Raf, MEKK phosphorylates and activates MEK (also known as MAPKK), providing an alternative route to MAPK cascade (Fig. 5.4). In addition, MEKK also phosphorylates and activates SEK (also known as SAPKK, stress activated protein kinase kinase). This initiates a divergent cascade leading to phosphorylation and activation of JNK/SAPK (Jun nuclear kinases or stress activated protein kinases) and of the transcription factor Jun, and ultimately to gene expression (Fig. 5.4). The activity of MEKK is regulated by receptors, i.e., EGFR, in a Ras-dependent manner. The kinase domain of MEKK has now been shown to bind to the effector domain of GTP-Ras and this binding promoted the ability of MEKK to phosphorylate MEK.[119]

Rap1

Finally, Rap1 is a small GTPase with an effector region identical to that of Ras. Rap-GTP binds the same effectors as Ras-GTP but does not activate them. Thus, Rap may regulate Ras-GAP and other effectors of Ras by competing with Ras for effectors and hence antagonizing Ras function (see ref. 120).

GTP-BINDING PROTEINS INVOLVED IN THE SECRETORY RESPONSE OF MAST CELLS

Recent evidence indicates that many GTP-binding proteins, both heterotrimeric and small, participate in the signaling pathway to secretion. Those implicated in secretion from mast cells will be described in this section.

THE GTP-BINDING PROTEIN G_i

Two G_i-like proteins in mast cells

Some heterotrimeric G proteins can be activated directly by compound 48/80 or by other positively charged secretagogues.[25,121,122] Since the response to 48/80 is Ptx sensitive,[9] it follows that G_i-like proteins must be involved. Studies of ATP_4^- permeabilized mast cells have helped to elucidate the function of these Ptx sensitive heterotrimeric G proteins. These cells were exposed to GTP-γ-S, re-sealed by addition of magnesium and then stimulated with 48/80. The results indicated a presence of two Ptx sensitive heterotrimeric G proteins. One of these proteins regulates phospholipase C while the other can induce degranulation in the absence of PLC activation. The latter protein therefore acts downstream of PLC and has been postulated as the putative G_e protein.[25]

G_{i3}

The two proteins, $G\alpha_{i2}$ and $G\alpha_{i3}$, identified as Ptx substrates in mast cells (see above), have been tested for their possible G_e-like function by using $G\alpha_i$ C-terminal peptides and specific antibodies directed against this region.[15] When a peptide, corresponding to the C-terminus of $G\alpha_{i3}$, was introduced into ATP_4^- permeabilized mast cells which were then resealed, 48/80-induced secretion was inhibited independently of PLC. Similarly, the antibody, that recognized the C-terminus of $G\alpha_{i3}$, inhibited Ca^{2+}/GTP-γ-S-induced secretion from Streptolysin-O permeabilized cells. On the other hand, $G\alpha_{i2}$-derived peptide or antibody to the $G\alpha_{i2}$ C-terminal region had no effects on secretion. Thus, G_{i3} may be a candidate for a G protein, G_e, acting downstream of PLC at a late stage in the secretory pathway. A model was suggested that places G_{i3} as G_e in series with a calcium binding protein, C_e. A PTK is also postulated downstream of PLC, but on a separate, G_e-independent pathway (Fig. 5.5) (see refs. 15 and 25; and R. Sagi-Eisenberg, personal communications).

Localization of G_{i3}

Immunofluorescence revealed that in mast cells, G_{i3} is present both on plasma membranes and in the Golgi compartment, while G_{i2} is exclusively on the plasma membranes. Thus, both proteins are accessible to the polycationic secretagogues.[15] The localization of G_{i3} is different in RBL-2H3 cells, that are normally unresponsive to 48/80: it is found only in the Golgi compartment. However, when RBL-2H3 cells were cocultured with 3T3 fibroblasts, they acquired responsiveness to 48/80 and G_{i3} was now additionally present on the plasma membranes.[123] Thus, the cell environment is also a critical regulatory

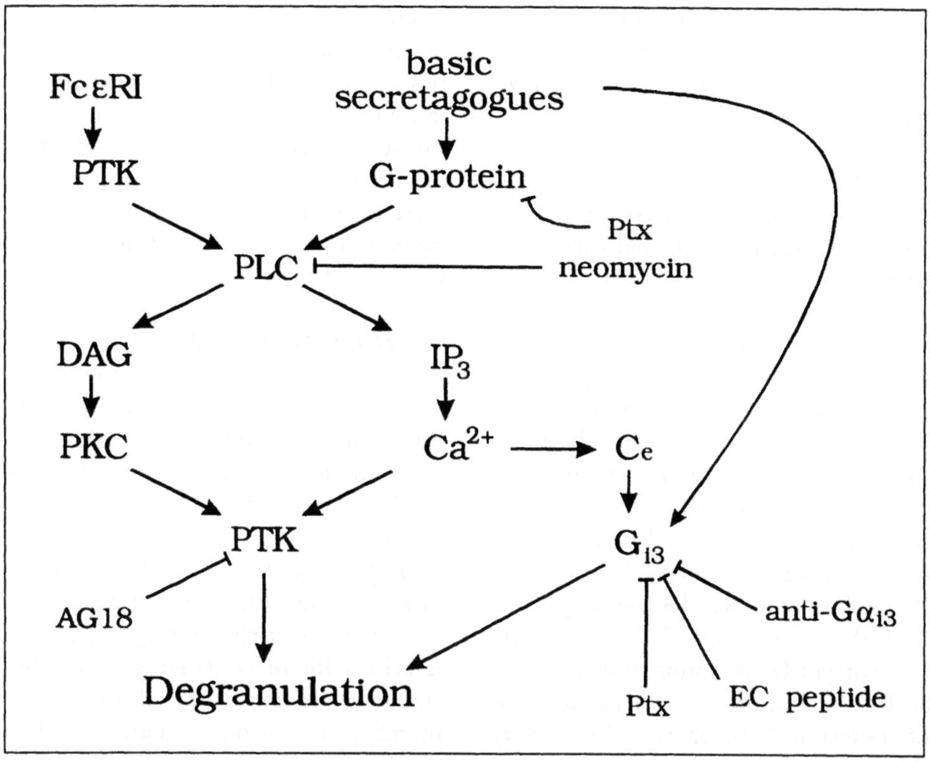

Fig. 5.5. Involvement of G_{i3} in exocytosis. Two Ptx sensitive G proteins act in series: one regulates PLC and the other, G_{i3} degranulation. Both are stimulated by basic secretagogues and inhibited by Ptx. The latter is thought to be G_{i3} because its activity can be specifically prevented by a peptide corresponding to the C-terminus of $G\alpha_{i3}$ and by anti-$G\alpha_{i3}$ antibody. G_{i3} acts independently of PLC but is activated by calcium via a calcium binding protein, C_e. PLC and G_e pathways are independent. A step, sensitive to a PTK inhibitor, AG18, is postulated downstream of calcium and PKC. The scheme is based on refs. 15, 25 and R. Sagi-Eisenberg, personal communication.

factor, affecting both the cell morphology and responsiveness.

Rab Family

Small GTPases of the Rab family regulate vesicle trafficking between membrane compartments. Over 25 different proteins with different specificities have been identified in yeast and mammalian cells.[43,46] GTP-γ-S, in contrast to its stimulatory effects on most GTP-binding protein-regulated pathways, inhibits Rab-mediated vesicle trafficking, but not vesicle formation. This indicates that the cycling between GDP- and GTP-bound states is essential for vesicle targeting and/or docking and/or fusion.[124]

Rab3a

Rab3a is expressed specifically in nervous and endocrine tissues and is associated with secretory vesicles. In mast cells, mRNAs with homology to Rab3b and Rab3d have been identified using PCR-cloning method.[125] Rab3b has also been identified in mast cells by immunoblotting and found to be phosphorylated by protein kinase A (PKA). This effect may be connected to the cAMP-induced inhibition of both secretion and IP_3 formation in response to 48/80.[126] Rab3a was first implicated in regulated secretion from mast cells by studies with a synthetic peptide, corresponding to its effector domain. Application of this peptide into patch-clamped cells stimulated secretion,[127] but later work with scrambled forms of this peptide showed that these effects were nonspecific.[128,129] More recent work in chromaffin cells indicates that Rab3a is a negative regulator of exocytosis.[130,131]

Rabphilin-3a

Rabphilin-3a is a putative target or regulator of Rab3a found in neurones and hormone-secreting cells. It inhibits the Rab-GAP-stimulated GTPase activity of Rab3a and stimulates nucleotide exchange on Rab3a.[132] Rabphilin-3a is composed of the N-terminal Rab3a-binding domain and the C-terminal calcium and phospholipid-binding domain.[133] The middle, regulatory, region is phosphorylated both by calmodulin-dependent protein kinase II and by PKA, offering a potential for Ca^{2+} and cAMP regulation of Rab3A activity.[134-136] Although the presence of rabphilin in mast cells has not yet been demonstrated, a role for the Rab3/rabphilin system in the control of secretion from these cells is very probable.

Rho Family

Rac, Rho and Cdc42 are the three subfamilies of the Ras-related GTPases belonging to the Rho (Ras homologous) family. Similarly to Ras, they are post-translationally lipid-modified and controlled by regulators such as GEFs and GAPs.[137]

Rac, Rho, Cdc42 and the cytoskeleton

The best established function of the Ras-related family is mediating signals from receptors to the cytoskeleton. In Swiss 3T3 fibroblasts, microinjections of constitutively active mutants of Rho and Rac result in the assembly of actin to produce membrane ruffles and stress fibres, respectively.[138,139] Cdc42, was recently reported to induce formation of filopodia.[140] These three GTPases also control, independently of actin polymerization, the assembly of multimolecular focal complexes that link actin networks to the plasma membrane. The three types of focal complexes are specific and distinct although they all contain focal adhesion proteins, vinculin, paxillin and $pp125^{FAK}$, as well as phospho-tyrosine. Microinjection studies revealed a hierarchical and coordinated control of the activities of these GTPases. Activation of Cdc42 in Swiss 3T3 cells leads to sequential activation of Rac and then Rho.[140] The molecular mechanism of the control of actin rearrangements is not understood but must involve recruitment and/or activation of various actin-binding proteins.

Other functions of Ras-related GTPases

In addition to their effects on cell morphology, other functions for the three Ras-related subfamilies have recently been revealed, some of them quite independent of the cytoskeletal effects. Some of these are discussed below. These functions often overlap in such a way that two or three different Rho-related GTPases seem to exert control over the same process, and there is still a lot of controversy over the position and the hierarchy of the Rho-family members in various signaling pathways.

Phospholipid metabolism

The "noncytoskeletal" functions include control of enzymes that regulate the turnover of phosphoinositides, such as PI3-K, PIP5-K and PLD.[111-114,141-143] Since activities of many actin binding proteins are regulated by phosphoinositides,[144] the effects that the GTPases exert on the cytoskeleton may well be, at least in part, a consequence of their effects on phospholipid metabolism. However, control of Rac by upstream PI3-K was also suggested and this could be a mechanism by which Ras, that directly activates PI3-K, communicates with Rac (Fig. 5.4).[115] Recently, the following mechanism for the Rac >> Rho hierarchy has been suggested:[145] Rac, in addition to its effect on actin, controls the release of arachidonic acid from membrane phospholipids, probably by activating $cPLA_2$. The subsequent generation of leukotrienes activates Rho and induces formation of stress fibers by a Rho-dependent mechanism. The authors suggest the existence of both leukotriene-dependent and independent routes to Rho activation.[145]

Focal adhesions formation

The Rho-dependent formation of focal adhesions requires activities of PTK(s) and of PKC,[146,147] indicating that such kinases may be another target for Rho-related GTPases. Recent experiments with the *Clostridium botulinum* C3 exoenzyme, which inactivates Rho by ADP-ribosylation, confirm that a Rho-PTK pathway may exist. Thus, in permeabilized Swiss 3T3 cells, GTP-γ-S stimulated tyrosine phosphorylation of two focal adhesion proteins, the tyrosine kinase pp125^{FAK} and paxillin. Pre-treatment with C3 or with the p21rho17-44 peptide, but not the p21ras17-44 peptide, markedly reduced this effect.[148] Therefore, signals can travel in both directions: from tyrosine kinases to GTPases and vice versa. Tyrosine phosphorylation of both pp125^{FAK} and paxillin is also induced by activation (clustering) of integrins.[149] It has been suggested that autophosphorylation of pp125^{FAK} (on Tyr397) is induced by direct binding of the N-terminal end of the pp125^{FAK} to the cytoplasmic tail of the activated β integrin. In turn, the autophosphorylated pp125^{FAK} may serve to recruit Src-related PTKs and Grb2, thus possibly linking integrins to the Ras pathway.[92,149] The C-terminal end of pp125^{FAK}, on the other hand, contains paxillin binding site.[149]

In RBL-2H3 cells, tyrosine phosphorylation of paxillin, as well as its redistribution, was observed as a relatively late event after the aggregation of FcεRI. Both these events were enhanced by cell adherence.[150] Moreover, Src-related PTK Lyn, expressed as a fusion protein, was found to bind paxillin from RBL-2H3 cell lysates and this binding was increased after cell activation.[151] Consequently, Lyn and other Src-related PTKs[152] may phosphorylate cytoskeletal proteins directly. The cross-talk between integrin- and Rho-dependent signaling pathways, although evident, is not yet understood and neither is it clear which kinase (pp125^{FAK}, Src-related PTK or other?) is responsible for paxillin phosphorylation. FcεRI and integrins may control phosphorylations of focal adhesion proteins synergistically, possibly via Rho in both cases. Yet another level of control may be exerted by seven transmembrane domain receptors since their activation also leads, in a Rho-dependent manner, to paxillin phosphorylation.[153] The sequencing of paxillin showed that this cytoskeletal protein contains a multitude of familiar signaling motifs,[154] and is presumably capable to participate in many alternative pathways.

NADPH oxidase regulation

In phagocytic leukocytes, Rac has been found to regulate the production of superoxide by NADPH oxidase by a mechanism that does not seem to involve the cytoskeleton.[155-157] Rac could activate the NADPH oxidase activity in a cell-free system containing only the five purified proteins: two cytochrome b558 subunits, two SH3 domain-containing phosphoproteins, p47phox and p67phox and the GTP-bound Rac. It appears that the native Rac in its GDP-bound form can also stimulate NADPH oxidase but only when complexed with Rho-GDI, a guanine nucleotide dissociation inhibitor.[157] Alternatively, Rho-GDI may also bind to Ras-GTP and protect it from GAP. At present, this is an unresolved discrepancy.

The recruitment of Rac to the plasma membrane requires a membrane bound protein, a putative guanine nucleotide exchange factor for Rac (Rac-GEF), that is different from p47phox and p67phox.[158] The translocation to the plasma membrane of the other components of the NADPH oxidase complex, the p47phox and p67phox proteins, occurs independently of that of Rac, but in a coordinated manner.[157] Therefore, Rac does not seem to regulate the assembly of the complex, but its enzymic activity. p67phox can bind to Rac directly in a GTP-dependent manner, and therefore may be the Rac effector protein in phagocytic cells.[159] Oxidase activation requires phosphorylation of both p67phox and p47phox proteins and it is possible that in intact cells, Rac activates a kinase that controls their phosphorylation state.

The recent finding that a serine/threonine kinase p65PAK (p21 activated kinase) binds to and is activated by the GTP-bound forms of Rac and Cdc42 may be relevant to the last point since p65PAK may be an effector of Rac.[160]

Muscle contraction

Rho has also been implicated in regulation of cytoplasmic division[161] and of smooth muscle contraction,[162] both via its effects on the actomyosin system. In smooth muscle, Rho decreases the calcium requirements for contraction, indicating a cooperative control by Rho and calcium. This is similar to the situation in T cells, where calcium and Ras synergistically control the translocation of the transcription factor NFAT to the nucleus.[81] Inhibition of receptor-mediated cortical actomyosin contractility by C3 transferase has also been observed in neural cell lines, indicating that Rho is required for this type of force generation and for the subsequent neurite retraction.[163]

Gene expression

As shown in Figure 5.4, Ras activation leads to the activation of MAPK and ultimately of transcription factors via a cascade of kinases. The MAPK kinase, MEK or MAPKK, is believed to be activated by both Raf-dependent and Raf-independent routes, the latter utilizing a kinase termed MEKK, a potential Ras effector. MEKK additionally initiates a divergent cascade leading to activation of JNK/SAPK and ultimately of a transcription factor Jun. Recent evidence points to Rac and Cdc42 involvement in the control of the MEKK-JNK pathway (reviewed in ref. 164). In many cells, this pathway is activated by agonists such as TNF-α or cytokines, and experiments with various transfected mutants of Cdc42, Rac and Rho showed that activities of Cdc42 or Rac1, but not Rho, are required for JNK activation. The serine/threonine kinase p65PAK has been suggested as a possible link between Cdc42/Rac1 and MEKK (Fig. 5.4).

Rac and Rho in mast cells

In secretory cells, the cytoskeleton may play a role in granule migration, in regulating the access of secretory vesicles to the plasma membrane and also in the endocytotic events following exocytosis. We have investigated the effects on actin organization of calcium, guanine nucleotides and ATP, three factors that in various combinations can elicit secretion from permeabilized mast cells. Because of their known effects on actin morphology, we have also examined the effects of Rac and Rho on mast cell cytoskeleton and on exocytosis. The results of this work have shown that both these GTPases are present in mast cells and that they regulate organization of actin as well as secretion.[32,40] The secretory and the cytoskeletal responses are strongly correlated but can occur independently of each other, indicating divergent signaling pathways (unpublished observations and ref. 165). This direct demonstration of the control of exocytosis by Rac and Rho assigns a new function for these GTPases and implies their participation in pathways initiated by activation of FcεRI.

Calcium-dependent cortical actin disassembly and secretion

After permeabilization, mast cells lose up to 70% of their total actin. The remaining F-actin is present mostly in the cell periphery, as a prominent subplasmalemmal network.[32,166] A reduction of this cortical F-actin by 40-50% is induced by exposure to calcium. This effect requires ATP in washed permeabilized cells,[33] but is ATP-independent when calcium is present at the time of permeabilization.[166] This may be due to the leakage of the endogenous ATP and possibly of some other factors[33] from the cells. However, permeabilized mast cells must retain a kinase that is presumably required for the calcium-dependent cortical disassembly. A possible candidate is PKC since pretreatment of cells with PMA prior to permeabilization causes a loss of the cortical actin.[166] The secretory response to calcium is obtained only from a certain proportion of cells, but the reduction in F-actin is apparent in both secreting and nonsecreting cell populations. Thus, the disassembly by itself is insufficient to induce secretion.[33,165]

Using antibodies raised against recombinant Rac and Rho, and C3-induced ADP-ribosylation of the endogenous Rho, we have shown that after permeabilization, only 30-40% of the total Rac or Rho leaks out of cells.[32,40] Pretreatment of the permeabilized cells with C3 transferase, a specific Rho inhibitor, reduces both the calcium/ATP induced cortical disassembly and secretion, indicating that Rho regulates both processes.[40] Moreover, when permeabilized mast cells were pretreated with the exogenous constitutively active mutant of Rho, V14RhoA, calcium/ATP-induced secretion was significantly increased and a small increase in cortical F-actin disassembly was also observed.[40,165] Thus, a Rho-regulated signal is additionally required for a step, possibly subsequent to the cortical disassembly, leading to exocytosis. The demonstration of Rho's involvement in the calcium/ATP induced responses raises an interesting possibility that Rho may be regulated by calcium.

Similarly to V14RhoA, the constitutively active mutant of Rac, V12Rac1, also enhanced calcium/ATP-induced secretion.[40] However, addition of a dominant negative mutant of Rac, N17Rac1, did not reduce either secretion or disassembly.[40,165] In contrast to C3, the effects of N17Rac1 are directed at the guanine nucleotide exchange factor rather than at the Rac itself. Therefore, this inhibitor would not affect the activity of Rac that may already be in the GTP-bound form. It is not yet clear whether Rac also has any effects on the cortical disassembly. Flow cytometry studies of cell populations showed that the enhancement of secretion by both V14RhoA and V12Rac1 was caused by an increased proportion of cells that are competent to respond to stimulation (Fig. 5.6).[40,165]

Calcium-independent cortical actin disassembly and secretion

Cortical actin disassembly can also be induced in a calcium-independent manner; the two mechanisms are different and may synergize. Treatment of permeabilized mast cells with GTP-γ-S/EGTA results in a centripetal redistribution of actin filaments due to selective decrease and increase of F-actin in the cortical and the interior region, respectively (Fig. 5.7C and D). The overall F-actin content after GTP-γ-S/EGTA treatment is increased.[32] In contrast to the calcium induced disassembly, that induced by GTP-γ-S/EGTA is inhibited by high concentrations (> 1 μM) of phalloidin, F-actin stabilizing agent, and is ATP independent. Again, cortical disassembly is exhibited by both populations of cells, secreting and nonsecreting; in this case, cell populations were distinguished by fluorescence microscopy (Norman et al, unpublished observations) (Fig. 5.8).

AlF_4^-, an activator of heterotrimeric G proteins, mimics one aspect of the GTP-γ-S induced response: the selective decrease of the cortical F-actin (Fig. 5.7E). Thus, in addition to Rho, a heterotrimeric G protein may regulate disassembly of cortical filaments and/or their detachment from the plasma membrane. C3 transferase did not prevent the GTP-γ-S induced disassembly[32] and therefore Rho may not be involved in this function unless calcium is also present. However, this effect needs to be re-examined: while the responses to calcium/ATP were studied in suspended cells, those to GTP-γ-S were performed in adherent cells. It is already clear that adherent cells are more resistant to GTPase inhibitors; the secretory response of suspended cells to GTP-γ-S/EGTA is strongly inhibited by either C3 or N17Rac1,[40] but that of adherent cells is inhibited only when both C3 and N17Rac1 are added together (Norman et al, unpublished observations).

Taken together, our results indicate that calcium-independent and calcium-dependent membrane-associated proteins are involved in cortical disassembly. The calcium-independent mechanism may include uncapping of gelsolin caps regulated by a heterotrimeric GTPase, probably via the phosphoinositide pathway. Gelsolin is an actin-regulatory protein that severs actin filaments in a calcium-dependent manner. It is also regulated by PIP_2 which, in the absence of calcium, decreases its affinity for the barbed ends of actin filaments.[167] Using immunofluorescence and confocal microscopy, we have observed GTP-γ-S/EGTA-induced loss of cortical gelsolin, suggesting that uncapping of gelsolin caps may play an important role in the cortical actin disassembly and in exocytosis.[33]

Rho and actin polymerization

The GTP-γ-S induced accumulation of F-actin in the interior of permeabilized

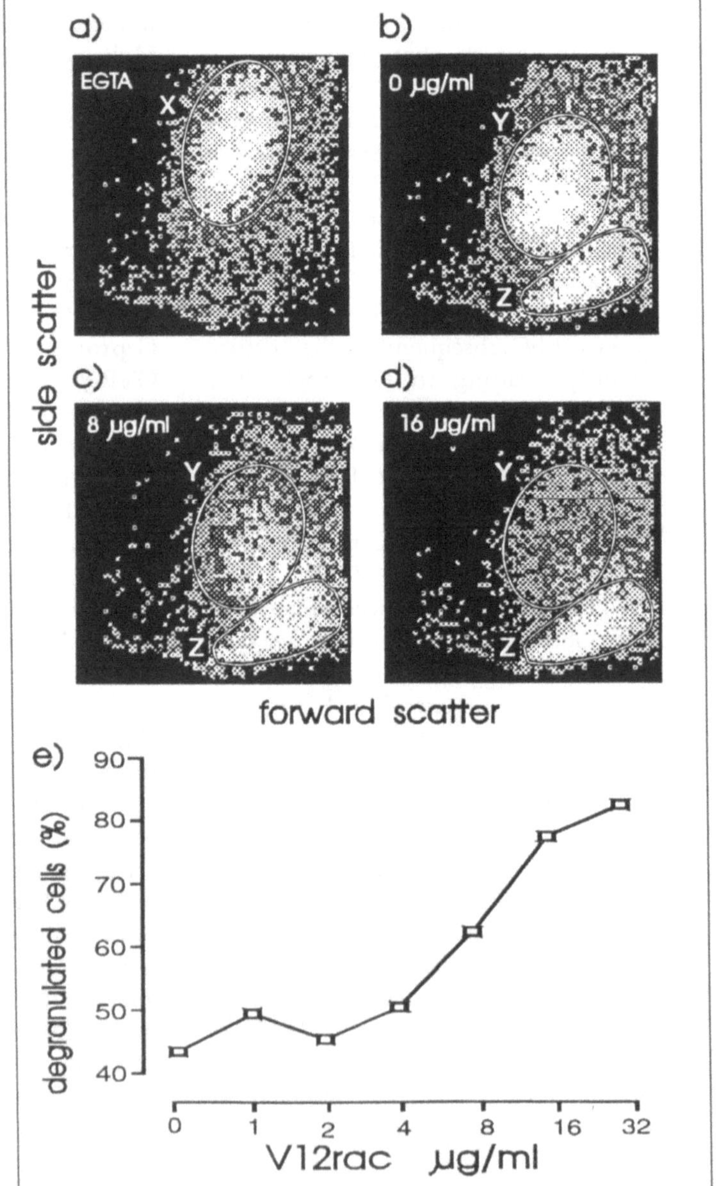

Fig. 5.6. V12Rac1 enhances the proportion of cells that degranulate in response to calcium/MgATP. Light scatter profiles of permeabilized mast cells that were preincubated with the indicated concentrations of V12Rac1 and then exposed for 30 minutes at 30°C to (a) 3 mM EGTA-GB (control), or (b, c and d) to calcium (pCa 4.75)/3 mM MgATP. Most of the control cells were contained within region X. Treatment with calcium/MgATP produced two discrete subpopulations of cells, contained in regions Y (nonsecreting cells) and Z (secreting cells). The number of cells present in regions Y and Z was reduced and enhanced, respectively, by addition of V12Rac1. (e) Increasing concentrations of V12Rac1 progressively enhance the percentage of cells in the secreting cell subpopulation. The percentage of cells in the secreting population (region Z) is plotted against increasing concentrations of V12Rac1. Reprinted with permission from Price LS et al, Current Biology 1995; 5:68-73.

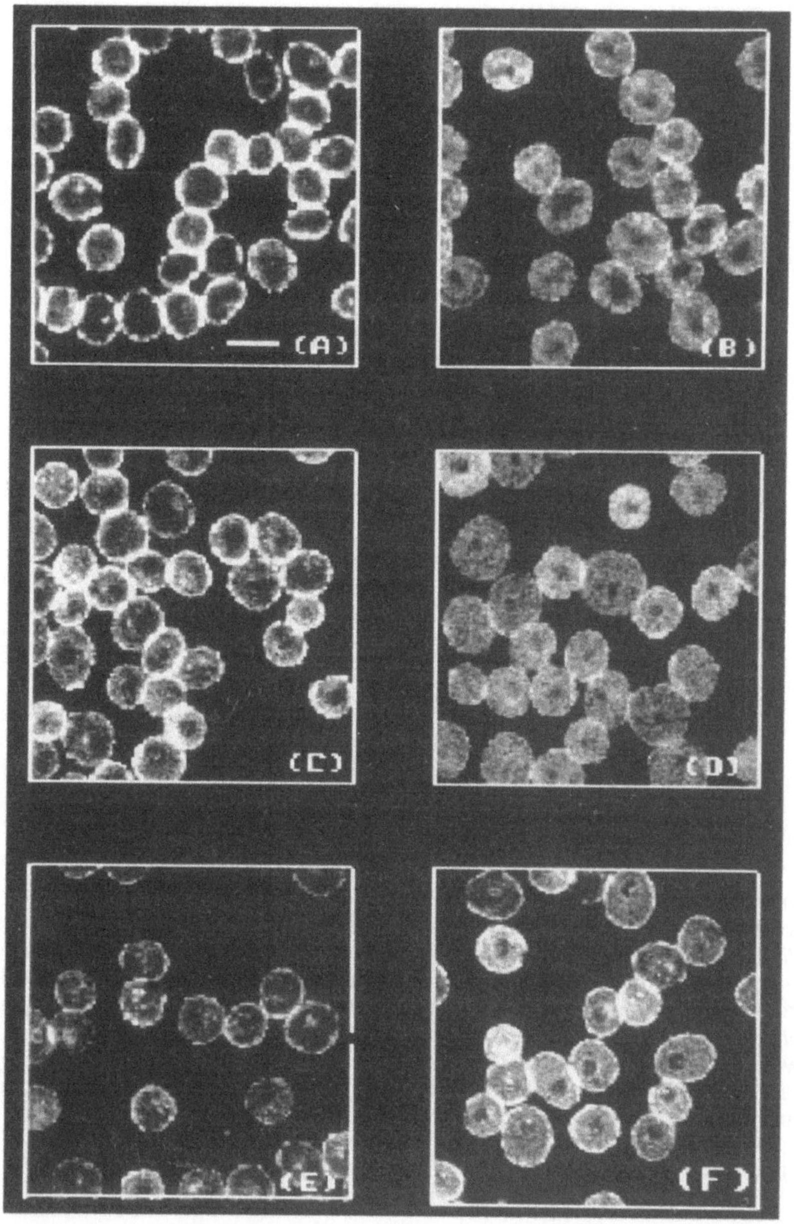

Fig. 5.7. F-actin distribution as shown in confocal micrographs of rhodamine phalloidin (RP) stained mast cells. The cells were stained with RP after fixation to visualize all the F-actin present. (A) unstimulated and (B) 48/80-treated intact mast cells. (C) unstimulated and (D) GTP-γS treated permeabilized cells showed similar patterns. Different responses were exhibited by permeabilized cells treated with (E) AlF_4^- and (F) V14RhoA. The bar represents 20 μm. Reproduced from Norman JC et al, The Journal of Cell Biology 1994; 126:1005-15 by copyright permission of The Rockefeller University Press.

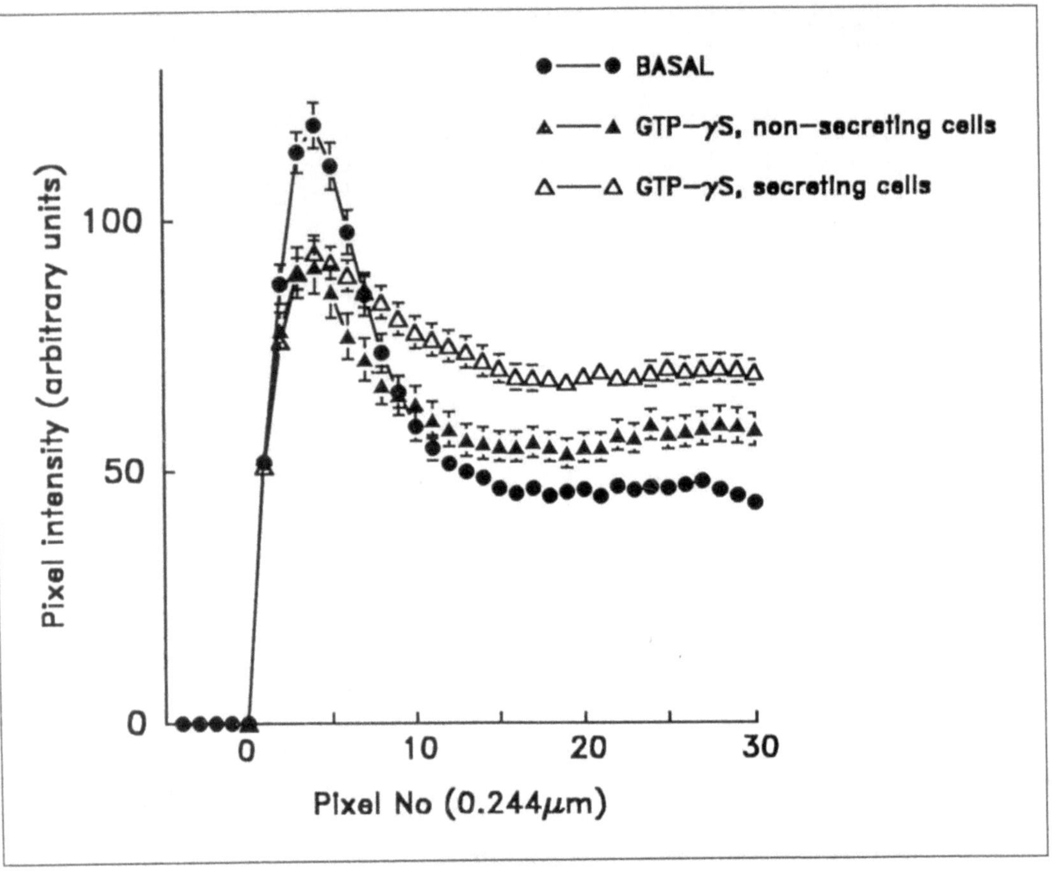

Fig. 5.8. Changes in the F-actin distribution induced by GTP-γ-S in permeabilized cells. Densitometric line scan analyses were performed on confocal images similar to those in Figures 5.7C and D. The intensity profiles of the radial lines (3 lines per each cell) were pooled to form a mean profile +/- sem, n > 50 cells. Secreting and nonsecreting cells were identified using a double label staining protocol (RP for F-actin and Bodipy-Con-A, BCA, for secreting cells). Following stimulation with GTP-γ-S, cells with BCA fluorescence greater than that of the control cells (exposed to 3 mM EGTA only) were considered to be secreting. Using this criterion, 73% of the cells were secreting in response to GTP-γ-S. Control cells (closed circles); GTP-γ-S stimulated cells (triangles). Cells with BCA fluorescence above and below control levels are represented by open and closed triangles, respectively. Both secreting and nonsecreting cells exhibit cortical actin disassembly but the level of F-actin in the cell interior is much greater in the secreting cell population (Norman et al, unpublished observations).

mast cells was attributed to polymerization and relocalization of actin filaments. The de novo actin polymerization was found to be regulated by Rho (Fig. 5.7F). This effect can be prevented by C3 transferase or by cytochalasin, indicating a Rho-dependent formation of actin filament barbed ends as nucleation sites for actin polymerization.[32] These sites may form a part of multiprotein complexes similar to focal adhesions. Newly polymerized filaments appear despite the loss of the soluble monomeric actin from permeabilized cells, suggesting the existence of a membrane-bound or slow-leaking G-actin pool. Another function of Rho may thus be to initiate the release of this sequestered G-actin, a 'ready-to-go' pool of actin monomer.

Rac and relocalization of cortical filaments

When F-actin in permeabilized mast cells is prelabeled with a low concentration of rhodamine-phalloidin, at least some of the labeled cortical filaments relocalize to the cell interior after exposure to

GTP-γ-S. This effect is inhibited by N17Rac1, but not by C3 or cytochalasin.[32] Conversely, N17Rac1 does not inhibit de novo actin polymerization. Therefore, Rac regulates the entrapment of cortical filaments and this effect is different and independent from that of Rho. The cytoskeletal effects of both Rac and Rho are ATP independent, making participation of any downstream kinases improbable.

Involvement of Rac and Rho in mast cell response to 48/80

Treatment of intact cells with compound 48/80 produced changes in microfilament morphology which were very similar to those induced in permeabilized cells by GTP-γ-S: a centripetal rearrangement of F-actin (Fig. 5.7A and B).[32] The similarity of these responses supports the evidence for a direct activation of plasma membrane heterotrimeric G proteins by compound 48/80[25,121,122] and indicates that such a G protein may activate Rac- and Rho-dependent responses. In permeabilized cells, however, AlF_4^--induced activation of G proteins resulted only in the cortical disassembly with no appearance of F-actin in the cell interior. This suggests that in intact cells, a cytosolic factor may be present that transduces signals from the 48/80 sensitive G protein to Rac and Rho and that this factor may leak out after permeabilization. It is notable that the increase in F-actin content of mast cells was quantitatively similar regardless of whether intact cells were stimulated by 48/80 or permeabilized cells by GTP-γ-S. Thus, the same membrane-bound pool of G-actin may be designated for recruitment in both cases.

GTP-γ-S induced cytoskeletal and secretory responses are correlated

This has been demonstrated by cell population studies using both flow cytometry and fluorescence microscopy (Norman et al, unpublished observations and ref. 165). Both de novo actin polymerization and the relocalization of filaments occur preferentially in secreting cells (Fig. 5.8). The secretory and cytoskeletal responses are therefore correlated and a common element must exist in their regulatory pathways. However, the relationship is not a simple one since the two activities can be separated from one another. For example in response to GTP-γ-S, cytochalasin inhibits de novo actin polymerization but not secretion. Low concentrations of GDP-β-S inhibit secretion but not the actin responses (Norman et al, unpublished observations). This indicates that the signaling pathways from these GTPases to secretion and to the cytoskeleton may act parallel rather than in series.

A model

We have therefore postulated another GDP-β-S sensitive GTP-binding protein required for exocytosis in addition to Rho, Rac and the AlF_4^--sensitive heterotrimeric G protein. A model for the control of exocytosis is proposed, with Rac and Rho occupying a central role in the signaling pathway to secretion. The model includes at least four GTP-binding proteins and shows both shared and divergent pathways from Rac and Rho (Fig. 5.9). Calcium has not been included in this model, but it is clear that it is relevant to both cytoskeletal and secretory responses.

The importance of Rac in the signaling pathway to secretion has recently been confirmed by experiments that examined fractions from a brain extract for their ability to reconstitute secretion from permeabilized mast cells: the active fraction contained Rac coupled to Rho GDI. Interestingly, Rac by itself did not have any effect on secretion while Rho GDI was inhibitory (A O'Sullivan et al, manuscript submitted).

CONCLUDING REMARKS

Crosslinking of FcεRI induces an increase in F-actin content and changes in actin morphology[168-170] that are discussed in chapter 8. How similar are these to the GTP-γ-S induced effects in permeabilized cells? How relevant are the effects induced by calcium- and GTP-γ-S in permeabilized

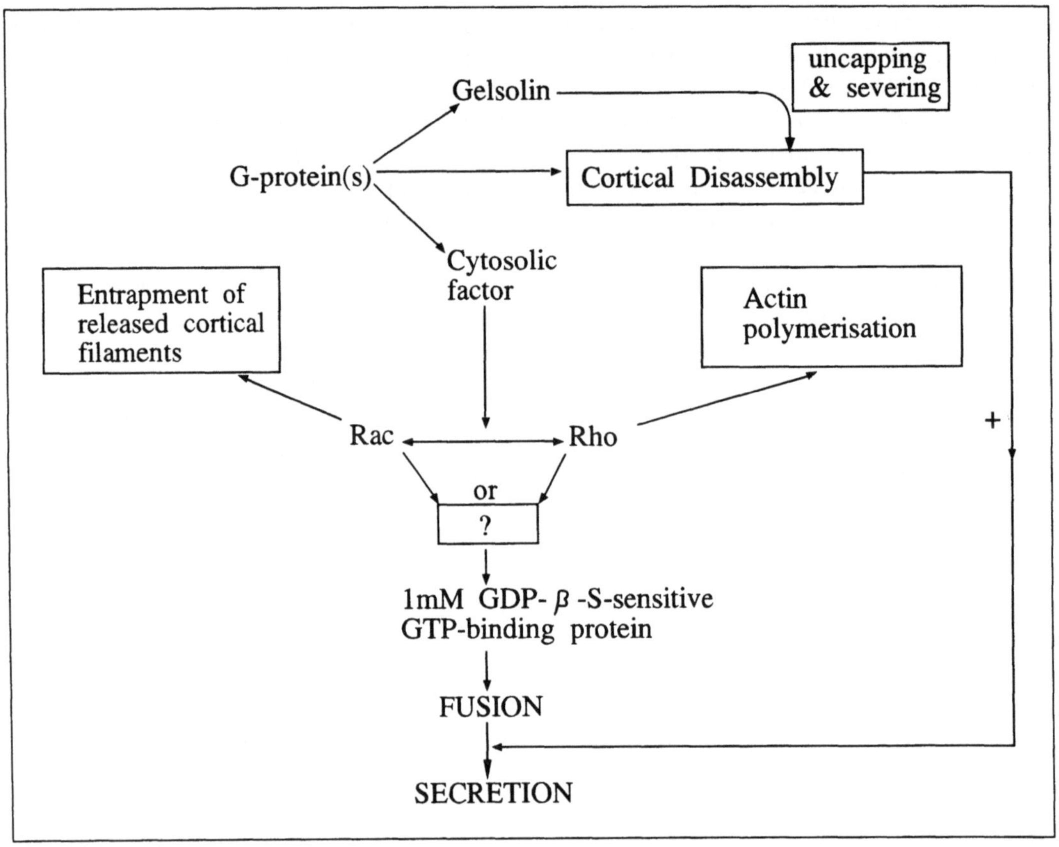

Fig. 5.9. Model showing interactions between four GTP-binding proteins which regulate cytoskeleton and exocytosis in mast cells. An AlF_4^- sensitive G protein leads to disassembly of cortical actin filaments. The disassembly facilitates secretion by allowing an increased number of granules to be recruited for exocytosis, but is not necessary for the fusion of (presumably docked) granules with the plasma membrane. Activation of Rac and Rho causes entrapment of released cortical filaments and de novo actin polymerization, respectively. The activity of either Rac or Rho is also necessary for exocytotic fusion. The pathways from Rac and Rho to the cytoskeleton and to secretion are probably divergent. Exocytosis requires the activity of another GTP-binding protein that is more sensitive to GDP-β-S than Rho, Rac and the AlF_4^- sensitive G protein. A cytosolic factor mediates communication between the G protein and Rac and Rho. The model is based on refs. 32, 33, 40 and on a Norman et al, unpublished observations.

cells to understanding the response to FcεRI activation? A nonselective activation of all GTP- or calcium-binding proteins will clearly produce a response different from that to the finely controlled physiological stimulus. However, studies with the permeabilized cells, in which conditions can be controlled and defined, have opened some pressing and clearly focused questions, particularly whether receptor crosslinking induces activation of Rac, Rho and G_{i3}.

The cytoskeletal responses to FcεRI activation include an increase in the total content of F-actin,[168] cytochalasin-insensitive association of the receptor with the detergent insoluble cytoskeleton[171] and formation of ruffles and plaques.[168,172] These changes are directed at the cell periphery and the appearance of any internal structures has not been reported. The response is dependent on both PKC and on PTKs, particularly Syk,[168,169,173] but cannot be mimicked simply by PKC activation.[172]

Thus, the characteristics are different from the responses of permeabilized cells to GTP-γ-S, which are apparently independent of kinases and do not include a formation of plaques.[32] It is interesting to note that the response of RBL-2H3 cells to adenosine receptor activation, which is Ptx sensitive, results in cytoskeletal changes in a PKC and PTK independent manner.[170] This type of response, known to potentiate secretion from RBL-2H3 cells,[11] may therefore be more similar to both the 48/80 induced response of intact cells and GTP-γ-S induced response of permeabilized cells. At this stage, it is very important to establish the relative positions of Rac, Rho and G_{i3} in signaling pathways from FcεRI and from other mast cell receptors. Figure 5.10, a purely speculative scheme, shows various pathways that need to be considered and tested.

Again, as is the case with other cellular functions, multiple levels of control over actin cytoskeleton are exerted by various receptors and stimuli. Specific GTP-binding proteins, activated differentially by different receptors, seem to contribute to the regulation of specific cell functions. In accordance with the state of play in the rest of the signaling network, their

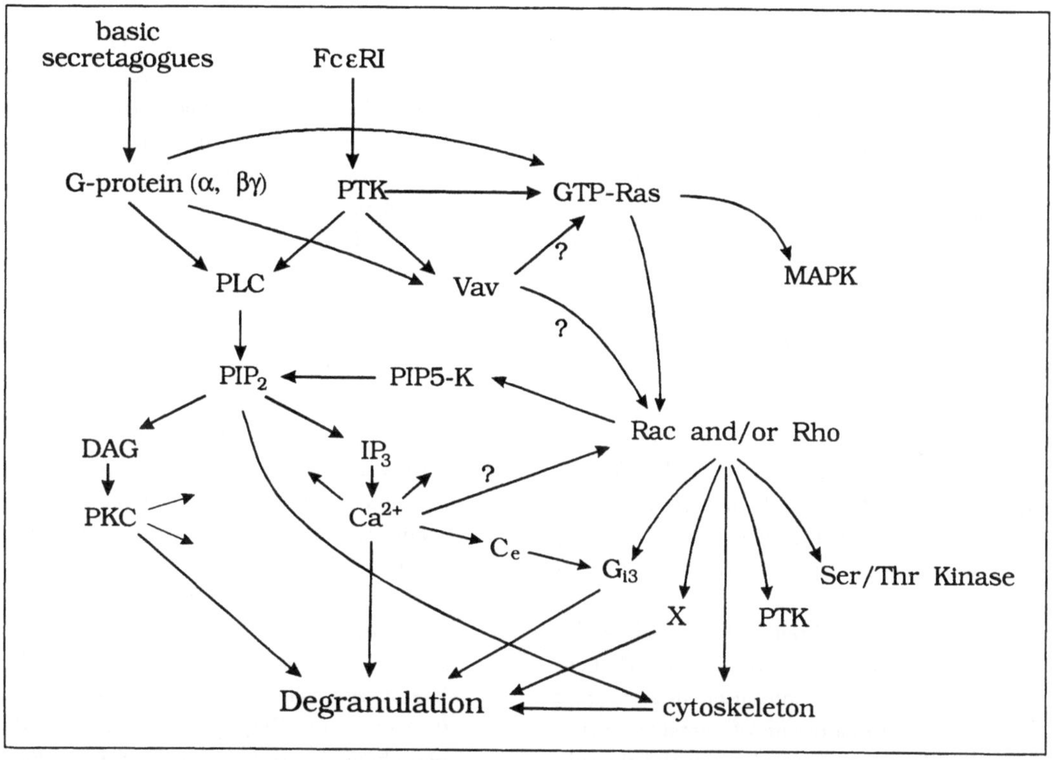

Fig. 5.10. Rac and/or Rho and their possible participation in the secretory response to FcεRI activation. Receptor activation leads to stimulation of Ras, PLC and Vav (via a PTK and/or α or βγ subunits of a G protein). GTP-Ras initiates a cascade of kinases leading to MAPK, that may exert modulatory effects on degranulation, and also activates Rac. PLC activation leads to calcium mobilization and activation of PKC that may induce degranulation independently of GTP-binding proteins. Changes in PIP_2 levels exert influence on the cytoskeleton via PIP_2 regulated actin-binding proteins. Both PKC and calcium regulate many other components of the signaling pathways, including Rac and/or Rho and G_{i3}. Vav may activate Ras as well as Rac and/or Rho dependent pathways. Rac and Rho control Ser/Thr and Tyr kinases, cytoskeletal reorganization, direct or indirect activation of G_{i3} (?), and activation of phosphoinositide kinases that regulate PIP_2 levels. The model is speculative and includes most of the points discussed in this chapter. The relationship between Rac and Rho or that between G_{i3} and Rac/Rho is not yet understood, although in this scheme G_{i3} is placed downstream of the small GTPases.

effects are either amplified or attenuated. The challenge is to understand this network in a broader, more integrated sense and to answer the questions such as how, when and why are the alternative pathways recruited and how is a specificity of response achieved.

References

1. Park DJ, Min HK, Rhee SG. IgE-induced tyrosine phosphorylation of phospholipase C-gamma1 in rat basophilic leukemia cells. J Biol Chem 1991; 266:24237-40.
2. Deanin GG, Martinez AM, Pfeiffer JR et al. Tyrosine kinase-dependent phosphatidylinostiol turnover and functional responses in the FcεR1 signaling pathway. Biochem Biophys Res Commun 1991; 179:551-7.
3. Beaven MA, Metzger H. Signal transduction by Fc receptors: the FcεRI case. Immunol Today 1993; 14:222-6.
4. DeFranco AL. Transmembrane signaling by antigen receptors of B and T lymphocytes. Current Biology 1995; 7:163-75.
5. Wilson BS, Deanin GG, Standefer JC et al. Depletion of guanine nucleotides with mycophenolic acid suppresses IgE receptor-mediated degranulation in rat basophilic leukemia cells. J Immunol 1989; 143: 259-65.
6. Casey PJ. Lipid modification of G proteins. Curr Opin Cell Biol 1994; 6:219-25.
7. Deanin GG, Cutts JL, Pfeiffer JR et al. Role of isoprenoid metabolism in IgE receptor-mediated signal transduction. J Immunol 1991; 146:3528-35.
8. Shakarjian MP, Eiseman E, Penhallow RC et al. 3-Hydroxy-3-methylglutaryl-coenzyme A reductase inhibition in a rat mast cell line. Impairment of tyrosine kinase-dependent signal transduction and the subsequent degranulation response. J Biol Chem 1993; 268:15252-9.
9. Saito H, Okajima F, Molski TFP et al. Effects of ADP-ribosylation of GTP-binding protein by pertussis toxin on immunoglobulin E-dependent and -independent histamine release from mast cells and basophils. J Immunol 1987; 138:3927-34.
10. Warner JA, Yancey KB, MacGlashan DW. The effect of pertussis toxin on mediator release from human basophils. J Immunol 1987; 139:161-5.
11. Ali H, Cunho-Melo JR, Beaven MA. Activation of phospholipase C via adenosine receptors provides synergistic signals for secretion in antigen stimulated RBL-2H3 cells. Evidence for a novel adenosine receptor. J Biol Chem 1990; 265:745-53.
12. McCloskey MA. Cholera toxin potentiates IgE-coupled inositol phospholipid hydrolysis and mediator secretion by RBL-2H3 cells. Proc Natl Acad Sci USA 1988; 85:7260-4.
13. Narasimhan V, Holowka D, Fewtrell C et al. Cholera toxin increases the rate of antigen-stimulated calcium influx in rat basophilic leukemia cells. J Biol Chem 1988; 263:19626-32X.
14. Hide M, Ali H, Price SR et al. GTP-binding protein G_E{aZ}: Its down-regulation by dexamethasone and its credentials as a mediator of antigen-induced responses in RBL-2H3 cells. Mol Pharmacol 1991; 40:473-9.
15. Aridor M, Rajmilevich G, Beaven MA et al. Activation of exocytosis by the heterotrimeric G protein Gi3. Science 1993; 262:1569-72.
16. Gomperts BD. Involvement of guanine nucleotide-binding proteins in the gating of Ca^{2+} by receptors. Nature 1983; 306:64-6.
17. Fernandez JM, Neher E, Gomperts BD. Capacitance measurements reveal stepwise fusion events in degranulating mast cells. Nature 1984; 312:453-5.
18. Cockcroft S, Howell TW, Gomperts BD. Two G proteins act in series to control stimulus-secretion coupling in mast cells: Use of neomycin to distinguish between G proteins controlling polyphosphoinositide phosphodiesterase and exocytosis. J Cell Biol 1987; 105:2745-50.
19. Howell TW, Cockcroft S, Gomperts BD. Essential synergy between Ca^{2+} and guanine nucleotides in exocytotic secretion from permeabilized mast cells. J Cell Biol 1987; 105:191-7.
20. Ali H, Collado-Escobar DM, Beaven MA. The rise in concentration of free Ca^{2+} and of pH provides sequential, synergistic signals

for secretion in antigen-stimulated RBL-2H3 cells. J Immunol 1989; 143:2626-33.
21. Luini A, De Matteis MA. Evidence that receptor-linked G protein inhibits exocytosis by a post-second-messenger mechanism in AtT-20 cells. J Neurochem 1990; 54:30-8.
22. Buccione R, Di Tullio G, Caretta M et al. Analysis of protein kinase C requirement for exocytosis in permeabilized RBL-2H3 cells: a GTP-binding protein(s) as a potential target for protein kinase C. Biochem J 1994; 298:149-56.
23. Ishizaka T. Role of GTP-binding protein in histamine release from mast cells. Clin Immunol Immunopathol 1989; 50:20-9.
24. Ali H, Cunha-Melo JR, Beaven MA. Receptor-mediated release of inositol 1,4,5-trisphosphate and inositol 1,4-bisphosphate in rat basophilic leukemia RBL-2H3 cells permeabilized with streptolysin O. Biochim Biophys Acta 1989; 1010:88-99.
25. Aridor M, Traub LM, Sagi-Eisenberg R. Exocytosis in mast cells by basic secretagogues: Evidence for direct activation of GTP-binding proteins. J Cell Biol 1990; 111:909-17.
26. Churcher Y, Allan D, Gomperts BD. Relationship between arachidonate generation and exocytosis in permeabilized mast cells. Biochem J 1990; 266:157-63.
27. Okano Y, Yamada K, Yano K et al. Guanosine 5'-(gamma-thio)triphosphate stimulates arachidonic acid liberation in permeabilized rat peritoneal mast cells. Biochem Biophys Res Commun 1987; 145:1267-75.
28. Truett AP, Snyderman R, Murray JJ. Stimulation of phosphorylcholine turnover and diacylglycerol production in human polymorphonuclear leukocytes. Novel assay for phosphorylcholine. Biochem J 1989; 260:909-13.
29. Hirasawa N, Santini F, Beaven MA. Activation of the mitogen-activated protein kinase/cytosolic PLA2 pathway in a rat mast cell line. J Immunol 1995; 154:5391-402.
30. Fasolato C, Hoth M, Penner R. A GTP-dependent step in the activation mechanism of capacitative calcium influx. J Biol Chem 1993; 268:20737-40.
31. Qian YX, McCloskey MA. Activation of mast cell K^+ channels through multiple G protein-linked receptors. Proc Natl Acad Sci USA 1993; 90:7844-8.
32. Norman JC, Price LS, Ridley AJ et al. Actin filament organization in activated mast cells is regulated by heterotrimeric and small GTP-binding proteins. J Cell Biol 1994; 126(4):1005-15.
33. Borovikov YS, Norman JC, Price LS et al. Secretion from permeabilized mast cells is enhanced by addition of gelsolin: contrasting effects of endogenous gelsolin. J Cell Sci 1995; 108:657-66.
34. Tatham PER, Gomperts BD. In: Siddle K et al, eds. Peptide Hormones-A Practical Approach. Oxford: IRL Press, 1990:257-69.
35. Koffer A, Gomperts BD. Soluble proteins as modulators of the exocytotic reaction of permeabilized rat mast cells. J Cell Sci 1989; 94:585-91.
36. Churcher Y, Gomperts BD. ATP dependent and ATP independent pathways of exocytosis revealed by interchanging glutamate and chloride as the major anion in permeabilized mast cells. Cell Regul 1990; 1:337-46.
37. Koffer A. Calcium-induced secretion from permeabilized rat mast cells: requirements for guanine nucleotides. Biochim Biophys Acta 1993; 1176:236-44.
38. Koffer A, Churcher Y. Calcium and GTP-gamma-S as single effectors of secretion from permeabilized rat mast cells: requirements for ATP. Biochim Biophys Acta 1993; 1176:222-30.
39. Lillie THW, Gomperts BD. Guanine nucleotide is essential, Ca^{2+} is a modulator, in the exocytotic reaction of permeabilized rat mast cells. Biochem J 1992; 288:181-7.
40. Price LS, Norman JC, Ridley AJ et al. Small GTPases, rac and rho, as regulators of secretion in mast cells. Current Biology 1995; 5(1):68-73.
41. Liao F, Shin HS, Rhee SG. Tyrosine phosphorylation of phospholipase C-gamma1 induced by crosslinking of the high affinity or low-affinity Fc receptor for IgG in U937 cells. Proc Nat Acad Sci USA 1992; 89:3659-63.

42. Lee SB, Rhee SG. Significance of PIP2 hydrolysis and regulation of phospholipase C isozymes. Curr Opin Cell Biol 1995; 7:183-9.
43. Takai Y, Kaibuchi K, Kikuchi A et al. Small GTP-binding proteins. Int Rev Cytol 1992; 133:187-230.
44. Hall A. Ras-related proteins. Curr Opin Cell Biol 1993; 5:265-8.
45. Kahn RA, der CJ, Bokoch GM. The ras superfamily of GTP-binding proteins: guidelines on nomenclature [news]. FASEB J 1992; 6:2512-3.
46. Boguski MS, McCormick F. Proteins regulating ras and its relatives. Nature 1993; 366:643-53.
47. Mayer BJ, Baltimore D. Signalling through SH2 and SH3 domains. Trends in Cell Biol 1993; 3:8-13.
48. Harlan JE, Hajduk PJ, Yoon HS et al. Pleckstrin homology domains bind to phosphatidylinositol-4,5-bisphosphate. Nature 1994; 371:168-70.
49. Quilliam LA, Khosravi FR, Huff SY et al. Guanine nucleotide exchange factors: activators of the Ras superfamily of proteins. Bio Essays 1995; 17:395-404.
50. McCormick F. How receptors turn ras on. Nature 1993; 363:15-7.
51. Rozakis AM, Fernley R, Wade J et al. The SH2 and SH3 domains of mammalian Grb2 couple the EGF receptor to the Ras activator mSos. Nature 1993; 363:83-5.
52. Gale NW, Kaplan S, Lowenstein EJ et al. Grb2 mediates the EGF-dependent activation of guanine nucleotide exchange on Ras. Nature 1993; 363:88-92.
53. Li N, Batzer A, Daly R et al. Guanine-nucleotide-releasing factor hSos1 binds to Grb2 and links receptor tyrosine kinases to Ras signalling. Nature 1993; 363:85-8.
54. Smit L, de Vries Smits AM, Bos JL et al. B cell antigen receptor stimulation induces formation of a Shc-Grb2 complex containing multiple tyrosine-phosphorylated proteins. J Biol Chem 1994; 269:20209-12.
55. Saxton TM, van Oostveen I, Bowtell D et al. B cell antigen receptor crosslinking induces phosphorylation of the p21ras oncoprotein activators SHC and mSOS1 as well as assembly of complexes containing SHC, GRB-2, mSOS1, and a 145 kDa tyrosine-phosphorylated protein. J Immunol 1994; 153:623-36.
56. Ravichandran KS, Lee KK, Songyang Z et al. Interaction of Shc with the zeta chain of the T cell receptor upon T cell activation. Science 1993; 262:902-5.
57. Buday L, Egan SE, Rodriguez Viciana P et al. A complex of Grb2 adaptor protein, Sos exchange factor, and a 36 kDa membrane-bound tyrosine phosphoprotein is implicated in ras activation in T cells. J Biol Chem 1994; 269:9019-23.
58. Sieh M, Batzer A, Schlessinger J et al. GRB2 and phospholipase C-gamma 1 associate with a 36- to 38-kilodalton phosphotyrosine protein after T cell receptor stimulation. Mol Cell Biol 1994; 14:4435-42.
59. Reif K, Buday L, Downward J et al. SH3 domains of the adapter molecule Grb2 complex with two proteins in T cells: the guanine nucleotide exchange protein Sos and a 75 kDa protein that is a substrate for T cell antigen receptor-activated tyrosine kinases. J Biol Chem 1994; 269:14081-7.
60. Downward J, Graves JD, Warne PH et al. Stimulation of $p21^{ras}$ upon T cell activation. Nature 1990; 346:719-23.
61. Izquierdo M, Downward J, Graves JD et al. Role of protein kinase C in T cell antigen receptor regulation of p21ras: evidence that two p21ras regulatory pathways coexist in T cells. Mol Cell Biol 1992; 12:3305-12.
62. Turner H, Reif K, Rivera J et al. Regulation of the adapter molecule Grb2 by the FcεR1 in the mast cell line RBL-2H3. J Biol Chem 1995; 270:9500-6.
63. Bar-Sagi D, Gomperts BD. Stimulation of exocytotic degranulation by microinjection of the *ras* oncogenic protein into rat mast cells. Oncogene 1988; 3:463-9.
64. Margolis B, Hu P, Katzav S et al. Tyrosine phosphorylation of vav proto-oncogene product containing SH2 domain and transcription factor motifs. Nature 1992; 356:71-4.
65. Katzav S, Martin ZD, Barbacid M. *vav*, a novel human oncogene derived from a locus ubiquitously expressed in hematopoietic cells. EMBO J 1989; 8:2283-90.
66. Adams JM, Houston H, Allen J et al. The hematopoietically expressed *vav* proto-

oncogene shares homology with the dbl GDP-GTP exchange factor, the bcr gene and a yeast gene (CDC24) involved in cytoskeletal organization. Oncogene 1992; 7:611-8.
67. Puil L, Pawson T. Vagaries of Vav. Current Biology 1992; 2(5):275-7.
68. Bustelo XR, Ledbetter JA, Barbacid M. Product of vav proto-oncogene defines a new class of tyrosine protein kinase substrates. Nature 1992; 356:68-71.
69. Hirasawa N, Scharenberg A, Yamamura H et al. A requirement for Syk in the activation of the microtubule-associated protein kinase/phospholipase A2 pathway by FcεRI is not shared by a G protein-coupled receptor. J Biol Chem 1995; 270:10960-7.
70. Gulbins E, Coggeshall KM, Baier G et al. Tyrosine kinase-stimulated guanine nucleotide exchange activity of Vav in T cell activation. Science 1993; 260:822-5.
71. Gulbins E, Langlet C, Baier G et al. Tyrosine phosphorylation and activation of Vav GTP/GDP exchange activity in antigen receptor-triggered B cells. J Immunol 1994; 152:2123-9.
72. Gulbins E, Coggeshall KM, Baier G et al. Direct stimulation of Vav guanine nucleotide exchange activity for Ras by phorbol esters and diglycerides. Mol Cell Biol 1994; 14:4749-58.
73. Bustelo XR, Suen KL, Leftheris K et al. Vav cooperates with Ras to transform rodent fibroblasts but is not a Ras GDP/GTP exchange factor. Oncogene 1994; 9: 2405-13.
74. Khosravi FR, Chrzanowska WM, Solski PA et al. Dbl and Vav mediate transformation via mitogen-activated protein kinase pathways that are distinct from those activated by oncogenic Ras. Mol Cell Biol 1994; 14:6848-57.
75. Kazanietz MG, Bustelo XR, Barbacid M et al. Zinc finger domains and phorbol ester pharmacophore. Analysis of binding to mutated form of protein kinase C zeta and the *vav* and c-raf proto-oncogene products. J Biol Chem 1994; 269:11590-4.
76. Ramos MF, Druker BJ, Fischer S. Vav binds to several SH2/SH3 containing proteins in activated lymphocytes. Oncogene 1994; 9:1917-23.
77. Avruch J, Zhang XF, Kyriakis JM. Raf meets Ras: completing the framework of a signal transduction pathway. Trends Biochem Sci 1994; 19:279-83.
78. Leevers SJ, Paterson HF, Marshall CJ. Requirement for Ras in Raf activation is overcome by targeting Raf to the plasma membrane. Nature 1994; 369:411-4.
79. Tordai A, Franklin RA, Patel H et al. Crosslinking of surface IgM stimulates the Ras/Raf-1/MEK/MAPK cascade in human B lymphocytes. J Biol Chem 1994; 269: 7538-43.
80. Franklin RA, Tordai A, Patel H et al. Ligation of the T cell receptor complex results in activation of the Ras/Raf-1/MEK/ MAPK cascade in human T lymphocytes. J Clin Invest 1994; 93:2134-40.
81. Woodrow M, Clipstone NA, Cantrell D. p21ras and calcineurin synergize to regulate the nuclear factor of activated T cells. J Exp Med 1993; 178:1517-22.
82. Santini F, Beaven MA. Tyrosine phosphorylation of a mitogen-activated protein kinase-like protein occurs at a late step in exocytosis. Studies with tyrosine phosphatase inhibitors and various secretagogues in rat RBL-2H3 cells. J Biol Chem 1993; 268:22716-22.
83. Offermanns S, Jones SV, Bombien E et al. Stimulation of mitogen-activated protein kinase activity by different secretory stimuli in rat basophilic leukemia cells. J Immunol 1994; 152:250-61.
84. Fruman DA, Bierer BE, Benes JE et al. The complex of FK506-binding protein 12 and FK506 inhibits calcineurin phosphatase activity and IgE activation-induced cytokine transcripts, but not exocytosis, in mouse mast cells. J Immunol 1995; 154:1846-51.
85. Franklin RA, Tordai A, Patel H et al. Ligation of the T cell receptor complex results in activation of the Ras/Raf-1/MEK/MAPK cascade in human T lymphocytes. J Clin Invest 1994; 93:2134-40.
86. Siegel JN, June CH, Yamada H et al. Rapid activation of C-Raf-1 after stimulation of the T cell receptor or the muscarinic recep-

tor type 1 in resting T cells. J Immunol 1993; 151:4116-27.

87. Burgering BM, Bos JL. Regulation of Ras-mediated signalling: more than one way to skin a cat. Trends Biochem Sci 1995; 20:18-22.
88. Crespo P, Xu N, Simonds WF, Gutkind JS. Ras-dependent activation of MAP kinase pathway mediated by G protein beta gamma subunits. Nature 1994; 369:418-20.
89. Egan SE, Weinberg RA. The pathway to signal achievement. Nature 1993; 365: 781-3.
90. Tanaka S, Morishita T, Hashimoto Y et al. C3G, a guanine nucleotide-releasing protein expressed ubiquitously, binds to the Src homology 3 domains of CRK and GRB2/ASH proteins. Proc Natl Acad Sci USA 1994; 91:3443-7.
91. Miki H, Miura K, Matuoka K et al. Association of Ash/Grb-2 with dynamin through the Src homology 3 domain. J Biol Chem 1994; 269:5489-92.
92. Schlaepfer DD, Hanks SK, Hunter T et al. Integrin-mediated signal transduction linked to Ras pathway by GRB2 binding to focal adhesion kinase. Nature 1994; 372:786-91.
93. Moran M, Polakis P, McCormick F et al. Protein-tyrosine kinases regulate the phosphorylation, protein interactions, subcellular distribution and activity of p21 ras GTPase-activating protein. Mol Cell Biol 1991; 11:1804-12.
94. Serth J, Weber W, Frech M et al. Binding of the H-ras p21 GTPase activating protein by the activated epidermal growth factor receptor leads to inhibition of the p21 GTPase activity in vitro. Biochemistry 1992; 31:6361-5.
95. Settleman J, Albright CF, Foster LC et al. Association between GTPase activators for Rho and Ras families. Nature 1992; 359:153-4.
96. Ridley AJ, Self AJ, Kasmi F et al. Rho family GTPase activating proteins p190, bcr and rhoGAP show distinct specificities in vitro and in vivo. EMBO J 1993; 12: 5151-60.
97. McGlade J, Brunkhorst B, Anderson D et al. The N-terminal region of GAP regulates cytoskeletal structure and cell adhesion. EMBO J 1993; 12:3073-81.
98. Lazarus AH, Kawauchi K, Rapoport MJ et al. Antigen-induced B lymphocyte activation involves the p21ras and ras GAP signaling pathway. J Exp Med 1993; 178: 1765-9.
99. Gold MR, Crowley MT, Martin GA et al. Targets of B lymphocyte antigen receptor signal transduction include the p21 ras GAP and two GAP-associated proteins. J Immunol 1993; 150:377-86.
100. Gold MR, Crowley MT, Martin GA et al. Targets of B lymphocyte antigen receptor signal transduction include the p21ras GTPase-activating protein (GAP) and two GAP-associated proteins. J Immunol 1993; 150:377-86.
101. Rodriguez-Viciana P, Warne PH, Dhand R et al. Phosphatidylinositol-3-OH kinase as a direct target of Ras. Nature 1994; 370:527-32.
102. Stephens L, Jackson T, Hawkins PT. Synthesis of phosphatidylinositol 3,4,5-trisphosphate in permeabilized neutrophils regulated by receptors and G proteins. J Biol Chem 1993; 268:17162-72.
103. Gold MR, Chan VW, Turck CW et al. Membrane Ig crosslinking regulates phosphatidylinositol 3-kinase in B lymphocytes. J Immunol 1992; 148:2012-22.
104. Cantrell DA, Izquierdo M, Reif K et al. Regulation of PtdIns-3-kinase and the guanine nucleotide binding proteins p21ras during signal transduction by the T cell antigen receptor and the interleukin-2 receptor. Semin Immunol 1993; 5:319-26.
105. Yano H, Nakanishi S, Kimura K et al. Inhibition of histamine secretion by wortmannin through the blockade of phosphatidylinositol 3-kinase in RBL-2H3 cells. J Biol Chem 1993; 268:25846-56.
106. Barker SA, Caldwell KK, Hall A et al. Wortmannin blocks lipid and protein kinase activities associated with PI 3-kinase and inhibits a subset of responses induced by FcεR1 receptor crosslinking. Mol Biol Cell 1995; 6:1145-58.
107. McGlade CJ, Ellis C, Reedijk M et al. SH2 domains of the p85 alpha subunit of phosphatidylinositol 3-kinase regulate bind-

ing to growth factor receptors. Mol Cell Biol 1992; 12:991-7.
108. Van Horn DJ, Myers Jr MG, Backer JM. Direct activation of the phosphatidylinositol 3'-kinase by the insulin receptor. J Biol Chem 1994; 269:29-32.
109. Reif K, Gout I, Waterfield MD et al. Divergent regulation of phosphatidylinositol 3-kinase P85 alpha and P85 beta isoforms upon T cell activation. J Biol Chem 1993; 268:10780-8.
110. Weng WK, Jarvis L, LeBien TW. Signaling through CD19 activates Vav/mitogen-activated protein kinase pathway and induces formation of a CD19/Vav/phosphatidylinositol 3-kinase complex in human B cell precursors. J Biol Chem 1994; 269:32514-21.
111. Zhang J, King WG, Dillons S et al. Activation of platelet phosphatidylinositide 3-kinase requires the small GTP-binding protein rho. J Biol Chem 1993; 268:22251-4.
112. Zheng Y, Bagrodia S, Cerione RA. Activation of phosphoinositide 3-kinase activity by Cdc42Hs binding to p85. J Biol Chem 1994; 269:18727-30.
113. Wennstrom S, Hawkins P, Cooke F et al. Activation of phosphoinositide 3-kinase is required for PDGF-stimulated membrane ruffling. Current Biol 1994; 4:385-93.
114. Kumagai N, Morii N, Fujisawa K et al. ADP-ribosylation of rho p21 inhibits lysophosphatidic acid-induced protein tyrosine phosphorylation and phosphatidylinositol 3-kinase activation in cultured Swiss 3T3 cells. J Biol Chem 1993; 268:24535-8.
115. Hawkins PT, Eguinoa A, Qui R et al. PDGF stimulates an increase in GTP-Rac via the activation of phosphoinositide 3OH kinase. Current Biol 1995; 5:393-400.
116. Stephens L, Smrcka A, Cooke FT et al. A novel phosphoinositide 3 kinase activity in myeloid-deprived cells is activated by G protein gamma subunits. Cell 1994; 77:83-93.
117. Spaargaren M, Bischoff JR. Identification of the guanine nucleotide dissociation stimulator for Ral as a putative effector molecule of R-ras, H-ras, K-ras, and Rap. Proc Natl Acad Sci USA 1994; 91:12609-13.
118. Hofer F, Fields S, Schneider C et al. Activated Ras interacts with the Ral guanine nucleotide dissociation stimulator. Proc Natl Acad Sci USA 1994; 91:11089-93.
119. Russel M, Lange Carter CA, Johnson GL. Direct interaction between Ras and the kinase domain of Mitogen-Activated Protein Kinase Kinase Kinase (MEKK1). J Biol Chem 1995; 270:11757-60.
120. Nassar N, Horn G, Herrmann C et al. The 2.2 A crystal structure of the Ras-binding domain of the ser/threo kinase c-Raf1 in complex with Rap1A and a GTP analogue. Nature 1995; 375:554-60.
121. Higashijima T, Uzu S, Nakajima T et al. A peptide toxin from wasp venom mimics receptors by activating GTP-regulatory proteins (G proteins). J Biol Chem 1988; 263:6491-4.
122. Mousli M, Bronner C, Landry Y et al. Direct activation of GTP-binding regulatory proteins (G proteins) by substance P and compound 48/80. FEBS Lett 1990; 259: 260-2.
123. Swieter M, Midura RJ, Nishikata H et al. Mouse 3T3 fibroblasts induce rat basophilic leukemia (RBL-2H3) cells to acquire responsiveness to compound 48/80. J Immunol 1993; 150:617-24.
124. Balch WE. Small GTP-binding proteins in vesicular transport. Trends in Biochem Sci 1990; 15:473-7.
125. Oberhauser AF, Balan V, Fernandez Badilla CL et al. RT-PCR cloning of Rab3 isoforms expressed in peritoneal mast cells. FEBS Lett 1994; 339:171-4.
126. Izushi K, Shirasaka T, Chokki M et al. Phosphorylation of smg p21B in rat peritoneal mast cells in association with histamine release inhibition by dibutyryl-cAMP. FEBS Lett 1992; 314:241-5.
127. Oberhauser AF, Monck JR, Balch WE et al. Exocytotic fusion is activated by Rab3a peptides. Nature 1992; 360:270-3.
128. MacLean CM, Law GJ, Edwardson JM. Stimulation of exocytotic membrane fusion by modified peptides of the rab3 effector domain: Re-evaluation of the role of rab3 in regulated exocytosis. Biochem J 1993; 294:325-8.

129. Law GN, AJ, Mason WT. Mastoparan like effects of rab3al peptide. FEBS Lett 1994; 333:56-60.
130. Ludger J, Lledo P, Roa M et al. The GTPase rab3A negatively controls calcium-dependent exocytosis in neuroendocrine cells. EMBO J 1994; 13:2029-37.
131. Holz RW, Brondyk WH, Senter RA et al. Evidence for the involvement of rab3A in Ca^{2+}-dependent exocytosis from adrenal chromaffin cells. J Biol Chem 1994; 269:10229-34.
132. Fujita Y, Sasaki T, Araki K et al. GDP/GTP exchange reaction-stimulating activity of Rabphilin-3A for Rab3A small GTP-binding protein. FEBS Lett 1994; 353: 67-70.
133. Yamaguchi T, Shirataki H, Kishida S et al. Two functionally different domains of rabphilin-3A, Rab3A p25/smg p25A-binding and phospholipid- and $Ca(^{2+})$-binding domains. J Biol Chem 1993; 268:27164-70.
134. Kato M, Sasaki T, Imazumi K et al. Phosphorylation of Rabphilin-3A by calmodulin-dependent protein kinase II. Biochem Biophys Res Commun 1994; 205:1776-84.
135. Numata S, Shirataki H, Hagi S et al. Phosphorylation of Rabphilin-3A, a putative target protein for Rab3A, by cyclic AMP-dependent protein kinase. Biochem Biophys Res Commun 1994; 203:1927-34.
136. Fykse EM, Li C, Sudhof TC. Phosphorylation of rabphilin-3A by Ca^{2+}/calmodulin- and cAMP-dependent protein kinases in vitro. J Neurosci 1995; 15:2385-95.
137. Takai Y, Sasaki T, Tanaka K et al. Rho as a regulator of the cytoskeleton. TIBS 1995; 20:227-231.
138. Ridley AJ, Hall A. The small GTP-binding protein rho regulates the assembly of focal adhesions and actin stress fibers in response to growth factors. Cell 1992; 70:389-99.
139. Ridley AJ, Paterson HF, Johnston CL et al. The small GTP-binding protein rac regulates growth factor-induced membrane ruffling. Cell 1992; 70:401-10.
140. Nobes CD, Hall A. Rho, Rac and Cdc42 regulate the assembly of multimolecular focal complexes associated with actin stress fibers, lamellipodia and filopodia. Cell 1995; 81:53-62.
141. Chong LD, Traynor KA, Bokoch GM et al. The small GTP-binding protein Rho regulates a phosphatidylinositol 4-phosphate 5-kinase in mammalian cells. Cell 1994; 79:507-13.
142. Bowman EP, Uhlinger DJ, Lambeth JD. Neutrophil phospholipase D is activated by a membrane-associated rho family small molecular weight GTP-binding protein. J Biol Chem 1993; 268:21509-215-12.
143. Malcolm CK, Ross AH, Qiu RG et al. Activation of rat liver phospholipase D by the small GTP-binding protein RhoA. J Biol Chem 1994; 269(42):25951-4.
144. Lassing I, Lindberg U. Evidence that the phosphatidylinositol cycle is linked to cell motility. Exp Cell Res 1988; 174:1-15.
145. Peppelenbosch MP, Qiu RG, de Vries Smits AM et al. Rac mediates growth factor-induced arachidonic acid release. Cell 1995; 81:849-56.
146. Ridley AJ, Hall A. Signal transduction pathways regulating Rho-mediated stress fibre formation: requirement for a tyrosine kinase. EMBO J 1994; 13:2600-10.
147. Morel F, Doussiere J, Vignais PV. The superoxide-generating oxidase of phagocytic cells. Physiological, molecular and pathological aspects. Eur J Biochem 1991; 201:523-46.
148. Seckl MJ, Morii N, Narumiya S et al. Guanosine 5'-3-O-(thio)triphosphate stimulates tyrosine phosphorylation of p125FAK and paxillin in permeabilized Swiss 3T3 cells. Role of p21rho. J Biol Chem 1995; 270:6984-90.
149. Richardson A, Parsons JT. Signal transduction through integrins: a central role for focal adhesion kinase? Bio Essays 1995; 17(3):229-236.
150. Hamawy MM, Swaim WD, Minoguchi K et al. The aggregation of the high affinity IgE receptor induces tyrosine phosphorylation of paxillin, a focal adhesion protein. J Immunol 1994; 153:4655-62.
151. Minoguchi K, Kihara H, Nishikata H et al. Src family tyrosine kinase Lyn binds several proteins including paxillin in rat basophilic leukemia cells. Mol Immunol 1994; 31:519-29.

152. Thomas SM, Soriano P, Imamoto A. Specific and redundant roles of Src and Fyn in organizing the cytoskeleton. Nature 1995; 376:267-271.
153. Rankin S, Morii N, Narumiya S et al. Botulinum C3 exoenzyme blocks the tyrosine phosphorylation of p125FAK and paxillin induced by bombesin and endothelin. FEBS Lett 1994; 354:315-9.
154. Turner CE, Miller JT. Primary sequence of paxillin contains putative SH2 and SH3 domain binding motifs and multiple LIM domains: identification of a vinculin and $pp125^{FAK}$-binding region. J Cell Sci 1994; 107:1583-91.
155. Abo A, Boyhan A, West I et al. Reconstitution of neutrophil NADPH oxidase activity in the cell free system by four components: p67-phox, p47-phox, p21-rac1 and cytochrome b245. J Biol Chem 1992; 267:16767-70.
156. Rotrosen D, Yeung CL, Leto TL et al. Cytochrome b558: the flavin-binding component of the phagocyte NADPH oxidase. Science 1992; 256:1459-62.
157. Bokoch-GM. Regulation of the phagocyte respiratory burst by small GTP-binding proteins. Trends Cell Biol 1995; 5:109-13.
158. Heyworth PG, Bohl BP, Bokoch GM et al. Rac translocates independently of the neutrophil NADPH oxidase components p47phox and p67phox. Evidence for its interaction with flavocytochrome b558. J Biol Chem 1994; 269:30749-52.
159. Diekmann D, Abo A, Johnston C et al. Interaction of Rac with p67phox and regulation of phagocytic NADPH oxidase activity. Science 1994; 265:531-3.
160. Manser E, Leung T, Salihuddin H et al. A brain serine/threonine protein kinase activated by cdc42 and rac1. Nature 1994; 367:40-6.
161. Kishi K, Sasaki T, Kuroda S et al. Regulation of cytoplasmic division of Xenopus embryo by rhop21 and its inhibitory GDP/GTP exchange protein (rho GDI). J Cell Biol 1993; 120:1187-95.
162. Hirata K, Kikuchi A, Sasaki T et al. Involvement of rho p21 in the GTP enhancement of calcium sensitivity in smooth muscle contraction. J Biol Chem 1992; 267:8719-22.
163. Jalink K, van Corven EJ, Hengeveld T et al. Inhibition of lysophosphatidate- and thrombin-induced neurite retraction and neuronal cell rounding by ADP ribosylation of the small GTP-binding protein Rho. J Cell Biol 1994; 126:801-10.
164. Vojtek AB, Cooper JA. Rho family members: Activators of MAP kinase cascades. Cell 1995; 82:527-9.
165. Price LS. Ph.D. Thesis 1995, University College London.
166. Koffer A, Tatham PER, Gomperts BD. Changes in the state of actin during the exocytotic reaction of permeabilized rat mast cells. J Cell Biol 1990; 111:919-27.
167. Janmey PA, Iida K, Yin HL et al. Polyphosphoinositide micelles and polyphosphoinositide-containing vesicles dissociate endogenous gelsolin-actin complexes and promote actin assembly from the fast-growing end of actin filaments blocked by gelsolin. J Biol Chem 1987; 262:12228-36.
168. Pfeiffer JR, Seagrave JC, Davis BH et al. Membrane and cytoskeletal changes associated with IgE-mediated serotonin release from rat basophilic leukemia cells. J Cell Biol 1985; 101:2145-55.
169. Apgar J. Regulation of the antigen-induced F-actin response in rat basophilic leukemia cells by protein kinase C. J Cell Biol 1991; 112:1157-63.
170. Apgar JR. Polymerization of actin in RBL-2H3 cells can be triggered through either the IgE receptor or the adenosine receptor but different signaling pathways are used. Mol Biol Cell 1994; 5:313-22.
171. Apgar JR. Association of the crosslinked IgE receptor with the membrane skeleton is independent of the known signaling mechanisms in rat basophilic leukemia cells. Cell Regul 1991; 2:181-91.
172. Pfeiffer JR, Oliver JM. Tyrosine kinase-dependent assembly of actin plaques linking FcεR1-crosslinking to increased cell substrate adhesion in RBL-2H3 tumor mast cells. J Immunol 1994; 152(1):270-9.
173. Oliver JM, Burg DL, Wilson BS et al. Inhibition of mast cell Fc epsilon R1-mediated signaling and effector function by the Syk-selective inhibitor, piceatannol. J Biol Chem 1994; 269:29697-703.

CHAPTER 6

Phosphorylation of Serines and Threonines in FcεRI-Mediated Responses in Mast Cells

Juan Rivera

INTRODUCTION

Cellular ontogeny, proliferation and differentiation are regulated by a complex interplay of extracellular signals and highly ordered intracellular events. The interaction of a ligand with its receptor is critical to the initiation of these processes and results in either movement of this complex into the cell where it directly influences intracellular events or in modification of the receptor and generation of second messengers. The latter process activates effector proteins and this is often accompanied by a shift in the localization of these proteins to a different microenvironment.[1,2] Some of these proteins are protein kinases. Protein kinase families are classified on the basis of the phospho-accepting amino acid residue in the substrate protein: those that phosphorylate serine and threonine residues and those that phosphorylate tyrosines. However some kinases with dual activity have been described.[3] The majority (>90%) of phosphorylations modify serine or threonine residues. In fact, a significant basal level of serine/threonine phosphorylation is found in most cells suggesting a role for these enzymes in homeostasis of the cell. Protein kinases with specificity towards serine/threonine residues are also important in cell proliferation and differentiation.[4] Serine/threonine kinases constitute a large and diverse family of proteins. Their diverse effects are mediated by their discrete specificities for target substrates, their sensitivity to various cofactors and by covalent modification of proteins.

IgE Receptor (FcεRI) Function in Mast Cells and Basophils,
edited by Majed M. Hamawy.

We know little about the physiological relevance of serine and threonine phosphorylation in the development of mast cells and basophils. However, significant gains have been made in understanding the role of serine/threonine phosphorylation in allergen-mediated activation of these cells. The serine/threonine kinases known to be found in mast cells are listed in Table 6.1. Of these the best characterized is the family of protein kinase C (PKC) which plays a central role in the release of histamine and of other mediators of allergic symptoms. Virtually no information is available on the role of calcium-calmodulin dependent protein kinases, cAMP-dependent protein kinases and protein kinase CK2 (casein kinase II), although all of these have been detected in the mast cell. Recent interest in regulation of gene expression in mast cells has advanced the knowledge on the role of mitogen-activated protein (MAP) kinases, also known as extracellular signal regulated (ERK) kinases, in these cells. Furthermore other serine/threonine kinases such as Raf and MEK, shown to be involved in activation of MAP kinase activity, are phosphorylated and presumably activated in response to aggregation of FcεRI.[5]

In this chapter the focus will be primarily on PKC-induced serine/threonine phosphorylation, but I will also discuss the FcεRI-mediated activation of serine/threonine kinases more generally and the role of this enzymatic activity in mast cell responses.

ACTIVATION OF SERINE AND THREONINE KINASES

ACTIVATION OF PKC

Aggregation of the FcεRI generates intracellular second messenger molecules that activate a second wave of protein phosphorylation primarily mediated by serine/threonine kinases. For PKC the activation of phospholipase C-γ (PLC-γ) in mast cells by tyrosine phosphorylation is a critical event in translocating cytosolic

Table 6.1. Serine/threonine kinases in mast cells

Kinase	Cell Detected	Reference
cAMP-PKA[a]	rat serosal mast cells	Holgate et al[18]
Casein Kinase II	rat mast (RBL-2H3)	J Rivera (unpublished)
MAP (p42/p44)	mouse mast cells (BMMC) rat mast (RBL-2H3)	Tsai et al[36] Offermanns et al[30]
MEK (MAPKK)	rat mast (RBL-2H3)	Hirasawa et al[5]
MLCK	rat mast (RBL-2H3)	Ludowyke et al[25]
PI-3 Kinase[a]	rat mast (RBL-2H3)	Barker et al[51]
PKC (α, β, δ, ε, ζ)	rat mast (RBL-2H3)	Ozawa et al[45]
pp90rsk	mouse mast (BMMC)	Tsai et al[36]
pp70-S6	mouse mast (BMMC)	Tsai et al[36]
Raf-1	rat mast (RBL-2H3)	Hirasawa et al[5]

[a] Abbreviations used: cAMP-PKA, cyclic AMP-dependent protein kinase; MAP, mitogen-activated protein kinase; MEK or MAPKK, mitogen-activated protein kinase kinase; MLCK, myosin light chain kinase; PI-3 Kinase, phosphatidylinositide-3 kinase; PKC, protein kinase C; pp90rsk, phosphoprotein 90 ribosomal S6 kinase; pp70-S6, phosphoprotein 70 S6 kinase.

[b] While not the first example of PI-3 Kinase in mast cells, this paper is the first to describe a potential serine/threonine kinase activity for PI-3 kinase in mast cells.

PKC towards the membrane.[2] PLC-γ hydrolyzes phosphatidylinositol-4,5-bisphosphate (PIP_2), generating the potent activator of PKC, 1,2-*sn*-diacylglycerol (DAG). It binds to the C1 domain of PKC which is characterized by highly conserved cysteine-rich repeats found in all isozymes. A second crucial product of PI hydrolysis is inositol-(1,4,5)-trisphosphate (IP_3), a second messenger which binds to receptors on the endoplasmic reticulum that trigger release of Ca^{2+} from these stores. The increase in intracellular calcium is required for the activity of PKC isozymes that contain the C2 domain, a glycine-rich region also found in other Ca^{2+}-dependent proteins such as phospholipase A_2 (PLA_2) and PLC-γ.[6,7] In principle two mechanisms are intertwined in the catalytic activation of PKC by diacylglycerols. The first is that DAG, in part due to its inherent hydrophobicity and its localization to the lipid environment of cellular membranes, interacts only with PKC that is membrane localized. The increase in DAG in the membrane shifts the localization of cellular PKC towards membranes (translocation), a process observed in mast cells.[8,9] The second is the functional role of DAG in activation of PKC. The presence of PKC in the membrane is in itself not sufficient for activation in the absence of DAG production.[10] Therefore, the binding of DAG in the C1 domain of PKC serves to sequester PKC in the membrane but also to activate its catalytic activity. In fact the two domain structure hypothesized for PKC would suggest that binding of DAG to the regulatory domain reduces the affinity of the pseudosubstrate site in the regulatory domain for the C-terminal catalytic domain.[11] This activation is mimicked by the DAG analog, phorbol 12-myristate-13-acetate (PMA). The initial rise in DAG in the membrane due to PLCγ activation is transient; however, a sustained activation of PKC is maintained by subsequent activation of PLD and generation of DAG predominantly from phosphatidylcholine.[12-14] Nevertheless, the critical event for generation of DAG and early activation of PKC maybe a transient calcium-independent rise in PI hydrolysis that is detected within 15 seconds of FcεRI aggregation.[15] Alternatively, recent studies suggest that several PKC isoforms can be activated in vitro by PIP_2 or by arachidonic acid, suggesting a need for activation of kinases of the PI pathway or activation of PLA_2 activity, respectively, rather than the production of DAG as critical for activation.[2,16]

Activation of cAMP-Dependent Protein Kinase (PKA)

A rapid rise in cAMP is initiated by the aggregation of the FcεRI on mast cells. The increase in cAMP is independent of extracellular calcium, peaks within 30 seconds after receptors are aggregated and returns to the basal level in 2 minutes.[17] In mast cells, increase in cAMP-dependent protein kinase activity coincides with the rise in cAMP.[18] Increases in cAMP levels and activation of adenylate cyclase seems to have an inhibitory effect on secretion in mast cells,[19,20] by a mechanism that inhibits mobilization of intracellular calcium.[21,22] However, the role for a rise in cAMP-dependent protein kinase activity in mast cell signaling and the intracellular targets of this activity are unknown.

Activation of Protein Kinase CK2

Mast cells express protein kinase CK2 (casein kinase II, Table 6.1). This enzyme, a complex tetrameric holoenzyme ($\alpha_2\beta_2$), can be found in the cytoplasm and nucleus of many cells. It is known to phosphorylate a large number of substrates that are involved in the control of cell division and in signal transduction. In vivo substrates for this enzyme include the protein phosphatase-1 inhibitor-2 (I-2), casein, calmodulin and phosvitin.[23] The activity of the catalytic α subunits is regulated by the association with the regulatory β subunits. The mechanism for its activation, however, is unclear although it is a substrate of the structurally and functionally similar cyclin-dependent protein kinase (p34cdc2).[23] Recent studies suggest that CK2 can be rapidly translocated to the nuclear matrix in response to a growth stimulus.[24] It remains to be seen whether in mast cells CK2 plays a role in FcεRI-mediated regulation of gene expression.

Activation of Ca^{2+}/Calmodulin-Dependent Protein Kinases

Calcium/calmodulin protein kinase activity has been described in the mast cell analog RBL-2H3.[25,26] Choi et al[26] found phosphorylation of myosin light chains at threonine 18 by the Ca^{2+}/calmodulin-dependent myosin light chain kinase (MLCK) when the FcεRI was aggregated. Calmodulin antagonists have inhibitory effects on histamine release from mast cells. Compounds such as triflouroperazine, W7 and 5-iodo-1-C8 show a dose-dependent inhibition of histamine release from mast cells.[27] While this evidence has been interpreted as indicating a role for Ca^{2+}/calmodulin-dependent protein kinases in secretion, the data should be interpreted cautiously since multiple calmodulin regulated proteins, such as calcineurin,[28] exist in these cells and these antagonists can affect PKC. Recent studies with a new calmodulin inhibitor (CGS 9343B) that has little influence on PKC have shown that low doses enhanced release,[29] while at high doses inhibition was observed. This biphasic effect may reflect the role of multiple proteins whose activities are modulated by calmodulin and are activated in response to aggregation of FcεRI.

Activation of Raf, MEK, MAP, pp90rsk and pp70-S6 Kinases

Phosphorylation of tyrosines and increase of MAP kinase activity in response to aggregation of FcεRI have been described.[30] Activation of MAP kinases requires the activity of Raf and MEK kinases.[31] Recent studies show that Raf 1 and MEK 1 are phosphorylated in response to aggregation of receptors in RBL cells[5] suggesting a similar pathway exists in mast cells. The mechanism by which Raf 1 is activated is a subject of intense investigation. In T cells, MAP kinases are activated when Raf-1 interacts with the membrane localized GTP-bound form of Ras.[32] Activation of the T cell antigen receptor (TCR) on the cell surface results in phosphorylation of the receptor and its association with a variety of molecules; one of which is the adapter molecule Grb2.[32] Grb2 is complexed via its SH3 domain with the guanine nucleotide exchange factor (GEF) Sos. The latter protein facilitates the shift in the Ras:GDP to the Ras:GTP form. In RBL-2H3 cells of all the above proteins are present[5,34] and some exist as multiprotein complexes that are recruited to receptor on antigen aggregation.[35] Furthermore, aggregation of FcεRI may regulate protein-protein interaction[34] and modulate Ras:GDP to Ras:GTP exchange activity (JS Song and J Rivera, unpublished observations). However, whether Ras is responsible for the activation of Raf 1 in mast cells remains unclear. Activation of pp90^{rsk} and pp70-S6 kinases in response to FcεRI has been described in mouse mast cells.[36] The mechanism of this activation is unclear although presumably these kinases may serve as targets for MAP kinase activity.

SEQUENCE MOTIFS IN FcεRI PHOSPHORYLATION

Sequence Targets for Phosphorylation of Receptor

The phosphorylation of the FcεRI-β and γ chains has been well documented.[37,38] In brief, all studies show phosphorylation of the β chain on serine and tyrosine while phosphorylation of the γ chain is on threonine and tyrosine. Tyrosine phosphorylation of both subunits occurs in the immunoreceptor tyrosine-based activation motif (ITAM), originally described by M. Reth,[39] (Table 6.3) although which of the two tyrosines in this motif, if not both, is phosphorylated remains to be defined.

Sequence Targets for Serine Phosphorylation of β Chain

The target serine residue on β chain phosphorylated in response to aggregation of FcεRI is presently unknown. The N-terminal cytoplasmic domain is rich in proline residues and contains serines as potential targets for proline-directed serine/threonine kinase activity (i.e., MAP kinase). However, peptide sequence analysis failed to identify a consensus motif.[40] The C-terminal cytoplasmic domain, which contains the ITAM, is not proline rich and contains no likely sites for PKC phosphorylation. However, sequence analysis reveals a potential site for phosphorylation by protein kinase CK2. The sequence is within the second YXXL motif of the ITAM and has a serine residue in an acidic amino acid environment (S-E/A-L-E-D). Human, rat and murine β chain, contain this consensus sequence (Table 6.2). CK2 phosphorylation sites on the in vivo substrate I-2 (S-G-E-E-D) and on the CK2 β subunit (S-S-E-E-V) also fit the consensus of S-X-X-E/D/Yp/Sp described by Marin et al[41] which requires an acidic residue at the +3 position. Interestingly, this consensus motif is also found on proteins homologous to FcεRI-β chain human and mouse CD20, and human HTm4 (Table 6.2). The presence of this site for CK2 proximal to a potentially phosphorylated tyrosine in the ITAM of the β chain may indicate that phosphorylation of serine regulates the phosphorylation or protein interactions of the ITAM. Whether the FcεRI-β chain is a substrate for CK2 activity in vivo is yet to be determined.

Sequence Targets for Phosphorylation of γ Chain Threonines

The γ chain contains four threonines that are potential targets for phosphorylation (Table 6.3). Since we previously showed that the γ chain is a target for PKC-δ activity (see details in the following section), we have analyzed the γ chain for generic consensus sites (K/R-X-X-S/T) of phosphorylation by PKC. By these criteria only one site was found; however, this site contained a serine therefore negating it as an in vivo target. Since PKC-δ was found to be the only isozyme capable of recognizing the γ chain as a substrate (see details below), we reasoned that a unique consensus site recognizable only to this isozyme might be present. The sequences

Table 6.2. Sequence alignment of a potential protein kinase CK2 phosphorylation site on the C-terminal cytoplasmic domain of FcεRI β chain and homologous proteins[a]

D R L Y E E L N H V Y S P I Y	S E L E D	Mouse β
D R L Y E E L H V Y S P I – Y	S A L E D	Rat β
D R V Y E E L N I Y S A T – Y	S E L E D	Human β
K R M C T R S K S N V V L – L	S A G E K	Mouse CD 20
K R T C S R P K S N I V L – L	S A E E K	Human CD 20
A M W C N A N C C – – – – – N	S R E E I	Human Htm4
	S X X E/D/Y_p/S_p	Consensus[b]

[a] Sequences are from the following sources: For human, rat and mouse β chain; Küster et al (1992), J Biol Chem, 267:12782; Kinet et al (1988), Proc Natl Acad Sci, 85:6483; and Ra et al (1989), J Biol Chem, 264:15323, respectively. For human and mouse CD 20; Tedder et al (1988), Proc Natl Acad Sci, 85:208; Tedder et al (1988), J Immunol, 141:4388, respectively. The human Htm4 sequence, a hematopoietic cell-specific molecule related to CD20 and the FcεRI β chain, was described by Adra et al (1994), Proc Natl Acad Sci, 91:10178.

[b] Consensus sequence is as described by Marin et al.[41] There is an absolute requirement for an acidic amino acid residue at position +3, additional acidic residues in positions +1, +2, +4, increase the phosphorylation rate. Identified physiological substrates have at least one acidic residue in these positions.

of human, rat and murine γ chain ITAM motifs as well as human and mouse ζ_3 chain ITAMs which contain threonine residues were aligned. This analysis revealed three highly conserved threonine residues between the γ and ζ chains (PG Swann and J Rivera, unpublished observations, Table 6.3). In γ chains two of the threonines are preceded by basic amino acids which is a prerequisite for PKC recognition. However, in the ζ_3 ITAM only one of these threonines is preceded by a basic amino acid. It seems reasonable to assume that the conservation might reflect the importance of these two potential phosphorylation sites to γ and ζ chain function, although phosphorylation of serine or threonine on the ζ chain has not been reported. A consensus sequence for the most conserved of the phosphorylation sites is R/N/K-X_n-D/E-T-Y-D/E, where n = 0-3. One obvious characteristic of this sequence is the mixed basic/acidic environment of this threonine residue, an unusual feature for recognition since target sites for PKC are normally found in basic environments. In a recent study Kielbassa et al[42] found that the PKC-δ isoform is uniquely able to phosphorylate weak basic sites. Possibly the specificity of PKC-δ is related to its ability to recognize such unique environments.

PHOSPHORYLATION OF THE FcεRI BY SERINE/THREONINE KINASES

ACTIVATION OF PKC RESULTS IN PHOSPHORYLATION OF FcεRI-γ CHAIN

Treatment of the RBL-2H3 cells with phorbol 12-myristate-13-acetate (PMA) results in the phosphorylation of FcεRI.[43,44] This phosphorylation modifies threonine residue(s) on the γ chain exclusively[44] analagous to the desired in vivo phosphorylation on threonine in response to antigen.[37,38] Adherent cells sensitized with mo-

Table 6.3. Sequence alignment of threonine residues in FcεRI γ and TCR ζ chains[a]

D X X Y X X L $X_{(6-8)}$	-	- - - - -	-	Y X X L	ITAM[39]
D A V Y T G L *n*	T	*r n* Q E	T	Y E T L	Mouse γ
D G L Y Q G L S	T	A T *k* D	T	Y D A L	Mouse ζ3
D A V Y T G L *n*	T	*r n* Q E	T	Y E T L	Rat γ
D G L Y Q G L S	T	A T *k* D	T	Y D A L	Rat ζ3
D G V Y T G L S	T	*r n* Q E	T	Y E T L	Human γ
D G L Y Q G L S	T	A T *k* D	T	Y D A L	Human ζ3
D R V Y E E L---*n*		I Y S A	T	Y S E L	Human β[b]
r/n/k		X_n D/E	T	YD/E	Consensus

[a] Primary sequences were derived from the following sources available through Genbank: For human, rat, and mouse γ sequences; Küster et al (1990), J Biol Chem, 265:6448.; Blank et al (1989), Nature, 337:187; and Ra et al (1989), J Biol Chem, 264:15323; respectively. For human, rat and mouse ζ sequences, Weissman et al (1988), Proc Natl Acad Sci USA, 85:9709; Itoh Y et al (1993), J Immunol, 151:4705; and Weissman et al (1988), Science, 239:1018. ; respectively. The ζ_3 ITAM has the highest homology to the γ ITAM and is the only ζ chain ITAM that contains threonine residues. *Italicized* small lettering indicates positively charged amino acid residues required for potential interaction with PKC. Framed in threonine residues (T) are conserved. Consensus sequence is derived from γ and ζ_3 sequences only.

[b] Human β chain sequence is included in alignment since this threonine located next to the terminal YXXL is conserved. This threonine is not found in rat or mouse β chain. Sequence derived from Küster et al (1992), J Biol Chem, 267:12782.

nomeric immunoglobulin E (IgE), but not stimulated with antigen showed a basal phosphorylation of threonine on the γ chain. Aggregation of the IgE increased the phosphorylation of threonine by an average of 3-fold and was detected within 15 seconds of antigen addition. The PKC specific inhibitor Ro31-7549 at doses that are not selective for individual isozymes of PKC completely inhibited the phosphorylation on threonine of γ without affecting phosphorylation of tyrosine. This compound also inhibited the basal level and PMA-induced phosphorylation of the γ chain. Since no effect on the phosphorylation of the β chain was observed, it appears that PKC activity is not required for serine phosphorylation of the β chain induced by aggregation of FcεRI.

Association of PKC-δ with FcεRI

We investigated whether any of the PKC isozymes found in RBL cells[44] would coimmunoprecipitate with the receptor. Immunoprecipitated receptors immunoblotted with antibodies to PKCs α, β, δ, ε and ζ revealed the presence of PKC-δ.[45] Increased association (1.5- to 2-fold) of PKC-δ with receptor was found in immunoprecipitates derived from stimulated cells. Relative quantitation of immunoblots suggested that 1-2% of the cellular PKC-δ associated with receptor. Glutathione-S-transferase (GST) fusion proteins of receptor cytoplasmic domains revealed the presence of PKC-δ with the C-terminal cytoplasmic domain of the β chain (βc).[44] Recent experiments with recombinant proteins suggest an indirect interaction since incubation of recombinant PKC-δ with the GST-βc did not result in the coimmunoprecipitation of these two proteins (H Haleem-Smith and J Rivera, unpublished observations) (Fig. 6.1).

Phosphorylation of the FcεRI-γ Chain Peptide by Recombinant PKC-δ

To assess if PKC-δ would phosphorylate γ chain as a substrate, the PKC isozymes present in RBL cells and that translocate to the membrane in response to antigen-activation (α, β, δ and ε)[45] were incubated with a synthetic peptide encoding the entire cytoplasmic domain of the γ chain in an in vitro kinase assay. Only PKC-δ phosphorylated the γ chain peptide as a substrate, even though all isozymes showed activity towards type III-S histones. In the presence of phosphatidylinositol (PI) the rate of phosphorylation of the peptide by PKC-δ increased by 2.5-fold confirming the preference of this isozyme for PI as a lipid cofactor.[44] These results and the present information about PKC-δ substrate specificity[42] support the suggestion of a target sequence in the γ chain as described in the above section.

Functional Effects of Phosphorylation of the γ Chain on Threonines

Little is known about the functional role of phosphorylation of threonine(s) on FcεRI-γ chain. Endocytosis of receptor as a consequence of PKC activation in RBL-2H3 cells was previously described.[46] More recent studies have shown that inhibition of PKC activity (with Ro31-7549) results in an early inhibition of the antigen-mediated endocytosis.[44] The $t^{1/2}$ of endocytosis in the absence and presence of the PKC inhibitor was 2.5 and 18 minutes, respectively. The PMA-induced endocytosis was completely inhibited by Ro31-7549 and correlated with inhibition of threonine phosphorylation of γ chain. Collectively these experiments suggest that PKC activity is required at an early step in the endocytosis and may reflect, as noted above, the requirement for threonine phosphorylation of the γ chain as part of the signal for this event. Recent studies by Bingham et al[43] with a low secreting variant of the RBL-2H3 cells suggest that threonine phosphorylation of the γ chain precedes the tyrosine phosphorylation. The defect in this variant line is unknown and tyrosine phosphorylation of both β and γ chains appeared defective. Nevertheless, the presence of threonine phosphorylation on the γ chain prior to antigen stimulation[44] and the

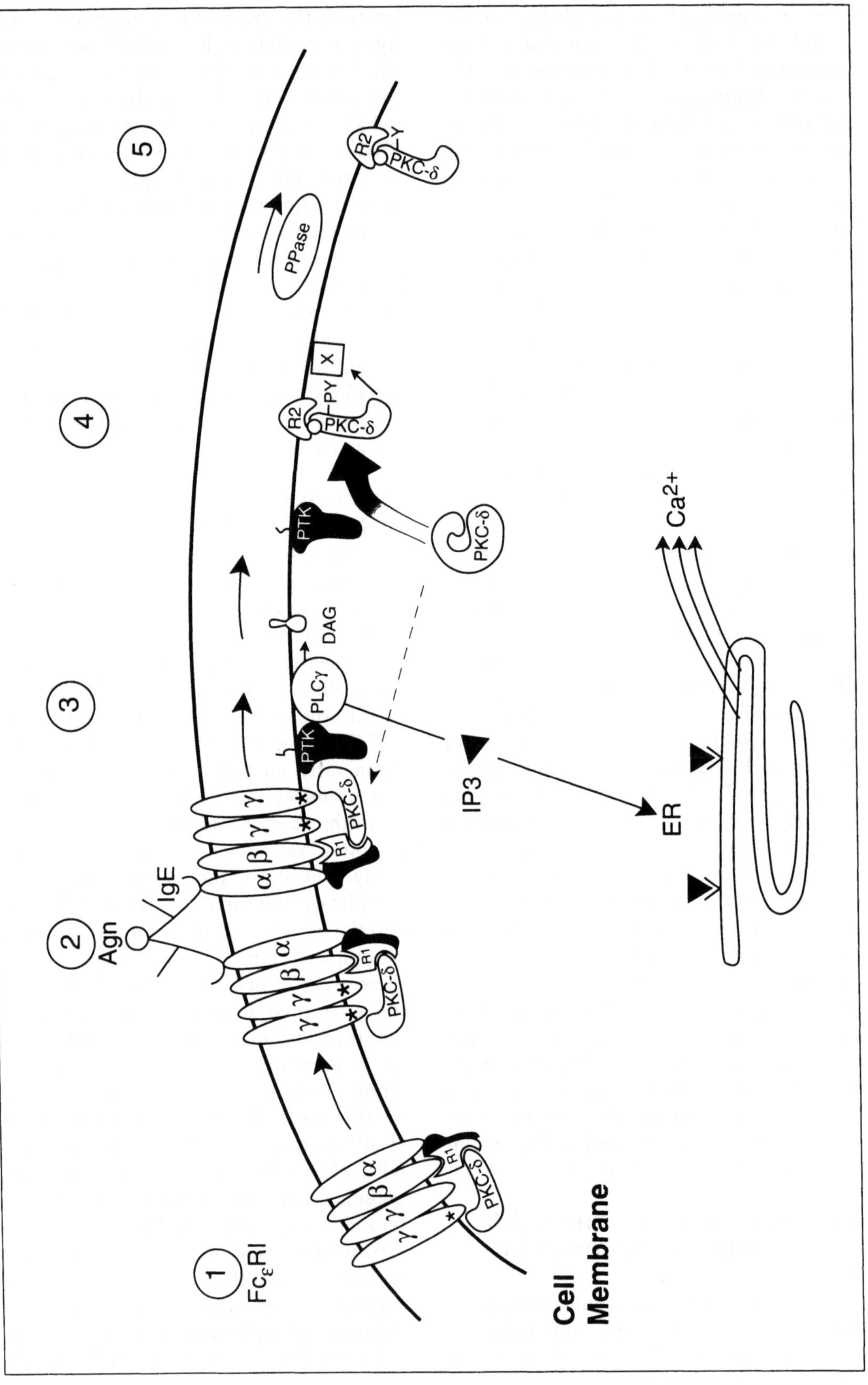
1
Fc$_\varepsilon$RI
Cell Membrane
2
Agn
IgE
3
4
5
α β γ γ
R1
PKC-δ
PTK
PLCγ
DAG
IP3
ER
Ca^{2+}
R2
PY
X
PPase
Y

Fig. 6.1 (above). Model of how PKC-δ functions on the membrane upon activation of FcεRI. (1) Some of the membrane-associated PKC-δ is associated with receptor in nonstimulated cells. This association occurs with the β chain of the FcεRI and preliminary evidence suggests an indirect interaction via a protein I call "R1." Low levels (*) of threonine phosphorylation of γ chain can be detected on the resting cells, indicating an active kinase. (2) Aggregation of the receptor by IgE-antigen (Agn) stimulates additional phosphorylation (*). This may be due to an increase in the amount of PKC-δ associated with receptor and/or an increase in kinase activity. (3) Aggregation of receptor also results in increased tyrosine kinase activity leading to activation of phospholipase C-γ(PLC-γ), hydrolysis of phosphatidyl inositol-(4,5)-bisphosphate (PIP_2), and generation of diacylglycerol (DAG) and inositol-(1,4,5)-trisphosphate (IP 3). The second messengers serve to activate PKC in the membrane (DAG) and to release calcium from intracellular stores (IP 3), although calcium is not required for translocation of PKC-δ. (4) This translocation results in a slight increase in association with receptor (2-fold) but most of the translocated enzyme is not associated with receptor and is tyrosine phosphorylated, suggesting this modification occurs by interaction with a membrane localized tyrosine kinase. (5) Tyrosine phosphorylation decreases the activity towards the receptor γ chain, but the enzyme is not inactive and therefore may phosphorylate other substrates (X). The tyrosine phosphorylation of PKC-δ is regulated by phosphatase activity (PPase) since the enzyme is dephosphorylated within 3 minutes of receptor aggregation. The latter process may regulate its activity towards the new substrate (X).

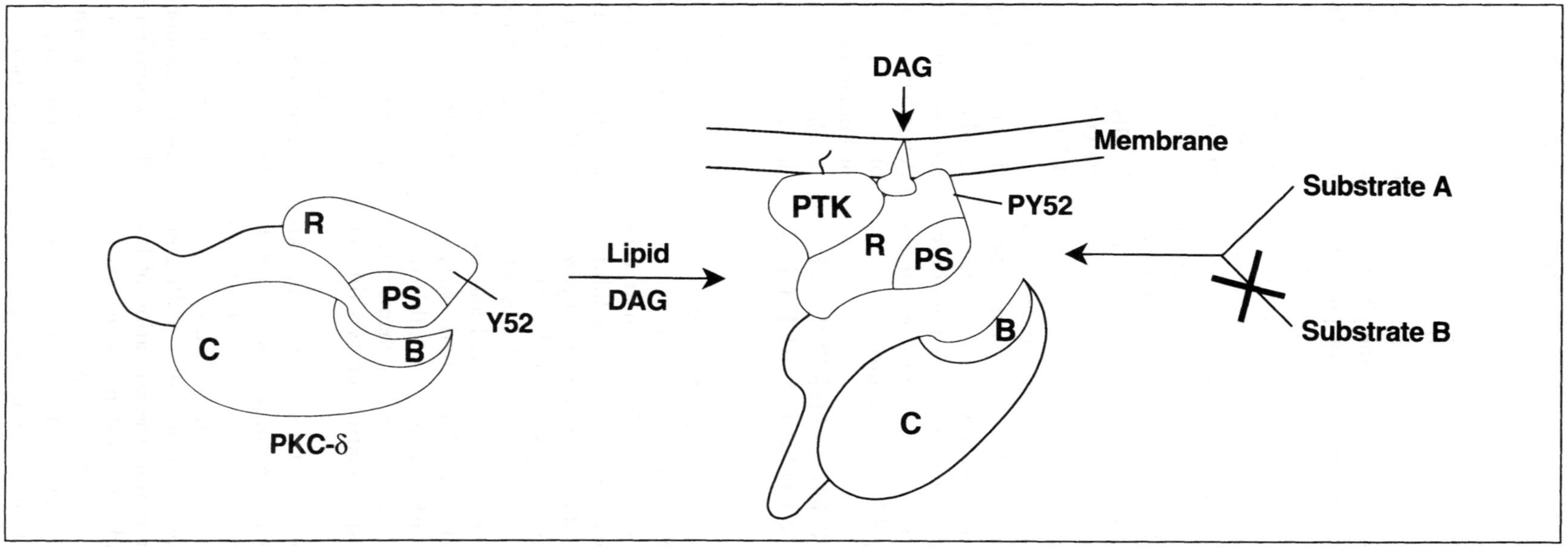

Fig. 6.2. Model of how phosphorylation of tyrosine regulates the substrate selectivity of PKC-δ. The two domain model of PKC suggests that the pseudosubstrate (PS) site of the regulatory (R) domain can interact with the substrate binding (B) site of the catalytic (C) domain with high affinity thereby blocking binding of substrate. The interaction of the enzyme with diacylglycerol (DAG) and lipid results in its translocation and in a conformational change which lowers the affinity of the pseudosubstrate site for the substrate binding site allowing substrates to interact. Phosphorylation at tyrosine 52 (Y52) by a tyrosine kinase may further reduce the interaction of the pseudosubstrate site with the substrate binding site, increasing the activity of the enzyme (Substrate A). However, substrates of PKC-δ with weak basic to slightly acidic phosphorylation sites (Substrate B) may interact weakly or not at all with the substrate binding domain due to charge interference following the phosphorylation of the Y52 (see refs. 42, 47).

threonine phosphorylation of γ chain prior to tyrosine phosphorylation[43] suggest an active PKC-δ prior to antigen-aggregation (Fig. 6.1). However, inhibition of threonine phosphorylation (with Ro31-7549) does not inhibit phosphorylation of tyrosine on the γ chain suggesting that threonine phosphorylation is not a prerequisite for phosphorylation on tyrosine. Recent studies on the regulation of PKC-δ[47] provide a potential clue for the importance of PKC-δ association with the FcεRI. Immuno-depletion of the PKC-δ from the membrane fraction of RBL-2H3 cells resulted in coimmunoprecipitation of more than 50% of activated receptors, as detected by antiphosphotyrosine immunoblots. This suggests that a large fraction of activated receptors is associated with PKC-δ. It is possible that the fraction of receptors which associate with PKC-δ also represents the fraction of receptors associated with tyrosine kinase activity and capable of signaling.

ROLE OF SERINE/THREONINE KINASES IN OTHER MAST CELL EVENTS

Regulation of PI Hydrolysis by Serine/Threonine Kinases

Treatment of basophils or mast cells (RBL-2H3) with PMA has an inhibitory effect on phosphatidylinositol (PI) hydrolysis without affecting secretion.[48,49] Studies with a permeabilized cell model in which all PKC isozymes were depleted and then replaced with each individual isozyme showed that PKC-α and ε inhibited PI hydrolysis presumably by a mechanism that inhibits phosphorylation of tyrosine on PLCγ.[50]

The diversity of serine/threonine kinases that regulate the phosphoinositide pathway is evident from recent studies by Barker et al.[51] Treatment of RBL-2H3 cells with wortmannin, an inhibitor of PI 3-kinase, did not inhibit the activation of Lyn and Syk, and only had minor effects on mobilization of Ca^{2+}, but inhibited production of inositol (1,4,5) P3, membrane ruffling and secretion. Immunoprecipitates of phosphatidylinositol 3-kinase (PI 3-kinase) from antigen-activated cells contained both a lipid and protein (serine/threonine) kinase activity as determined by in vitro assay. Moreover, FcεRI immunoprecipitates from antigen-activated cells contained a serine/threonine activity inhibited by wortmannin. This suggests the possibility that PI 3-kinase may not only regulate lipid metabolism in response to FcεRI but serve to regulate protein components of this pathway by phosphorylation of serines or threonines or both.

Serine/Threonine Phosphorylation and the Calcium Response

A number of studies show that serine/threonine phosphorylation plays a role in regulation of the calcium response. Indirect evidence exists for a role of cAMP-dependent protein kinases in the regulation of intracellular calcium levels. Treatment of human mast cells with isobutylmethylxanthine, an agent that activates cAMP-dependent protein kinase activity, inhibits the Ca^{2+} response.[22] Treatment of normal rat mast cells with dibutyryl cyclic AMP inhibits antigen-induced secretion and $^{45}Ca^{2+}$ uptake whereas secretion induced by calcium ionophore A23187 is not affected.[52] On the other hand, treatment of RBL cells with dibutyryl cyclic AMP, at concentrations that inhibit secretion in normal mast cells, has a minimal effect on secretion from these cells.[53] Whether cAMP-dependent protein kinase activity is critical in regulating the Ca^{2+} response is unclear, although the ability of RBL cells to secrete in the presence of agents that increase intracellular cAMP suggests that cAMP-dependent activity is not essential for regulation of FcεRI-mediated release. One report suggests that in mast cells the elevation of cAMP in conjunction with calcium ionophore is a potent stimulus for the expression of IL-3.[54] This is due to an increase in the half-life of the interleukin mRNA probably through a serine/threonine phosphorylation-dependent mechanism.

H. Ali and colleagues have analyzed the Ca^{2+} response triggered by antigen-activation or thapsigargin in the presence or absence of extracellular calcium and PMA.[55] Thapsigargin treatment depleted intracellular Ca^{2+} stores and replenishment of these stores was unaffected by the addition of PMA. In contrast, the calcium response to antigen-activation, which mimicked that of thapsigargin in the absence of PMA, was more transient and intermittent in the presence of PMA. This shift from a sustained to an intermittent response suggests that PMA induces a modification in a PKC isozyme that regulates the calcium response. R. Penner's laboratory has demonstrated, by whole cell patch-clamp recordings and single-cell calcium measurements, that the intracellular calcium response activated-channel (I_{crac}) is sensitive to the activation of PKC.[56] Treatment of RBL-2H3 cells with PMA suppresses entry of Ca^{2+} through I_{crac}. Inhibitors of protein kinase C activity (staurosporine and bisindolylmaleimide) reversed this suppression and restored Ca^{2+} influx. Collectively, the studies implicate phosphorylation of serine or threonine in regulating the influx of calcium rather than the mobilization from intracellular stores. They also suggest that the mechanism by which thapsigargin activates influx is potentially different than that activated by aggregation of FcεRI.

An additional level of complexity in the regulation of the calcium response by serine/threonine kinases is revealed by the studies of Wolfe and colleagues. These studies analyzed the effects of PMA on the Ca^{2+} response and secretion utilizing adherent or nonadherent cells.[57] In nonadherent cells PMA inhibited antigen-induced Ca^{2+} responses[49,57] and the PKC inhibitor bisindolylmaleimide (GF 109203X) reversed the inhibitory effect. However, in adherent cells PMA had little effect on the Ca^{2+} response. These results suggest that adhesion of cells may modify the action of PKC towards a substrate that is involved in the influx of calcium, so it becomes refractive to the effects of PMA.

Regulation of Secretion by Serine/Threonine Kinase Activity

In addition to affecting the sensitivity of cells to PMA,[57] adhesion enhances secretion by RBL-2H3 cells.[58] Treatment of nonadherent cells with PMA can be stimulatory (at concentrations <10 nM) or inhibitory (at concentrations >10 nM) whereas secretion by adherent cells is minimally affected. On the other hand, the PKC inhibitor GF 109203X inhibits antigen-stimulated secretion from both adherent and nonadherent cells while thapsigargin alone induces secretion only in the adherent cells.[57] This secretion requires the catalytic activity of PKC since secretion is inhibited in adherent cells when PKC is down-regulated with PMA or by treatment with GF 109203X.[57] Calphostin C, a PKC inhibitor that binds to the regulatory domain of PKC fails to inhibit secretion in adherent cells although it had previously been shown to inhibit secretion in the RBL-2H3 cells.[59] These studies raise the possibility that a PKC activity required for secretion is constitutively present in adherent cells whereas in nonadherent cells the PKC activity must be activated and may never achieve the levels observed in adherent cells. One explanation proposed is that adhesion of the cells may alter the regulatory domain of PKC.[57] This is consistent with the inability of calphostin C to inhibit secretion from adherent cells and with recent studies showing that at least one isoform of PKC is tyrosine phosphorylated[47] in the regulatory domain[60] and that this may influence its activity (see below). An alternate possibility is that the localization of PKC isozymes differs in adherent versus nonadherent cells. Preliminary studies suggest that while differences are observed in the ability of specific isozymes to translocate in adherent versus nonadherent cells no immediate conclusions could be drawn as to how these differences contribute to the refractive nature of PKC in adherent cells.[57]

Role of Serine/Threonine Kinases in Gene Transcription

The production of cytokines by mast cells in response to aggregation of FcεRI[61,62] is probably the single most important finding that has stimulated research on the molecular signals involved in mast cell gene transcription. The focus of this effort has been on two constituents implicated in other systems,[63] namely MAP kinases and PKC. While FcεRI-mediated activation of MAP kinase is well documented,[5,30] there is little information about its role in regulating mast cell gene transcription. In contrast, a role for PKC in the regulation of gene expression in mast cells has been documented since the early studies of cytokine induction in mast cells.[64] Over the past 5 years studies from the laboratory of E. Razin have implicated PKC activity in regulating cell growth, gene expression and the DNA-binding activity of transcription factors in response to aggregation of FcεRI.[65,66] These studies suggest that depletion of PKC leads to loss of DNA-bound AP-1 complexes by reducing the expression of c-fos and c-jun.[67] A straightforward interpretation of these results is that aggregation of the receptor induces PKC activation resulting in expression of c-fos and c-jun thus leading to an increase in DNA-bound complexes of AP-1 and initiation of gene transcription. However, an intriguing finding was the observation that the newly synthesized c-fos did not associate with c-jun, to increase levels of AP-1, but instead interacted with a newly described transcription factor of the USF family, *Fos*-interacting protein (FIP).[67] This suggests that aggregation of FcεRI initiates not only the expression of transcription factor mRNA but also the formation of specific complexes that may regulate specific genes.

Depletion of PKC isozymes by prolonged pretreatment of the cells with PMA, followed by permeabilization and washing of the permeabilized cells, results in almost complete loss of the FcεRI-dependent induction of mRNA for c-fos and c-jun.[68] Addition of the individual PKC isozymes to the permeabilized cells showed that PKC-β and ε were the only isozymes capable of restoring the response to antigen-activation of the FcεRI.[68] Further studies now implicate PKC-β in the regulation of the FIP-*Fos* complex in the murine mast cell line MC-9.[69] Antibodies to a C-terminal peptide of FIP and to c-fos were used to identify these proteins in MC-9 cells and to confirm their presence in DNA-binding complexes by gel shift assays using the DNA target sequence for binding of FIP. Short-term treatment of the cells with PMA followed by incubation with antisense oligonucleotide to the C2 domain of PKC specifically downregulated PKC-β. Aggregation of FcεRI on the PKC-β-depleted cells reduced FIP-*Fos* DNA binding activity whereas AP-1 DNA binding activity was unaffected. As already noted, PKC-β regulates the synthesis of c-fos[68] in response to aggregation of FcεRI. This may explain the loss of FIP-*Fos* binding. However, aggregation of FcεRI was also found to regulate the synthesis of FIP. Whether the latter is mediated via the activity of PKC-β remains to be determined. Together these studies indicate that serine/threonine phosphorylation mediated by PKC isozymes, in response to the FcεRI, play a critical role in regulating the synthesis of transcription factors as well as in secretion of effectors (see previous section).

TARGETS OF SERINE/THREONINE KINASE ACTIVITY

Targets of MAP Kinase Activity

Enhanced MAP kinase activity in mast cells in response to aggregation of FcεRI has been described.[5,30] Candidates for target molecules of this activity include microtubule-associated protein, Myc, Jun, Fos, pp90rsk, pp70-S6 kinase and cytosolic phospholipase A_2 ($cPLA_2$). All of these are present in mast cells, contain the P-X-S/T-P phosphorylation consensus site[41] and have been shown to be targets in other systems.[70] It is known that MAP kinase activity in the mast cell is required for $cPLA_2$ activity.[5] Presumably this increase results from the direct phosphorylation of $cPLA_2$ at serine 505, as previously described.[71] The p42 form of MAP kinase, demonstrated to phosphorylate $cPLA_2$, is the predominant isoform of MAP kinase phosphorylated in mast cells,[5,36] and therefore presumably activated in response to aggregation of FcεRI.[5]

Targets of PKC Activity

In a previous section (Phosphorylation of the FcεRI by Serine/Threonine Kinases) FcεRI-γ chain is described as a target for PKC-δ activity. This section will focus on other targets of PKC activity and the known or presumed functional consequences.

Phosphorylation of myosin light chains and heavy chains by PKC in response to FcεRI was shown to temporally correlate with exocytosis of granule contents from RBL-2H3 cells.[25] In fact both PKC and Ca^{2+}/calmodulin-dependent MLCK phosphorylate myosin light chains in response to receptor activation.[26] MLCK is responsible for basal levels of phosphorylation at serine 19 and aggregation of receptors induces an additional MLCK phosphorylation at threonine 18. The PKC activity induced by aggregation of receptor phosphorylated serine 1 or serine 2. Secretion occurred when both diphosphorylation by MLCK and monophosphorylation by PKC of myosin light chains was stimulated by antigen or by the combination of PMA and calcium ionophore (A23187). The role, if any, these phosphorylation events play in activating myosin molecules and in the secretory process in mast cells remains unclear.

T. Kawakami's laboratory has explored the physical interaction of PKC with members of the Tec subfamily of tyrosine kinases expressed in mast cells. A member of this family, Bruton's tyrosine kinase (BTK), is defective in human X-linked agammaglobulinemia and in X-linked immunodeficient (xid) mice. Initial studies show that BTK is tyrosine phosphorylated and activated in response to aggregation of FcεRI although its role in mast cell function is unclear.[72] Immunoprecipitation of BTK from mast cells and immunoblots with antibodies to PKC revealed the presence of PKC-βI in association with BTK. This interaction of PKC with BTK mapped to the pleckstrin homology domain (PH) of BTK.[73] Protein kinase C phosphorylated BTK in vitro and diminished its activity as determined by the ability of the latter to autophosphorylate. Prolonged treatment of mast cells with PMA- or calphostin C enhanced the phosphorylation of tyrosine on BTK in response to stimulation. This may indicate an increased ability of BTK to autophosphorylate in the absence of PKC activity. Alternatively BTK may be a better substrate for other tyrosine kinases in the absence of PKC activity. Studies in mast cells with Emt, another member of the Tec subfamily, show that the PH domain of this tyrosine kinase interacts with multiple isoforms of PKC in vitro as well as in vivo.[74] The coexpression of Emt and Ca^{2+}-dependent isoforms of PKC in COS cells increased the phosphorylation of tyrosine on Emt, suggesting that PKC coexpression activates the autophosphorylating activity of Emt or of another tyrosine kinase present in COS cells which phosphorylates Emt. Depletion of PKC by prolonged treatment with PMA, or inhibition of PKC activity by calphostin C, decreased the phosphorylation of tyrosine on Emt. The role of Tec subfamily members in the function of mast cells and the effect which phosphorylation of these kinases by PKC has on their ability to recognize substrates remains to be elucidated.

Targets of Other Serine/ Threonine Kinase Activities

The targets for Raf-1 and MEK activities are not yet defined in mast cells but it is likely that the activated Raf-1 phosphorylates MEK and that the increased MEK activity results in phosphorylation of MAP kinase as observed in other systems.[70] What is certain is that increase in activity of MAP kinases, in response to aggregation of FcεRI, parallels the increase in phosphorylation of Raf and MEK.[5]

Studies from the laboratory of Baird and Holowka have suggested a role for serine phosphorylation in the desensitization of the FcεRI in RBL-2H3 cells.[75] In these experiments two distinct populations of FcεRI were created by sensitizing cells with two murine IgEs of distinct antigenic specificity. The calcium or secretory response was measured to assess the responsiveness of receptors. Aggregation of one receptor population with specific antigen reduced the responsiveness of the second receptor population to its specific antigen (heterologous desensitization). Pretreatment of cells with agents that inhibit serine/threonine phosphatases, such as okadaic acid, enhanced the desensitization. Heterologous desensitization was not affected by long or short-term exposure of cells to PMA implying that a serine/threonine kinase other than a phorbol ester-sensitive isoform of PKC was involved in this process. A potential target for the serine/threonine activity was identified as a 55 kDa receptor-associated protein whose serine phosphorylation correlated with desensitization. The phosphorylation increased in both receptor populations in response to prolonged aggregation of FcεRI (30 minutes) by a specific antigen. These results suggest that unidentified 55 kDa receptor-associated protein plays a role in the desensitization of the FcεRI.

REGULATION OF SERINE/ THREONINE KINASE ACTIVITY

Phosphatases

What information we have on the role of serine /threonine phosphatases in the regulation of mast cell function comes from studies on the inhibition of the Ca^{2+}/calmodulin-dependent protein phosphatase, calcineurin, by cyclosporin A (CsA), FK506 or rapamycin. CsA or FK506 bind to immunophilins and these complexes interact with calcineurin. Other serine/threonine phosphatases such as PP1, PP2A and PP2C are unaffected.[76] In bone marrow-derived mast cells (BMMC) these inhibitors diminish FcεRI-induced increase in IL-1β, IL-6 and TNF-α at doses where exocytosis is unaffected.[77] The target(s) of calcineurin activity in the mast cells are still undefined; however, potential candidates may be members of the S6 kinase family since immunosuppressant drugs such as rapamycin, which may indirectly target calcineurin in mast cells[36], inhibit pp70-S6 kinase and may subsequently affect gene transcription as shown in other systems.[78]

Okadaic acid, a potent cell permeant inhibitor of the serine/threonine protein phosphatase PP2A and to a lesser extent of PP1 and PP2B, promotes FcεRI-induced responses in mast cells at nM doses and inhibits them at μM doses. Treatment of human lung mast cells with μM concentrations of this agent prior to aggregation of FcεRI attenuated the release of histamine and the synthesis of leukotriene C4.[79] Calyculin A, another serine/threonine phosphatase inhibitor with potent activity towards PP2A and PP1, gave similar results suggesting that serine/threonine phosphorylation regulated by these phosphatases is critical for mast cell function. In the RBL-2H3 cells, μM okadaic acid enhances the heterologous antigen-mediated desensitization, as described above, suggesting an active role for serine/threonine phosphatases in regulating the desensitization of receptor.[75]

Tyrosine Phosphorylation

There is increasing interest in "cross-talk" between tyrosine and serine/threonine kinases. Serine/Threonine phosphorylation of tyrosine kinases[80,81] has been documented for several years, but only in the past 2 years has evidence been presented for serine/threonine kinases as targets of tyrosine kinase activity. The new evidence has come primarily from studies on the regulation of PKC isozymes. Denning and colleagues[82] reported that tyrosine phosphorylation of PKC-δ in cells overexpressing oncogenic Ras inactivated PKC-δs activity. In contrast Li and co-workers[83] found increased activity of PKC-δ after it was tyrosine phosphorylated in response to PMA. In all studies tyrosine phosphorylation of only membrane-associated PKC-δ was observed (Fig. 6.1). In RBL-2H3 cells, FcεRI-induced tyrosine phosphorylation of PKC-δ did not affect its activity towards substrates such as myelin basic protein (MBP) peptide or histone type III but decreased the ability of this enzyme to phosphorylate the FcεRI-γ chain (Fig. 6.2).[47] These results suggest that modification of PKC-δ, and perhaps of other PKCs, by tyrosine phosphorylation, can modulate substrate specificity. To test this hypothesis we recently defined the tyrosine on PKC-δ that is modified in response to aggregation of FcεRI.[60] Mutation of the tyrosine residue at position 52 to phenylalanine completely ablated the phosphorylation of PKC-δ in response to activation of FcεRI. The molecular mechanism for this effect remains to be determined; however, at minimum the tyrosine phosphorylation of PKC-δ in mast cells seems to influence its catalytic activity and may serve to regulate its function in vivo (Fig. 6.2). In fact, overexpression of the tyrosine 52 mutant PKC-δ in transfected RBL-2H3 cells results in a cell line with a slower growth rate in comparison to RBL transfectants overexpressing wild-type PKC-δ (E-Y Chang, Z Szallasi, PM Blumberg et al, unpublished observations).

Localization of Serine/Threonine Kinases

The redistribution of serine/threonine kinases in response to FcεRI may regulate its enzymatic activity. Sequestration or presentation of enzymes to their substrates is an effective mechanism by which cells regulate enzymatic cascades. Evidence for this cellular mechanism is abundant in diverse pathways ranging from activation of MAP kinase via interaction of membrane bound Ras with cytosolic Raf[84] to PKC's interaction with DAG and with various proteins in the membrane.[85] Evidence that targeting of Raf to the membrane is sufficient to activate MAP kinase[86] and that targeting of Sos to the membrane activates Ras[87] provides the most compelling examples of how re-localization in response to a stimulus can profoundly regulate enzymatic activity. Many of the constituents of signaling pathways exist as multicomponent signaling particles[88] which are at equilibrium with the target membrane and are sequestered for signal transduction upon induction of a phosphorylation event or generation of second messengers. For example, in the RBL-2H3 cells the guanine nucleotide exchange factor Vav is found in a complex with Grb2, Raf-1 and p42 mapk prior to activation of FcεRI.[35] Vav is also found in association with the γ chain of the FcεRI and aggregation of receptor-induced increased association of the receptor with Vav and Grb2.[35] The recruitment of this multicomponent signaling particle may serve to ensure that the serine/threonine activity of the constituent kinases are directed as a response of receptor aggregation. This level of regulation along with its multiple components of phosphorylation and dephosphorylation appears to be critical for regulated cell response.

CONCLUDING REMARKS

Our knowledge of serine/threonine phosphorylation in response to activation of FcεRI activation of mast cells is at its inception. Its importance in mast cell function is undeniable since exocytosis is absolutely dependent on the catalytic activity of at least one serine/threonine kinase, namely PKC. Determining the role of serine/threonine kinases in the mast cell, whose activities they regulate, and determining which cellular proteins they modify are a daunting yet exciting challenge. Our gain in knowledge may open new areas as potential therapeutic targets.

Acknowledgments

I gratefully acknowledge the contribution of manuscripts made available to me prior to publication by Drs. B. Baird, C. Fewtrell, D. Holowka, T. Kawakami, J. Oliver, R. Penner and B. Wilson. I thank Drs. E-Y. Chang, H. Metzger, E. Razin and P. Swann for their critical reviews of this chapter.

References

1. Koch CA, Anderson D, Moran MF et al. SH2 and SH3 domains: Elements that control interactions of cytoplasmic signaling proteins. Science 1991; 252:668-74.
2. Nishizuka Y. Intracellular signaling by hydrolysis of phospholipids and activation of protein kinase C. Science 1992; 258: 607-14.
3. Wu J, Harrison JK, Vincent LA et al. Molecular structure of a protein-tyrosine/threonine kinase activating p42 mitogen-activated protein (MAP) kinase: MAP kinase kinase. Proc Natl Acad Sci USA 1993; 90:173-77.
4. Nishizuka Y. The molecular heterogeneity of protein kinase C and its implications for cellular regulation. Nature 1988; 334: 661-65.
5. Hirasawa N, Santini F, Beaven MA. Activation of the mitogen-activated protein kinase/cytosolic phospholipase A_2 pathway in a rat mast cell line. J Immunol 1995; 154:5391-402.
6. Clark JD, Lin LL, Kriz RW et al. A novel arachidonic acid-selective cytosolic PLA_2 contains a Ca^{2+}-dependent translocation domain with homology to PKC and GAP. Cell 1991; 65:1043-048.
7. Baker ME. Similarity between phospholipase C and the regulatory domain of protein kinase C. Mol Cell Endocrinol 1989; 61:129-31.
8. White JR, Pluznik DH, Ishizaka K et al. Antigen-induced increase in protein kinase C activity in plasma membrane of mast cells. Proc Natl Acad Sci USA 1985; 82:8193-98.
9. White KN, Metzger H. Translocation of protein kinase C in rat basophilic leukemic cells induced by phorbol ester or by aggregation of IgE receptors. J Immunol 1988; 141:942-47.
10. Chakravarthy BR, Whitfield JF, Durkin JP. Inactive membrane protein kinase Cs: a possible target for receptor signalling. Biochem J 1994; 304:809-16.
11. Dekker LV, Parker PJ. Protein kinase C-a question of specificity. Trends Biochem Sci 1994; 19:73-77.
12. Dinh TT, Kennerly DA. Assessment of receptor-dependent activation of phosphatidylcholine hydrolysis by both phospholipase D and phospholipase C. Cell Regul 1991; 2:299-309.
13. Lin P, Wiggan GA, Gilfillan AM. Activation of phospholipase D in a rat mast (RBL 2H3) cell line; a possible unifying mechanism for IgE-dependent degranulation and arachidonic acid metabolite release. J Immunol 1991; 146:1609-616.
14. Lin P, Fung WJC, Gilfillan AM. Phosphatidylcholine-specific phospholipase D-derived 1,2-diacylglycerol does not initiate protein kinase C activation in the RBL 2H3 mast-cell line. Biochem J 1992; 287:325-31.
15. Pribluda VS, Metzger H. Calcium-independent phosphoinositide breakdown in rat basophilic leukemia cells. Evidence for an early rise in inositol 1,4,5-trisphosphate which precedes the rise in other inositol phosphates and in cytoplasmic calcium. J Biol Chem 1987; 262:11449-54.
16. Kochs G, Hummel R, Fiebich B et al. Activation of purified human protein kinase Cα and β_1 in vitro by Ca^{2+}, phosphatidylinositol and phosphatidylinositol 4,5-bisphosphate. Biochem J 1993; 291:627-33.
17. Ishizaka T, Ishizaka K. Activation of mast cells for mediator release through IgE receptors. Prog Allergy 1984; 34:188-35.
18. Holgate ST, Lewis RA, Austen KF. 3',5'-cyclic-AMP-dependent protein kinase of the rat serosal mast cell and its immunologic activation. J Immunol 1980; 124:2093-99.
19. Sullivan T J, Parker KL, Kulczycki Jr A et al. Modulation of cyclic AMP in purified rat mast cells. III. Studies on the effects of concanavalin A and anti-IgE on cyclic AMP concentrations during histamine release. J Immunol 1976; 117:713-16.
20. Diamant B, Kazimierczak W, Patkar SA. Does cyclic AMP play any role in histamine release from rat mast cells. Allergy 1978; 33:50-51.
21. Foreman JC, Hallet MB, Mongar JL. The relationship between histamine secretion and ^{45}calcium uptake by mast cells. J Physiol 1977; 271:193-14.
22. Columbo M, Botana LM, Horowitz EM et al. Studies of the intracellular Ca^{2+} levels in human adult skin mast cells activated by

the ligand for the human c-kit receptor and anti-IgE. Biochem Pharmacol 1994; 47: 2137-45.

23. Allende JE, Allende CC. Protein kinase CK2: an enzyme with multiple substrates and a puzzling regulation. FASEB J 1995; 9:313-23.
24. Tawfic S, Ahmed K. Growth stimulus-mediated differential translocation of casein kinase 2 to the nuclear matrix. Evidence based on androgen action in the prostate. J Biol Chem 1994; 269:24615-20.
25. Ludowyke R I, Peleg I, Beaven MA et al. Antigen-induced secretion of histamine and the phosphorylation of myosin by protein kinase c in rat basophilic leukemia cells. J Biol Chem 1989; 264:12492-501.
26. Choi OH, Adelstein RS, Beaven MA. Secretion from rat basophilic RBL-2H3 cells is associated with diphosphorylation of myosin light chains by myosin light chain kinase as well as phosphorylation by protein kinase C. J Biol Chem 1994; 269: 536-41.
27. Chakravarty N. The role of calmodulin and protein kinase C in histamine secretion from mast cells. Agents Actions 1992; 36:183-91.
28. Kaye RE, Fruman DA, Bierer BE et al. Effects of cyclosporin A and FK506 on Fc epsilon receptor type I-initiated increases in cytokine mRNA in mouse bone marrow-derived progenitor mast cells: resistance to FK506 is associated with a deficiency in FK506-binding protein FKBP12. Proc Natl Acad Sci USA 1992; 89:8542-46.
29. Grosman N. Influence of CGS 9343B, an inhibitor of calmodulin activity, on histamine release from isolated rat mast cells. Agents Actions 1992; 36:192-99.
30. Offermanns S, Jones SVP, Bombien E et al. Stimulation of mitogen-activated protein kinase activity by different secretory stimuli in rat basophilic leukemia cells. J Immunol 1994; 152:250-61.
31. Crews CM, Erikson RL. Extracellular signals and reversible protein phosphorylation: What to MEK of it all. Cell 1993; 74:215-17.
32. Cantrell D. G proteins in lymphocyte signalling. Curr Op Immunol 1994; 6:380-84.
33. Buday L, Egan SE, Rodriguez-Viciana P et al. A complex of Grb-2 adaptor protein, Sos exchange factor and a 36 kDa membrane bound tyrosine phosphoprotein is implicated in ras activation in T cells. J Biol Chem 1994; 269:9019-23.
34. Turner H, Reif K, Rivera J et al. Regulation of the adapter molecule Grb2 by the FcεRI in the mast cell line RBL 2H3. J Biol Chem 1995; 270:9500-506.
35. Song JS, Gomez J, Rivera J. Regulation of p95 Vav association with the FcεRI γ chain in RBL-2H3 mast cells: Evidence for a constitutive in vivo association of Vav with Grb2, Raf-1 and ERK2 in an active complex. 1996; (submitted).
36. Tsai M, Chen RY, Tam SY et al. Activation of MAP kinases, pp90rsk and pp70-S6 kinases in mouse mast cells by signaling through the c-kit receptor tyrosine kinase or FcεRI: rapamycin inhibits activation of pp70-S6 kinase and proliferation in mouse mast cells. Eur J Immunol 1993; 23: 3286-91.
37. Quarto R, Metzger H. The receptor for immunoglobulin E: examination for kinase activity and as a substrate for kinases. Mol Immunol 1986; 23:1215-23.
38. Paolini R, Jouvin MH, Kinet JP. Phosphorylation and dephosphorylation of the high-affinity receptor for immunoglobulin E immediately after receptor engagement and disengagement. Nature 1991; 353:855-58.
39. Reth M. Antigen receptor tail clue (letter). Nature 1989; 338:383-84.
40. Alvarez E, Northwood IC, Gonzalez FA et al. Pro-Leu-Ser/Thr-Pro is a consensus primary sequence for substrate protein phosphorylation. J Biol Chem 1991; 266: 15277-85.
41. Marin O, Meggio F, Sarno S et al. Phosphorylation of synthetic fragments of inhibitor-2 of protein phosphatase-1 by casein kinase-1 and -2. Evidence that phosphorylated residues are not strictly required for efficient targeting by casein kinase-1. Eur J Biochem 1994; 223:647-53.
42. Kielbassa K, Müller HJ, Meyer HE et al. Protein kinase Cδ-specific phosphorylation of the elongation factor eEF-1α and an eEF-1α peptide at threonine 431. J Biol Chem 1995; 270:6156-62.

43. Bingham BR, Monk PN, Helm BA. Defective phosphorylation and Ca^{2+} mobilization in a low secreting variant of the rat basophilic leukemia cell line. J Biol Chem 1994; 269:19300-306.
44. Germano P, Gomez J, Kazanietz MG et al. Phosphorylation of the γ chain of the high affinity receptor for immunoglobulin E by receptor-associated protein kinase C-δ. J Biol Chem 1994; 269:23102-107.
45. Ozawa K, Szallasi Z, Kazanietz MG et al. Ca^{2+}-dependent and Ca^{2+}-independent isozymes of protein kinase C mediate exocytosis in antigen-stimulated rat basophilic RBL-2H3 cells: reconstitution of secretory responses with Ca^{2+} and purified isozymes in washed permeabilized cells. J Biol Chem 1993; 268:1749-55.
46. Ra C, Furuichi K, Rivera J et al. Internalization of IgE receptors on rat basophilic leukemic cells by phorbol ester. Comparison with endocytosis induced by receptor aggregation. Eur J Immunol 1989; 19: 1771-77.
47. Haleem-Smith H, Chang EY, Szallasi Z et al. Tyrosine phosphorylation of protein kinase C-δ in response to the activation of the high affinity receptor for immunoglobulin E modifies its substrate recognition. Proc Natl Acad Sci USA 1995; 92:9112-9116.
48. Sagi-Eisenberg R, Pecht I. Protein kinase C, a coupling element between stimulus and secretion of basophils. Immunol Lett 1984; 8:237-41.
49. Beaven MA, Guthrie DF, Moore JP et al. Synergistic signals in the mechanism of antigen-induced exocytosis in 2H3 cells: evidence for an unidentified signal required for histamine release. J Cell Biol 1987; 105:1129-35.
50. Ozawa K, Yamada K, Kazanietz MG et al. Different isozymes of protein kinase C mediate feedback inhibition of phospholipase C and stimulatory signals for exocytosis in rat RBL-2H3 cells. J Biol Chem 1993; 268:2280-85.
51. Barker SA, Caldwell KK, Hall A et al. Wortmannin blocks lipid and protein kinase activities associated with PI-3 Kinase and inhibits a subset of responses induced by FcεRI crosslinking. Mol Biol Cell 1995; 6:1145-58.
52. Foreman J, Mongar JL, Gomperts BD et al. A possible role for cyclic AMP in the regulation of histamine secretion and the action of cromoglycate. Biochem Pharmacol 1975; 24:538-40.
53. Fewtrell C, Metzger H. Stimulus-secretion coupling in rat basophilic leukemia cells. In: Becker EL, Simon AS, Austen KF, eds. Biochemistry of actue allergic reactions. New York: Alan R Liss, Inc., 1981:295-14.
54. Hahn S, Moroni C. Modulation of cytokine expression in PB-3c mastocytes by IBMX and PMA. Lymphokine Cytokine Res 1994; 13:247-52.
55. Ali H, Maeyama K, Sagi-Eisenberg R et al. Antigen and thapsigargin promote influx of Ca^{2+} in rat basophilic RBL-2H3 cells by ostensibly similar mechanisms that allow filling of inositol 1,4,5-trisphosphate-sensitive and mitochondrial Ca^{2+} stores. Biochem J 1994; 304:431-40.
56. Parekh AB, Penner R. Depletion-activated Ca^{2+} current is inhibited by protein kinase in RBL-2H3 Cells. Proc Natl Acad Sci USA 1995; 92:7907-11.
57. Wolfe PC, Chang E-Y, Rivera J et al. Differential effects of the protein kinase C activator PMA on calcium responses and secretion in adherent and suspended RBL-2H3 mucosal mast cells. J Biol Chem 1996; 271:6658-6665.
58. Hamawy M, Oliver C, Mergenhagen SE et al. Adherence of rat basophilic leukemia (RBL-2H3) cells to fibronectin-coated surfaces enhances secretion. J Immunol 1992; 149:615-21.
59. Yamada K, Jelsema CL, Beaven MA. Certain inhibitors of protein serine/threonine kinases also inhibit tyrosine phosphorylation of phospholipase Cγ1 and other proteins and reveal distinct roles for tyrosine kinase(s) and protein kinase C in stimulated, rat basophilic RBL-2H3 cells. J Immunol 1992; 149:1031-37.
60. Szallasi Z, Denning MF, Chang E-Y et al. Determination of tyrosine phosphorylation sites of PKC-δ upon activation of the high affinity receptor for IgE in rat basophilic leukemia cells. Biochem Biophys Res Com

1995; 214:888-94.

61. Plaut M, Pierce JH, Watson CJ et al. Mast cell lines produce lymphokines in response to crosslinkage of FcεRI or to calcium ionophores. Nature 1989; 339:64-67.
62. Wodnar-Filipowicz A, Heusser CH, Moroni C. Production of the haemopoietic growth factors GM-CSF and interleukin-3 by mast cells in response to IgE receptor-mediated activation. Nature 1989; 339:150-52.
63. Ullman KS, Northrop JP, Verweij CL et aL. Transmission of signals from the T lymphocyte antigen receptor to the genes responsible for cell proliferation and immune function: The missing link. Annu Rev Immunol 1990; 8:421-52.
64. Burd PR, Rogers HW, Gordon JR et al. Interleukin-3 dependent and -independent mast cells stimulated with IgE and antigen express multiple cytokines. J Exp Med 1989; 170:245-57.
65. Chaiken E, Ziltener HJ, Razin E. Protein kinase C plays an inhibitory role in IL-3 and IL-4 mediated mast cell proliferation. J Biol Chem 1990; 265:22109-16.
66. Baranes D, Razin E. Protein kinase C regulates proliferation of mast cells and the expression of the mRNAs of fos and jun protooncogenes during activation by IgE-Ag or calcium ionophore A23187. Blood 1991; 78:2354-61.
67. Lewin I, Nechushtan H, Qingen K et al. Regulation of AP-1 expression and activity in antigen-stimulated mast cells: The role played by protein kinase C and the possible involvement of Fos interacting protein. Blood 1993; 82:3746-51.
68. Razin E, Szallasi Z, Kazanietz MG et al. Protein kinases C-β and C-ε link the mast cell high-affinity receptor for IgE to the expression of c-fos and c-jun. Proc Natl Acad Sci USA 1994; 91:7722-26.
69. Lewin I, Jacob-Hirsch J, Zang ZC et al. PKC-β regulates the FIP-Fos but not AP-1 DNA binding activity in FcεRI stimulated mast cells. J Biol Chem 1996; 371:1514-1519.
70. Blumer KJ, Johnson GL. Diversity in function and regulation of MAP kinase pathways. Trends Biochem Sci 1994; 19:236-41.
71. Lin LL, Wartmann M, Lin AY et al. cPLA2 is phosphorylated and activated by MAP kinase. Cell 1993; 72:269-78.
72. Kawakami Y, Yao L, Miura T et al. Tyrosine phosphorylation and activation of Bruton Tyrosine Kinase upon FcεRI crosslinking. Mol Cell Biol 1994; 14: 5108-13.
73. Yao L, Kawakami Y, Kawakami T. The pleckstrin homology domain of Bruton tyrosine kinase interacts with protein kinase C. Proc Natl Acad Sci USA 1994; 91: 9175-79.
74. Kawakami Y, Yao L, Tashiro M et al. Activation and interaction with protein kinase C of a cytoplasmic tyrosine kinase, Itk/Tsk/Emt, on Fc epsilon RI crosslinking on mast cells. J Immunol 1995; 155:3556-62.
75. Weetall M. Studies on the high affinity receptor for IgE (FcεRI): Binding of a chimeric IgE/IgE and desensitization of cellular responses. Ph.D. Thesis (1992), Cornell University, Ithaca, NY.
76. Liu J, Albers MW, Luan S et al. Inhibition of T cell signaling by immunophilin-ligand complexes correlates with loss of calcineurin phosphatase activity. Biochemistry 1992; 31:3896-901.
77. Fruman DA, Bierer BE, Benes JE et al. The complex of FK506-binding protein 12 and FK506 inhibits calcineurin phospatase activity and IgE activation-induced cytokine transcripts, but not exocytosis, in mouse mast cells. J Immunol 1995; 154:1846-51.
78. Blenis J. Signal transduction via the MAP kinases: proceed at your own rsk. Proc Natl Acad Sci USA 1993; 90:5889-94.
79. Peachell PT, Munday MR. Regulation of human lung mast cell function by phosphatase inhibitors. J Immunol 1993; 151:3808-16.
80. Morgan DO, Kaplan JM, Bishop JM et al. Mitosis-specific phosphorylation of p60c-src by p34cdc2-associated protein kinase. Cell 1989; 57:775-86.
81. Winkler DG, Park I, Kim T et al. Phosphorylation of Ser-42 and Ser-59 in the N-terminal region of the tyrosine kinase $p56^{lck}$. Proc Natl Acad Sci USA 1993; 90:5176-80.
82. Denning MF, Dlugosz AA, Howett MK.

Expression of an oncogenic ras^{Ha} gene in murine keratinocytes induces tyrosine phosphorylation and reduced activity of protein kinase C δ. J Biol Chem 1993; 268: 26079-81.
83. Li W, Mischak H, Yu JC et al. Tyrosine phosphorylation of protein kinase C-δ in response to its activation. J Biol Chem 1994; 269:2349-52.
84. Vojtek AB, Hollenberg SM, Cooper JA. Mammalian ras interacts directly with the serine/threonine kinase raf. Cell 1993; 74:205-14.
85. Ron D, Chen CH, Caldwell J et al. Cloning of an intracellular receptor for protein kinase C: A homolog of the β subunit of G proteins. Proc Natl Acad Sci USA 1994; 91:839-43.
86. Stokoe D, Macdonald SG, Cadwallader K et al. Activation of raf as a result of recruitment to the plasma membrane. Science 1994; 264:1463-67.
87. Aronheim A, Engelberg D, Li N et al. Membrane targeting of the nucleotide exchange factor Sos is sufficient for activating the ras signaling pathway. Cell 1994; 78:949-61.
88. Egan SE, Giddings BW, Brooks MW et al. Association of Sos ras exchange protein with Grb2 is implicated in tyrosine kinase signal transduction and transformation. Nature 1993; 363:45-51.

CHAPTER 7

Protein Tyrosine Phosphatases in Signaling Through the FcεRI on Mast Cells and Basophils

Mark Swieter

INTRODUCTION

The process of reversible tyrosine phosphorylation is a key regulatory mechanism in eukaryotic cells for controlling events in cell growth, differentiation and activation.[1,2] The level of protein tyrosine phosphorylation within a cell reflects a dynamic balance between the often competing activities of the protein tyrosine kinases that phosphorylate proteins and the protein tyrosine phosphatases (PTPs) that dephosphorylate them. It is the shift in this balance brought on by microenvironmental cues that determines the phosphorylation status of a protein at any given time. In this chapter, an analysis is provided of the PTPs that are likely to be involved in the signaling pathways initiated by aggregating the high affinity IgE receptors (FcεRI) on mast cells and basophils. To be sure, relatively little work has been done in this area up to now. But that should be interpreted as an indication of just how open to exploration this area of investigation is.

PTPs can be categorized on the basis of their structural characteristics into transmembrane and cytosolic subsets. As the name implies, transmembrane PTPs are inserted into cell membranes by virtue of transmembrane segments. An example of this group of PTP is the leukocyte common antigen, CD45. In addition to their transmembrane domains, most of the transmembrane PTPs have large extracellular domains and an intracellular domain containing dual tandem PTP catalytic subdomains.

IgE Receptor (FcεRI) Function in Mast Cells and Basophils, edited by Majed M. Hamawy.

In contrast, the other group of PTPs lack extracellular and transmembrane regions and contain only a single catalytic subdomain. They are called cytosolic PTPs. In some cases, the noncatalytic regions of the cytosolic PTPs have been shown to control the localization of the PTPs within the cell or the phosphatase activity of the PTPs.[3,4] Multiple transmembrane and cytosolic PTPs have been cloned, each consisting of structurally distinct noncatalytic domains fused to a conserved phosphatase domain. Such findings highlight the diversity of this group of regulatory proteins and support the notion that individual members of the PTP family will be involved in distinct cellular functions.[5]

FcεRI-ASSOCIATED PTP

Protein tyrosine phosphorylation is an early event in the signaling cascade initiated by FcεRI engagement. Several proteins become phosphorylated on tyrosine residues including the FcεRI-β and -γ subunits, the protein kinases Syk and Fak, phospholipase C-γ1, paxillin, Nck, Vav and a number of other proteins in the 70 kDa and 105 kDa to 115 kDa ranges.[6-20] Interruption or reversal of FcεRI crosslinking by the addition of hapten results in rapid protein dephosphorylation.[6,7] However, none of the PTPs involved in regulating the phosphorylation/dephosphorylation status of these or any other tyrosine phosphorylated proteins in mast cells or basophils have been identified.

The tyrosine kinase Lyn coimmunoprecipitates with the FcεRI in both activated and unstimulated RBL-2H3 mast cells.[10,20] Accordingly, it has been proposed that Lyn tyrosine phosphorylates the FcεRI-β and -γ subunits following FcεRI aggregation.[20] However, the Lyn that coimmunoprecipitates with the FcεRI is catalytically active regardless of whether the cells from which it was derived were stimulated or not.[10] Thus, it would be expected that the β and γ subunits of the FcεRI are tyrosine phosphorylated even in the unstimulated cell. This does not happen. Swieter et al[21] recently showed that PTP activity coisolates with the FcεRI. Using a combination of immunoblotting, enzyme depletion and studies with RBL-2H3 cell variants, they were able to show that the PTP was not CD45, PTP1C or PTP1D. In in vitro phosphatase assays the FcεRI-associated PTP dephosphorylated tyrosine phosphorylated β and γ subunits but not Lyn or Syk. These findings led them to speculate that the PTP may function to maintain the receptor subunits in their dephosphorylated state. Clearly, the identity of this PTP needs to determined.

CD45

STRUCTURE AND STRUCTURE/FUNCTION RELATIONSHIPS

CD45 is a transmembrane glycoprotein expressed exclusively by hematopoietic cells (see also chapter 12). It is also known as leukocyte common antigen and constitutes one of the most abundant proteins in the leukocyte plasmalemma.[1] Due to its exclusive distribution and its apparently critical role in signaling through the multisubunit antigen recognition receptors on T cells,[22,23] B cells[24] and mast cells,[25,26] CD45 is one of the most extensively studied PTPs.

As with the other transmembrane PTPs, CD45 has extracellular, transmembrane and intracellular domains. The extracellular domain is large and heavily glycosylated and, in T cells, undergoes variable exon splicing that is regulated in cell type-specific[27] and developmental[28] manners as well as depend on the activation state of the cell.[29,30] There is a single putative transmembrane domain and it appears to confer some specificity on CD45 in its interactions with other membrane-spanning proteins.[31] However, the functional center of CD45 lies in its intracellular domain that comprises a pair of tandem protein tyrosine phosphatase subdomains, both of which are required for full enzymatic activity of the phosphatase.

The CD45 gene maps to chromosome one in humans and mice and is comprised of 33 exons in both species.[32,33] Exons 3 through 21 code for the extracellular domain, exon 22 encodes the transmembrane domain and exons 23 through 33 code for the intracellular domain. Exons 4, 5 and 6 undergo alternative splicing giving rise to the structural diversity of CD45. An additional CD45 cDNA that lacks exons 4,5,6 and 7, has been identified by polymerase chain reaction,[34] but whether a corresponding protein is expressed is unknown. The molecular size of CD45 varies from 180 kDa to 220 kDa in T cells,[33,35-37] whereas the CD45 isoform expressed in B cells is 220 kDa.[35] The isoforms present in mast cells and basophils appear to be somewhat smaller.[25,38,39]

A role for the extracellular domain in the functioning of CD45 is implied by the regulated alternative splicing of RNA transcripts of exons 4 through 7. In addition, it may be significant that different isoforms of CD45 are found on T cells, B cells, mast cells and basophils, all of which bear distinctly different antigen recognition receptors. Although specific ligands for CD45 have yet to be identified, it is envisioned that the structural heterogeneity of the extracellular domain will serve in a variety of capacities, including homotypic cell binding,[40] responses to antigen-specific activation,[34] as well as influencing the proximity of the PTP to intracellular substrates. Experiments in which CD45-deficient T cells were transfected with either the shortest or the longest CD45 isoform suggested that variations in the structure of the extracellular domain impact on intracellular signaling.[41] After activation with anti-TCR (anti-CD3) antibodies, transfected cells exhibited isoform-dependent differences in interleukin-2 production and in protein tyrosine phosphorylation of cellular substrates including Vav.

The transmembrane domain of CD45 may serve in ways other than merely anchoring the PTP into the cell surface membrane.[31] CD45AP (CD45 associated protein) is a 36 kDa putative adaptor protein that coimmunoprecipitates with CD45 from lymphocytes. To examine what parts of the proteins were necessary for their interaction, HeLa cells were transfected with a variety of chimeric constructs involving CD45 and the T cell coreceptor CD8 as well as CD45AP and the tyrosine kinase Src. Only when CD45/CD8 constructs contained the transmembrane domain of CD45 and when CD45AP/Src constructs contained the transmembrane domain of CD45AP did the two reciprocally coimmunoprecipitate. Thus, the transmembrane domains of CD45 and CD45AP were essential for the physical interaction between the two molecules. How critical the transmembrane domain of CD45 is for functional interactions with other surface-associated molecules (such as antigen receptors, see above and below) awaits future study.

The business end of CD45 is the intracellular domain with the tandem protein tyrosine phosphatase segments. This is perhaps best illustrated by the observation that, so long as the intracellular domain is catalytically active,[42,43] the extracellular and transmembrane domains of CD45 can be replaced with those of other surface molecules[43,44] or eliminated altogether[45,46] and still T cells will undergo TCR-dependent signaling events. It should be kept in mind, however, that such an observation is not meant to suggest that the extracellular and transmembrane domains are unnecessary for CD45 to function in vivo. In the context of the quantity and quality of microenvironmental factors that undoubtedly influence cell behavior in intact organisms, the extracellular and transmembrane domains are certain to be important.

The Role of CD45 in Signaling Through the FcεRI

The initial observation that CD45 might be involved in FcεRI-mediated signaling was by Hook et al.[25] In the process of testing mouse mAbs which they had produced against human basophils, they identified two that bound to CD45. The mAbs did not stimulate histamine release from human basophils either directly or indirectly with anti-mouse IgG at concentrations from 0.1-100 μg/ml. However, when incubated for 18 hours at 37°C with basophils, the mAbs inhibited goat anti-IgE induced histamine release in a dose-dependent manner. The higher the basophil concentration in the cell preparation, the more effective the mAbs were at inhibiting histamine release. Optimal temperature for inhibition was 37°C and maximal inhibition of release was reached after 3-5 hours when optimal concentrations of anti-CD45 mAbs were used. When using suboptimal concentrations of the mAbs, 18-hour incubations were needed to achieve a similar magnitude of inhibition. Importantly, neither IgE nor anti-CD45 mAbs were lost after overnight incubations. Because the inhibition required a period of preincubation of the basophils with the anti-CD45 mAbs, it seems likely that metabolic events must take place for the inhibition to occur. Interestingly, mediator release induced by other mechanisms including by the calcium ionophore A23187 was not affected by anti-CD45 mAbs. Thus, CD45 in human basophils probably influences FcεRI-mediated secretion at one or more steps preceding calcium entry into the cytosol from outside the cell.

Not all FcεRI-bearing cells appear to require CD45 for receptor-dependent activation.[38] Using a rat mast cell line (RBL-2H3), CD45-containing and CD45-deficient cells were obtained by cell sorting using a pan-specific anti-CD45 mAb. Both populations, which expressed similar levels of FcεRI by fluorescence intensity, released β-hexosaminidase in response to IgE/antigen provocation. Similar dose-response curves were generated for the two cell lines, although the CD45-deficient cells released about 20% less β-hexosaminidase (40-45% vs. 55-60%) at optimal antigen concentration than the CD45-containing cells did. However, the difference in mediator release is probably not important. It may be more a reflection of the secretory capability of the cells rather than an effect of CD45 deficiency since release in response to calcium ionophore (ionomycin) gave similar results (45 vs. 70%). Moreover, production of inositol triphosphates, which are responsible for releasing calcium from internal stores, was the same in the two cell lines in response to antigen stimulation. Support for these findings comes from a recent report by Swieter et al.[21] They showed that variants of RBL-2H3 cells that express no detectable CD45 release similar amounts of histamine in response to FcεRI crosslinking as do cells with low or high levels of the PTP. Thus, CD45 may not be critical for FcεRI-mediated secretion in these cells. Whether CD45 affects other signaling pathways initiated by receptor aggregation has not been reported.

The most convincing evidence that CD45 is involved in signaling through the FcεRI comes from the experiments of Berger et al.[26] They used CD45 "knock-out" mice generated by replacing through homologous recombination exon 6 of the CD45 gene with cDNA for neomycin resistance.[24] Primary mast cell cultures were grown from bone marrow of normal (+/+) and gene targeted (-/-) mice in the presence of IL-3. By week 3 more than 99% of the cells were mast cells as determined by alcian blue staining or IgE binding. By FACS analysis, anti-CD45 staining was barely, if at all, detectable in the -/- mast cells whereas +/+ mast cells were intensely positive. Importantly, mast cells from both kinds of mice expressed similar levels of FcεRI. After passive sensitization with monoclonal anti-DNP IgE, mast cells from +/+ animals released about 45% of the β-hexosaminidase stored in their cytoplasmic granules, whereas mast cells from -/- mice did not respond at all to antigenic stimulation. Because the calcium ionophore A23187 induced similar levels of mediator release from both +/+ and -/- mast cells, the cellular machinery downstream from the CD45 defect was intact. Thus, CD45 probably exerts its influence on FcεRI-mediated degranulation at some point before calcium enters the cell from the extracellular environment, and CD45 probably does not influence mediator release once calcium has fluxed across the surface membrane.

Although no data were presented regarding calcium flux or tyrosine phosphorylation events in the CD45-deficient mast cells, there is evidence in genetically manipulated T cells that CD45 may be involved in regulating calcium flux in FcεRI-mediated signaling pathways. The α, β and γ subunits of the rat FcεRI were cotransfected into both the CD45-deficient Jurkat T cell line J45.01 (J45ε) and the CD45 containing parental line (JRε).[47] Both sets of transfected cells mobilized calcium in response to ionophore. By contrast, activation of these cells through the FcεRI (with IgE/DNP-HSA or with mouse anti-FcεRI mAbs plus goat anti-mouse antibodies) resulted in calcium influx in the Jkε cells, but not in the J45ε line. Thus, CD45 seems to play a role in FcεRI-induced calcium mobilization.

Such findings are further supported by signaling studies using T cells and B cells. The B cell line J558L µm³ lacks CD45 and fails to mount a calcium mobilization response to BCR aggregation. Transfection of the cells with full length CD45 cDNA is sufficient to restore a normal calcium response following receptor ligation.[48] In other studies, B cells from CD45 knockout mice were examined for the intactness of signaling pathways involving the B cell antigen receptor (BCR).[49] Splenic B cells from -/- mice failed to proliferate in response to BCR crosslinking with anti-IgM antibodies, whereas those from +/+ mice did. Intriguingly, both basal and anti-BCR-induced tyrosine phosphorylations of the Igα component of the BCR complex and of phospholipase C-γ2 were indistinguishable in the B cells from both types of animals. Thus, early signaling events appeared normal. However, when calcium mobilization was measured, calcium from internal stores was released into the cytosol upon BCR ligation, but calcium from the extracellular environment failed to cross the surface membrane. This implicated CD45 in the regulation of calcium influx from extracellular sources during BCR-induced cell activation.

Similar findings have been reported in CD45-deficient T cells.[50,51] Peyron et al[50] describe a CD45-negative Jurkat T cell line in which aggregation of the T cell receptor resulted in some increase in PTK activity and a small, but detectable, increase in intracellular free calcium. However, complete activation leading to the production of lymphokines was not seen. In addition, Volarevic et al[51] found that crosslinking the T cell receptor in a CD45-deficient variant of the YAC-1 mouse thymoma cell line resulted in a delayed, gradual increase in intracellular calcium when compared to wild-type cells. Taken together, the results of experiments with CD45-deficient mast cells, B cells and T cells suggest that one role of CD45 in signaling through multisubunit antigen recognition receptors is to regulate calcium mobilization from the extracellular environment.

Mechanisms of CD45 Action

CD45 has been found to coimmunoprecipitate from quiescent T cells and B cells with the ζ chain of the TCR complex[52] and with the Igα/Igβ heterodimer of the BCR complex,[53] respectively. Because these three receptor components are tyrosine phosphorylated in response to receptor aggregation,[54,55] they are probably substrates of CD45. Such an arrangement would implicate CD45 as a negative regulator of receptor-mediated cell activation in T cells and B cells.[52,53] The FcεRIs on RBL-2H3 cells are also associated with a phosphatase that may function to maintain the receptor subunits in a dephosphorylated state, but the PTP does not appear to be CD45.[21] Using RBL-2H3 cells and several variants that expressed levels of CD45 from undetectable to high, FcεRIs were isolated from unstimulated populations of each cell line and subjected to in vitro phosphatase assays. Regardless of the amount of CD45 expressed in the RBL cells, there were no differences in the levels of receptor-associated PTP activity. Thus, the CD45 does not seem to physically interact with the FcεRI in these cells.

Considering the results of the experiments with CD45-deficient mast cells, T cells and B cells alluded to in the previous section, it is clear that CD45 does not act solely as a negative regulator of cellular activation through the antigen recognition receptors. Rather, one of the major functions of CD45 must be that of a positive regulator. Otherwise the responses of CD45-deficient cells to receptor ligation would be heightened instead of diminished.

The absence of CD45 from T cells impacts at a very early stage on cell activation through the antigen receptor since protein tyrosine phosphorylations are blocked.[56,57] In contrast, CD45 deficiency in B cells does not hinder early signaling events such as the basal or anti-IgM-induced protein tyrosine phosphorylations of Igα or phospholipase C-γ2, an enzyme required for the generation of second messengers that bring about calcium flux.[49] Thus, there appears to be more than one mechanism of action of CD45 on signaling through antigen recognition receptors.

Src kinases have a tyrosine residue at their C-termini that, when phosphorylated, results in downregulation of their catalytic activity. Lck and Fyn are Src family members found in T cells. When Lck is absent from T cells, TCR signaling is impaired and there is a failure to mobilize calcium.[58] Because CD45 dephosphorylates Lck and Fyn in vitro,[59,60] CD45 may function to regulate the activity of Lck and Fyn in T cells. This is supported by the observations that in CD45-deficient T cells the regulatory tyrosines on Lck and Fyn are hyperphosphorylated and become dephosphorylated after transfection of the cells with CD45.[61,62] Fyn, Lyn and Blk are Src kinases found in B cells. Lyn and CD45 both coimmunoprecipitate with the Igα/Igβ heterodimer of the BCR[53] and CD45 dephosphorylates tyrosine phosphorylated Igα/Igβ in vitro.[48] In CD45-deficient immature B cells only Lyn was dysregulated; it was hyperphosphorylated and activated.[63] Hence, CD45 may exert its influence on BCR-induced signaling pathways via Lyn. Last, in RBL-2H3 cells Lyn is constitutively activated and can be coimmunoprecipitated with the FcεRI in both unstimulated and activated cells.[10] It is hyperphosphorylated, perhaps due to the lack of CD45 in these cells.[21] These observations are consistent with a role for CD45 in the regulation of Lyn kinase activity in mast cells. But, clearly, the mechanism by which CD45 controls FcεRI-dependent cell activation requires further investigation.

In summary, CD45 is essential for signaling through the FcεRI. It probably functions as an activator of members of the Src family of kinases and in some way regulates the entry of calcium from the extracellular environment into the cytosol. In these capacities, CD45 functions as a positive modulator of cell activation.

HePTP

Hematopoietic protein tyrosine phosphatase, also known as leucocyte protein tyrosine phosphatase (LC-PTP), is a 40 kDa cytosolic PTP that appears to be predominantly expressed in mast cells and lymphocytes.[39,64,65] In T cells, its mRNA levels increase following activation with any of a number of mitogenic agents including anti-TCR antibodies,[64,66] and in RBL-2H3 mast cells, the HePTP protein becomes tyrosine phosphorylated after FcεRI aggregation.[39] Therefore, HePTP may play a role in regulating signaling pathways initiated at the antigen recognition receptors in these cells.

The intron/exon organization of the HePTP gene has been examined and found to contain 11 exons, nine of which encode protein.[67] Interestingly, the gene maps to the same locus on human chromosome 1 (1q32.1) as does CD45.[67,68] The amino acid sequence of HePTP reveals two domains: a C-terminal consensus protein tyrosine phosphatase domain and an N-terminal noncatalytic end of unknown function. It is likely that the N-terminal half of the PTP serves to localize HePTP to intracellular sites where it performs its functions, or serves as a regulatory domain for PTP activity.[69]

Swieter and co-workers[39] recently showed that HePTP was present in RBL-2H3 mast cells and that it localized to distinct cytoplasmic compartments. Although it failed to coimmunoprecipitate with the FcεRI, HePTP did become tyrosine phosphorylated after aggregation of the FcεRI with anti-receptor antibodies. Therefore, HePTP is implicated as an effector in FcεRI-mediated signaling. HePTP also became phosphorylated in cells stimulated with calcium ionophore so long as calcium was in the extracellular medium, but not in cells treated with the protein kinase C activator, phorbol ester. Thus, tyrosine phosphorylation of HePTP is a relatively late (i.e., calcium requiring) event in FcεRI-mediated signaling and it can occur independent of protein kinase C activation.

OTHER PTP

Hampe and Pecht[70] also found PTP activity in RBL-2H3 cells. They described a PTP activity in membrane fractions that displayed a pH optimum of 5-6 and that increased in activity 2- to 3-fold after FcεRI aggregation. The activity increase was dependent on a rise in intracellular calcium concentration, and could be brought about by stimulating the cells with calcium ionophore. All of these characteristics distinguish this PTP from the one described by Swieter et al.[21] However, it is not known whether the membrane-associated PTP of Hampe and Pecht is one that has already been identified.

Two additional cytosolic PTPs bearing dual tandem Src homology region 2 (SH2) domains have been shown to be present in RBL-2H3 cells by immunoblotting.[21] These are PTP1C (also known as SH-PTP1, hematopoietic cell phosphatase (HCP), or SHP) and PTP1D (also known as SH-PTP2 or Syp). Given that one of the functions of SH2 domains is to bind tyrosylphosphates, it seems likely that these two PTPs would be involved in FcεRI-mediated signaling pathways. However, they do not appear to physically interact with the FcεRI.[21] On the other hand, PTP1C may be involved in down-regulating signals initiated by FcεRI engagement through interactions with an Fcγ receptor.[71] Daëron et al have shown that coaggregation of FcεRI and Fcγ receptors in primary mouse mast cell cultures or in RBL-2H3 cells transfected with murine FcγRIIB1 cDNA inhibits IgE-dependent mediator release. This sort of occurrence is envisioned to happen when immune complexes containing IgE and IgG reach the surfaces of mast cells. But now that it has been shown that certain subclasses of IgG can also activate mast cells through the FcεRI,[72] it is conceivable that aggregates containing IgG alone may be effective. A parallel in the B cell was recently described.[73] Following coaggregation of the BCR and the FcγRIIB1, PTP1C bound to the Fcγ receptor and dephosphorylated the BCR components. In this way, BCR induced signaling was downregulated. Perhaps PTP1C in mast cells performs similar functions.

CONCLUDING REMARKS

Tyrosine phosphorylation of numerous proteins is a characteristic feature of FcεRI aggregation in mast cells and basophils. The phosphorylations come about as a result of a shift in the balance of the orchestrated actions of protein tyrosine kinases and phosphatases, which sometimes work in concert with one another and which at other times work at cross-purposes. Thus, to gain a comprehensive understanding of the signaling pathways that are initiated by aggregating FcεRI in mast cells and basophils, it is necessary to learn the identities and capabilities of the enzymes involved. Over the last few years progress has been made in identifying relevant phosphatases. As more is learned about these and other PTP, the means for controlling undesirable behavior of mast cells and basophils comes closer to our grasp.

REFERENCES

1. Trowbridge IS, Thomas ML. CD45: an emerging role as a protein tyrosine phosphatase required for lymphocyte activation and development. Ann Rev Immunol 1994; 12:85-16.
2. Stone RL, Dixon JE. Protein-tyrosine phosphatases. J Biol Chem 1994; 269:31323-26.

3. Frangioni JV, Beahm PH, Shifrin V et al. The nontransmembrane tyrosine phosphatase PTP-1B localizes to the endoplasmic reticulum via its 35 amino acid C-terminal sequence. Cell 1992; 68:545-60.
4. McLaughlin S, Dixon JE. Alternative splicing gives rise to a nuclear protein tyrosine phosphatase in *Drosophila*. J Biol Chem 1993; 268:6839-42.
5. Chan AC, Desai DM, Weiss A. The role of protein tyrosine kinases and protein tyrosine phosphatases in T cell antigen receptor signal transduction. Ann Rev Immunol 1994; 12:555-92.
6. Benhamou M, Gutkind JS, Robbins KC et al. Tyrosine phosphorylation coupled to IgE receptor-mediated signal transduction and histamine release. Proc Nat Acad Sci USA 1990; 87:5327-30.
7. Paolini R, Jouvin MH, Kinet JP. Phosphorylation and dephosphorylation of the high-affinity receptor for immunoglobulin E immediately after receptor engagement and disengagement. Nature 1991; 353:855-58.
8. Park DJ, Min HK, Rhee SG. IgE-induced tyrosine phosphorylation of phospholipase C-γ1 in rat basophilic leukemia cells. J Biol Chem 1991; 266:24237-40.
9. Benhamou M, Stephan V, Robbins KC et al. High-affinity IgE receptor-mediated stimulation of rat basophilic leukemia (RBL-2H3) cells induces early and late protein-tyrosine phosphorylations. J Biol Chem 1992; 267:7310-14.
10. Eiseman E, Bolen JB. Engagement of the high-affinity IgE receptor activates *src* protein-related tyrosine kinases. Nature 1992; 355:78-80.
11. Hutchcroft JE, Geahlen RL, Deanin GG et al. FcεRI-mediated tyrosine phosphorylation and activation of the 72 kDa protein-tyrosine kinase, PTK72, in RBL-2H3 rat tumor mast cells. Proc Nat Acad Sci USA 1992; 89:9107-11.
12. Kawakami T, Inagaki N, Takei M et al. Tyrosine phosphorylation is required for mast cell activation by FcεRI crosslinking. J Immunol 1992; 148:3513-19.
13. Li W, Deanin GG, Margolis B et al. FcεRI-mediated tyrosine phosphorylation of multiple proteins, including phospholipase Cγ1 and the receptor βγ complex, in RBL-2H3 rat basophilic leukemia cells. Mol Cell Biol 1992; 12:3176-82.
14. Margolis, B, Hu P, Katzav S et al. Tyrosine phosphorylation of vav proto-oncogene product containing SH2 domain and transcription factor motifs. Nature 1992; 356:71-74.
15. Benhamou M, Ryba NJP, Kihara H et al. Protein-tyrosine kinase $p72^{syk}$ in high affinity IgE receptor signaling. J Biol Chem 1993; 268:23318-24.
16. Hamawy MM, Mergenhagen SE, Siraganian RP. Cell adherence to fibronectin and the aggregation of the high affinity immunoglobulin E receptor synergistically regulate tyrosine phosphorylation of 105-115 kDa proteins. J Biol Chem 1993; 268:5227-33.
17. Hamawy MM, Mergenhagen SE, Siraganian RP. Tyrosine phosphorylation of $pp125^{FAK}$ by the aggregation of high affinity immunoglobulin E receptors requires cell adherence. J Biol Chem 1993; 268:6851-54.
18. Minoguchi K, Benhamou M, Swaim WD et al. Activation of protein tyrosine kinase $p72^{syk}$ by FcεRI aggregation in rat basophilic leukemia cells. $p72^{syk}$ is a minor component but the major protein tyrosine kinase of pp72. J Biol Chem 1994; 269:16902-908.
19. Swaim WD, Minoguchi K, Oliver C et al. The anti-ganglioside monoclonal antibody AA4 induces protein tyrosine phosphorylations, but not degranulation, in rat basophilic leukemia cells. J Biol Chem 1994; 269:19466-73.
20. Jouvin MHE, Adamczewski M, Numerof R et al. Differential control of the tyrosine kinases Lyn and Syk by the two signaling chains of the high affinity immunoglobulin E receptor. J Biol Chem 1994; 269:5918-25.
21. Swieter M, Berenstein EH, Siraganian RP. Protein tyrosine phosphatase activity associates with the high affinity IgE receptor and dephosphorylates the receptor subunits, but not Lyn or Syk. J Immunol 1995; 155:5330-36.
22. Pingel JT, Thomas ML. Evidence that the leukocyte-common antigen is required for antigen-induced T lymphocyte proliferation. Cell 1989; 58:1055-65.
23. Koretzky GA. Role of the CD45 tyrosine

phosphatase in signal transduction in the immune system. FASEB J 1993; 7:420-26.
24. Kishihara K, Penninger J, Wallace VA et al. Normal B lymphocyte development but impaired T cell maturation in CD45-exon6 protein tyrosine phosphatase-deficient mice. Cell 1993; 74:143-56.
25. Hook WA, Berenstein EH, Zinsser FU et al. Monoclonal antibodies to the leukocyte common antigen (CD45) inhibit IgE-mediated histamine release from human basophils. J Immunol 1991; 147:2670-76.
26. Berger SA, Mak TW, Paige CJ. Leukocyte common antigen (CD45) is required for immunoglobulin E-mediated degranulation of mast cells. J Exp Med 1994; 180:471-76.
27. Thomas ML. The leukocyte common antigen family. Ann Rev Immunol 1989; 7:339-69.
28. Hathcock KS, Laszlo G, Dickler HB et al. Expression of variable exon A-, B-, C-specific CD45 determinants on peripheral and thymic T cell populations. J Immunol 1992; 148:19-28.
29. Birkeland ML, Johnson P, Trowbridge IS et al. Changes in CD45 isoform expression accompany antigen-induced murine T cell activation. Proc Nat Acad Sci USA 1989; 86:6734-38.
30. Rothstein DM, Yamada A, Schlossman SF et al. Cyclic regulation of CD45 isoform expression in a long-term human CD4+ CD45RA+ T cell line. J Immunol 1991; 146:1175-83.
31. Cahir McFarland ED, Thomas ML. CD45 protein-tyrosine phosphatase associates with the WW domain-containing protein, CD45AP, through the transmembrane region. J Biol Chem 1995; 270:28103-107.
32. Shen FW, Tung JS, Boyse EA. Further definition of the Ly-5 system. Immunogenetics 1986; 24:146-49.
33. Ralph SJ, Thomas ML, Morton C et al. Structural variants of human T200 glycoprotein (leukocyte-common antigen). EMBO J 1987; 6:1251-57.
34. Leitenberg D, Novak TJ, Farber D et al. The extracellular domain of CD45 controls association with the CD4-T cell receptor complex and the response to antigen-specific stimulation. J Exp Med 1996; 183:249-59.
35. Thomas ML, Reynolds PJ, Chain A et al. B cell variant of mouse T200 (Ly-5): evidence for alternative mRNA splicing. Proc Nat Acad Sci USA 1987; 84:5360-63.
36. Barclay AN, Jackson DI, Willis AC et al. Lymphocyte specific heterogeneity in the rat leucocyte common antigen (T200) is due to differences in polypeptide sequences near the NH_2-terminus. EMBO J 1987; 6:1259-64.
37. Saga Y, Tung JS, Shen FW et al. Alternative use of 5' exons in the specification of Ly-5 isoforms distinguishing hematopoietic cell lineages. Proc Nat Acad Sci USA 1987; 84:5364-68.
38. Schneider H, Korn M, Haustein D. CD45-deficient RBL-2H3 cells. Cellular responses to FcεR-induced stimulation. Immunol Invest 1993; 22:503-15.
39. Swieter M, Berenstein EH, Swaim WD et al. Aggregation of IgE receptors in rat basophilic leukemia 2H3 cells induces tyrosine phosphorylation of the cytosolic protein-tyrosine phosphatase HePTP. J Biol Chem 1995; 270:21902-906.
40. Spertini F, Wang AVT, Chatila T et al. Engagement of the common leukocyte antigen CD45 induces homotypic adhesion of activated human T cells. J Immunol 1994; 153:1593-602.
41. McKenney DW, Onodera H, Gorman L et al. Distinct isoforms of the CD45 protein-tyrosine phosphatase differentially regulate interleukin 2 secretion and activation signal pathways involving vav in T cells. J Biol Chem 1995; 270:24949-54.
42. Desai DM, Sap J, Silvennoinen O et al. The catalytic activity of the CD45 membrane-proximal phosphatase domain is required for TCR signaling and regulation. EMBO J 1994; 13:4002-10.
43. Motto DG, Musci MA, Koretzky GA. Surface expression of a heterologous phosphatase complements CD45 deficiency in a T cell clone. J Exp Med 1994; 180:1359-66.
44. Desai DM, Sap J, Schlessinger J et al. Ligand-mediated negative regulation of a chimeric transmembrane receptor tyrosine phosphatase. Cell 1993; 73:541-54.
45. Volarevic S, Niklinska BB, Burns CM et al. Regulation of TCR signaling by CD45 lacking transmembrane and extracellular do-

mains. Science 1993; 260:541-44.
46. Niklinska BB, Hou D, June C et al. CD45 tyrosine phosphatase activity and membrane anchoring are required for T cell antigen receptor signaling. Mol Cell Biol 1994; 14:8078-84.
47. Adamczewski M, Numerof RP, Koretzky GA et al. Regulation by CD45 of the tyrosine phosphorylation of high affinity IgE receptor β- and γ-chains. J Immunol 1995; 154:3047-55.
48. Justement LB, Campbell KS, Chien NC et al. Regulation of B cell antigen receptor signal transduction and phosphorylation by CD45. Science 1991; 252:1839-42.
49. Benatar T, Carsetti R, Furlonger C et al. Immunoglobulin-mediated signal transduction in B cells from CD45-deficient mice. J Exp Med 1996; 183:329-34.
50. Peyron JF, Verma S, Malefyt RD et al. The CD45 protein tyrosine phosphatase is required for the completion of the activation program leading to lymphokine production in the Jurkat human T cell line. Int Immunol 1991; 3:1357-66.
51. Volarevic S, Niklinska BB, Burns CM et al. The CD45 tyrosine phosphatase regulates phosphotyrosine homeostasis and its loss reveals a novel pattern of late T cell receptor-induced Ca^{2+} oscillations. J Exp Med 1992; 176:835-44.
52. Furukawa T, Itoh M, Krueger NX et al. Specific interaction of the CD45 protein-tyrosine phosphatase with tyrosine-phosphorylated CD3 ζ chain. Proc Nat Acad Sci USA 1994; 91:10928-32.
53. Brown VK, Ogle EW, Burkhardt AL et al. Multiple components of the B cell antigen receptor complex associate with the protein tyrosine phosphatase, CD45. J Biol Chem 1994; 269:17238-44.
54. June CH, Fletcher MC, Ledbetter JA et al. Increases in tyrosine phosphorylation are detectable before phospholipase C activation after T cell receptor stimulation. J Immunol 1990; 144:1591-99.
55. Gold MR, Matsuchi L, Kelly RB et al. Tyrosine phosphorylation of components of the B cell antigen receptors following receptor crosslinking. Proc Nat Acad Sci USA 1991; 88:3436-40.
56. Koretzky GA, Picus J, Schultz Tet al. Tyrosine phosphatase CD45 is required for T cell antigen receptor and CD2-mediated activation of a protein tyrosine kinase and interleukin 2 production. Proc Nat Acad Sci USA 1991; 88:2037-41.
57. Koretzky GA, Picus J, Thomas ML et al. Tyrosine phosphatase CD45 is essential for coupling T cell antigen receptor to the phosphatidylinositol pathway. Nature 1990; 346:66-68.
58. Straus D, Weiss A. Genetic evidence for the involvement of the lck tyrosine kinase in signal transduction through the T cell antigen receptor. Cell 1992; 70:585-93.
59. Mustelin T, Altman A. Dephosphorylation and activation of the T cell tyrosine kinase $p56^{lck}$ by the leukocyte common antigen (CD45). Oncogene 1990; 5:809-13.
60. Mustelin T, Pessa-Morikawa T, Autero M et al. Regulation of the $p59^{fyn}$ protein tyrosine kinase by the CD45 phosphotyrosine phosphatase. Eur J Immunol 1992; 22: 1173-78.
61. Cahir McFarland ED, Hurley TR, Pingel JT et al. Correlation between Src family member regulation by the protein-tyrosine-phosphatase CD45 and transmembrane signaling through the T cell receptor. Proc Nat Acad Sci USA 1993; 90:1402-6.
62. Burns CM, Sakaguchi K, Appella E et al. CD45 regulation of tyrosine phosphorylation and enzyme activity of *src* family kinases. J Biol Chem 1994; 269:13594-600.
63. Katagiri T, Ogimoto M, Hasegawa K et al. Selective regulation of Lyn tyrosine kinase by CD45 in immature B cells. J Biol Chem 1995; 270:27987-90.
64. Zanke B, Suzuki H, Kishihara K et al. Cloning and expression of an inducible lymphoid-specific, protein tyrosine phosphatase (HePTPase). Eur J Immunol 1992; 22: 235-39.
65. Adachi M, Sekiya M, Isobe M et al. Molecular cloning and chromosomal mapping of a human protein-tyrosine phosphatase LC-PTP. Biochem Biophys Res Commun 1992; 186:1607-15.
66. Adachi M, Sekiya M, Ishino M et al. Induction of protein-tyrosine phosphatase LC-PTP by IL-2 in human T cells. LC-PTP is an

early response gene. FEBS Lett 1994; 338:47-52.

67. Adachi M, Miyachi T, Sekiya M et al. Structure of the human LC-PTP (HePTP) gene: similarity in genomic organization within protein-tyrosine phosphatase genes. Oncogene 1994; 9:3031-35.
68. Zanke B, Squire J, Griesser H et al. A hematopoietic protein tyrosine phosphatase (HePTP) gene that is amplified and overexpressed in myeloid malignancies maps to chromosome 1q32.1. Leukemia 1994; 8:236-44.
69. Mauro LJ, Dixon JE. 'Zip codes' direct intracellular protein tyrosine phosphatases to the correct cellular 'address'. TIBS 1994; 19:151-55.
70. Hampe CS, Pecht I. Protein tyrosine phosphatase activity enhancement is induced upon Fcε receptor activation of mast cells. FEBS Lett 1994; 346:194-98.
71. Daëron M, Malbec O, Latour S. Regulation of high-affinity IgE receptor-mediated mast cell activation by murine low-affinity IgG receptors. J Clin Invest 1995; 95:577-85.
72. Benhamou M, Berenstein EH, Jouvin, MH et al. The receptor with high affinity for IgE on rat mast cells is a functional receptor for rat IgG2a. Mol Immunol 1994; 31:1089-97.
73. D'Ambrosio D, Hippen KL, Minskoff SA et al. Recruitment and activation of PTP1C in negative regulation of antigen receptor signaling by FcγRIIB1. Science 1995; 268:293-97.

CHAPTER 8

Regulation and Roles of the Membrane, Cytoskeletal and Adhesive Responses of RBL-2H3 Rat Tumor Mast Cells to FcεRI Crosslinking

Janet M. Oliver, Janet R. Pfeiffer and Bridget S. Wilson

INTRODUCTION

Mast cells, basophils and RBL-2H3 rat tumor mast cells are activated by treatments that crosslink their high affinity IgE receptor, FcεRI. Receptor crosslinking causes degranulation, releasing inflammatory agents that initiate the immediate hypersensitivity diseases including allergy, asthma and anaphylaxis. Receptor crosslinking also stimulates the production of lipid mediators that amplify the immediate inflammatory response and of cytokines that maintain inflammation through the recruitment and activation of other pro-inflammatory cells. Additionally, FcεRI crosslinking induces a series of immediate and dramatic changes in the membrane, cytoskeletal and adhesive properties of basophils and mast cells that are the subject of this chapter.

We begin by reviewing and illustrating the membrane, cytoskeletal and adhesive responses of RBL-2H3 mast cells to FcεRI crosslinking. We then discuss the biochemical pathways activated by FcεRI crosslinking that may initiate and regulate these responses. Finally, we present evidence linking membrane, cytoskeletal and adhesive responses not only to the regulation of mast cell and basophil migration and localization but also to the control of the physical properties and topographical distribution

IgE Receptor (FcεRI) Function in Mast Cells and Basophils,
edited by Majed M. Hamawy.

of the crosslinked receptors and to the modulation of FcεRI-mediated secretion. This subject was previously reviewed by Oliver et al.[1]

THE MEMBRANE, CYTOSKELETAL AND ADHESIVE RESPONSES OF RBL-2H3 CELLS TO FcεRI CROSSLINKING

ANTIGEN-INDUCED MEMBRANE RESPONSES

Ruffling

The FcεRI of IgE-primed mast cells and basophils can be crosslinked by specific antigen or anti-IgE antibody. The FcεRI can also be crosslinked directly by IgE polymers or antibody specific for the FcεRI-α subunit. The most dramatic response of RBL-2H3 cells to these various crosslinking agents occurs at the level of cell surface topography.[1,2] Before crosslinking, RBL-2H3 cells grown in suspension or on a surface maintain a randomly microvillous membrane architecture (Figs. 8.1A and 8.2A). FcεRI priming, usually accomplished in RBL-2H3 cells by the addition of anti-DNP-IgE, has no effect on membrane topography. However, ligands (typically DNP-BSA or anti-IgE) that crosslink the IgE-FcεRI complexes initiate a sequence of events in which microvilli are first aligned into rows and are then transformed into lamellae by a process that involves the formation first of fringed lamellae and then of smooth-surfaced lamellae. This remarkable process is illustrated in Figures 8.1 and 8.2. Aligned microvilli are typically seen within 30 seconds after the addition of crosslinking agent (Fig. 8.2B), and fringed lamellae are present between 1-3 minutes of crosslinking (Figs. 8.1B and 8.2C). By 5 minutes, only smooth-surfaced lamellae are visible at the cell surface (Figs. 8.1C,D and 8.2D). The same antigen-induced transformation of surface topography occurs in suspended and adherent cells. In adherent cells, the continued remodeling of the cell surface associated with spreading (below) results in a further transition from a rose-like to a honeycombed surface topography (Fig. 8.1D).

It is clear from video analyses that the lamellae are highly dynamic. They are also persistent. Thus, antigen concentrations that stimulate maximum secretion in a 20-minute assay also induce ruffling responses that persist for at least 20 minutes. Like secretion, membrane ruffling induced by multivalent ligand (DNP-BSA, anti-IgE) can be reversed at any time by adding an excess of monovalent hapten (DNP-lysine, Fab fragments of anti-IgE) to halt continued crosslinking. Fringed lamellae and aligned microvilli reappear during this morphological reversal.

Antigen-induced ruffling is accompanied by cell stiffening, demonstrated by Liu et al[3] as an increased resistance of cells grown on poly-HEMA-coated surfaces (to prevent attachment and spreading) to deformation by the glass probe of the Cell Poker developed by Elson and colleagues. This antigen-induced stiffening, interpreted as a contractile response, is reversed by the addition of monovalent hapten. We speculate that the contractile response to FcεRI crosslinking may be responsible for the initial alignment of microvilli. The aligned microvilli may then provide a scaffold that is filled by the antigen-induced polymerization of actin (described below), generating first fringed and then smooth lamellae.

Ligand-induced ruffling is not unique to RBL-2H3 cells. PDGF-treated fibroblasts, EGF-treated A431 cells, MDCK and PTK2 cells treated with hepatocyte growth factor/scatter factor (HGF/SF) and neutrophils treated with formyl peptide all show membrane ruffling responses.[2,4-6] Nevertheless, the sequence of events involved in this transformation of surface topography from randomly aligned microvilli, to aligned microvilli, to fringed and ultimately smooth-surfaced lamellae has only been described in the RBL-2H3 cell system.

Macropinocytosis

Antigen-induced ruffling in RBL-2H3 cells is associated with a dramatic increase in fluid pinocytosis (macropinocytosis), measured in the flow cytometer from the increased uptake of high molecular weight fluorescein-conjugated dextrans by antigen-activated RBL-2H3 cells.[2,7,8] Like antigen-induced ruffling, macropinocytosis depends on continued FcεRI crosslinking and returns rapidly to basal rates when an excess of competing monovalent hapten is added. Increased fluid pinocytosis has been linked to the onset of receptor-mediated membrane ruffling in other cell types;[1,2,4] Dowrick et al[4] have presented evidence that it may occur by a process involving

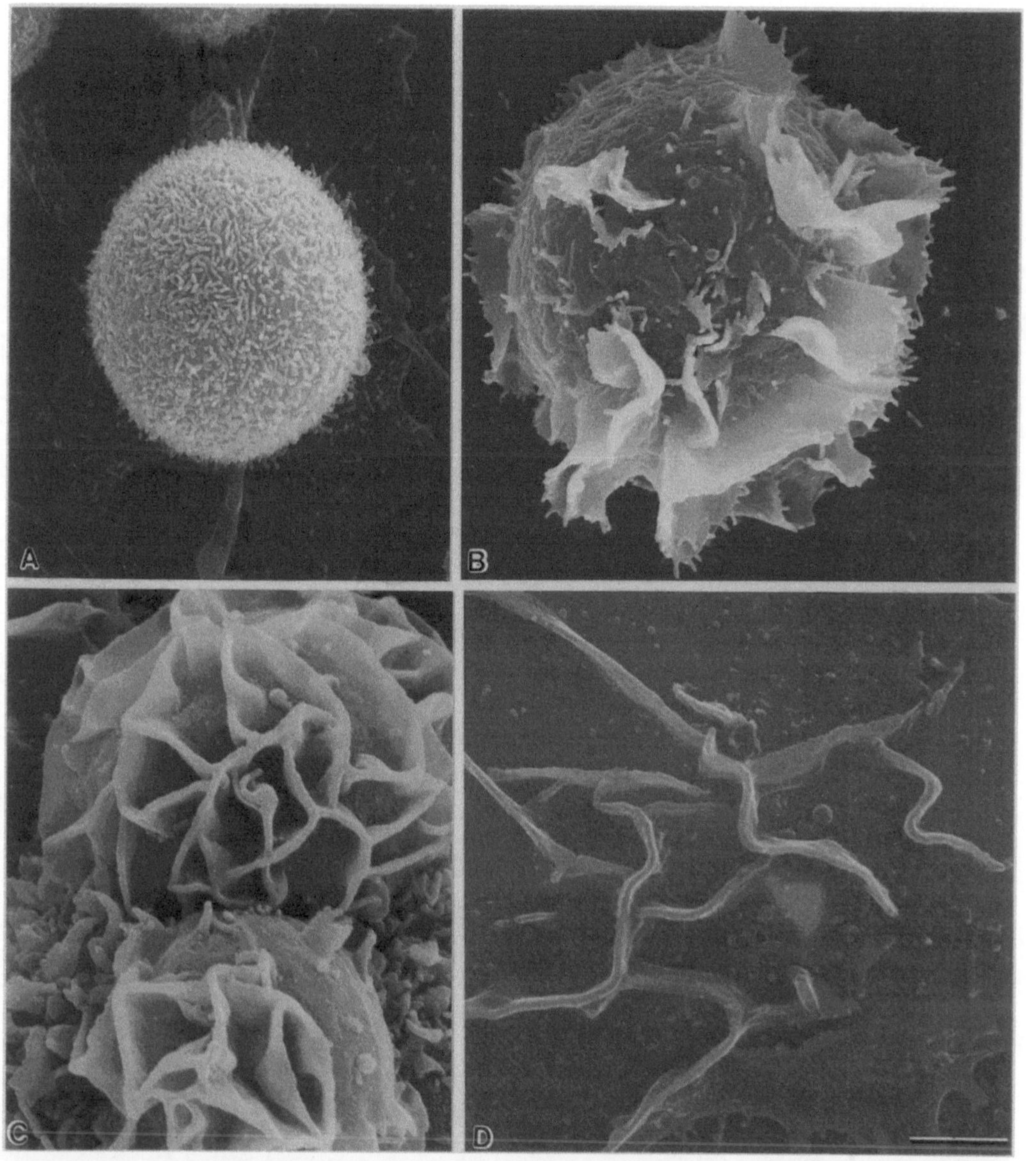

Fig. 8.1. FcεRI crosslinking transforms cell surface topography. Monolayers of IgE-primed RBL-2H3 cells were incubated without crosslinker (A) or with 1.0 μg/ml anti-IgE for 3 minutes (B), 7 minutes (C) or 15 minutes (D) at 37°C, then fixed and processed for scanning electron microscopy. Unstimulated RBL-2H3 cells maintain a microvillous surface architecture (A). After FcεRI crosslinking, the cell surface is characterized by the presence first of fringed lamellae (B), then a complex pattern of smooth-surfaced lamellae (C), and eventually by a simplified lamellar architecture (D). Bar = 5 μm.

the contraction and closure of ruffles. It is clear from recent work[8] that fluid uptake coupled to the ruffling response of antigen-treated RBL-2H3 cells is regulated separately from the coated pit-mediated endocytic pathway, described below, that mediates the internalization of crosslinked receptors.

Antigen-Induced Cytoskeletal Responses

Actin assembly

Not surprisingly, the antigen-induced reorganization of cell surface architecture is accompanied by changes in the levels of F-actin in RBL-2H3 cells. F-actin levels

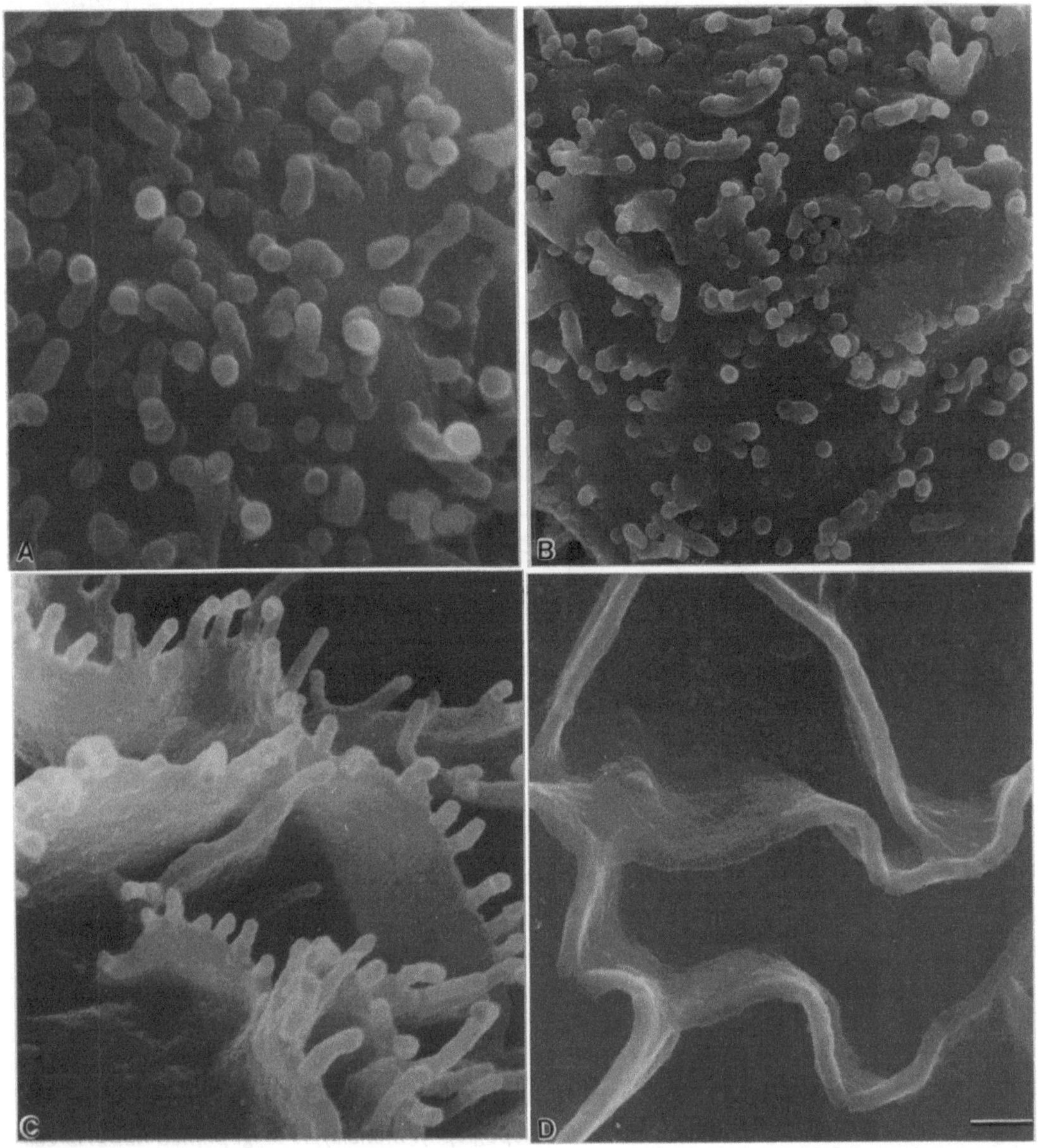

Fig. 8.2. Stages in the transformation of surface topography. Cells were activated and observed as described in Figure 8.1. Photography at higher magnification demonstrates the transition of surface architecture from randomly distributed microvilli (A; unstimulated cells), to aligned microvilli (B; 30 sec activation), to fringed lamellae (C; 3-minute activation) to smooth-surfaced lamellae (D; 7-minute activation). Bar = 1 μm.

can be measured by flow cytometry following incubation of fixed, permeabilized RBL-2H3 cells with rhodamine- or fluorescein-conjugated phalloidin.[2,8-11] FcεRI crosslinking induces an increase in F-actin levels to between 30-50% above resting levels. The increase is measurable within 15 seconds of antigen binding, reaches an initial peak by 30 seconds to 1 minute, and then increases slowly to a plateau level over the next 2-4 minutes (Fig. 8.3). A transient phase of actin disassembly separates the rapid actin assembly response (0-1 minute) from the slower increase to a plateau level (2-4 minutes)[8] (illustrated in Fig. 8.3). This phase coincides in time with the transition from fringed to smooth lamellae that was illustrated in Figs. 8.1 and 8.2.

Brief incubation of RBL-2H3 cells with dihydrocytochalasin B (dHCB) and other cytochalasin derivatives that cap actin filaments has little effect on resting levels of F-actin but prevents the antigen-induced increase in actin polymerization. We showed previously[1] that dHCB treatment also has little effect on the surface morphology of unstimulated cells. However, dHCB inhibits ruffling in response to FcεRI crosslinking and causes the microvilli of adherent cells to redistribute to one pole of the cell, resulting in a cap-like morphology.

Actin distribution

The ruffling response to FcεRI crosslinking is also associated with striking changes in the organization of the membrane-associated microfilament system. We showed that the microvilli of unstimulated RBL-2H3 cells are supported by actin bundles that originate in the cell cortex and extend to the tips of the microvilli. In contrast, the lamellae are supported by a microfilament-rich meshwork that excludes organelles and granules.[1] The concentration of F-actin in the lamellae of activated RBL-2H3 cells is readily illustrated by conventional fluorescence microscopy of

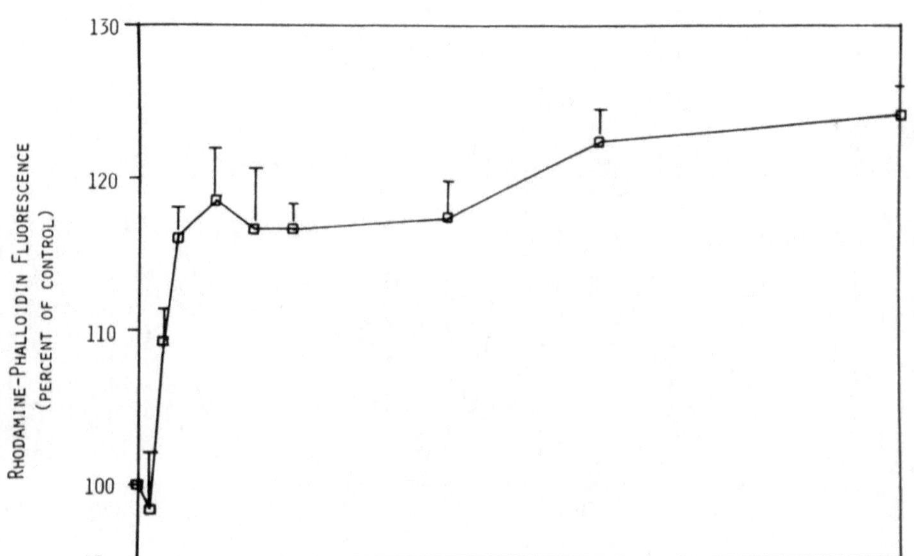

Fig. 8.3. FcεRI crosslinking stimulates actin polymerization. IgE-primed cells were fixed at various times after the addition of antigen and F-actin levels were measured by flow cytometry following incubation with rhodamine-phalloidin. Reproduced from Pfeiffer et al, The Journal of Cell Biology 1985; 101:2145-2155 by copyright permission of The Rockefeller University Press.

the dorsal surface of rhodamine-phalloidin-labelled cells (Fig. 8.4E,F). The results of whole mount electron microscopy by Spudich[12] showed elegantly that these lamellae consist of aligned bundles of microfilaments connected by microfilament meshworks. Using confocal microscopy, Ludowyke et al[13,14] have observed an additional accumulation of F-actin in the perinuclear region of activated RBL-2H3 cells. This antigen-induced centripetal accumulation of F-actin was also observed in primary peritoneal mast cells[15,16] (see also chapter 5).

Additional actin-containing structures, that we named actin plaques,[17] form at the cell-substrate interface of antigen-activated RBL-2H3 cells. By TEM (Fig. 8.5A), they consist of a dense meshwork of microfilaments associated with sites of close cell-

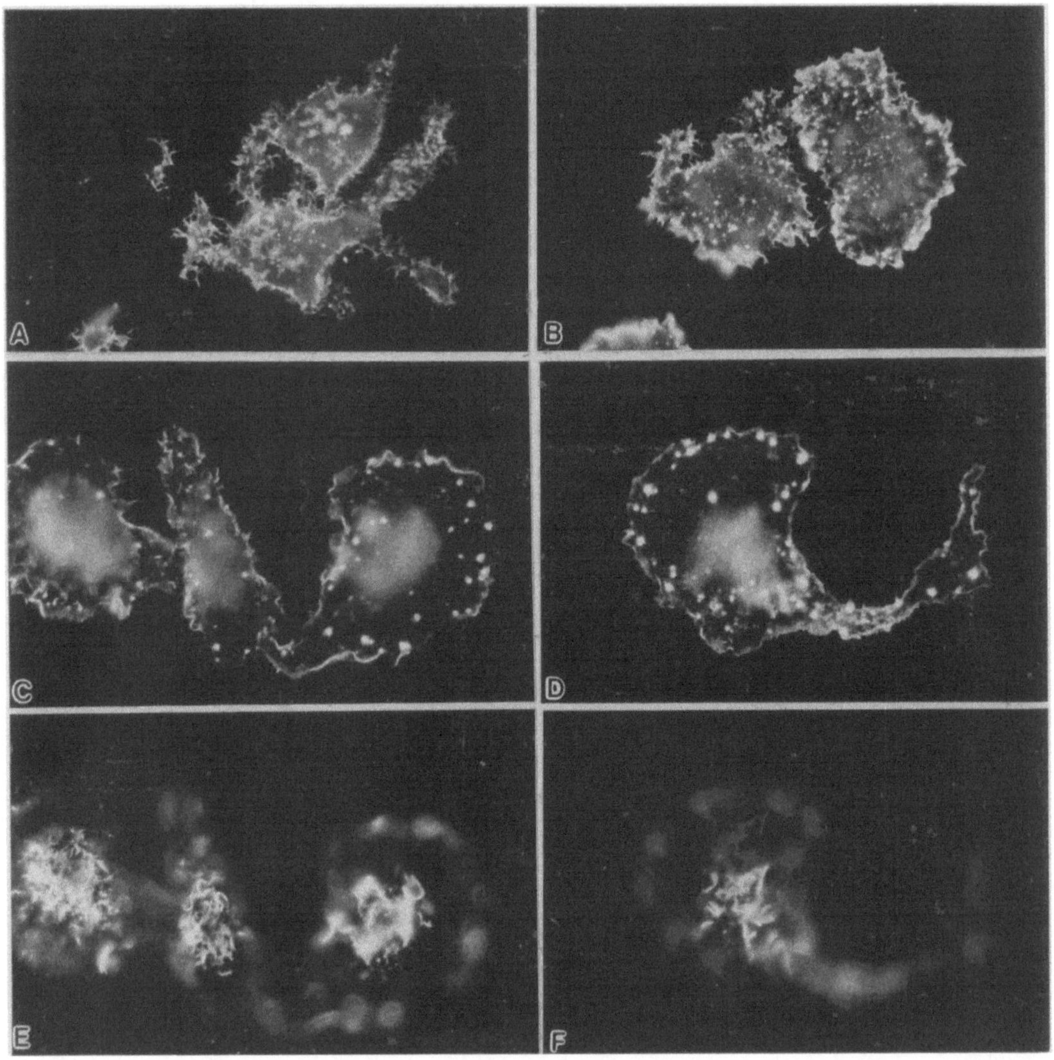

Fig. 8.4. FcεRI crosslinking induces actin plaque assembly in adherent cells. IgE-primed cell monolayers were fixed without antigen treatment (A) and following activation with 1 µg/ml DNP-BSA for 2 minutes (B), 5 minutes (C,E) and 15 minutes (D,F). F-actin was labeled with rhodamine-phalloidin and its association with actin plaques and with ruffles was photographed by focusing at either the ventral cell surface (plaques; A-D) or the dorsal cell surface (ruffles; E,F). Reproduced with permission from Pfeiffer et al, J Cell Biol 1994; 152:270-279. Copyright 1994, The American Association of Immunologists.

substrate contact.[1,17] By fluorescence microscopy of rhodamine-phalloidin-labelled cells (Fig. 8.5B), they appear as intensely fluorescent dots that are distributed over the entire ventral cell surface but reach their highest concentration near the cell periphery. Actin plaques assemble rapidly, being visible after 2 minutes of FcεRI crosslinking (Fig. 8.4B) and fully developed within 5 minutes (Fig. 8.4C). They develop in antigen-activated cell monolayers formed on glass, plastic and tissue culture surfaces, but are most prominent in cells plated onto fibronectin-coated surfaces. They are highly dynamic, disassembling within minutes after the addition of competing monovalent hapten to antigen-stimulated cell monolayers. Actin filament bundles

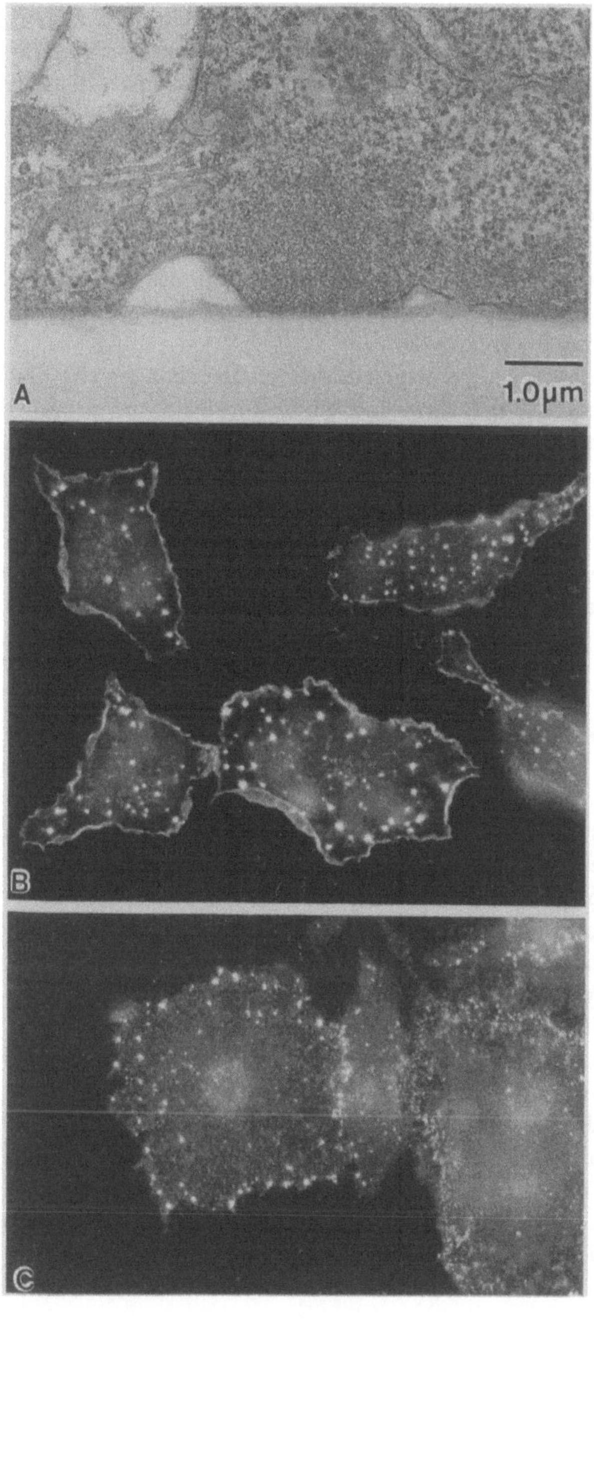

Fig. 8.5. The morphology and composition of actin plaques. IgE-primed adherent cell monolayers were incubated with 1 μg/ml DNP-BSA for 15 minutes. Panels A and B show the concentration of F-actin in actin plaques respectively by TEM and by fluorescence microscopy of rhodamine-phalloidin-labelled cells. Panel C shows the accumulation of anti-phosphotyrosine antibody in actin plaques. Modified with permission from Pfeiffer et al, J Cell Biol 1994; 152:270-279. Copyright 1994, The American Association of Immunologists.

interconnect the plaques. No analogous structures form when cells are activated in suspension, indicating that actin plaque assembly depends both on FcεRI crosslinking and on cell-substratum interaction.

The discovery of actin plaques in antigen-stimulated RBL-2H3 cells adds to the list of actin-based structures implicated in cell adherence to surfaces. These include focal adhesions,[18-20] that subtend the substratum at the periphery of serum-grown fibroblasts; rosettes (also called podosomes);[19,21-23] F-actin aggregates that replace focal adhesions when monolayers of fibroblasts are transformed with a range of viruses that encode active protein-tyrosine kinases; podosomes (also called actin dots),[24-25] that occur in a variety of adherent cells of the myeloid lineage, including monocytes, macrophages and osteoclasts; and point contacts[27] found at the cell-substratum interface in neural cells. Structurally, actin plaques are most like the plug-like podosomes of macrophages and least like focal adhesions that maintain a tangential rather than a perpendicular orientation with respect to the adherent membrane. The rapid assembly and disassembly of actin plaques differentiates them from both focal adhesions and macrophage podosomes, that appear to be slow to assemble and disassemble.

Biochemical and topographical changes in other cytoskeletal proteins

Work by Beaven and colleagues[28,29] has shown that FcεRI crosslinking results in the increased phosphorylation of the heavy and light chains of RBL-2H3 myosin II. Both protein kinase C (PKC) and myosin light chain kinase were implicated in these responses. Antigen-induced phosphorylation is accompanied by a rapid redistribution of myosin II away from the actin meshwork-enriched cell cortex and into the perinuclear area of RBL-2H3 cells.[12] Ludowyke et al[13,14] have reported a similar redistribution of myosin away from the cell cortex and into a cytoplasmic network concentrated perinuclearly in permeabilized RBL-2H3 cells stimulated with PMA and Ca^{2+} ionophore and with ATPγS. The relationship of the biochemical and morphological observations has not been studied in detail. However, recent data suggest that myosin light chain phosphorylation can inhibit actin-myosin interaction[30] and that myosin heavy chain phosphorylation causes thick filament disassembly.[31] It is thus possible that the antigen-induced phosphorylation of myosin localized near the cell periphery could disrupt membrane-associated actin-myosin complexes, simultaneously providing a pool of cortical actin needed for the assembly of lamellae and a pool of cytoplasmic myosin that could, following dephosphorylation and interaction with newly polymerized F-actin in the perinuclear region, bring about the stiffening or contractile response postulated above to contribute to the alignment of microvilli.

Antigen binding also changes the distribution and phosphorylation of proteins previously detected in other adhesion structures. Thus, we[17] have observed by immunofluorescence microscopy that actin plaques are weakly reactive with antibodies to talin and vinculin and strongly reactive with antiphosphotyrosine antibody (Fig. 8.5C). Talin, vinculin and anti-phosphotyrosine-reactive proteins were found previously in the focal adhesions of fibroblasts, in the rosette structures of transformed fibroblasts and in macrophage podosomes.[19,27] Actin plaques also bind antibody to the p80/85 protein, cortactin, previously identified as a component of fibroblast rosettes.[32]

The protein-tyrosine kinases, focal adhesion kinase (FAK) and Src, have been implicated in the assembly of focal adhesions in fibroblasts.[33,34] Biochemical studies by Hamawy et al[35,36] have demonstrated that FcεRI crosslinking in adherent RBL-2H3 cells induces the tyrosine phosphorylation of FAK and also of paxillin, a component of fibroblast focal adhesions.[37] However FAK has not yet been localized to actin plaques.

Antigen-Induced Adhesive Responses

Footpads

Unstimulated RBL-2H3 cells attach loosely to surfaces, forming monolayers that are readily detached in a buffer stream. FcεRI crosslinking induces a marked spreading response in adherent RBL-2H3 cells, accompanied by a substantial increase in their resistance to detachment by shear forces. The especially tight apposition of the ventral membrane to the substratum at actin plaques (Fig. 8.5A) strongly suggested a role for these specialized structures in both the spreading and adhesion responses of RBL-2H3 cells to FcεRI crosslinking. This suggestion was reinforced by the results of experiments in which monolayers of resting and activated cells were incubated with a $ZnCl_2$-containing buffer intended to "stiffen" the membrane, followed by shearing in a buffer stream. Most unstimulated cells were detached completely by this treatment, although occasional cells left remnants like those illustrated in Figure 8.6A. In contrast, the cell bodies were detached from monolayers of activated cells, leaving behind sheets of ventral membrane bearing actin plaques ("footpads") on the substrate (Fig. 8.6B).[17] These preparations link actin plaque assembly to the antigen-induced adhesive and spreading response. They provide a starting point for biochemical analyses of these unique structures.

The role of integrin activation in FcεRI-stimulated adhesion

Current studies suggest that integrins are the principal molecules mediating mast cell and basophil adhesion to other cells and to matrix molecules. In particular, Valent and Bettelheim,[38] Hamawy et al[39] and Ra et al[40] have reviewed evidence that mast cells, including RBL-2H3, express substantial amounts of the β1 integrins, VLA-4 (CD49d/CD29; α4β1) and VLA-5 (CD49e/CD29; α5β1) that bind fibronectin (VLA-4 also binds VCAM-1) and of the β3 integrin, VNR (CD51/CD61), that binds vitronectin. Basophils express the same β1 integrins and additionally express the β2 integrins, LFA-1 (CD11a/CD18), Mac1 (CD11b/CD18) and p150/95 (CD11c/CD18). Basophils and mast cells also express ICAM-1, the immunoglobulin superfamily counter-receptor for LFA-1. Other surface molecules with adhesive functions include receptor tyrosine kinase, c-Kit, found at high levels on the mast cell surface, that can serve as an adhesion receptor though interaction with its ligand, stem cell factor (SCF), that is expressed in a transmembrane form on endothelial cells and fibroblasts; the selectin LAM-1 (also known as Mel-14 and LECAM-1) expressed on basophils; and the selectin

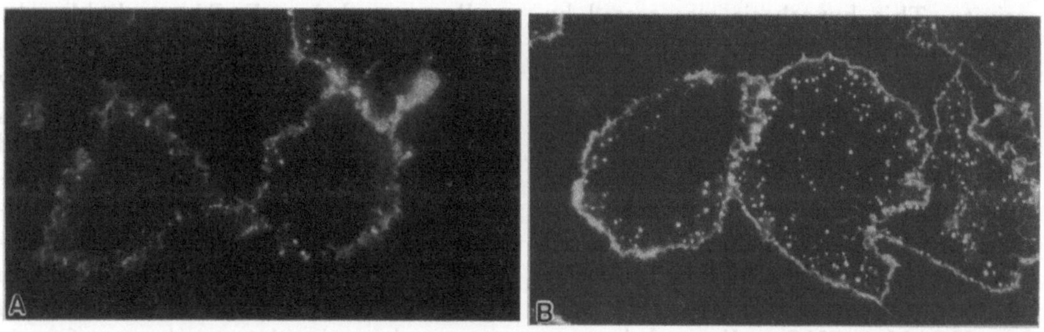

Fig. 8.6. Antigen-activated adherent cells leave footpads on fibronectin-coated substrates. IgE-primed RBL-2H3 cells were incubated on fibronectin-coated coverslips for 10 minutes without (A) or with (B) 1 µg/ml DNP-BSA, then rinsed and displaced by incubation with buffer containing $ZnCl_2$, followed by shearing in a stream of buffer. Membrane fragments were fixed, labeled with rhodamine-phalloidin and observed by fluorescence microscopy. Modified with permission from Pfeiffer et al, J Cell Biol 1994; 152:270-279. Copyright 1994, The American Association of Immunologists.

counter-receptor, sialyl-Lewis **X**, also expressed on basophils.

Although antigen-stimulated RBL-2H3 cells flatten, spread and assemble actin plaques on most surfaces, including clean glass coverslips and tissue culture plastic, these responses are more marked on fibronectin-coated surfaces. These observations, together with evidence from Hamawy et al[35] that RBL-2H3 cells express the β1 integrin subunit, raised the possibility that FcεRI crosslinking may upregulate the expression levels or binding activity of one or both of the principal fibronectin-binding β1 integrins, VLA-4 and VLA-5, of mast cells.

Flow cytometric studies confirmed the expression of the α4 subunit of VLA-4 at the RBL-2H3 cell surface and also established that the increased adhesion of antigen-activated cells is not associated with increased VLA-4 expression at the cell surface. Previously, Masumoto and Hemler[41] reported that VLA-4 can exist on the surfaces of leukocytes in three activation states, one with little or no ligand binding activity, one with little activity towards fibronectin but with moderate activity towards a second specific ligand, vascular cell adhesion molecule-1 (VCAM-1), and one with high activity towards both VCAM-1 and fibronectin. We therefore postulated that the increased adhesiveness of antigen-stimulated RBL-2H3 cells might result in part from an increase in VLA-4 binding activity. This hypothesis was tested by modification of a new assay for integrin activity[42] in which IgE-primed RBL-2H3 cells were labeled with the red fluorescent dye, PKH-26, and incubated with rolling with FITC-labeled CHO cells expressing VCAM-1, a specific ligand for VLA-4. After incubation, paraformaldehyde-fixed cell mixtures were analyzed by flow cytometry for their content of free RBL-2H3 cells, free VCAM-1-CHO cells and heteroconjugates. Typical results are illustrated in Figure 8.7. Essentially 100% of unstimulated or antigen-stimulated RBL-2H3 cells remain in the free state during incubation with VCAM-1-CHO cells in the presence of EDTA, of antibody 49D7 that binds the VLA-4 α4 subunit (Fig. 8.7A,B), or of antibody that recognizes and blocks VCAM-1 on the transfected CHO cells (not shown). These data establish that there is little or no nonspecific interaction between RBL-2H3 and VCAM-1-CHO cells. Interactions between RBL-2H3 and VCAM-1-CHO cells in complete medium without inhibitors reduce the free RBL-2H3 population to around 50% of the input population when unstimulated RBL-2H3 are incubated with VCAM-1-transfected CHO cells (Fig. 8.7C). These results establish that RBL-2H3 cells bind VCAM-1-CHO cells by a VLA-4-dependent mechanism. Importantly, the percent of free RBL-2H3 cells declines further, to 25% of their input number, when antigen is added simultaneously with, or 5 minutes before, the addition of VCAM-1-CHO cells (Fig. 8.7D). These results establish that FcεRI crosslinking enhances the binding activity of VLA-4 on RBL-2H3 cells.

Further studies have implicated the antigen-induced increase in VLA-4 binding activity in the induction of RBL-2H3 adhesion and actin plaque assembly. In Figure 8.8, IgE-primed cells were allowed to settle for 20 minutes on fibronectin-coated coverslips in medium with or without anti-VLA-4 antibody, then incubated for 10 minutes with antigen, followed by fixation and labeling with rhodamine-phalloidin. In the absence of anti-VLA-4, cells responded to FcεRI crosslinking by ruffling, spreading and actin plaque assembly (Fig. 8.8A). In the presence of antibody specific for the α4 subunit of rat VLA-4, cells still attached to fibronectin-coated coverslips and showed antigen-induced ruffling. However, antigen-induced spreading and actin plaque assembly were inhibited (Fig. 8.8B). Experiments are still in progress to determine if VLA-4 itself accumulates in the membrane of actin plaques or, alternatively, if enhanced VLA-4-ligand interaction is responsible for adhesion strengthening and spreading, events that may precede and prepare for actin plaque assembly.

THE BIOCHEMICAL BASIS OF THE FcεRI-MEDIATED CYTOSKELETAL, MEMBRANE AND ADHESIVE RESPONSES

EVENTS AT THE LEVEL OF THE FcεRI

Biochemical studies by Metzger and colleagues, reviewed in Metzger et al[43] and in chapter one of this book, established that the FcεRI is expressed at the mast cell surface as an $\alpha\beta\gamma_2$ heterotrimer. In 1989, Kinet, Metzger and their associates completed the task of cloning and sequencing all three receptor subunits.[44,45] Briefly, the FcεRI consists of an IgE-binding α subunit, a β subunit related to the recently recognized transmembrane-4 family,[46] and a disulfide-linked γ subunit dimer that is

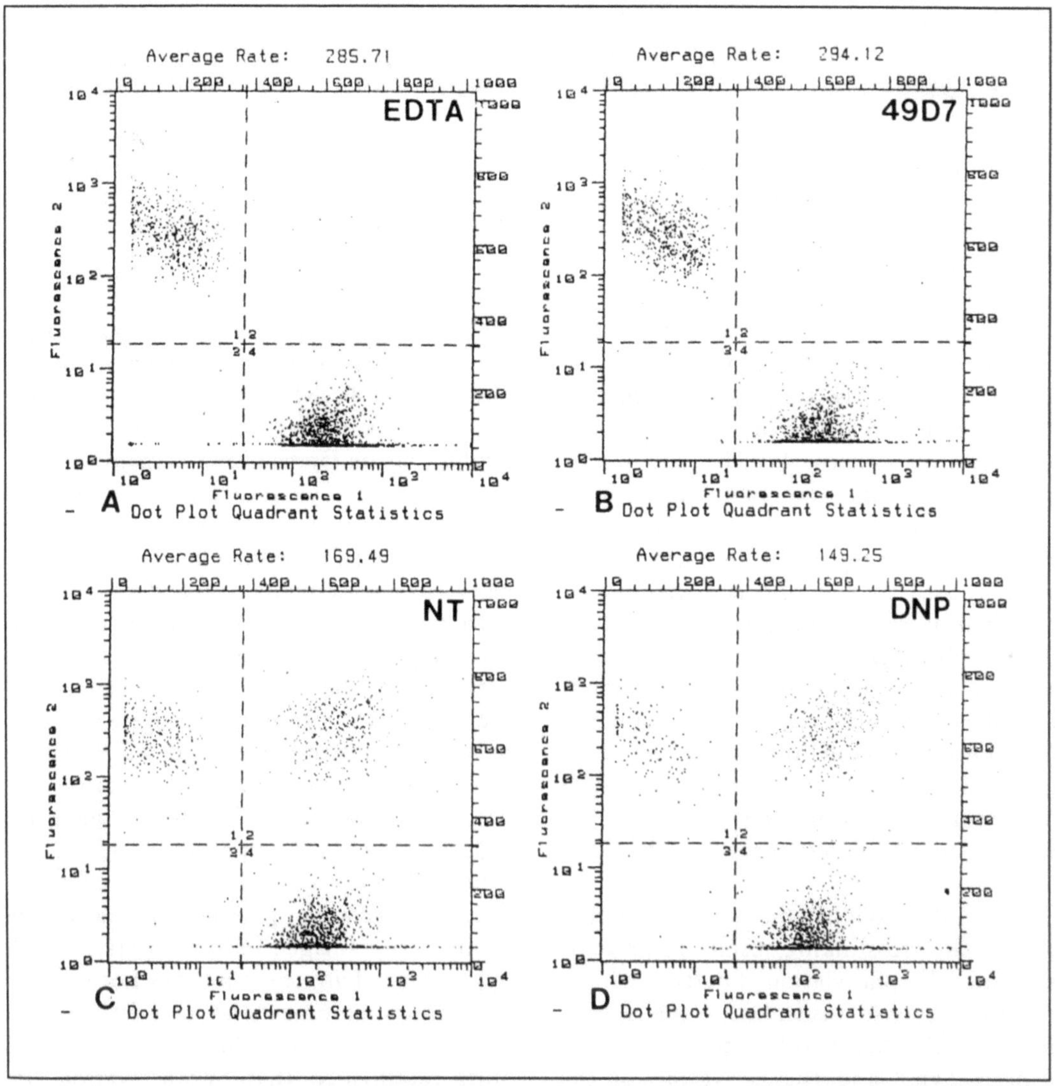

Fig. 8.7. FcεRI crosslinking activates VLA-4 binding activity. IgE-primed PKH-26-labeled RBL-2H3 cells (2 x 10⁴ cells in 100 μl Hanks'-BSA medium) were preincubated for 5 minutes with: (A), 2 mM EDTA; (B), 5 μg/ml Mab 49D7 (anti-VLA-4); (C), no addition; and (D), 0.1 μg/ml DNP-BSA. After 5 minutes, FITC-labelled VCAM-1-CHO cells (6 x 10⁴ in 100 μl) were added and the cells rolled together for 5 minutes. Cell suspensions were fixed and 10,000 events analyzed by flow cytometry. Numbers of free RBL-2H3 cells recovered from an input of 2,500 RBL-2H3 cells/10,000 total cells analyzed were (A), 2490 (100%); (B), 2695 (108%); (C), 1337 (53%); (D), 627 (25%). The lower proportion of free cells in the antigen-activated samples indicates the greater binding activity of their cell surface VLA-4 towards its ligand, VCAM-1.

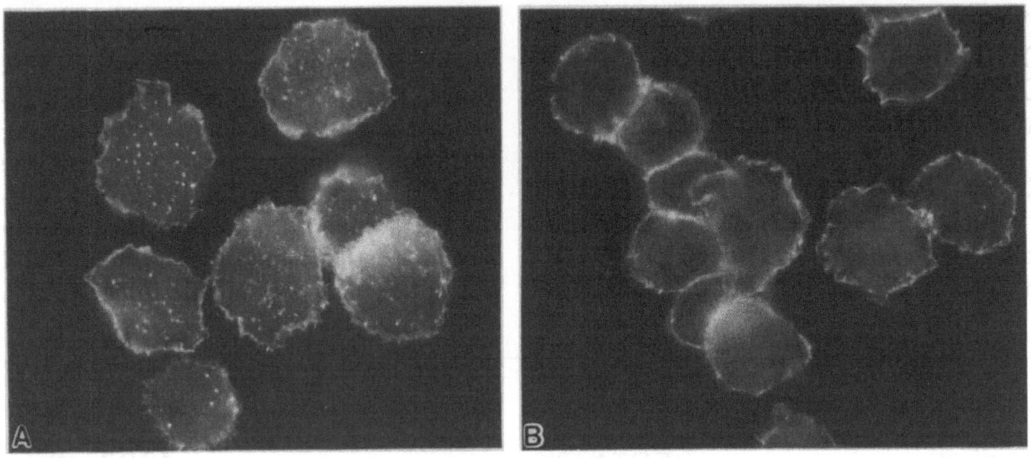

Fig. 8.8. Anti-VLA-4 inhibits antigen-induced spreading and actin plaque assembly. IgE-primed cells were incubated for 10 minutes in suspension with no addition (A) or 5 μg/ml Mab 49D7 (anti-VLA-4) (B). Cells were allowed to settle for 20 minutes on fibronectin-coated coverslips, then stimulated for 10 minutes with 1 μg/ml DNP-BSA. Cells were fixed and observed by fluorescence microscopy after rhodamine-phalloidin labeling.

closely related to the ζ subunit of the T cell receptor. Immunoreceptor tyrosine activation motifs (ITAMs) are present in the carboxyl-terminal domains of the cytoplasmic sequences of both the FcεRI-β and γ subunits.[47] Long before the complete receptor structure was known, it was clear that FcεRI crosslinking is the initiating event leading both to secretion and membrane/cytoskeletal responses. Recent work (unpublished) in collaboration with E. Ortega and colleagues using monoclonal antibodies directed against the FcεRI-α subunit[48] has shown that FcεRI dimers are the minimal signaling-competent complexes capable of stimulating membrane/cytoskeletal as well as secretory responses in RBL-2H3 cells.

In 1990, Benhamou et al[49] reported that FcεRI crosslinking results in protein-tyrosine phosphorylation. Shortly thereafter, this response was linked to the recruitment of the Src family protein-tyrosine kinase, Lyn[50] and the activation of the 72 kDa protein-tyrosine kinase, PTK72 or Syk.[51] Subsequent studies (see chapter 3) have suggested that Lyn is partially precoupled to the FcεRI,[52] that Lyn is specifically responsible for the phosphorylation of the FcεRI β and γ subunits,[53] and that Lyn-mediated transphosphorylation is responsible for receptor subunit phosphorylation.[54] Very recent studies have shown directly that FcεRI-γ subunit phosphorylation in turn creates docking sites for the binding of the tandem SH2 domains of Syk that facilitates Syk activation.[55-57]

We reported in 1991 that the relatively nonspecific tyrosine kinase inhibitors, genistein and tyrphostins 8 and 11, inhibit both secretion and membrane/cytoskeletal responses to FcεRI crosslinking and, conversely, that the phosphatase inhibitor, sodium vanadate, can mimic antigen-induced secretion and membrane responses.[9,59] These were the first results linking tyrosine kinase activation to the control of antigen-induced functional responses in RBL-2H3 cells. Subsequent studies showed that the stilbene derivative, piceatannol, selectively inhibits the antigen-induced activation of Syk but not Lyn, and abolishes both secretion and membrane and cytoskeletal responses to FcεRI crosslinking.[53] These results established that Syk activation, in particular, is essential for the initiation of the known FcεRI-mediated signaling responses.

More recent studies have further narrowed the biochemical events required to initiate FcεRI-coupled functional responses. In particular, we showed that crosslinking transfected chimeric receptors consisting of the cytoplasmic tail of the FcεRI-γ subunit coupled to the transmembrane and extracellular domains of the IL2 receptor α subunit (the Tac antigen; hence the construct is abbreviated TTγ) results in the recruitment and activation of both Lyn and Syk and in the induction of both secretory and membrane/cytoskeletal responses.[60] In contrast, crosslinking the chimeric receptor, TTβ, that expresses the carboxyl terminus of the FcεRI β subunit, not only elicits no signaling responses, but also inhibits secretion induced by FcεRI crosslinking. These data indicate that full signaling activity leading to both secretion and membrane/cytoskeletal responses resides in the FcεRI-γ subunit and that the β subunit plays a regulatory role in FcεRI-mediated cell activation.

Other work from our laboratory[59,61] has demonstrated a requirement for protein isoprenylation at a very early step in the FcεRI-mediated secretory and membrane/cytoskeletal response pathways. Prenylation refers to the postsynthetic modification of proteins on N-terminal cysteine residues with farnesyl and geranylgeranyl groups, a lipid modification that permits their association with membranes.[62] The principal families of prenylated proteins identified to date are the γ subunits of the heterotrimeric GTP-binding proteins (G proteins), all of the Ras-related small GTP-binding proteins (GTPases) and the nuclear lamins. The isoprenoid pathway inhibitor, lovastatin, causes the rounding of unstimulated RBL-2H3 cells and it blocks both the secretory and membrane/cytoskeletal responses of RBL-2H3 cells to FcεRI crosslinking. Ionomycin and phorbol myristate acetate (PMA) were able to elicit secretory and ruffling responses respectively in lovastatin-treated cells, indicating that signaling can be restored in lovastatin-treated cells by reagents that bypass the FcεRI. These data, and subsequent studies by Shakarjian et al,[63] implicate protein prenylation in the assembly of the signaling complex that initiates both secretory and membrane responses. Further studies revealed a small reduction in antigen-induced tyrosine phosphorylation[63] and a substantial reduction in protein serine/threonine phosphorylation (GG Deanin and JM Oliver, unpublished observations) in lovastatin-treated cells. It is thus possible that a prenylated protein regulates antigen-induced protein-serine/threonine phosphorylation that was shown before[64] to accompany antigen-induced protein-tyrosine phosphorylation. Although no prenylated kinase has yet been identified in the FcεRI signaling complex, it is striking that a prenylated serine/threonine kinase, rhodopsin kinase, has been identified as part of the rhodopsin pathway in the visual system.[65]

Post-Receptor Events in the Pathway to Membrane, Cytoskeletal and Adhesive Responses

Overview of the antigen-activated signaling pathways

Syk activation results in the phosphorylation of a series of receptor-associated proteins including phospholipase C-γ1 and -γ2,[64,66] the 85 kDa subunit of phosphatidylinositol 3-kinase (PI 3-kinase),[8,67] the guanine nucleotide exchange protein, Vav[68] the adaptor protein, Grb2, that forms complexes with another guanine nucleotide exchange protein, Sos, in RBL-2H3 cells[69] and an additional protein-tyrosine kinase, Bruton kinase.[70] Other receptor-associated proteins include protein-tyrosine phosphatases, very likely including CD45[71] and/or HePTP,[72] responsible for the rapid dephosphorylation of receptor subunits and receptor-associated kinases following treatments that prevent further FcεRI crosslinking[73,74] (see chapter 7) and protein serine/threonine kinases, possibly including an isoprenylated species (discussed above) responsible for the increased phosphorylation of tyrosine kinase substrates on serine and threonine residues in antigen-stimulated cells[64] (see chapter 6).

These initial biochemical events induce the mobilization of Ca^{2+}, mediated both by the 1,4,5-inositol trisphosphate (1,3,4-IP_3)-

induced release of Ca^{2+} stores and Ca^{2+} influx.[75] They also induce further biochemical responses including the tyrosine phosphorylation of multiple Grb2-binding proteins;[69] the activation of members of the Ras superfamily of GTPases, including species of Ras itself (T Graham and BS Wilson, in preparation), and, very likely, members of the Rab and Rac families (see below); and the activation of downstream effector molecules including at least three serine/threonine kinases (protein kinase C (PKC), the dual function serine and lipid kinase, PI 3-kinase and Raf-1), the dual function protein serine and tyrosine kinases, mitogen-activated protein kinases I and II (MAPK1 and MAPK2; ERK1 and ERK2) and phospholipase A_2 responsible for arachidonate release.[8,69,76-78]

Figure 8.9 depicts some of the signaling proteins that are activated by FcεRI crosslinking in RBL-2H3 cells. Their possible sequence of activation in RBL-2H3 cells and their likely relationship to antigen-induced functional responses is also indicated. The proliferation of dotted lines and question marks in the downstream signaling pathway indicates the difficulty of linking these biochemical responses to each other, let alone of tracing a clear pathway from receptor crosslinking to specific membrane, cytoskeletal and adhesive responses. Our goal in the remainder of this section is to make connections between antigen-activated biochemical events and functional responses that can, in turn, provide directions for more detailed analysis.

The roles of Ca^{2+} mobilization and targets of PMA in antigen-induced membrane and cytoskeletal responses

One established property of antigen-induced secretion is its dependence on the presence of extracellular Ca^{2+}.[1,43] The central role of Ca^{2+} mobilization in secretion is further indicated by the ability of Ca^{2+} ionophores such as ionomycin to mimic antigen-induced secretion.[1,43] In contrast, antigen-induced protein-tyrosine phosphorylation, ruffling, F-actin assembly, fluid pinocytosis and actin plaque assembly all occur normally in nominally Ca^{2+}-free medium, that does not support Ca^{2+} influx or secretion. Furthermore, ionomycin does not induce ruffling or spreading responses resembling those induced by FcεRI crosslinking.[1,2,64] This differential dependence on high extracellular Ca^{2+} was one of the first results suggesting that the pathways to secretion and to membrane/cytoskeletal responses diverge at the early postreceptor level.

Other early work showed that ruffling and spreading (although not actin plaque assembly; see below) are induced by phorbol esters, including PMA, and diacylglycerol analogs that do not cause secretion.[1,2] It was originally considered that this result links protein kinase C activation to the membrane and cytoskeletal responses. However, recent studies showed that PMA activates a range of proteins that express a common cysteine-rich diacylglycerol-binding motif:[79] proteins with this motif in the FcεRI-coupled signaling cascade include not only members of the PKC family, that can activate multiple downstream proteins through serine/threonine phosphorylation (see chapter 6), but also the guanine nucleotide exchange protein, Vav,[68] that can activate Ras-related GTPases.

The role of GTPases in antigen-induced membrane and cytoskeletal responses

We postulate that the activation of members of the Ras superfamily of GTPases is an intermediate step leading to antigen-induced secretion and to antigen-induced membrane, cytoskeletal and adhesive interactions. A role for GTPases in FcεRI-mediated functional responses was first indicated by evidence that the nonhydrolyzable GTP analog, GTPγS, induces actin assembly, ruffling and spreading as well as secretion in streptolysin O-permeabilized RBL-2H3 cells.[59] Further experiments showed that the inosine monophosphate (IMP) dehydrogenase inhibitor, mycophenolic acid (MPA), could reduce GTP levels by approximately 70% and abolish the antigen-induced secretory response while having only a small effect on the

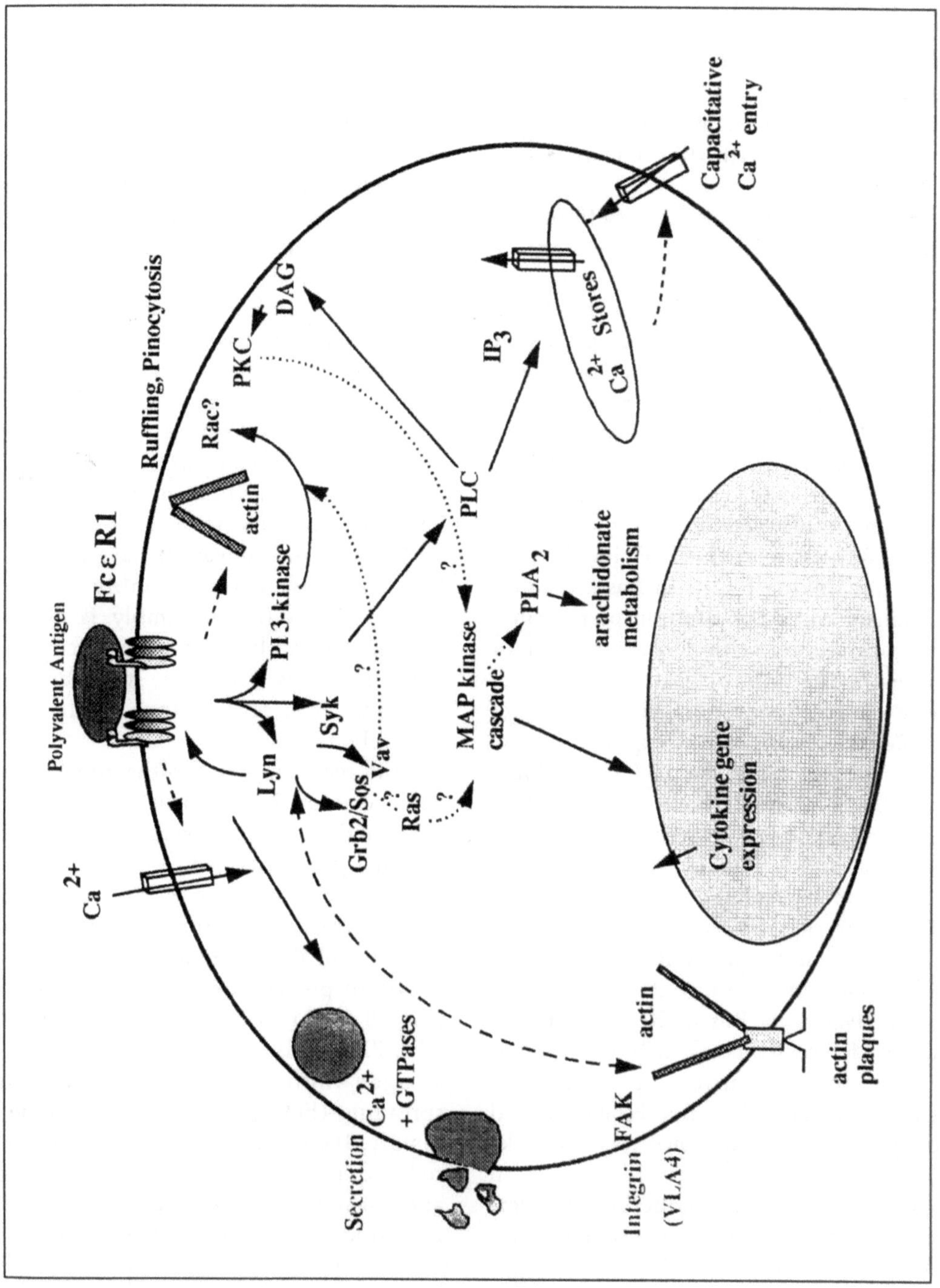

Fig. 8.9. FcεRI-coupled signaling pathways in RBL-2H3 cells. This cartoon illustrates some of the signaling events that have been identified in antigen-activated RBL-2H3 cells and indicates their possible sequence based on studies in RBL-2H3 and other cells.

antigen-induced membrane/cytoskeletal responses.[7] We speculated from this result that the signaling cascades leading to these different responses include distinct GTP-binding proteins with different affinities for GTP.

A possible role for Ras itself in the control of the FcεRI-coupled secretory or membrane/cytoskeletal response pathways was explored following evidence that Grb2 exists in RBL-2H3 cells in a complex with the Ras-directed guanine nucleotide exchange factor, Sos, and is tyrosine phosphorylated in responses to FcεRI crosslinking.[67] We identified three forms of Ras, H-Ras, K-Ras and R-Ras, in RBL-2H3 cells[80] and confirmed by assays of H-Ras and K-Ras-associated GTP levels that FcεRI crosslinking results in Ras activation (T Graham and BS Wilson, in preparation; analogous studies with R-Ras are still in progress). The role of Ras activation in antigen-induced secretion and membrane/cytoskeletal responses was explored by use of the specific farnesyl-transferase inhibitor, BZA-5B. After 72 hour treatment with 100 µg/ml BZA-5B, essentially 100% of the H-Ras and K-Ras in RBL-2H3 cells is in the nonfarnesylated form and FcεRI crosslinking causes a reduction, rather than an increase, in the amount of Ras in the GTP-bound form. Despite this evidence for the inactivation of Ras in BZA-5B-treated cells, FcεRI crosslinking still induces apparently normal ruffling, spreading and actin plaque assembly and FcεRI-induced secretion is enhanced. These results make it unlikely that Ras activation plays an important role in the induction of membrane and cytoskeletal responses. They suggest that activated Ras (or another farnesylated protein) may actually inhibit secretion. Since lovastatin inhibits FcεRI-coupled signaling, and BZA-5B does not, these data also suggest that the prenylated protein(s) implicated above in the FcεRI signaling complex is a geranylgeranylated species.

Current studies on the regulation of FcεRI-coupled signaling by small G proteins are focused on possible roles for members of the Rab3 family of GTPases in the control of secretion and of members of the Rho family of GTPases, including Rac, in the regulation of antigen-induced membrane and cytoskeletal responses. Rab3 proteins are being studied as potential regulators of secretion because at least two members of this geranylgeranylated family, Rab3B and Rab3D, were recently identified in RBL-2H3 cells[81] and because Rab family members have been linked to the control of secretion in other cells.[82] Rac is being studied as a potential regulator of membrane/cytoskeletal responses because of evidence that it controls F-actin assembly and membrane ruffling in many cell types, including primary mast cells[5] (see chapter 5). Hall and colleagues have implicated Rho in stress fiber assembly in fibroblasts.[5,6] Stress fiber assembly is observed in PMA-treated but not antigen-stimulated RBL-2H3 cells. However, it remains possible that Rho contributes to actin plaque assembly.

Indirect support for the proposed involvement of Rac in antigen-induced ruffling and fluid pinocytosis in RBL-2H3 cells has been obtained recently from studies using the selective PI 3-kinase inhibitor, wortmannin.[8] Wortmannin, a known inhibitor of FcεRI-induced secretion,[67] was shown to prevent antigen-induced ruffling (Fig. 8.10D) and fluid pinocytosis in RBL-2H3 cells, while having no effect on antigen-induced receptor endocytosis, spreading (Fig. 8.10D) or actin plaque assembly. Basal F-actin levels were somewhat reduced in wortmannin-treated cells, but antigen-induced F-actin assembly was not impaired. Wortmannin-treated cells retained the ability to assemble stress fibers in response to PMA. These results suggest strongly that PI 3-kinase controls a branch point downstream of actin polymerization that regulates antigen-induced ruffling and fluid pinocytosis, without affecting the assembly of stress fibers or actin plaques. Previous studies established that wortmannin also inhibits ruffling induced by PDGF or insulin in fibroblasts. In the studies with fibroblasts, it was shown further that the inhibition by wortmannin of ruffling could be relieved by microinjection of activated Rac.[6,83] Based on these results, it was suggested that PI 3-kinase activation may be an essential upstream event in the receptor-activated signaling pathway leading to Rac activation and ruffling responses. There is as yet very little information on mechanisms linking PI 3-kinase to Rac activation. Members of the Ras superfamily are regulated not only by guanine nucleotide exchange factors (GEFs) like Vav (see below), but also by GTPase activating proteins (GAPs) and by guanine nucleotide dissociation inhibitors (GDIs). Based on sequence homology, Hall has raised the possibility that the p85 subunit of PI 3-kinase could downregulate GAP activity, thus activating Rac or other GTPase targets by increasing the amount of G protein in the GTP-bound (active) form. Very recent results[84] raised the additional possibility that PI 3-kinase and other phosphoinositide kinases may interact directly with Rho family members to regulate their localization and activity.

We hypothesize that the guanine nucleotide exchange protein, Vav, may also modulate Rac activation in antigen-stimulated RBL-2H3 cells. Vav is suggested as an intermediate in the putative pathway from FcεRI crosslinking to Rac activation to ruffling in RBL-2H3 cells because it can be activated both by antigen-induced tyrosine phosphorylation[68] and by PMA.[79] Thus, it is attractive to propose a cascade of responses that begins with the antigen-induced tyrosine phosphorylation and

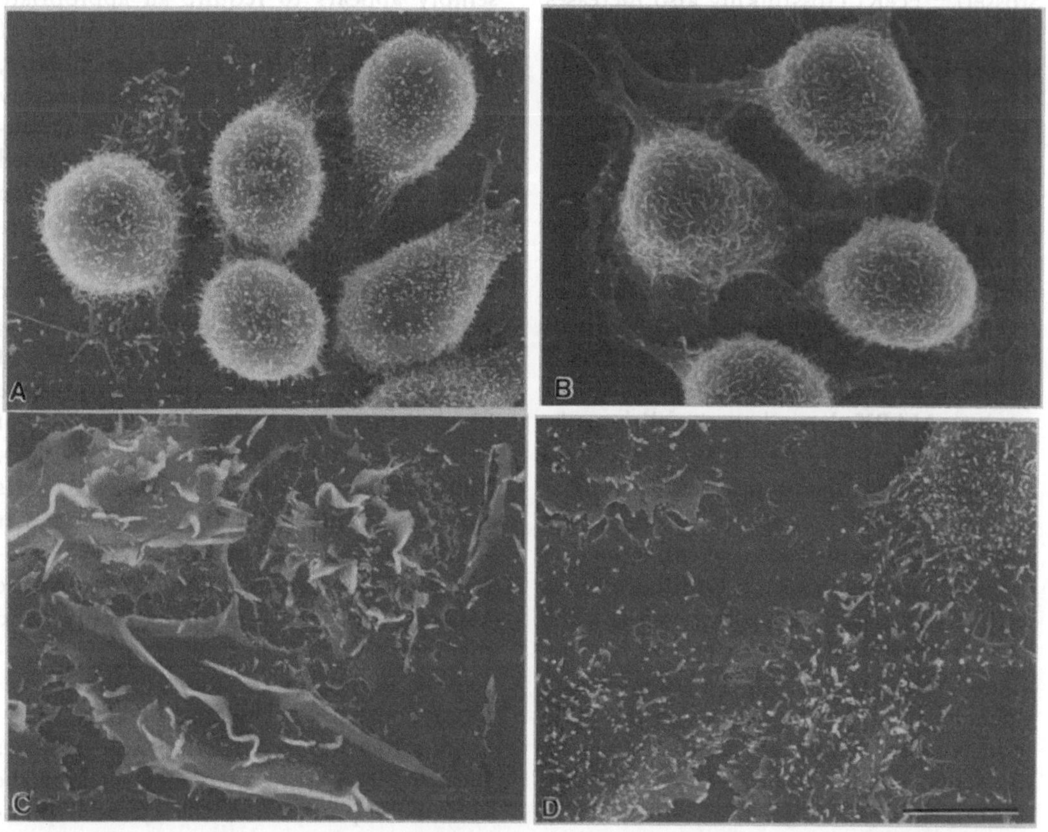

Fig. 8.10. Wortmannin is a selective inhibitor of FcεRI-mediated ruffling. IgE-primed RBL-2H3 cell monolayers were incubated for 15 minutes without (A,C) or with (B,D) 10 nM wortmannin. The cells in C and D were activated for 10 minutes with 1 µg/ml DNP-BSA. Coverslips were fixed and processed for SEM analysis. Bar = 10 µm. Reproduced from Barker et al, Molecular Biology of the Cell 1995; 6:1145-1158, by copyright permission of the American Society for Cell Biology.

activation of Syk, that in turn causes the tyrosine phosphorylation and activation of Vav (mimicked by PMA), that in turn activates Rac (mimicked by GTPγS and perhaps facilitated by PI 3-kinase), that finally results in membrane ruffling.

Roles for lipid metabolites

An extensive literature connects lipid metabolism to cytoskeletal remodeling. Thus, FcεRI crosslinking alters the levels and phosphorylation states of inositol phospholipids through the activation of PLC-γ, PI 3-kinase and other enzymes. The binding of polyphosphoinositides to microfilament-associated proteins such as gelsolin, profilin and villin has been linked to the regulation of actin monomer levels, the sites of actin filament nucleation and the rates of actin polymerization (reviewed by Stossel).[82] FcεRI crosslinking also increases the levels of phosphatidic acid (PA) and lysophosphatidic acid (LPA) in RBL-2H3 cells through the activation of PLC, PLD and PLA_2 (reviewed by Beaven and Metzger).[86] In fibroblasts, PA has been implicated directly in the regulation of actin polymerization.[87] LPA has been shown to modulate cytoskeletal organization indirectly by binding to a G protein-coupled receptor and activating a signaling pathways leading to Ras and Rho activation.[88]

PLA_2 activation, recently shown to occur by a MAPK-dependent signaling pathway in antigen-activated RBL-2H3 cells,[78] also yields arachidonic acid, the precursor in mast cells of a range of prostaglandin and leukotriene mediators of inflammation (reviewed in this book by Gilfillan). Studies in A431 cells by Peppelenbosch et al[89,90] have demonstrated roles for cyclooxygenase, a key enzyme regulating the conversion of arachidonate to prostaglandins, in the EGF-induced disassembly of cytoplasmic actin filaments and for lipoxygenase, that regulates leukotriene synthesis from arachidonate, in EGF and Rho-induced stress fiber assembly. By analogy with the results in A431 cells, it is possible that the activation of PLA_2 and the resulting increase in cyclooxygenase and lipoxygenase activity in antigen-activated RBL-2H3 cells may contribute to the reorganization of the mast cell cytoskeleton as well as to the synthesis of inflammatory mediators.

Until recently, there was no clear connection between the putative effects of lipid metabolites and GTPase activation on cytoskeletal organization. Bokoch and collaborators[91] have now linked Rho to the activation of phosphatidylinosital 4-phosphate s-kinase, leading to increased levels of phosphatidylinositol 4,5-bisphosphate. These results raise the possibility that some of the effects of lipid metabolites on cytoskeletal organization occur in conjunction with Rac and Rho.

The control of actin plaque assembly

In addition to signals permitting increased F-actin assembly, actin plaque assembly appears to require an appropriate surface for cell adhesion and activation of protein-tyrosine kinases. As noted above, crosslinker-induced actin plaque assembly occurs on many surfaces, but is most dramatic in cells plated onto surfaces coated with the VLA-4 ligand, fibronectin. Even on fibronectin, actin plaque assembly depends additionally on tyrosine kinase activation. Although PMA and GTPγS mimic the antigen-induced F-actin assembly, spreading and ruffling responses, both of these reagents induce the assembly of stress fibers rather than actin plaques at sites of cell-substrate interaction (Fig. 8.11A). Similarly, stimulating RBL-2H3 cells through transfected fMLP receptors, that signal through the activation of heterotrimeric G proteins instead of tyrosine kinases, induces F-actin assembly, and membrane ruffling but not actin plaque assembly (unpublished observations in collaboration with A Hall and L Sklar). However, structures resembling actin plaques form in PMA-treated, adherent cells in the additional presence of the tyrosine phosphatase inhibitor, sodium pervanadate[17] (Fig. 8.11B). We speculate that FAK may provide the tyrosine phosphorylation signal(s) that is required for actin plaque assembly. This enzyme was first detected in

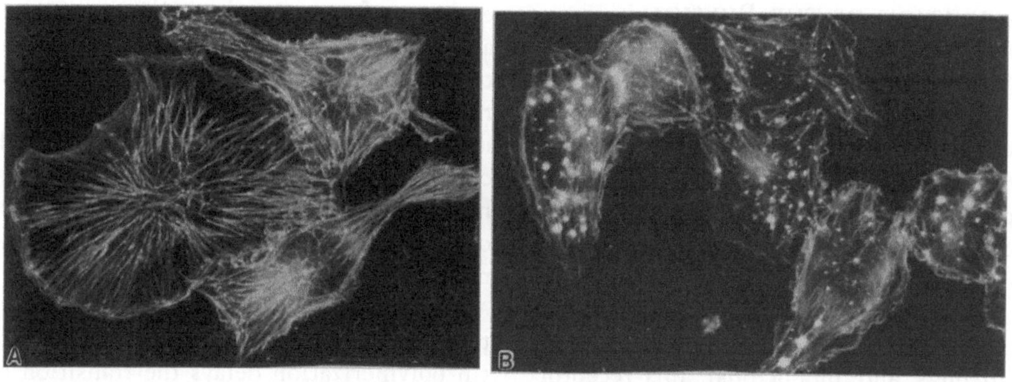

Fig. 8.11. PMA plus pervanadate stimulates assembly of structures resembling actin plaques. RBL-2H3 cells were incubated for 15 minutes with 25 nM PMA (A) or with PMA plus 0.1 mM pervanadate (B). The cells were fixed, incubated with rhodamine-phallodin and actin stress fibers and plaques observed by fluorescence microscopy. Modified with permission from Pfeiffer et al, J Cell Biol 1994; 152:270-279. Copyright 1994, The American Association of Immunologists.

RBL-2H3 cells by Hamawy et al,[35] who established that its activation requires both FcεRI crosslinking and cell adhesion. There is precedent in platelets for FAK activation through the coordinate activation of integrin and agonist receptors.[92]

CONSEQUENCES OF FcεRI INDUCED MEMBRANE, CYTOSKELETAL AND ADHESIVE RESPONSES IN BASOPHILS AND MAST CELLS

Regulation of Cell Migration, Location and Antigen Presentation

It is obvious that the capacity of basophils and mast cells to modulate their membrane, cytoskeletal and adhesive properties in response to FcεRI crosslinking and to other ligand-receptor interactions is likely to determine both their inherent distribution and their recruitment to sites of inflammation. Primary basophils are circulating cells that can, upon engaging the FcεRI or other receptors, especially receptors for cytokines, interact with endothelial cells and migrate across the endothelium to sites of inflammation. Primary mast cells are inherently localized to tissues but possess chemotactic activity that permits their redistribution to inflammatory sites.

The ability to upregulate the binding activity of VLA-4 may be a particularly important property in the determination of mast cell and basophil migration. Studies in other hematopoietic cells, especially neutrophils, monocytes and lymphocytes, have implicated VLA-4-VCAM-1 interactions in the control of leukocyte rolling, adhesion to vasular endothelium and extravasation.[93-95] We speculate that the additional capacity of activated cells to bind tightly to surfaces through the assembly of a novel adhesion structure, the actin plaque, may play a critical role in determining mast cell and basophil localization in tissues. Additional signals regulating mast cell migration and localization are undoubtedly generated through the multiple other adhesion molecules expressed on mast cells and basophils.[38,39]

In addition to migration and adhesion, basophils and mast cells are capable of the microfilament-mediated process of phagocytosis.[96] Recent evidence showed that mast cells can also present antigen to T cells by a MHC class II-dependent mechanism.[97,98] These data suggest that antigen uptake, processing and presentation may be an additional, important function of the dynamic membrane and cytoskeletal properties of mast cells and basophils.

Regulation of the Physical and Topographical Properties of the Crosslinked Receptors

As well as determining the motile and invasive properties of whole cells, antigen-induced membrane and cytoskeletal responses have been shown to regulate a subset of the physical and topographical properties of the FcεRI itself.

Biophysical analyses established that crosslinking agents, including antigen, IgE multimers and monoclonal anti-receptor antibodies, cause a rapid decrease in FcεRI lateral and rotational mobility in the RBL-2H3 cell membrane.[99-102] Additionally, receptor mapping studies by immunogold labeling and backscattered electron imaging in the SEM showed that crosslinked receptors redistribute rapidly from regions of membrane curvature, including both microvilli and lamellae, to smooth membrane regions.[103] All of these properties are relatively insensitive to modulation by ligand concentration or valency, temperature and time, suggesting that they reflect inherent physical properties of the receptors and are not extensively influenced by membrane or cytoskeletal properties.

In contrast, another series of receptor physical and topographical properties was found to be highly sensitive to modulation by antigen concentration, antigen valency, temperature and time. Thus, it was shown by flow cytometry that FcεRI crosslinking agents cause a time and antigen-concentration-dependent transformation of fluorescein-IgE-labelled receptors from a detergent (Triton-X-100)-soluble to a detergent-insoluble form (Fig. 8.12).[104-106] Other flow cytometric assays established that this transition is associated with a progressive increase in the resistance of bound antigen to displacement by competing monovalent hapten (Fig. 8.13).[107] It is also associated with a predictable time and antigen concentration-dependent redistribution of crosslinked receptors, visible by colloidal gold labeling of anti-IgE-IgE-receptor complexes[103] from an inherently uniform distribution on the cell surface (except for exclusion from coated pits)[108] into small and then progressively larger aggregates on the cell surface (Fig. 8.14A-C). The fluorescent lipid analog, 3,3'-dihexadecylindocarbocyanine (diI) coaggregates with crosslinked IgE receptors, suggesting that highly crosslinked receptor aggregates occupy or induce the formation of specific membrane domains.[109]

Treatment with dHCB and other cytochalasins to prevent antigen-induced actin polymerization delays the transition of crosslinked receptors to the detergent-insoluble form;[105] preserves the ability of competing monovalent hapten to displace bound antigen from IgE-receptor complexes (Fig. 8.13);[107] and maintains crosslinked IgE-receptor complexes in small clusters rather than large aggregates at the cell surface (Fig. 8.14D-F).[103] Together, these results strongly implicate crosslinker-induced changes in actin assembly and distribution in the control of selected receptor physical and topographical properties.

The mechanisms underlying the cytochalasin-sensitive transformation of the crosslinked FcεRI from a dispersed, detergent-soluble form that permits reversible antigen binding to a highly aggregated, detergent-insoluble form that binds antigen irreversibly are not clear. Hypotheses include the stabilization of receptor aggregates through the association of cytoskeletal elements with specific binding sites on the crosslinked receptor; the coassociation of crosslinked receptors and cytoskeletal elements with the detergent-insoluble membrane skeleton, more recently interpreted as the caveolin-enriched fraction of the plasma membrane; and a relatively nonspecific entrapment of receptor clusters in regions of cytoskeletal-membrane interaction. Evidence from Mao et al[110] that antigen-induced receptor insolubilization occurs normally in cells expressing receptors systematically depleted of subunit cytoplasmic domains makes it unlikely that the first hypothesis is correct. Although RBL-2H3 cells contain caveolin, detected by immunofluorescence microscopy, we observe no overlap between the distribu-

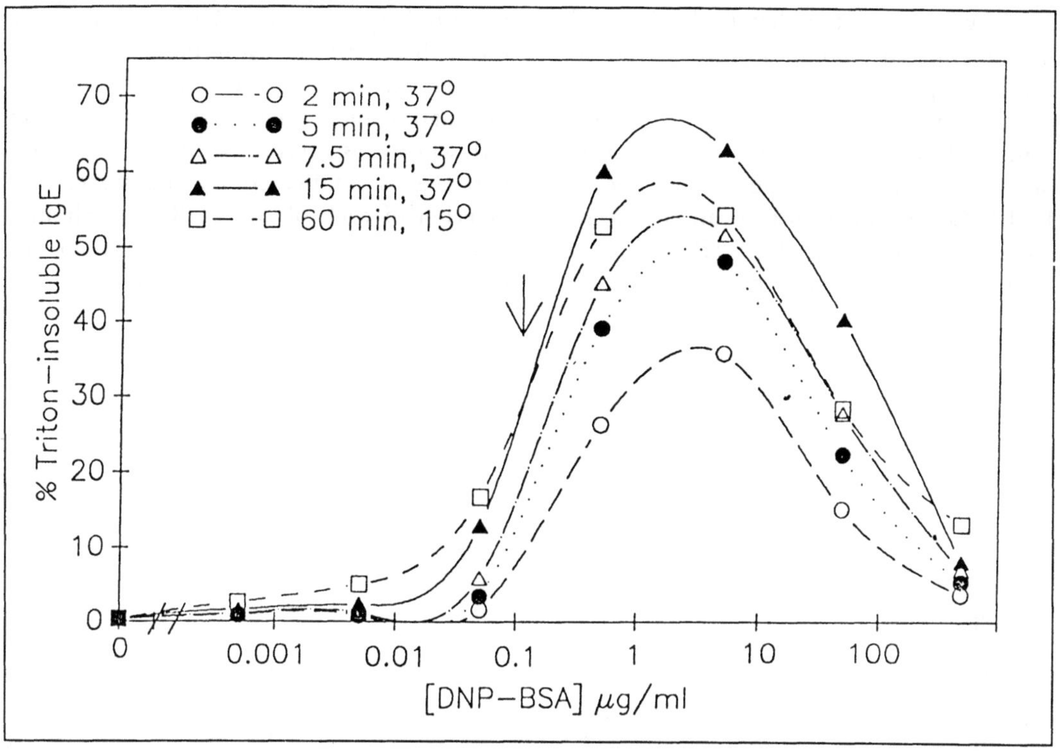

Fig. 8.12. Time and antigen concentration-dependence of the association of IgE-receptor complexes with the detergent-insoluble fraction of RBL-2H3 cells. Fluorescein-IgE-primed cells were incubated with DNP-BSA over a concentration range from 0.5 ng/ml to 500 μg/ml. After selected times, cells were extracted with Triton-X-100, fixed with paraformaldehyde, and analyzed by flow cytometry. The amount of fluorescein-IgE that remains associated with the detergent-insoluble cytoskeletons is expressed as the percent of fluorescein-IgE bound to intact cells. The arrow is centered in the antigen concentration range that gives optimal secretion. Reproduced from Seagrave et al. J Cell Physiol 1991; 148:139-51. Reprinted by permission of John Wiley & Sons, Inc.

tion of detergent-insoluble receptors and of caveolin (unpublished observations) and cultured mast cells anyway lack morphologically distinct caveolae.[108] These results are inconsistent with the second hypothesis. We therefore favor the hypothesis[105,110] that receptor clusters may become detergent-resistant through relatively nonspecific interactions with each other and with the cortical actin meshwork of antigen-stimulated RBL-2H3 cells.

Clustering is typically followed by receptor internalization through coated pits (Fig. 8.15). In our hands, small to medium sized clusters formed by crosslinking anti-DNP-IgE-receptor complexes with DNP-BSA or DNP-phycoerythrin particles or with anti-IgE precoupled to 15 nm colloidal gold particles are rapidly internalized. In contrast, dimers formed by crosslinking unprimed receptors with monoclonal anti-receptor antibodies are not significantly internalized (JM Oliver, JR Pfeiffer, C Santana et al, unpublished observations) and very large FcεRI aggregates are also not internalized.[103] It thus appears that some minimum aggregate size must be achieved before receptors are engaged by coated pits and, further, that internalization is not possible beyond some maximum aggregate size, probably due to physical properties of both the aggregate and the pits. Receptor internalization is inhibited in cytochalasin-treated cells, as predicted for a microfilament-dependent process.

It was postulated that FcεRI internalization, like the internalization of the mitogen receptors,[111] may depend on interactions of cytoskeletal elements with motifs contained in the cytoplasmic domain of one or other subunit of the heterotrimeric FcεRI. Consistent with this, an engineered receptor consisting of the FcεRI-α subunit linked to the plasma membrane via a glycerophosphoinositol (GPI) tail was not internalized after crosslinking.[108] However, further experiments have shown that the internalization of heterotrimeric receptors following crosslinking with anti-IgE-gold particles is unaffected by the systematic deletion of each of the FcεRI cytoplasmic domains.[108] Additionally, crosslinking agents drive internalization of chimeric

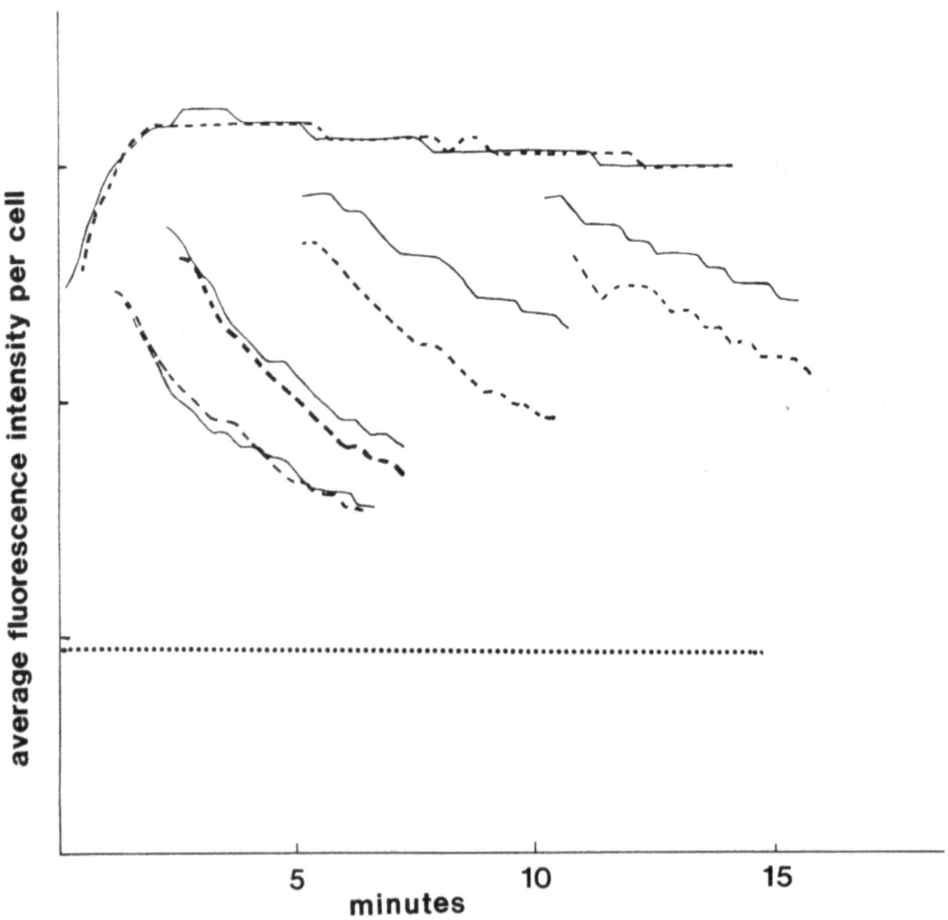

Fig. 8.13. The effect of incubation time on the displacement of antigen from IgE-receptor complexes. IgE-primed RBL-2H3 cells were incubated with DNP-coupled B-phycoerythrin (DNP-PhE; 0.25 μg/ml) without (solid lines) or with (dashed lines) 5 μg/ml dHCB. Binding was followed continuously in the flow cytometer to generate the upper curve. The lower curves were obtained by the addition of 10 μM DNP-lysine at the indicated times. The dotted lines show the fluorescence of cells that were incubated with DNP-B-PhE without prior IgE priming. Results were the same in the presence of an endocytosis inhibitor. Reproduced from Seagrave JC et al, J Cell Physiol 1990; 144:128-136. Reprinted by permission of John Wiley & Sons, Inc.

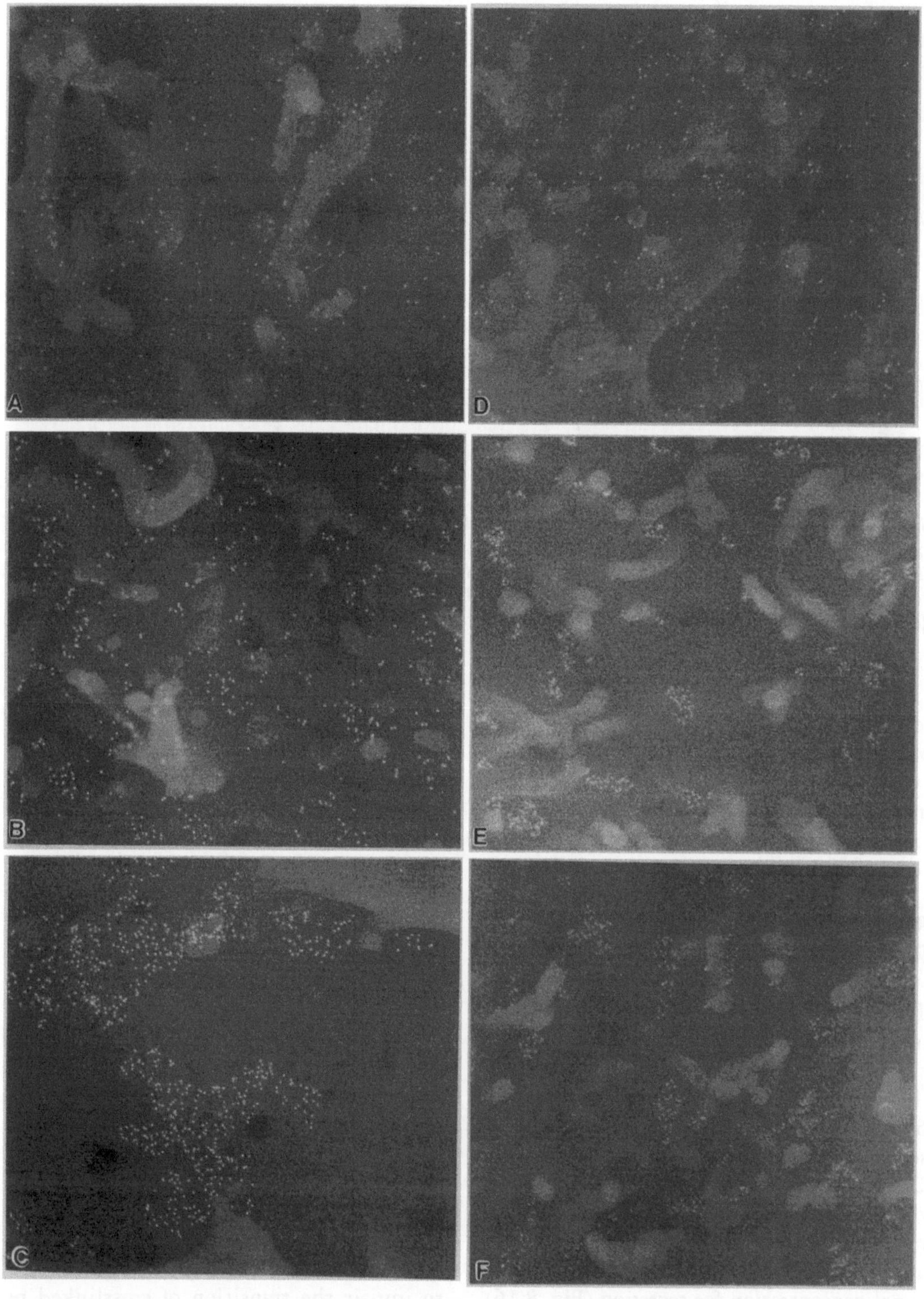

Fig. 8.14. The effects of antigen and dHCB on FcεRI topography. IgE-primed RBL-2H3 cells were incubated with 1 μg/ml anti-IgE without (A,B,C) or with (D,E,F) 5 μg/ml dHCB. Incubation times at 37°C were 2 minutes, (A,D), 5 minutes (B,E) and 15 minutes (C,F).

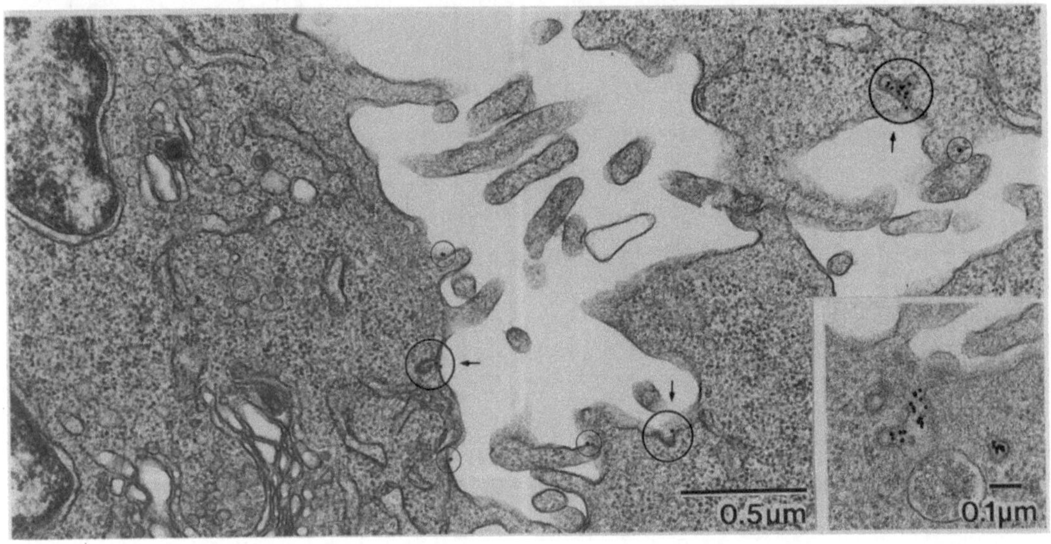

Fig. 8.15. Crosslinked IgE-receptor complexes are endocytosed through coated pits. IgE-primed cells were incubated for 2 minutes with 20 nm DNP-conjugated colloidal gold particles, then fixed and processed for TEM analysis. The redistribution of gold particles (circled) to coated pits (arrows) is indicated. Reproduced from Pfeiffer et al, The Journal of Cell Biology 1985; 101:2145-2155 by copyright permission of The Rockefeller University Press.

receptors consisting of just the β or γ carboxyl terminal domains coupled to irrelevant transmembrane and extracellular domains.[60] Thus internalization motifs, if they exist, are not limited to one specific receptor subunit.

Regulation of Signaling Responses

The role of receptor-cytoskeletal interactions in signal termination

In early publications, it was proposed that receptor-cytoskeletal interaction may be a required step in the stimulation of secretion.[104] However, detailed work by J. C. Seagrave in this laboratory[105] established that the optimal concentration of antigen for receptor insolubilization (Fig. 8.12) is approximately 10-fold higher than the optimal concentration for secretion (Fig. 8.16) and coincides with antigen concentrations that induce submaximal secretion, a phenomenon called high-dose inhibition. It was also shown that concentrations of anti-IgE that induce optimal secretion cause the rapid redistribution of IgE-receptor complexes into chains and small aggregates, but only slowly induce the appearance of large receptor clusters at the cell surface. In contrast, large receptor clusters form rapidly when cells are incubated with the higher anti-IgE concentrations that induce submaximal secretion.[103] These data strongly implicated the conversion of crosslinked receptors to a detergent-insoluble, highly aggregated form in the termination of signaling.

Further work by Seagrave and others has continued to support the view that the conversion of crosslinked receptors to the highly aggregated, detergent-insoluble form is associated with signal termination. In particular, dHCB treatment, reported above to reduce the conversion of bound antigen to the form that resists displacement by monovalent hapten (Fig. 8.13),[107] to impair the transition of crosslinked receptors to the detergent-insoluble form[105] and to inhibit the formation of large clusters of crosslinked receptors on the RBL-2H3 cell surface (Fig. 8.14),[103] also increases the rate and extent of secretion.[105,107] This enhancement of secretion by dHCB,

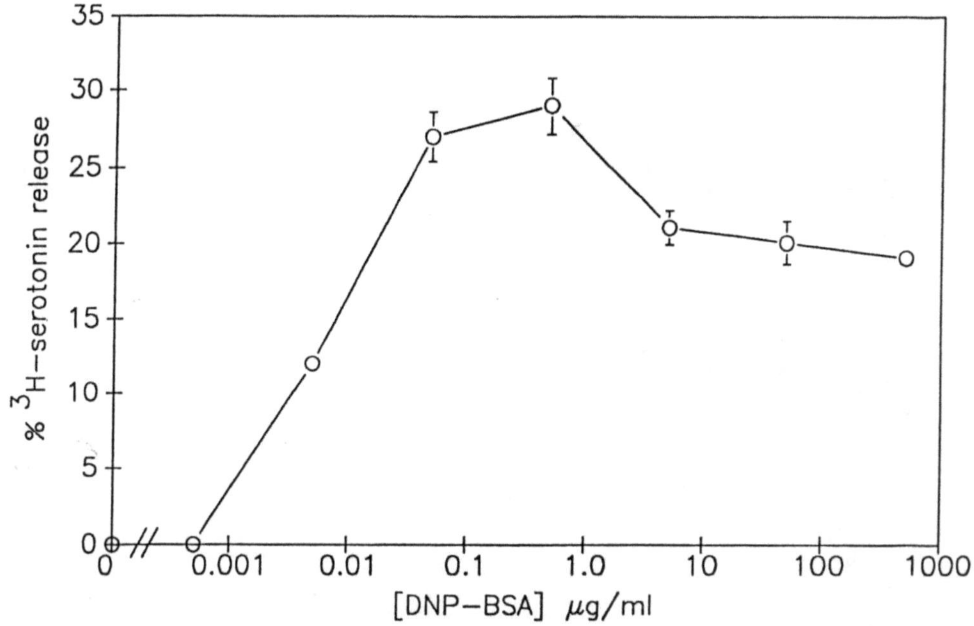

Fig. 8.16. The effect of antigen concentration on FcεRI-mediated secretion. IgE-primed, ³H-serotonin-labelled RBL-2H3 cells were incubated with concentrations of antigen ranging from 0.05 ng/ml to 500 µg/ml DNP-BSA. Data are presented as percent of total mediator released in 20 minutes and are corrected for spontaneous release and for quenching of scintillation counting at very high antigen concentrations. Ranges of the duplicate samples are given where they exceed the size of the symbols. Reproduced from Seagrave et al, J Cell Physiol 1991; 148:139-151. Reprinted by permission of John Wiley & Sons, Inc.

illustrated in Figure 8.17, is observed whether cells are activated with antigen concentrations that result in optimal secretion or with higher concentrations of antigen that cause submaximal secretion (high-dose inhibition).

There is continued speculation about mechanisms connecting early antigen-induced events such as receptor-cytoskeleton interaction, receptor insolubilization and receptor aggregation to the termination of secretion, usually measured after 20-30 minutes. We have approached this issue recently by studies of the time course of antigen-induced protein-tyrosine phosphorylation as a function of antigen concentration. The results in Fig. 8.18 shows these time courses measured flow cytometrically from the binding of anti-phosphotyrosine antibody to tyrosine phosphorylated proteins in fixed, permeabilized cells (Fig. 8.18A) and biochemically from the incorporation of ³²P-orthophosphate into antiphosphotyrosine-reactive proteins (Fig. 8.18B). In both assays, the initial tyrosine phosphorylation response measured after 1-2 minutes of crosslinking increased with increasing antigen concentration over a very broad range of antigen concentrations, including those that cause submaximal secretion in a 20 minute assay. Total tyrosine phosphorylation observed in both assays persisted at similar levels over 10 minutes in cells activated with antigen concentrations at or below the concentrations (0.1-1 µg/ml DNP-BSA in these experiments) that elicit maximum secretion in a 20 minute assay. In contrast, total tyrosine phosphorylation decreased rapidly from the initial peak value after activation with the high antigen concentrations (5 or 10 µg/ml DNP-BSA) that cause the high-dose inhibition of

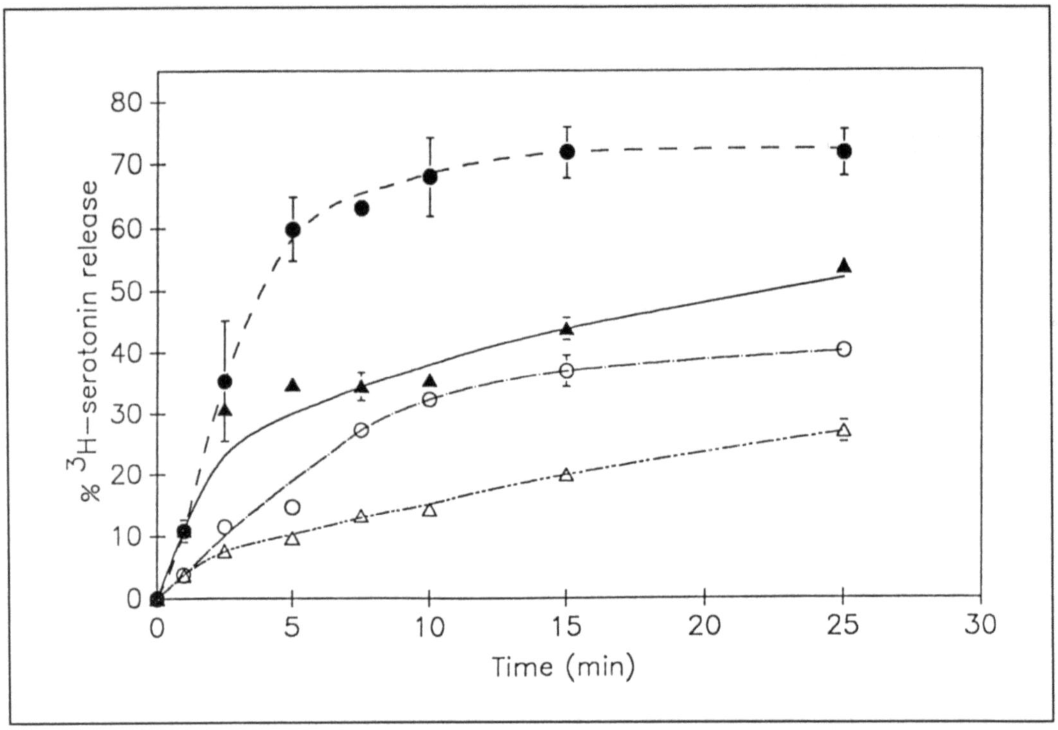

Fig. 8.17. The effects of cytochalasin on secretion. The release of ³H-serotonin from IgE-primed RBL-2H3 cells was measured in cells treated with 0.25 μg/ml DNP-BSA (circles) and 5 μg/ml DNP-BSA (triangles), without (open symbols) or with (closed symbols) the addition of 5 μg/ml. Reproduced from Seagrave et al, J Cell Physiol 1991; 148:139-151. Reprinted by permission of John Wiley & Sons, Inc.

secretion. Results obtained using the biochemical assay showed additionally that Syk, identified above as a critical component of the FcεRI-coupled signaling pathway, is particularly sensitive to the initial burst and rapid decline of phosphorylation in cells treated with high antigen concentrations. From these data, we infer that there is a direct relationship between the crosslinking rate and signal generation during the earliest phases of cell activation. We speculate that highly crosslinked receptors may segregate rapidly to membrane domains that are enriched in F-actin and isolated physically from tyrosine kinases, especially Syk, and other signaling elements, resulting in a rapid decrease in tyrosine phosphorylation levels and an early termination of signaling responses.

The role of antigen-stimulated adhesion in signal amplification

Recent results indicate that the FcεRI-mediated adhesion and spreading of adherent mast cells may serve not only to localize cells to sites of inflammation but also to modulate their signaling activity. Hamawy et al[35,112] were the first to report the amplification of antigen-induced secretion induced by permitting the adhesion of RBL-2H3 cells to surfaces and to link this response to the activation of FAK. We have confirmed the amplification of secretion in antigen-activated adherent cells and have demonstrated that antigen binding to adherent cells induces the in vivo phosphorylation of a 120 kDa protein, very likely FAK, that is not seen when cells are stimulated with antigen in suspension. A protein

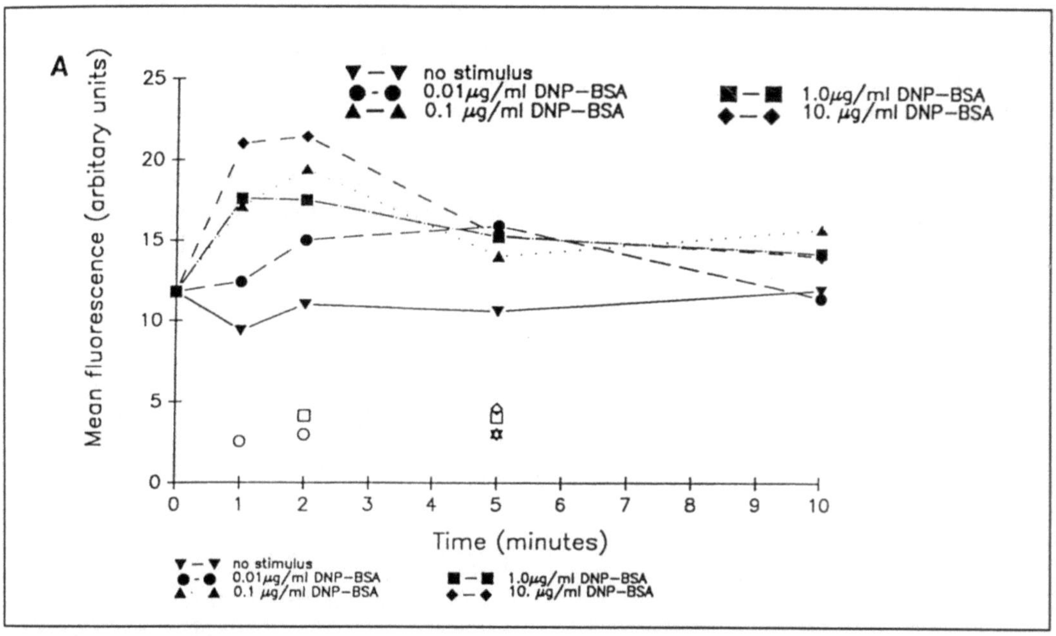

Fig. 8.18. The effect of antigen concentration on protein-tyrosine phosphorylation. In (A), suspensions of IgE-primed RBL-2H3 cells were activated with 0.01-10 µg/ml DNP-BSA and antigen activation was terminated by fixation at the times indicated. Cells were permeabilized and incubated sequentially with polyclonal anti-phosphotyrosine antibody followed by fluorescein-anti-rabbit IgG as described in Wilson et al (1995). Total levels of protein-tyrosine phosphorylation were compared between samples by flow cytometry. In (B), suspensions of ^{32}P-orthophosphate-labeled RBL-2H3 cells were activated with 0.1, 1.0 or 5.0 µg/ml DNP-BSA for the times indicated, then lysed and the antiphosphotyrosine-reactive proteins were isolated, separated by SDS-PAGE and detected by autoradiography as described in Li et al (1992). Arrowheads point to bands corresponding to Syk and the FcεRI β and γ subunits.

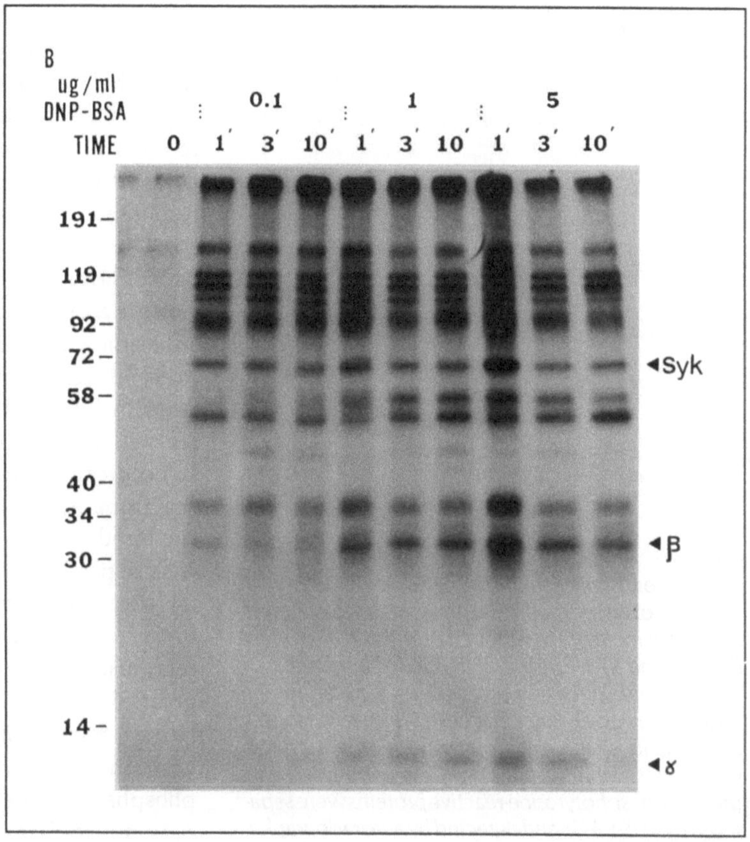

of around 80 kDa is also uniquely phosphorylated in antigen-stimulated adherent cells. Both proteins are marked by stars in Figure 8.19, lane 2. The experiment illustrated in Figure 8.19 additionally confirms that adhesion alone has little effect on tyrosine phosphorylation (compare lanes 1 and 4) and that the extent of antigen-induced phosphorylation of virtually every tyrosine phosphorylated protein is greater in the adherent, as compared to the suspended cell populations (compare lanes 2 and 5).

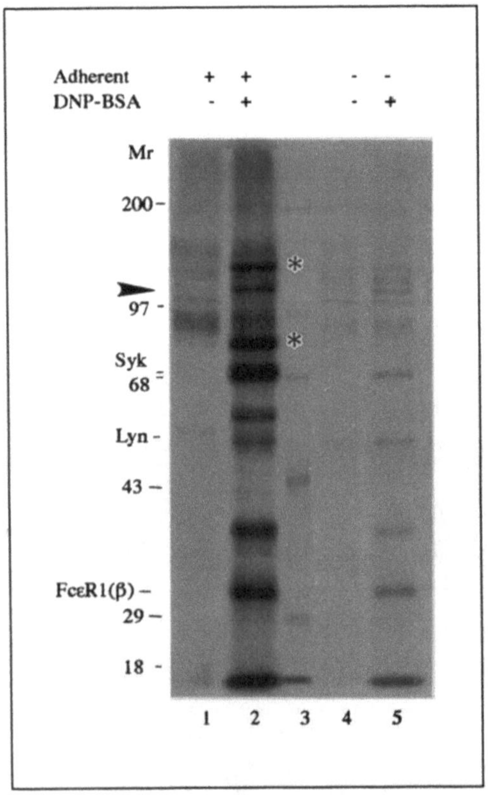

Fig. 8.19. Antigen-induced protein-tyrosine phosphorylation is amplified in adherent cells. Identical numbers of IgE-primed, ^{32}P-orthophosphate-labeled cells were incubated without (lanes 1 and 4) or with (lanes 2 and 5) 0.1 µg/ml DNP-BSA for 2 minutes. In lanes 1 and 2, cells were activated as adherent monolayers on fibronectin-coated dishes. In lanes 4 and 5, cells were grown on suspension plates and were activated in suspension. Cells were lysed and the anti-phosphotyrosine-reactive proteins were separated by SDS-PAGE and detected by autoradography. Asterisks indicate proteins visible in activated adherent but not suspension cells. The arrowhead points to a relatively invariant band that can be used to normalize between lanes. Lane 3 contains ^{14}C-labeled molecular weight markers.

We established above that FcεRI crosslinking upregulates the ligand binding activity of VLA-4 (Fig. 8.8) and suggested that activated VLA-4 with FAK may directly or indirectly control actin plaque assembly. Here, we suggest that activated VLA-4 may additionally contribute to the increased tyrosine phosphorylation response of antigen-activated adherent cells, perhaps by controlling the activation of FAK. The role of adhesion in signal amplification in RBL-2H3 cells is discussed further in chapters 9 and 12.

CONCLUDING REMARKS

The goal of this chapter was to draw attention to the dynamic membrane, cytoskeletal and adhesive properties of mast cells and basophils by demonstrating their esthetic appeal, their biochemical regulation, their ability to influence cellular activities such as extravasation and migration to sites of inflammation, and their ability to modulate receptor-mediated signaling responses.

Our own continuing studies will focus particularly on the assembly and composition of actin plaques, that may be general adhesion structures in mast cells, just as focal adhesions are general adhesion structures in fibroblasts. Other continuing studies will address the role of these dynamic structures in the control of functions ranging from cell migration and localization to the modulation of secretion. In addition, it is a particularly exciting time for studies that integrate the roles of tyrosine kinases, lipid metabolism and GTPases in the regulation of cytoskeletal organization. A great deal of information has been published about the modulation by FcεRI crosslinking of tyrosine phosphorylation, phosphatidylinositol turnover and arachidonate metabolism. We propose mast cells as an outstanding model for further studies to connect these biochemical pathways to each other and to cytoskeletal responses.

Acknowledgments

Original results presented here were obtained with support from ACS grant IM675 to JMO and NIH grants GM49814 (to JMO) and GM50562 (to BSW). Some of the previously unpublished micrographs in Figures 8.1 and 8.2 were generated in this laboratory by M. C. Schneebeck. The previously unpublished protein-tyrosine phosphorylation data were generated in collaboration with W. Li (Fig. 8.18) and A. M. Martinez (Fig. 8.19).

References

1. Oliver JM, Seagrave JC, Stump, RF et al. Signal transduction and cellular response in RBL-2H3 mast cells. Prog Allergy 1988; 42:185-45.
2. Pfeiffer JR, Seagrave JC, Davis BH et al. Membrane and cytoskeletal changes associated with IgE-mediated serotonin release from rat basophilic leukemia cells. J Cell Biol 1985; 101:2145-55.
3. Liu, Z-Y, Young, J-I, Elson EL. Rat basophilic leukemia cells stiffen when they secrete. J Cell Biol 1987; 105:2933-43.
4. Dowrick P, Kenworthy P, McCann B et al. Circular ruffle formation and closure lead to macropinocytosis in hepatocyte growth factor/scatter factor-treated cells. Eur J Cell Biology 1993; 61:44-53.
5. Hall A. Small GTP-binding proteins and the regulation of the actin cytoskeleton. Ann Rev Cell Biol 1994; 10:31-54.
6. Nobes CD, Hawkins P, Stephens L et al. Activation of the small GTP-binding proteins rho and rac by growth factor receptors. J Cell Sci 1995; 108:225-33.
7. Wilson BS, Deanin GG, Standefer JC et al. Depletion of guanine nucleotides with mycophenolic acid suppresses IgE receptor-mediated degranulation in rat basophilic leukemia cells. J Immunol 1989; 143: 259-65.
8. Barker SA, Caldwell KK, Hall A et al. Wortmannin blocks lipid and protein kinase activities associated with PI 3-kinase and inhibits a subset of responses induced by FcεRI crosslinking. Molec Biol Cell 1995; 6:1145-58.
9. Deanin GG, Martinez AM, Pfeiffer JR et al. Tyrosine kinase-dependent phosphatidylinositol turnover and functional responses in the FcεRI signalling pathway. Biochem Biophys Res Comm 1991; 179:551-57.
10. Apgar JR. Association of the crosslinked IgE receptor with the membrane skeleton is independent of the known signaling mechanisms in rat basophilic leukemia cells. Cell Reg 1991; 2:181-91.
11. Apgar JR. Polymerization of actin in RBL-2H3 cells can be triggered through either the IgE receptor or the adenosine receptor but different signaling pathways are used. Molec Biol Cell 1994; 5:313-32.
12. Spudich A. Myosin reorganization in activated RBL cells correlates temporally with stimulated secretion. Cell Motil Cytoskel 1994; 29:345-53.
13. Ludowyke RI, Kawasugi K, French PW. ATPγS induces actin and myosin rearrangement during histamine secretion in a rat basophilic leukemia cell line (RBL-2H3). Eur J Cell Biol 1994; 64:357-67.
14. Ludowyke RI, Kawasugi K, French PW. PMA and calcium ionophore induce myosin and F-actin rearrangement during histamine secretion from RBL-2H3 cells. Cell Motil Cytoskel 1994; 29:354-65.
15. Norman JC, Price LS, Ridley AJ et al. Actin filament organization in activated mast cells is regulated by heterotrimeric and small GTP-binding proteins. J Cell Biol 1994; 126:1005-15.
16. Price LS, Norman JC, Ridley AJ et al. The small GTPases Rac and Rho as regulators of secretion in mast cells. Current Biology 1995; 5:68-73.
17. Pfeiffer JR, Oliver JM. Tyrosine kinase-dependent assembly of actin plaques linking FcεRI crosslinking to increased cell substrate adhesion in RBL-2H3 tumor mast cells. J Immunol 1994; 152:270-79.
18. Maher PA, Pasquale EB, Wang JY et al. Phosphotyrosine-containing proteins are concentrated in focal adhesions and intracellular junctions in normal cells. Proc Natl Acad Sci USA 1985; 82:6576-80.
19. Burridge K, Fath K, Kelly T et al. Focal adhesions: transmembrane junctions between

the extracellular matrix and the cytoskeleton. Ann Rev Cell Biol 1988; 4:487-25.

20. Burridge K, Fath K. Focal contacts: transmembrane links between the extracellular matrix and the cytoskeleton. Bio Essays 1989; 10:104-08.
21. Nigg EA, Sefton BM, Hunter T et al. Immunofluorescent localization of the transforming protein of Rous sarcoma virus with antibodies against a synthetic src peptide. Proc Natl Acad Sci USA 1982; 79:5322-26.
22. Tarone G, Cirillo D, Giancotti FG et al. Rous sarcoma virus-transformed fibroblasts adhere primarily at discrete protrusions of the ventral membrane called podosomes. Exp Cell Res 1985; 159:141-57.
23. Mueller SC, Yey Y, Chen W-T. Tyrosine phosphorylation of membrane proteins mediates cellular invasion by transformed cells. J Cell Biol 1992; 119; 1309-25.
24. Trotter JA. The organization of actin in spreading macrophages. Exp Cell Res 1981; 132:235-48.
25. Zambonin-Zallone A, Teti A, Primavera MV et al. Cytoskeleton and adhesion of osteoclasts and monocytes. J Anat 1983; 137:57.
26. Marchisio PC, Cirillo D, Teti A et al. Rous sarcoma virus-transformed fibroblasts and cells of monocytic origin display a peculiar dot-like organization of cytoskeletal proteins involved in microfilament-membrane interactions. Exp Cell Res 1987; 169:202-14.
27. Tawil N, Wilson P, Carbonetto S. Integrins in point contacts mediate cell spreading: factors that regulate integrin accumulation in point contacts vs. focal contacts. J Cell Biol 1993; 120:261-71.
28. Ludowyke RI, Peleg I, Beaven MA et al. Antigen-induced secretion of histamine and the phosphorylation of myosin by protein kinase C in rat basophilic leukemia cells. J Biol Chem 1989; 264:12492-501.
29. Choi OH, Adelstein RS, Beaven MA. Secretion from rat basophilic RBL-2H3 cells is associated with diphosphorylation of myosin light chains by myosin light chain kinase as well as phosphorylation by protein kinase C. J Biol Chem 1994; 269: 536-41.
30. Nishikawa M, Seller JR, Adelstein RS et al. Protein kinase C mediates in vitro phosphorylation of the smooth muscle heavy meromyosin by myosin light chain kinase. J Biol Chem 1984; 259:8808-14.
31. Egelhoff TT, Brown SS, Spudich JA. Spatial and temporal control of nonmuscle myosin localization: identification of a domain that is necessary for myosin filment disassembly in vivo. J Cell Biol 1991; 112:677-88.
32. Wu H, Parsons JT. Cortactin, an 80/85 kilodalton $pp60^{src}$ substrate, is a filamentous actin-binding enriched in the cell cortex. J Cell Biol 1993; 120:1417-26.
33. Schaller MD, Borgman CA, Cobb BS et al. $pp125^{FAK}$, a structurally distinctive protein-tyrosine kinase associated with focal adhesions. Proc Natl Acad Sci USA 1992; 89:5192-96.
34. Hanks SK, Calalb MB, Harper MC et al. Focal adhesion protein-tyrosine kinase phosphorylated in response to cell attachment to fibronectin. Proc Natl Acad Sci USA 1992; 8487-91.
35. Hamawy MJ, Mergenhagen SE, Siraganian RP. Tyrosine phosphorylation of $pp125^{fak}$ by the aggregation of high affinity immunoglobulin E receptors requires cell adherence. J Biol Chem 1993; 268:6851-54.
36. Hamawy MM, Swaim WD, Minoguchi K et al. The aggregation of the high affinity IgE receptor induces tyrosine phosphorylation of paxillin, a focal adhesion protein. J Immunol 1994; 153:4655-62.
37. Burridge K, Turner CE, Romer LH. Tyrosine phosphorylation of paxillin and $pp125^{FAK}$ accompanies cell adhesion to extracellular matrix: a role in cytoskeletal assembly. J Cell Biol 1992; 119:893-03.
38. Valent P, Bettelheim P. Cell surface structures on human basophils and mast cells: biochemical and functional characterization. Adv Immunol 1992:52:333-23.
39. Hamawy MM, Mergenhagen SE, Siraganian RP. Adhesion molecules as regulators of mast-cell and basophil function. Immunol Today 1994:15:62-66.
40. Ra C, Yasuda M, Yagita H, Okumura K. Fibronectin receptor integrins are involved in mast cell activation. J Allergy Clin Immunol 1994:94:625-28.

41. Masumoto A, Hemler ME. Multiple activation states of VLA-4. J Biol Chem 1993; 268:228-34.
42. Leavesley DI, Oliver JM, Swart BW et al. Signals from platelet/endothelial cell adhesion molecule enhance the adhesive activity of the very late antigen-4 integrin of human CD34+ hemopoietic progenitor cells. J Immunol 1994:153:4673-83.
43. Metzger H, Alcarez R, Hohman R et al. The receptor with high affinity for immunoglobulin E. Ann Rev Immunol 1986; 4:419-70.
44. Blank U, Ra C, Miller L et al. Complete structure and expression in transfected cells of high affinity IgE receptor. Nature 1987; 337:187-89.
45. Metzger H. The receptor with high affinity for IgE. Immunol Rev 1992; 125:37-48.
46. Wright MD, Tomlinson MG. The ins and outs of the transmembrane 4 superfamily. Immunol Today 1994; 15:588-95.
47. Reth M. Antigen receptor tail clue [letter]. Nature 1989; 338:383.
48. Ortega E, Schweitzer-Stenner R, Pecht I. Possible orientational constraints determine secretory signals induced by aggregation of IgE receptors on mast cells. EMBO J 1988; 7:4101-09.
49. Benhamou M, Gutkind JS, Robbins KD. Tyrosine phosphorylation coupled to IgE receptor-mediated signal transduction and histamine release. Proc Natl Acad Sci USA 87:5327-30.
50. Eiseman E, Bolen JB. Engagement of the high-affinity IgE receptor activates src protein-related tyrosine kinases. Nature 1992; 355:78-80.
51. Hutchcroft JE, Geahlen RL, Deanin GG et al. FcεRI-mediated tyrosine phosphorylation and activation of the 72 kDa protein-tyrosine kinase, PTK72, in RBL-2H3 rat tumor mast cells. Proc Natl Acad Sci USA 1992; 89:9107-11.
52. Jouvin M-HE, Adamczewski M, Numerof R et al. Differential control of the tyrosine kinases lyn and syk by the two signaling chains of the high affinity immunoglobulin E receptor. J Biol Chem 1994; 269:5918-25.
53. Oliver JM, Burg DL, Wilson BS et al. Inhibition of mast cell FcεRI-mediated signaling and effector function by the Syk-selective inhibitor, piceatannol. J Biol Chem 1994; 269:29697-703.
54. Pribluda VS, Pribluda C, Metzger H. Transphosphorylation as the mechanism by which the high-affinity receptor for IgE is phosphorylated upon aggregation. Proc Natl Acad Sci USA 1994; 91:11246-50.
55. Kihara H, Siraganian RP. Src homology 2 domains of Syk and Lyn bind to tyrosine-phosphorylated subunits of the high affinity IgE receptor. J Biol Chem 1994; 269:22427-32.
56. Shiue L, Zoller MJ, Brugge JS. Syk is activated by phophotyrosine-containing peptides representing the tyrosine-based activation motifs of the high affinity receptor for IgE. J Biol Chem 1995; 270:10498-02.
57. Shiue L, Green J, Green OM et al. Interaction of $p72^{syk}$ with the γ and β subunits of the high-affinity receptor for immunoglobulin E, FcεRI. Molec Cell Biol 1995; 15:272-81.
58. Taylor JA, Karas JL, Ram MK et al. Activation of the high-affinity immunoglobulin E receptor FcεRI in RBL-2H3 cells is inhibited by syk SH2 domains. Molec Cell Biol 1995; 15:4149-57.
59. Deanin GG, Pfeiffer JR, Cutts JL et al. Isoprenoid pathway activity is required for IgE receptor-mediated, tyrosine kinase-coupled transmembrane signaling in permeabilized RBL-2H3 rat basophilic leukemia cells. Cell Reg 1991; 2:627-40.
60. Wilson BS, Kapp N, Lee RJ et al. Distinct functions of the FcεRI γ and β subunits in the control of FcεRI-mediated tyrosine kinase activation and signaling responses in RBL-2H3 mast cells. J Biol Chem 1995; 270:4013-22.
61. Deanin GG, Cutts JL, Pfeiffer JR et al. Role of isoprenoid metabolism in IgE receptor-mediated signal transduction. J Immunol 1991; 146:3528-35.
62. Clarke S. Protein isoprenylation and methylation of carboxyl-terminal cysteine residues. Ann Rev Biochem 1992; 61:355-86.
63. Shakarjian MP, Eiseman E. Penhallow RC et al. 3-Hydroxy-3-methylglutaryl-coenzyme A reductase inhibition in a rat mast cell line. J Biol Chem 1993; 268:15252-59.

64. Li W, Deanin GG, Margolis B et al. FcεRI-mediated tyrosine phosphorylation of multiple proteins, including phospholipase Cγ1 and the Receptor $\beta\gamma_2$ complex, in RBL-2H3 rat basophilic leukemia cells. Molec Cell Biol 1992; 12:3176-82.
65. Anant JS, Fung BK-K. In vivo farnesylation of rat rhodopsin kinase. Biochem Biophys Res Commun 1992; 183:468-73.
66. Park DJ, Min HK, Rhee SG. IgE-induced tyrosine phosphorylation of phospholipase Cγ1 in rat basophilic leukemia cells. J Biol Chem 1991; 266:24237-40.
67. Yano H, Nakanishi S, Kimura K et al. Inhibition of histamine secretion by wortmannin through the blockade of phosphatidylinositol 3-kinase in RBL-2H3 cells. J Biol Chem 1993; 268:25846-56.
68. Margolis B, Hu P, Katzav S et al. Tyrosine phosphorylation of the *vav* protooncogene product, a potential transcriptional regulator with SH2 and SH3 domains. Nature 1992; 356:71-74.
69. Turner H, Reif K, Rivera J et al. Regulation of the adapter molecule Grb2 by the FcεRI in the mast cell line RBL2H3. J Biol Chem 1995; 270:9500-06.
70. Kawakami Y, Yao L, Miura T et al. Tyrosine phosphorylation and activation of Bruton tyrosine kinase upon FcεRI cross-linking. Molec Cell Biol 1994; 14:5108-13.
71. Adamczewski M, Numerof RP, Koretzky GA et al. Regulation by CD45 of the tyrosine phosphorylation of high affinity IgE receptor β- and γ-chains. J Immunol 1995; 154:3047-55.
72. Swieter M, Berenstein EH, Swaim WD, Siraganian RP. Aggregation of IgE receptors in rat basophilic leukemia 2H3 cells induces tyrosine phosphorylation of the cytosolic protein-tyrosine phosphatase HePTP. J Biol Chem 1995; 270:21902-06.
73. Paolini R, Jouvin M-H, Kinet J-P. Phosphorylation and dephosphorylation of the high-affinity receptor for immunoglobulin E immediately after receptor engagement and disengagement. Nature 1991; 353: 855-58.
74. Paolini R, Numerof R, Kinet J-P. Phosphorylation/dephosphorylation of high-affinity IgE receptors: a mechanism for coupling/uncoupling a large signaling complex. Proc Natl Acad Sci USA 1992; 89:10733-37.
75. Lee RJ, Oliver JM. Roles for Ca^{2+} stores release and two Ca^{2+} influx pathways in the FcεRI-activated Ca^{2+} responses of RBL-2H3 mast cells. Mol Biol Cell 1995; 6:825-39.
76. Offermanns S, Jones SVP, Bombien E et al. Stimulation of mitogen-activated protein kinase activity by different secretory stimuli in rat basophilic leukemia cells. J Immunol 1994; 152:250-61.
77. Hirasawa N, Scharenberg A, Yamamura H et al. A requirement for syk in the activation of the microtubule-associated protein kinase/phospholipase A_2 pathway by FcεRI is not shared by a G protein-coupled receptor. J Biol Chem 1995; 270:10960-67.
78. Hirasawa N, Santini F, Beaven MA. Activation of the mitogen-activated protein kinase/cytosolic phospholipase A_2 pathway in a rat mast cell line. J Immunol 1995; 154:5391-02.
79. Gulbins E, Coggeshall KM, Gottfried B et al. Director stimulation of Vav guanine nucleotide exchange activity for Ras by phorbol esters and diglycerides. J Cell Biol 1994; 14:4749-58.
80. Graham T, Oliver JM, Wilson BS. Differential inhibitors of protein isoprenylation distinguish positive and negative regulatory pathways in FcεRI-mediated signaling in RBL-2H3 cells. Molec Biol Cell 1994; 5:321a (abstract).
81. Kapp N, Barker S, Wilson BS. Identification of Rab3D in rat basophilic leukemia cells. Molec Biol Cell 1994; 5:447a (abstract).
82. Fischer von Mollard G, Stahl B, Li C et al. Rab proteins in regulated exocytosis. TIBS 1994; 19:164-68.
83. Kotani K, Hara, K, Kotani et al. Phosphoinositide 3-kinase as an upstream regulator of the small GTP-binding protein Rac in the insulin signaling of membrane ruffling. Biochem Biophys Res Comm 1995: 208:985-90.
84. Tolias KF, Cantleuy LC, Carpenter CL. Rho family GTPases bind to phosphoinositide kinases. J Biol Chem 1995; 270:17656-59.
85. Stossel TP. On the crawling of animal cells. Science 1993; 260:1086-94.

86. Beaven MA, Metzger H. Signal transduction by Fc receptors: the FcεRI case. Immunol. Today 1993; 220:222-26.
87. Ha K-S, Exton JH. Activation of actin polymerization by phosphatidic acid derived from phosphatidylcholine in IIC9 fibroblasts. J Cell Biol 1993; 123:1789-96.
88. Moolenaar, WH. LPA: a novel lipid mediator with diverse biological actions. Trends in Cell Biology 1994; 4:213-19.
89. Peppelenbosch MP, Qiu R-G, de Vries-Smits, AMM et al. Rac mediates growth factor-induced arachidonic acid release. Cell 1995; 81:849-56.
90. Peppelenbosch MP, Tertoolen LG, Hage GJ et al. Epidermal growth factor-induced actin remodeling is regulated by 5-lipoxygenase and cyclooxygenase products. Cell 1993; 74:565-75.
91. Chong LD, Traynor-Kaplan A, Bokoch GM, Schwartz MA. The small GTP-binding protein Rho regulates a phosphatidylinositol 4-phosphate 5-kinase in mammalian cells. Cell 1994; 79:507-13.
92. Shattil SJ, Haimovich B, Cunningham M et al. Tyrosine phosphorylation of $pp125^{FAK}$ in platelets requires coordinated signaling through integrin and agonist receptors. J Biol Chem 1994; 269:14738-45.
93. Springer TA. Traffic signals for lymphocyte recirculation and leukocyte emigration: the multistep paradigm. Cell 1994; 76:301-14.
94. Alon R, Kassner PD, Carr MW et al. The integrin VLA-4 supports tethering and rolling in flow on VCAM-1. J Cell Biol 1995; 128:1243-53.
95. Berlin C, Bargatze RF, Campbell JJ et al. α4 integrins mediate lymphocyte attachment and rolling under physiologic flow. Cell 1995; 80:413-22.
96. Otani I, Conrad DH, Carlo JR et al. Phagocytosis by rat peritoneal mast cells: independence of IgG Fc-mediated and C3-mediated signals. J Immunol 1982; 129: 2109-12.
97. Fox CC, Jewell SD, Whitacre CC. Rat peritoneal mast cells present antigen to a PPD-specific T cell line. Cell Immunol 1994; 158:253-64.
98. Frandji P, Tkaczyk C, Oskeritziana C et al. Presentation of soluble antigens by mast cells: upregulation by interleukin-4 and granulocyte/macrophage colony-stimulating factor and downregulation by interferon-γ. Cell Immunol 1995; 163:37-46.
99. Menon AK, Holowka D, Webb WW et al. Crosslinking of receptor-bound IgE to aggregates larger than dimers lead to rapid immobilization. J Cell Biol 1986; 102: 541-50.
100. Menon AK, Holowka D, Webb WW et al. Clustering, mobility, and triggering activity of small oligomers of immunoglobulin E on rat basophilic leukemia cells. J Cell Biol 1986; 102:534-40.
101. Pecht I, Ortega E, Jovin TM. Rotational dynamics of the Fc_{ε} receptor on mast cells monitored by specific monoclonal antibodies and IgE. Biochemistry 1991; 30: 3449-58.
102. Myers, JN, Holowka D, Baird B. Rotational motion of monomeric and dimeric immunoglobulin E-receptor complexes. Biochemistry 1992; 31:567-75.
103. Seagrave JC, Pfeiffer JR, Wofsy C et al. Relationship of IgE receptor topography to secretion in RBL-2H3 mast cells. J Cell Physiol 1991; 148:139-51.
104. Robertson D, Holowka D, Baird B. Crosslinking of immunoglobulin E-receptor complexes induces their interaction with the cytoskeleton of rat basophilic leukemia cells. J Immunol 1986; 136:1-9.
105. Seagrave JC, Oliver JM. Antigen-dependent transition of IgE to a detergent-insoluble form is associated with reduced IgE receptor-dependent secretion from RBL-2H3 mast cells. J Cell Physiol 1990; 144:128-36.
106. Apgar JR. Antigen-induced crosslinking of the IgE receptor leads to an association with the detergent-insoluble membrane skeleton of rat basophilic leukemia (RBL-2H3) cells. J Immunol 1990; 145:3814-22.
107. Seagrave JC, Deanin GG, Martin JC et al. DNP-phycobiliproteins, fluorescent antigens to study dynamic properties of antigen-IgE-receptor complexes on RBL-2H3 rat mast cells. Cytometry 1987; 8:287-95.
108. Mao S-Y, Pfeiffer JR, Oliver JM et al. Effects of subunit mutation on the localization to coated pits and internalization of crosslinked IgE-receptor complexes. J

Immunol 1993; 151:2760-74.

109. Thomas JL, Holowka D, Baird B et al. Large-scale coaggregation of fluorescent lipid probes with cell surface proteins. J Cell Biol 1994; 125:795-02.

110. Mao S-Y, Alber G, Rivera J et al. Interaction of aggregated native and mutant IgE receptors with the cellular skeleton. Proc Natl Acad Sci USA 1992; 89:222-26.

111. Ktistakis NT, Thomas D, Roth MG. Characteristics of the tyrosine recognition signal for internalization of transmembrane surface glycoproteins. J Cell Biol 1990; 111:1393-07.

112. Hamawy MM, Oliver C, Mergenhagen SE et al. Adherence of rat basophilic leukemia (RBL-2H3) cells to fibronectin-coated surfaces enhances secretion. J Immunol 1992; 149:615-21.

CHAPTER 9

FcεRI-Mediated Cell Degranulation, Proliferation and Adhesion

Majed M. Hamawy and William D. Swaim

INTRODUCTION

FcεRI on mast cells and basophils plays a central role in allergic diseases. Crosslinking FcεRI triggers mast cells and basophils to release an array of potent inflammatory mediators. In addition, in vitro studies strongly suggest that FcεRI aggregation modulates the proliferation of mast cells and basophils and controls migration and localization. Mediators released from mast cells and basophils upon FcεRI aggregation fall into two groups: preformed and newly synthesized mediators. Preformed mediators, stored in secretory granules, include histamine, sulfated proteoglycans, proteases, cytokines (TNF-α), chemotactic factors for neutrophils and eosinophils including platelet activating factor, and adenosine. Chapters 10 and 11 in this book discuss in detail FcεRI-mediated release of newly synthesized mediators. In this chapter, FcεRI-mediated cell degranulation and other aspects of FcεRI functions are reviewed and discussed.

PREFORMED MEDIATORS RELEASED BY FcεRI AGGREGATION

Aggregating IgE on the surface of mast cells and basophils by a multivalent antigen triggers cell degranulation, leading to the release of several inflammatory mediators. (For details on mast cell and basophil preformed mediators see reference 1.) Only some of these mediators are discussed in this manuscript.

IgE Receptor (FcεRI) Function in Mast Cells and Basophils,
edited by Majed M. Hamawy.

Histamine (β-imidazolylethylamine)

This is a biogenic amine derived from the amino acid histidine by histidine decarboxylase.[2] In humans, mast cells and basophils are the major source of this amine. However, other cells including T lymphocytes, myelomonocytic cells and macrophages produce low levels of histamine. In the human, histamine is the major biogenic amine found in mast cell and basophil granules. Histamine is found together with other amines such as serotonin in rodent mast cells.[3] Serotonin is not found in human mast cells, but it is stored in human platelets. Upon degranulation, histamine is released from a proteoglycan/protein complex by cation exchange with extracellular sodium at neutral pH. Upon release, histamine is rapidly metabolized to either methylhistamine, methylimidazole acetic acid or imidazole acetic acid. Histamine produces its effects by acting on at least three distinct surface receptors known as H1, H2 and H3.[4]

Neutral Proteases

These enzymes are the dominant protein components of granules in human and rodent mast cells. They catalyze the cleavage of peptide bonds and are most active near neutral pH. Such enzymes have proven useful as markers that distinguish mast cells from basophils and other cells, and different mast cell subpopulations from one another. Some mast cell proteases are discussed below.

Tryptase

This is a serine endoprotease that is the major enzyme accounting for trypsin-like activity in human skin and lung mast cells.[5] Tryptase is a tetramer with a molecular mass of 134 kDa with subunits of 31 and 34 kDa, each with an active enzymatic site.[6] Five very homologous isoforms of tryptase have been identified in humans and have been divided into α and β groups.[7] Tryptase accounts for about 50% of the entire protein in the human mast cell.[8] In human mast cells the enzyme resides in secretory granules in fully active form by binding to heparin.[9] When free in solution, tryptase subunits dissociate from one another into inactive monomers.[9,10] Importantly, tryptase is absent for other types of cells including eosinophils, neutrophils, monocytes, lymphocytes and is found in very low amounts in basophils. Thus, this enzyme may serve as a selective marker for human mast cells. In rats, tryptase is found in small amounts in skin mast cells.[11,12] Unlike human tryptase, the rat enzyme is inhibited by inhibitors of serine proteases and does not bind heparin in a physiologic medium.

In murine mast cells, two tryptases have been reported (MMCP-6 and -7).[13] MMCP-6 is found in connective tissue mast cells (CTMC), whereas MMCP-7 is not expressed in mature mast cells, but has been found in bone marrow-derived IL-3 dependent mast cells in culture (BMMC).[13,14]

Chymase

This is a serine endopeptidase that is the principal enzyme accounting for the chymotrypsin-like activity in human skin, lung and bowel mast cells.[15-17] It is a 30 kDa monomer that is stored fully active in mast cell secretory granules.[18] Unlike tryptase, chymase stability does not require heparin, and its enzymatic activity is inhibited by classical biologic inhibitors of serine proteinase. In contrast to human tryptase, it is found only in a subpopulation of human mast cells, called MC_{TC}, that are preferentially located in skin and bowel submucosa.[19]

In rats, there are two forms of chymase: rat chymase I (rat mast cell protease I, RMCP-I) that is found exclusively in CTMC, and rat chymase II (rat mast cell protease II, RMCP II) that is found only in the mucosal mast cell.[20] It accounts for approximately 25% of total mast cell proteins.

In murine mast cells, at least five chymases have been reported (MMCP-1, -2,

-3, -4, -5).[13,21,22] MMCP-1 and -2 are present in mucosal mast cells while MMCP-3, -4 and -5 are present in CTMC.

Cathepsin G

This a serine endoprotease with a molecular mass of approximately 30 kDa.[23] This enzyme is also found in neutrophils and monocytes. Like chymase, cathepsin G appears to be found only in a subpopulation of human skin mast cells, i.e., MC_{TC}.

Carboxypeptidase

This is a zinc metallo-exopeptidase that removes carboxyterminal Phe and Leu residues.[24] It localizes to secretory granules from human skin and lung MC_{TC}.[25] It is a monomer with a molecular mass of 36 kDa.[26] Interestingly, chymase and carboxypeptidase reside in a complex with proteoglycan that is different from the tryptase:proteoglycan complex.[27]

In rats, carboxypeptidase A removes amino acids with aromatic residues from the C-termini of peptides. It accounts for approximately 25% of CTMC protein.[28,29]

Proteoglycans

These are composed of glycosaminoglycan side chains covalently linked to a single chain protein core (approximately 18 kDa) via a specific trisaccharide-protein linkage region.[30,31] The same protein core is associated with heparin and chondroitin sulfate proteoglycans. Because different types of mast cells have the machinery to produce the protein core, the type of proteoglycan present in mast cell is determined by the enzyme available to synthesis the glycosaminoglycan and not by the peptide core. The granules of human mast cells contain two major proteoglycans, heparin and chondroitin sulfate E, whereas basophil granules contain mainly chondroitin sulfate A.[30,32,33]

FcεRI-Induced Cell Growth and Proliferation

Few studies have examined the mitogenic effects of FcεRI crosslinking. However, these studies report contradictory data that could be due to the use of different mast cell populations and lines. Takagi et al[34] reported that FcεRI crosslinking on CTMC can cause cell proliferation. They showed that when anti-DNP IgE-sensitized mouse CTMC were stimulated by DNP and then cultured in medium containing IL-3, pure mast cell colonies developed. No mast cell colonies developed when CTMC were stimulated with the secretagogue 48/80, suggesting that the proliferation induced by FcεRI-stimulation is not due to mediators released from CTMC. Neither DNP nor IL-3 alone stimulated clonal growth of sensitized CTMC. Furthermore, neither accessory cells nor serum factors were required for IgE-mediated cell proliferation. Continuous presence of anti-DNP, IgE and IL-3 in culture was required for proliferation of CTMC. Notably, the concentrations of DNP and IgE for optimal proliferation of CTMC were similar to those for maximal degranulation. The addition of IL-4 to CTMC sensitized with anti-DNP IgE in the presence of IL-3 increased the number of the colonies and greatly enlarged both the colony size and the cell volume. However, the number of colonies greatly exceeded the maximal level supported by the combination of IL-4 and IL-3 in the absence of DNP. In contrast to these studies, Kuriu et al[35] reported no change in the proliferation of murine peritoneal mast cells after cell degranulation.

More recent studies have shown that short-term stimulation (up to 16 hours) of IL-3 dependent BMMC, Abelson-transformed mouse liver-derived mast cells (ABFTL) or RBL-2H3 cells by crosslinking FcεRI resulted in an inhibition of their proliferation as measured by ^{3}H-thymidine incorporation.[36,37] This was also associated with the accumulation of the mRNAs of the early response genes (proto-oncogenes) c-*fos*, c-*jun*, *jun*B and *jun*D. These proto-oncogenes encode nuclear proteins that interact with regulatory DNA elements, and thereby regulate many aspects of cell proliferation, differentiation and function. The kinetics of the elevation of the mRNA level of c-*jun* and *jun*B were similar, starting

5 minutes poststimulation, reaching maximum levels by 25 minutes and then gradually declining. The increase in c-*fos* mRNA was more transient than that of c-*jun* and *jun*B, reaching a maximum by 5 minutes poststimulation and then immediately decreasing. However, the relationship between the increase in the mRNAs of the proto-oncogenes and the FcεRI-mediated inhibition of mast cell proliferation is not clear. Strikingly, the pattern and the kinetics of early response gene expression induced by FcεRI aggregation were similar to that induced by stimulation of the Kit receptors. Crosslinking FcεRI in the presence of protein kinase C (PKC) inhibitors significantly reduced the inhibition of ^{3}H-thymidine uptake. Prolonged treatment of the cells with phorbol esters (to degrade PKC) blocked completely the inhibitory effect of FcεRI aggregation on mast cell proliferation. Together, the data suggest a role for PKC in mediating FcεRI-induced inhibition of mast cell proliferation. In contrast to these studies, long-term stimulation of BMMC (up to 9 days) resulted in an increase in ^{3}H-thymidine uptake and also in an increase in the number of cells.[38] However, in these studies the cells started to die after day 9.

The studies on the role of FcεRI in cell proliferation are further complicated by the fact the FcεRI aggregation initiates the release of several cytokines and growth factors that are known to influence mast cell and basophil growth (see chapter 10 in this book). Thus, it is feasible that many of the effects produced by FcεRI on proliferation are secondary to the release of growth factors.

FcεRI-MEDIATED CELL ADHESION AND CHEMOTAXIS

Adhesion of cells to other cells and to the extracellular matrix (ECM) plays a critical role in developmental, as well as cellular and immunological processes.[39-41] This attachment is mediated by specialized cell surface receptors and results in activation of intracellular signals that not only evoke new cellular responses but also modulate the response of cells to other stimuli. The specificity and affinity of receptors for adhesion molecules appear to be regulated by cell activation (see chapter 12). Thus, although cells can adhere spontaneously to other cells and to the ECM, cell activation usually enhances such binding. The enhanced binding of the cells may be a consequence of an increase in the binding affinity, an increase in the number of receptors or due to the expression of a new type of receptors.

Basophils and mast cells from different species adhere to other cells and to the ECM.[42-55] Both cells have surface receptors that can mediate binding to other cells and to ECM glycoproteins (see chapter 12). Recent studies indicate that FcεRI aggregation not only modulates cell adhesion but also regulates the expression and the affinity of adhesion molecules on mast cells and basophils.

BMMC adhere spontaneously to vitronectin through their surface integrins, whereas binding to laminin and fibronectin is markedly enhanced by FcεRI aggregation.[48,49,56] Significant cell adherence to laminin and fibronectin occurred with concentrations of Ag that did not induce cell degranulation, suggesting that FcεRI-induced adhesion was not mediated by mediators released from activated mast cells. This was confirmed by the failure of supernatants derived from activated mast cells to induce the adhesion of the cells to laminin. Notably, FcεRI aggregation also enhanced mast cell chemotaxis to laminin, which was independent of secretion.[57] Whether crosslinking FcεRI increases the expression and/or modulates the affinity of laminin receptors on the cells is still not determined.

RBL-2H3 cells adhere spontaneously to some ECM glycoproteins.[58] In these cells, FcεRI aggregation induces massive cell spreading and the redistribution of cytoskeletal proteins. This is accompanied by an increase in the strength of the adherence of these cells to the substratum. In chapter 8 of this book, Oliver et al show that the increase in the adhesiveness of

RBL-2H3 cells upon FcεRI aggregation is due to an increase in VLA-4 binding activity and is not due to an increase in the expression of this integrin on RBL-2H3 cells.

Mature mast cells are intimately associated with tissues, whereas basophils circulate in the blood and migrate into tissue only after activation.[50-52,54,55,59] The failure of inactivated basophils to adhere to the lining of the blood vessels, i.e., endothelium, can be attributed to many factors including the low expression of adhesion molecules on basophils and their counterparts on the endothelium and/or to the low affinity of these molecules in nonactivated cells. Notably, many of the mediators released as a result of allergic and inflammatory responses markedly increase both the number and the affinity of adhesion molecules on basophils and on the endothelium which, in turn, initiate the migration of basophils into tissues.[50,52] FcεRI aggregation regulates the expression of adhesion molecules on basophils.[54] For example, treatment of basophils with anti-IgE induces a rapid and sustained increase in CD11b.[54] In contrast, FcεRI aggregation causes a decrease in Leu 8 expression. Both effects were apparent within minutes of FcεRI aggregation. As with mast cells, FcεRI-effects were independent of cell degranulation. Thus, treatment of basophils with anti-IgE in the absence of Ca^{2+} resulted in similar alterations in CD11b and Leu 8. Furthermore, kinetic studies indicated that the changes in the level of these adhesion molecules did not correlate with that of secretion. In a different set of studies, within minutes, FcεRI aggregation induced basophil adhesion to vascular endothelium.[55] In these studies, coincubation of basophils and endothelial cells with anti-IgE resulted in time- and dose-dependent increases in basophils adhesion. These effects required both Mg^{2+} and Ca^{2+} and occurred in the absence of basophil degranulation. FcεRI-induced adhesion was inhibited upon incubating the cells with mAb to CD18, suggesting the involvement of this integrin in FcεRI-dependent adhesion of basophils to the endothelium.

CONCLUDING REMARKS

The aggregation of FcεRI result in the release of potent inflammatory mediators. These mediators recruit to sites of injury different types of cells leading, in many instances, to the elimination of the causative agent. The ability of FcεRI aggregation to induce or enhance the expression of adhesion molecules on mast cells and basophils and to increase the chemotaxis of these cells may be important in the recruitment and localization of mast cells and basophils to the site of injury. These FcεRI-induced effects, together with promotion of cell proliferation, may be crucial for launching and sustaining a successful inflammatory response.

REFERENCES

1. Schwartz L, Huff T. Biology of mast cells and basophils. In: Middleton E, Reed CE, Ellis EF, eds. Allergy: Principles and Practice. 4th ed. St. Louis: Mosby, 1993:135-68.
2. Schayer RW. Histidine decarboxylase in mast cells. Ann NY Acad Sci 1963; 103:164-72.
3. Weitzman G, Galli SJ, Dvorak AM et al. Cloned mouse mast cells and normal mouse peritoneal mast cells. Determination of serotonin content and ability to synthesize serotonin in vitro. Int Arch Allergy Appl Immunol 1985; 77:189-91.
4. Tedeschi A, Lorini M, Arquati M et al. Regulation of histamine release from human basophil leucocytes: role of H1, H2 and H3 receptors. Allergy 1991; 46:626-31.
5. Schwartz LB, Lewis RA, Austen KF. Tryptase from human pulmonary mast cells. Purification and characterization. J Biol Chem 1981; 256:11939-43.
6. Schwartz LB. Monoclonal antibodies against human mast cell tryptase demonstrate shared antigenic sites on subunits of tryptase and selective localization of the enzyme to mast cells. J Immunol 1985; 134:526-31.
7. Miller JS, Westin EH, Schwartz LB. Cloning and characterization of complementary DNA for human tryptase. J Clin Invest

1989; 84:1188-95.
8. Schwartz LB, Lewis RA, Seldin D et al. Acid hydrolases and tryptase from secretory granules of dispersed human lung mast cells. J Immunol 1981; 126:1290-94.
9. Schwartz LB, Bradford TR. Regulation of tryptase from human lung mast cells by heparin. Stabilization of the active tetramer. J Biol Chem 1986; 261:7372-79.
10. Alter SC, Metcalfe DD, Bradford TR et al. Regulation of human mast cell tryptase. Effects of enzyme concentration, ionic strength and the structure and negative charge density of polysaccharides. Biochem J 1987; 248:821-27.
11. Kido H, Yokogoshi Y, Katunuma N. Kunitz-type protease inhibitor found in rat mast cells. Purification, properties and amino acid sequence. J Biol Chem 1988; 263:18104-207.
12. Chen Z, Irani AA, Bradford TR et al. Localization of rat tryptase to a subset of the connective tissue type of mast cell. J Histochem Cytochem 1993; 41:961-69.
13. Reynolds DS, Stevens RL, Lane WS et al. Different mouse mast cell populations express various combinations of at least six distinct mast cell serine proteases. Proc Natl Acad Sci USA 1990; 87:3230-34.
14. McNeil HP, Reynolds DS, Schiller V et al. Isolation, characterization and transcription of the gene encoding mouse mast cell protease 7. Proc Natl Acad Sci USA 1992; 89:11174-78.
15. Schechter NM, Fraki JE, Geesin JC et al. Human skin chymotryptic proteinase. Isolation and relation to cathepsin G and rat mast cell proteinase I. J Biol Chem 1983; 258:2973-78.
16. Caughey GH, Zerweck EH, Vanderslice P. Structure, chromosomal assignment and deduced amino acid sequence of a human gene for mast cell chymase. J Biol Chem 1991; 266:12956-63.
17. Irani AA, Schechter NM, Craig SS et al. Two types of human mast cells that have distinct neutral protease compositions. Proc Natl Acad Sci USA 1986; 83:4464-68.
18. Schechter NM, Choi JK, Slavin DA et al. Identification of a chymotrypsin-like proteinase in human mast cells. J Immunol 1986; 137:962-70.
19. Wintroub BU, Kaempfer CE, Schechter NM et al. A human lung mast cell chymotrypsin-like enzyme. Identification and partial characterization. J Clin Invest 1986; 77:196-201.
20. Miller HR, Huntley JF, Newlands GF et al. Granule chymases and the characterization of mast cell phenotype and function in rat and mouse. Monogr Allergy 1990; 27:1-30.
21. Serafin WE, Reynolds DS, Rogelj S et al. Identification and molecular cloning of a novel mouse mucosal mast cell serine protease. J Biol Chem 1990; 265:423-29.
22. Huang RY, Blom T, Hellman L. Cloning and structural analysis of MMCP-1, MMCP-4 and MMCP-5, three mouse mast cell-specific serine proteases. Eur J Immunol 1991; 21:1611-21.
23. Schechter NM, Irani AM, Sprows JL et al. Identification of a cathepsin G-like proteinase in the MCTC type of human mast cell. J Immunol 1990; 145:2652-61.
24. Goldstein SM, Kaempfer CE, Proud D et al. Detection and partial characterization of a human mast cell carboxypeptidase. J Immunol 1987; 139:2724-29.
25. Irani AM, Goldstein SM, Wintroub BU et al. Human mast cell carboxypeptidase. Selective localization to MCTC cells. J Immunol 1991; 147:247-53.
26. Reynolds DS, Gurley DS, Stevens RL et al. Cloning of cDNAs that encode human mast cell carboxypeptidase A, and comparison of the protein with mouse mast cell carboxypeptidase A and rat pancreatic carboxypeptidases. Proc Natl Acad Sci USA 1989; 86:9480-84.
27. Goldstein SM, Leong J, Schwartz LB et al. Protease composition of exocytosed human skin mast cell protease-proteoglycan complexes. Tryptase resides in a complex distinct from chymase and carboxypeptidase. J Immunol 1992; 148:2475-82.
28. Everitt MT, Neurath H. Rat peritoneal mast cell carboxypeptidase: localization, purification and enzymatic properties. FEBS Lett 1980; 110:292-96.
29. Schwartz LB, Riedel C, Schratz JJ et al. Localization of carboxypeptidase A to the

macromolecular heparin proteoglycan-protein complex in secretory granules of rat serosal mast cells. J Immunol 1982; 128:1128-33.

30. Stevens RL, Fox CC, Lichtenstein LM et al. Identification of chondroitin sulfate E proteoglycans and heparin proteoglycans in the secretory granules of human lung mast cells. Proc Natl Acad Sci USA 1988; 85:2284-87.
31. Stevens RL, Avraham S, Gartner MC et al. Isolation and characterization of a cDNA that encodes the peptide core of the secretory granule proteoglycan of human promyelocytic leukemia HL-60 cells. J Biol Chem 1988; 263:7287-91.
32. Thompson HL, Schulman ES, Metcalfe DD. Identification of chondroitin sulfate E in human lung mast cells. J Immunol 1988; 140:2708-13.
33. Metcalfe DD, Bland CE, Wasserman SI. Biochemical and functional characterization of proteoglycans isolated from basophils of patients with chronic myelogenous leukemia. J Immunol 1984; 132:1943-50.
34. Takagi M, Nakahata T, Koike K et al. Stimulation of connective tissue-type mast cell proliferation by crosslinking of cell-bound IgE. J Exp Med 1989; 170:233-44.
35. Kuriu A, Sonoda S, Kanakura Y et al. Proliferative potential of degranulated murine peritoneal mast cells. Blood 1989; 74: 925-29.
36. Baranes D, Razin E. Protein kinase C regulates proliferation of mast cells and the expression of the mRNAs of fos and jun protooncogenes during activation by IgE-Ag or calcium ionophore A23187. Blood 1991; 78:2354-64.
37. Lewin I, Nechushtan H, Ke Q et al. Regulation of AP-1 expression and activity in antigen-stimulated mast cells: the role played by protein kinase C and the possible involvement of Fos interacting protein. Blood 1993; 82:3745-51.
38. Levi-Schaffer F, Shalit M. Proliferation and functional responses of bone marrow-derived mast cells after activation. Cell Immunol 1993; 148:435-43.
39. Shimizu Y, Shaw S. Lymphocyte interactions with extracellular matrix. FASEB J 1991; 5:2292-99.
40. Springer TA. Adhesion receptors of the immune system. Nature 1990; 346:425-34.
41. Hamawy MM, Siraganian RP, Mergenhagen SE. The role of the extracellular matrix in inflammation. In: Genco R, Hamada S, Lehner T, eds. Molecular Pathogenesis of Periodontal Disease. Washington, D.C.: American Society of Microbiology, 1994: 235-46.
42. Hamawy MM, Mergenhagen SE, Siraganian RP. Adhesion molecules as regulators of mast cell and basophil function. Immunol Today 1994; 15:62-66.
43. Hamawy MM, Mergenhagen SE, Siraganian RP. Cell adherence to fibronectin and the aggregation of the high affinity IgE receptor synergistically regulate tyrosine phosphorylation of 105-115 kDa proteins. J Biol Chem 1993; 268:5227-33.
44. Ducharme LA, Weis JH. Modulation of integrin expression during mast cell differentiation. Eur J Immunol 1992; 222603-07.
45. Gurish MF, Bell AF, Smith TJ et al. Expression of murine beta 7, alpha 4 and beta 1 integrin genes by rodent mast cells. J Immunol 1992; 149:1964-72.
46. Shimizu Y, Irani AM, Brown EJ et al. Human mast cells derived from fetal liver cells cultured with stem cell factor express a functional CD51/CD61 (alpha v beta 3) integrin. Blood 1995; 86:930-39.
47. Thompson HL, Burbelo PD, Yamada Y et al. Identification of an amino acid sequence in the laminin A chain mediating mast cell attachment and spreading. Immunology 1991; 72:144-49.
48. Thompson HL, Burbelo PD, Segui-Real B et al. Laminin promotes mast cell attachment. J Immunol 1989; 143:2323-27.
49. Dastych J, Costa JJ, Thompson HL et al. Mast cell adhesion to fibronectin. Immunology 1991; 73:478-84.
50. Bochner BS, Luscinskas FW, Gimbrone MA et al. Adhesion of human basophils, eosinophils and neutrophils to interleukin 1-activated human vascular endothelial cells: contributions of endothelial cell adhesion molecules. J Exp Med 1991; 173:1553-57.
51. Walsh LJ, Kaminer MS, Lazarus GS et al. Role of laminin in localization of human

dermal mast cells. Lab Invest 1991; 65:433-40.

52. Bochner BS, McKelvey AA, Sterbinsky SA et al. IL-3 augments adhesiveness for endothelium and CD11b expression in human basophils but not neutrophils. J Immunol 1990; 145:1832-37.
53. Yasuda M, Hasunuma Y, Adachi H et al. Expression and function of fibronectin binding integrins on rat mast cells. Int Immunol 1995; 7:251-58.
54. Bochner BS, Sterbinsky SA. Altered surface expression of CD11 and Leu 8 during human basophil degranulation. J Immunol 1991; 146:2367-73.
55. Bochner BS, MacGlashan Jr DW, Marcotte GV et al. IgE-dependent regulation of human basophil adherence to vascular endothelium. J Immunol 1989; 142:3180-86.
56. Thompson HL, Burbelo PD, Metcalfe DD. Regulation of adhesion of mouse bone marrow-derived mast cells to laminin. J Immunol 1990; 145:3425-31.
57. Thompson HL, Burbelo PD, Yamada Y et al. Mast cells chemotax to laminin with enhancement after IgE-mediated activation. J Immunol 1989; 143:4188-92.
58. Hamawy MM, Oliver C, Mergenhagen SE et al. Adherence of rat basophilic leukemia (RBL-2H3) cells to fibronectin-coated surfaces enhances secretion. J Immunol 1992; 149:615-21.

CHAPTER 10

Signal Transduction Pathways Regulating Arachidonic Acid Metabolite Generation Following FcεRI Aggregation

Alasdair M. Gilfillan

INTRODUCTION

Aggregation of high affinity IgE receptors (FcεRI) on the surface of mast cells and basophils results in the initiation of specific signal transduction events that lead to degranulation, cytokine gene expression and the de novo synthesis and release of phospholipid-derived pro-inflammatory mediators. The phospholipid-derived category of inflammatory mediators includes cyclooxygenase-derived products such as prostaglandin (PG) D_2, 6-keto-PGF1α, PGE_2 and thromboxane B_2; and 5-lipoxygenase-derived products such as leukotriene (LT) C_4, LTB_4 and LTE_4.[1-17] Two major steps are required for the generation of these inflammatory mediators from phospholipids: (a) the liberation of substrate, i.e., arachidonic acid from phospholipids, and (b) the generation of the active metabolites from arachidonic acid (Fig. 10.1). Although it is likely that the early signal transduction events in the FcεRI-mediated production of these molecules are similar to those leading to degranulation and perhaps cytokine gene expression, there is emerging evidence indicating that there is also divergency in these pathways. The aim of this chapter

IgE Receptor (FcεRI) Function in Mast Cells and Basophils,
edited by Majed M. Hamawy.

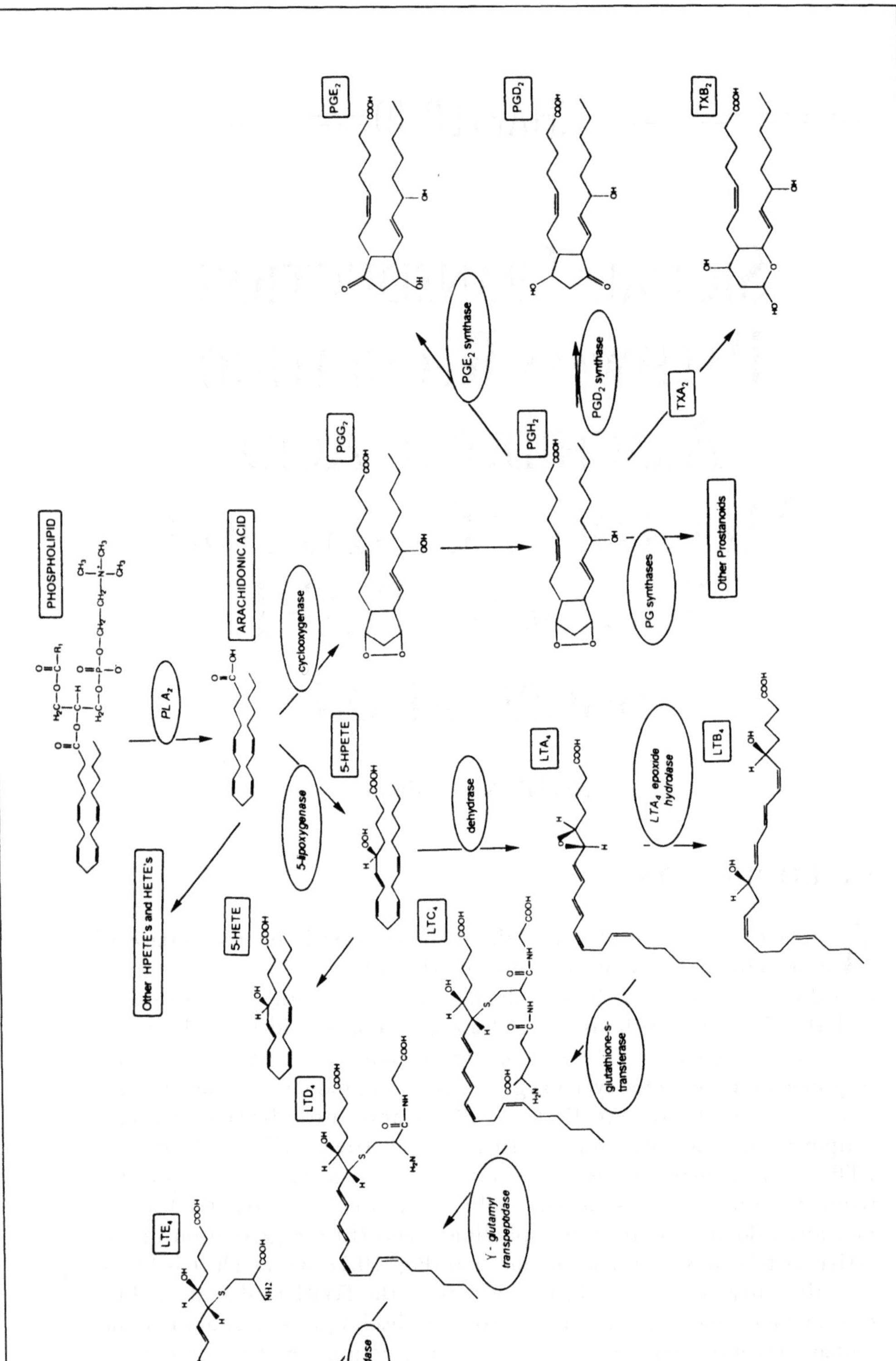

Fig. 10.1. The metabolic pathways that lead to the generation of prostaglandins and leukotrienes from phospholipids. PL: phospholipase; LT: leukotriene; PC: prostaglandin; TX: thromboxane, HETE: hydroxy eicosatetraenoic acid; HPETE: dihydroxy eicosatetraenoic acid.

is to describe the characteristics of the key enzymes that regulate the production of the phospholipid-derived inflammatory mediators and to review the current literature describing the signal transduction pathways that link FcεRI aggregation to the activation of these enzymes.

SECRETION OF BIOACTIVE ARACHIDONIC ACID METABOLITES

Studies utilizing mast cell deficient mice have revealed that mast cells are the major source of certain classes of arachidonic acid metabolites released following exposure to antigen.[12] The relative amounts of cyclooxygenase and lipoxygenase products released are largely governed by the species, the location of the cells and the nature of the stimulus. The major arachidonic acid metabolite generated following antigen challenge of mast cells recovered from human lung tissue,[7,17] monkey lung tissue[15] and human skin is PGD_2.[14] LTC_4, LTB_4 and the bioactive phospholipid, platelet activating factor (PAF) are also generated following antigen challenge, but these products are present in significantly lower concentrations.[2,7,14,16,17] In contrast, the major arachidonic acid metabolite released from human basophils[18,19] and rodent lung mast cells[6-8,10,11] is LTC_4 although significant amounts of PGD_2 and other metabolites are also released.

Generally, with most classes of stimuli including antigen challenge, the concentration-response relationships for the release of these arachidonic acid metabolites are virtually identical to those for degranulation. Where differences are observed, these are primarily due to long-term differential effects on the expression of the enzymes required for arachidonic acid metabolism and are independent of the signal transduction cascade leading to mediator release immediately following FcεRI aggregation. Certain neurohormones such as substance P, bradykinin and neurotensin, however, release significant amounts of histamine in the absence of PGD_2 release[8] indicating differential regulation of the respective signal transduction pathways leading to the release of these individual classes of mediators.

Following antigen challenge of mast cells from human lung,[12] human foreskin,[14] guinea pig lung[1] and monkey lung[15] there is a delay before any evidence of LTC_4 and PGD_2 release is observed. This contrasts the release of granule associated mediators which are detected almost immediately following antigen challenge. Similar responses are seen following FcεRI aggregation in rat basophilc leukemia (RBL-2H3) cells.[20,21] These differences in the kinetics of arachidonic acid metabolite release compared to degranulation likely reflect requirements for posttranslational modification of regulatory enzymes required for arachidonic acid metabolism and divergency in the signal transduction pathways leading to the release of both categories of mediators.

PHOSPHOLIPIDS AS A SOURCE OF PRO-INFLAMMATORY MEDIATORS

Structurally, phospholipids consist of a glycerol backbone to which specific fatty acid moieties are esterified at the sn_1 and sn_2 positions (Fig. 10.2). At the sn_3 position, phosphorylated polar head groups such as phosphorylcholine, phosphorylethanolamine, phosphorylserine or phosphorylinositol are attached via phosphodiester linkages forming phosphatidylcholine (PC), phosphatidylethanolamine (PE), phosphatidylserine (PS) and phosphatidylinositol (PI) respectively (Fig. 10.2).

The majority of studies describing the liberation of arachidonic acid/arachidonic acid metabolites from phospholipids following FcεRI aggregation have utilized exogenous [^{14}C]- or [^{3}H]-arachidonic acid to metabolically radiolabel the endogenous phospholipid pool(s). If the radiolabeled substrate is not allowed to equilibrate with the endogenous arachidonic acid-containing pools, however, the results can be substantially different from those observed under equilibrium labeling conditions. For example, labeling mouse bone marrow derived mast cell (BMMC) phospholipids or

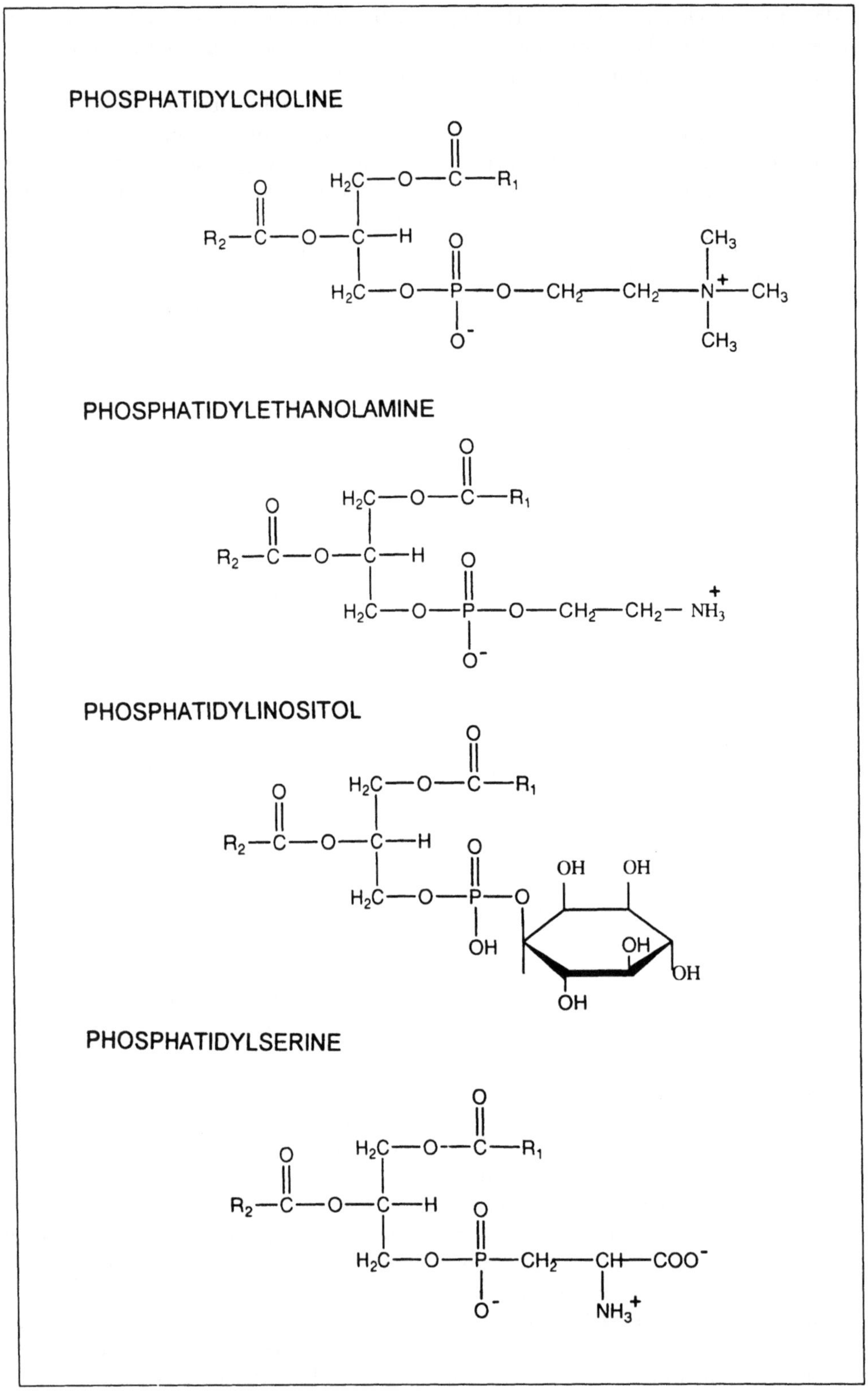

Fig. 10.2. The primary structures of the major phospholipids that provide arachidonic acid substrate for the production of leukotrienes and prostaglandins following FcεRI aggregation in mast cells. R1 and R2 represent fatty acids.

RBL-2H3 cell phospholipids with [^{3}H]- or [^{14}C]-arachidonic acid under nonequilibrium conditions (0.5-2 hours) results in the majority of the label being incorporated into PC (44-50%). Less label is associated with PE (26-38%), and PI and PS (15-16%).[20,22,23] As equilibrium labeling conditions are approached in these cells (24-48 hours), the amount of label associated with PC decreases concurrently with an increase in label associated with PE, primarily 1-alk-1-enyl-2-arachidonyl-sn-(glycero-3-phosphoethanolamine).[22] By 48 hours less than 20% of the label in RBL-2H3 cells is found in PC, approximately 20% in PI and PS, and almost 60% is found in PE.[23]

These differences are reflected in the apparent phospholipid sources for arachidonic acid released upon FcεRI aggregation. When RBL-2H3 cells,[20,23] rat peritoneal mast cells[24] or BMMCs[22] are incubated with [^{3}H]- or [^{14}C]-arachidonic acid under nonequilibrium conditions, more than 80% of the label released upon receptor aggregation is derived from PC. Only limited amounts of label are lost from PI/PS and PE. In contrast, when RBL-2H3 cells[23] and BMMCs[22] are incubated with labeled arachidonic acid to equilibrium, less than 10% of the secreted label is derived from PC. The majority of the label is derived from either PE or PI and PS. To circumvent the interpretational problems presented by radiolabeling protocols, Fonteh et al[25] utilized mass measurements to determine the source of arachidonic acid released from BMMCs in response to antigen, phospholipase A_2 (PLA_2) (naja naja venom), or a combination of both. Under these conditions, the major phospholipid source for liberated arachidonic acid was found to be PE with lesser quantities derived from PI, PS and PC.

Based on the studies described above and on data provided by double labeling experiments,[22] it appears that initial incorporation of free arachidonic acid into phospholipids is primarily into PC. This incorporated arachidonic acid is subsequently redistributed into PE. The movement of arachidonic acid from PC to PE is dramatically stimulated upon FcεRI aggregation, a mechanism that may serve to replenish the releasable stores of arachidonic acid.[22] The production of PAF following FcεRI aggregation may also be regulated by this process.[26] This remodeling pathway does not require CoA, therefore, the enzyme responsible has been termed CoA independent transacylase.[27]

Following FcεRI aggregation, free arachidonic acid is primarily released from 1-ether linked phospholipids such as 1-alk-1-enyl-2-arachidonyl-PE and to a lesser extent 1-alkyl-2-arachidonyl-PC.[27] Double labeling approaches have provided evidence for the existence of more than one pool of arachidonic acid which may be utilized for the generation of different classes of arachidonic acid metabolites.[28] Free arachidonic acid and perhaps other classes of arachidonic acid metabolites are released as a consequence of hydrolysis of PE; however, these studies suggest that LTC_4 and LTB_4 may be produced as a consequence of hydrolysis of PC and PI.

MECHANISMS OF ARACHIDONIC ACID LIBERATION FROM PHOSPHOLIPIDS

Arachidonic acid is liberated from phospholipids by the action of phospholipase(s) either functioning alone or in conjunction with other lipases. Phospholipase A_1 (PLA_1) and phospholipase A_2 (PLA_2) hydrolyze the sn_1 and sn_2 ester linkages respectively to yield lysophospholipids and free fatty acids including arachidonic acid (Fig. 10.3). PLA_2 is also responsible for the production of PAF. The sn_3 phosphodiester linkage is the site of hydrolysis by phospholipase C (PLC) and phospholipase D (PLD) (Fig. 10.3). PLC produces 1,2-diacylglycerol (DAG) and the phosphorylated polar head group. In addition, PLC is responsible for the liberation of inositol 1,4,5-trisphosphate and DAG from phosphatidylinositol 4,5-bisphosphate providing an important signaling pathway for many receptor subtypes including FcεRI.[29-31]

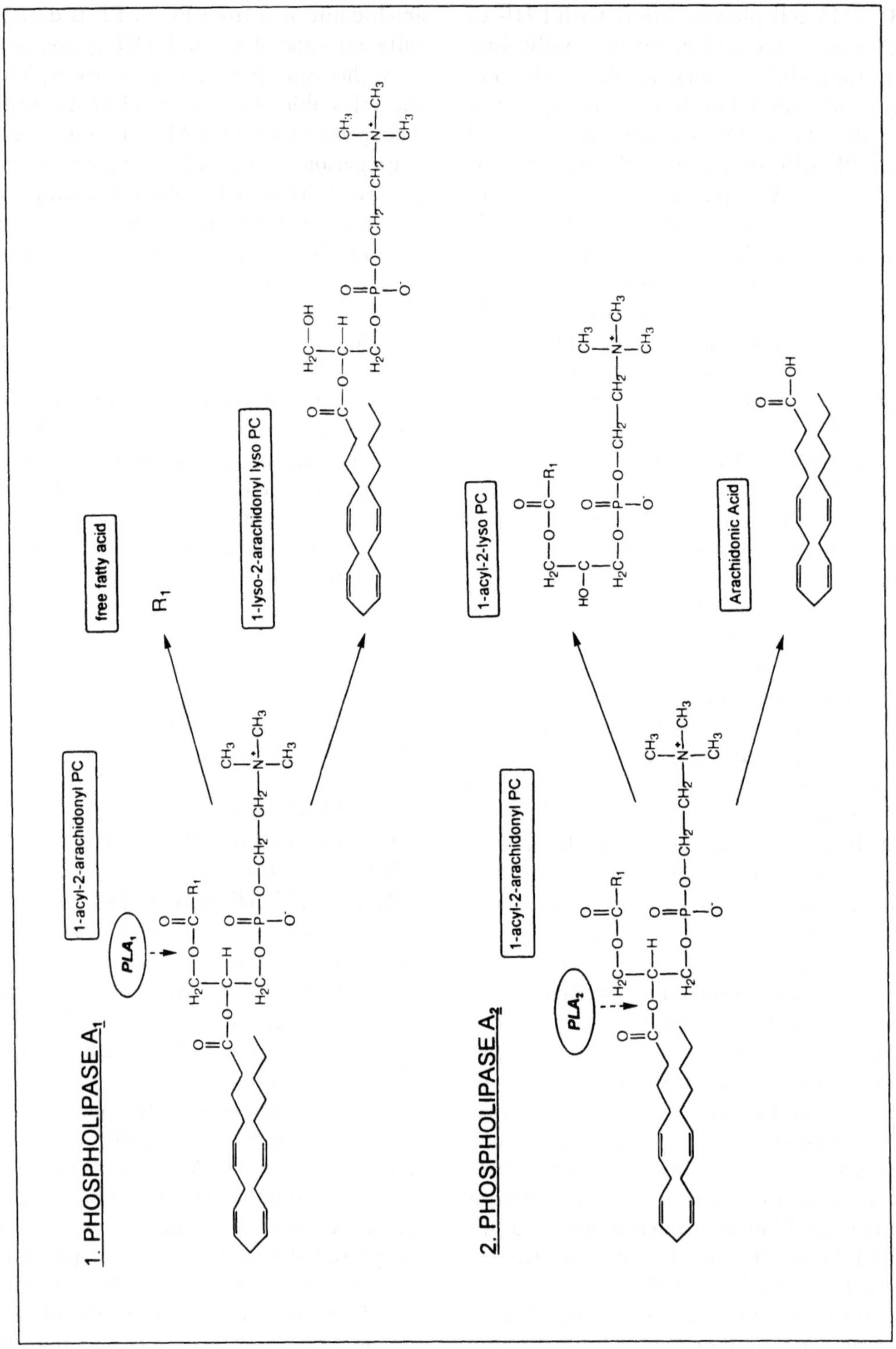
1. PHOSPHOLIPASE A1
PLA1
1-acyl-2-arachidonyl PC
free fatty acid
R1
1-lyso-2-arachidonyl lyso PC
2. PHOSPHOLIPASE A2
PLA2
1-acyl-2-arachidonyl PC
1-acyl-2-lyso PC
Arachidonic Acid

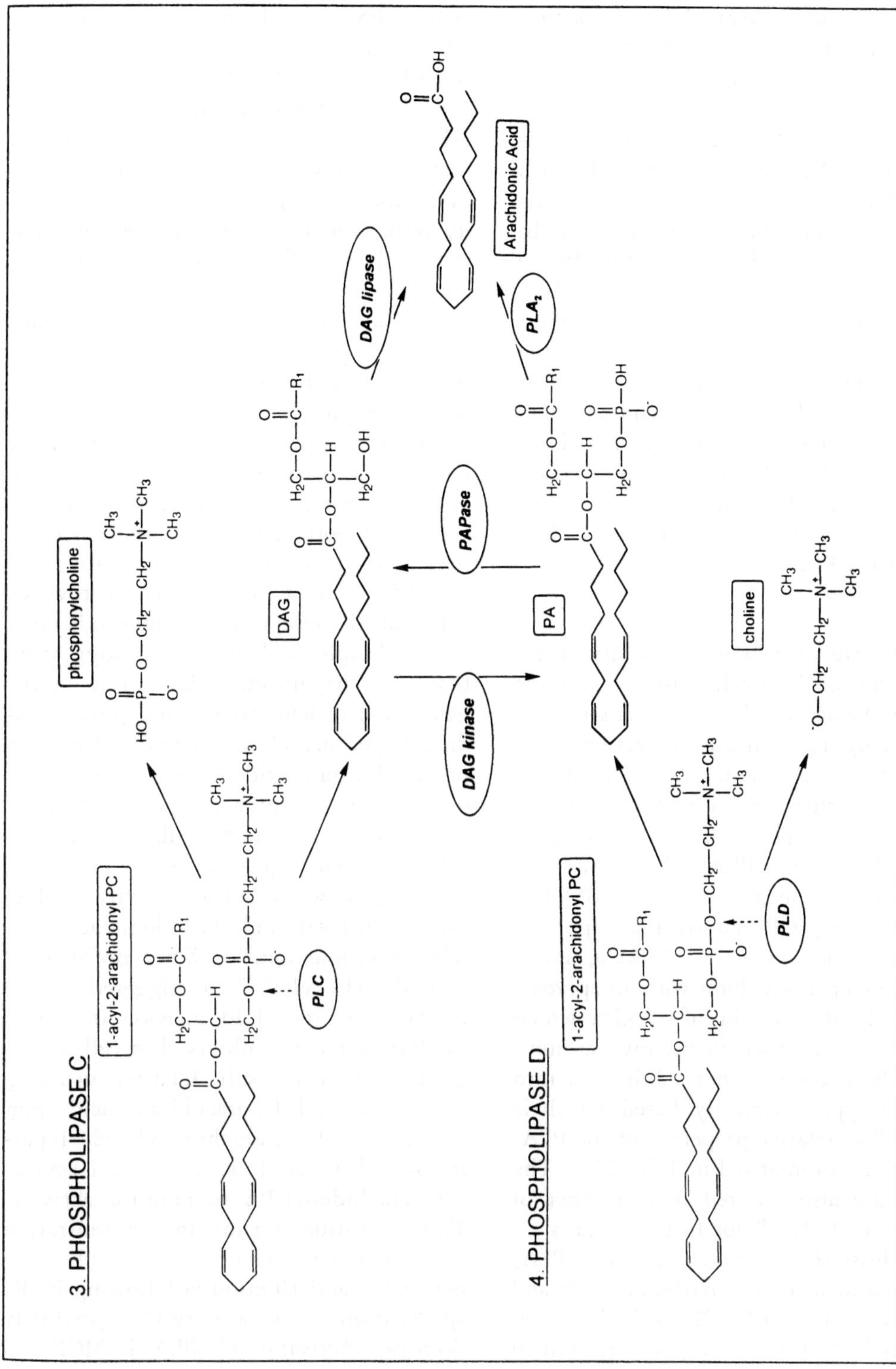

Fig. 10.3. The sites of PLA_1, PLA_2, PLC and PLD action and the products of their catalytic hydrolysis of phosphatidylcholine. PL: phospholipase; PC: phosphatidylcholine acid; PA: phosphatidic acid; DAG: 1,2 diacylglycerol; PAPase: phosphatidic acid phosphohydrolase.

The products liberated by the action of PLD are phosphatidic acid and the nonphosphorylated polar head group (Fig. 10.3). As 1-acyl-2-arachidonyl DAG can provide arachidonic acid following hydrolysis by DAG lipase, theoretically both PLC and PLD may play a role in the liberation of arachidonic acid metabolites. In addition, free arachidonic acid may also be liberated from PA by the action of PLA_2.[32] Mass and radiolabeling experiments suggest that the amount of DAG produced by the action of PLC would be insufficient to provide adequate substrate for arachidonic acid metabolite production following FcεRI aggregation.[20,33,34] The major phospholipases, therefore, responsible for the liberation of free arachidonic acid are likely to be PLA_2 and perhaps PLD.

PLA_2

PLA_2s are a family of genetically, structurally and biochemically distinct proteins that are characterized on the basis of molecular weight, substrate utilization, subcellular location, calcium requirements for activity and their ability to bind heparin. Three major categories of PLA_2s have been described: cytosolic PLA_2 ($cPLA_2$) with an approximate molecular weight of 85 kDa; secretory PLA_2 ($sPLA_2$) with an approximate molecular weight of 14 kDa; and a calcium independent form with an approximate molecular weight of 45 kDa which is primarily expressed in the myocardium. $sPLA_2$s have been further subdivided into three subtypes primarily based on their sources. The relative properties of the PLA_2 families are summarized in Table 10.1. The readers are also referred to a number of reviews on PLA_2[35-38] for further details.

At least three different types of PLA_2 have been identified in RBL-2H3 cells and BMMCs.[39-41] Both RBL-2H3 cells[21,39-41] and BMMCs[39-41] express $cPLA_2$ as determined by molecular weight, calcium requirements for activation, substrate utilization and recognition by $cPLA_2$ specific antibodies. This $cPLA_2$ hydrolyzes both arachidonyl-containing PE and PC at an optimal pH between 8.9-9.0.[39-41] Additional PLA_2 activity which utilizes PS as a substrate was detected in BMMCs and RBL-2H3 cells with an optimal pH between 5.5-7.4.[41] When purified by column chromatography followed by HPLC, this activity resolved into two peaks, one of which was identified as $sPLA_2$ by antibody recognition. The other activity peak, which was much more efficient at hydrolyzing PS than PE, PC and PI, was not recognized by antibodies against $cPLA_2$ and $sPLA_2$, suggesting a novel, perhaps mast cell specific, PS-utilizing PLA_2.[41] It is interesting to note that lysoPS, which would be produced by this reaction is a potent activator/coactivator for mediator release from mast cells.[42,43] Thus, activation of a PS specific PLA_2 may lead to the liberation of both arachidonic acid metabolites and granule associated mediators.

cPLA2 has a high selectivity towards arachidonyl containing phospholipids especially phosphatidylcholine,[44-46] suggesting that this enzyme may play a role in the generation of leukotrienes and prostaglandins. Experimental approaches[47-50] have also provided strong evidence to support a role of $cPLA_2$ in receptor mediated release of free arachidonic acid. A similar role in mast cells has been suggested by the close correlations between $cPLA_2$ activity, $cPLA_2$ expression levels and arachidonic acid metabolite production.[51-53] When BMMCs[53,54] or RBL-2H3 cells[21,55] are triggered via the FcεRI, a 2- to 3-fold increase in $cPLA_2$ catalytic activity is observed in cell homogenates and/or cytosolic fractions utilizing the release of [^{14}C]-arachidonic acid from 1-stearoyl-2-[^{14}C]-arachidonyl-PC, 1-palmitoyl-2-[^{14}C]-arachidonyl-PC or 1-acyl-2-[^{14}C]-arachidonyl-PE to monitor activity. This activation is time and concentration dependent with maximal activity observed between 5 and 10 minutes following FcεRI aggregation.[21] The activity then gradually decreases. Activation of $cPLA_2$ in MCII and RBL cells preceeds both arachidonic acid metabolite release and degranulation.[21,54] Prolonged aggregation of FcεRI on the surface of BMMCs results in a further enhancement of $cPLA_2$ activity due to enhanced expression of $cPLA_2$ protein.[53] This

Table 10.1. Characteristics of the major classes of PLA_2

Properties	$sPLA_2$	$cPLA_2$	Ca^{2+} independent PLA_2
Mass	~14 kDa	~85 kDa	~45 kDa
Location	Secreted	Cellular	Cellular
Tissue	type I: human pancreas snake venom type II: human synovial fluid/platelets, snake venom type III: bee venom	platelets, macrophages, neutrophils, mast cells and other cell types	myocardium
Calcium Requirements	mM	µM	none
Substrate Specificity	PE and PS greater than PC nonspecific for fatty acids	arachidonyl-containing phospholipids	arachidonyl-containing phospholipids
Regulation	Induced expression and secretion	Ca^{2+}-dependent translocation post-translational modification	?
Receptor	Yes	No	No
Heparin binding	Yes	No	No

This data is based on the information in refs. 35-37.

induction correlated with an enhanced ability of the cells to release archidonic acid metabolites in response to a second stimulus.[53]

Following subcellular fractionation of sonicates taken from triggered RBL-2H3 cells, the increase in $cPLA_2$ activity was observed in both membrane and cytosolic fractions.[21] The increase in membrane-associated $cPLA_2$ activity preceded the increase in cytosolic-associated $cPLA_2$ suggesting that there is both a net increase in $cPLA_2$ activity and membrane translocation. Translocation of $cPLA_2$ is necessary to allow access of the $cPLA_2$ catalytic domain to the membrane-associated phospholipids. Translocation and subsequent binding of $cPLA_2$ to the cell membrane is calcium dependent and is a consequence of the presence of a calcium dependent phospholipid binding consensus motif: the calB or C2 domain.[47,48,50] This domain is expressed in a number of signal transducing proteins such as the GTPase activating protein GAP, PLC-γ1 and calcium-dependent isoforms of PKC.[56] The importance of the $cPLA_2$ calB domain and membrane translocation for arachidonic acid metabolite release following FcεRI aggregation was demonstrated by overexpressing this domain in RBL cells in the absence of the $cPLA_2$ catalytic domain. This resulted in a reduction in [^{14}C]-arachidonic acid release compared to wild-type responses.[47] Recent studies have demonstrated that, following activation of RBL-2H3 cells via FcεRI, almost 30% of the $cPLA_2$ translocates to the nuclear envelope[57] suggesting that nuclear membrane phospholipids are a source of arachidonic acid following cell activation. It is also possible that phospholipase-derived products in some way influence transcription (also see later sections). Free arachidonic acid, for example, stimulates the growth of murine P815 mast cells.[58] This response appears to be a consequence of activation of protein kinase(s) with subsequent phosphorylation of 44 kDa and 135 kDa proteins.

The data supporting a role for $sPLA_2$ in FcεRI-dependent arachidonic acid metabolite generation has been somewhat conflicting. In contrast to the studies described above, our group found no evidence of expression of $sPLA_2$ in RBL-2H3 cells as assessed by PCR (J Kochan, personal communication) or activity measurements (T Chen et al, unpublished observations). $sPLA_2$ was also not detected in MCII bone marrow derived mast cells utilizing $sPLA_2$ specific monoclonal antibodies.[54] The 14 kDa group II $sPLA_2$, however, has been reported to be expressed in rat peritoneal mast cells[59,60] and in BMMCs.[61] Furthermore, when BMMCs were triggered with antigen, they were reported to release a protein with PLA_2 activity that was recognized and blocked by antibodies against group II $sPLA_2$.[22,61] This release preceded degranulation indicating that the secreted $sPLA_2$ did not originate within the secretory granules.

Cell surface receptors for $sPLA_2$ have recently been cloned[62-64] and binding of $sPLA_2$ to this receptor is independent of its catalytic property.[65] A binding site for $sPLA_2$ also appears to be expressed on the surface of mast cells and it has been suggested that the $sPLA_2$ released from activated mast cells may influence mast cell secretion by either binding to this receptor or liberating arachidonic acid from the external surface of the cell membrane.[66] Exogenously added $sPLA_2$ can produce histamine release from rat peritoneal mast cells[66] and BMMCs.[25] This effect is calcium independent and can be blocked by an anti-$sPLA_2$ antibody or by heparin binding.[66] In contrast to BMMCs,[25] however, exogenously added $sPLA_2$ does not appear to influence the release of arachidonic acid metabolites from peritoneal mast cells.[66] The ability of porcine $sPLA_2$ to liberate histamine from rat peritoneal mast cells was blocked by the nonspecific PLA_2 inhibitor para-bromophenacyl bromide[67] and the lipoxygenase inhibitor ETYA[68] suggesting that a lipoxygenase product may influence degranulation.[68-70] Parabromophenacyl bromide also inhibits the antigen dependent release of histamine and arachidonic acid metabolites from rat peritoneal mast cells[67,71,72] and human basophils.[73,74] In contrast, inhibitors that show greater specificity towards $sPLA_2$ such as thielocin A1, mepacrine, 2.4'-dibromoacetophenone and an anti-rat 14 kDa group II $sPLA_2$ antibody, inhibit degranulation but not PGD_2 release from rat peritoneal mast cells evoked by immunological and nonimmunological stimuli.[59,60]

The mM calcium requirements for $sPLA_2$ catalytic activity makes a precise regulation of the catalytic activity of this enzyme by receptor aggregation highly unlikely. The intracellular calcium concentrations following FcεRI aggregation are around 1 μM or less.[20] This would mean that $sPLA_2$ is not active inside mast cells. In contrast, $cPLA_2$ would be active under these conditions.[35] Although these results suggest that $sPLA_2$ does not influence arachidonic acid metabolism in mast cells, it has been reported that there is a closer correlation of expression levels of $sPLA_2$ than $cPLA_2$ to arachidonic acid metabolism when BMMCs are incubated with c-kit ligand.[61] For this to be the case, $sPLA_2$ would have to operate in an extracellular environment where calcium concentrations would be sufficient for activity.

Taken together, above data suggest that $cPLA_2$ is the major form of PLA_2 contributing to the generation of arachidonic acid metabolites. Although $sPLA_2$ may play a role in degranulation, its contribution to the generation of arachidonic acid metabolites is unclear. Certainly, liberation of arachidonic acid metabolites following FcεRI aggregation can proceed in the absence of $sPLA_2$ suggesting that, if $sPLA_2$ does contribute to the release of arachidonic acid metabolites, it is only in a limited capacity. The presence of a PS-specific

PLA_2 activity in mast cells is of interest, especially in light of the ability of lysoPS to influence degranulation. The role of this enzyme in mast cell activation warrants further investigation.

PLD

Activation of PC-PLD may contribute to the release of arachidonic acid metabolites by two distinct mechanisms: (a) the PC-PLD-catalyzed liberation of phosphatidic acid from PC followed by the secondary production of lysoPA and DAG by the actions of PLA_2 and PAPase respectively. PA, lysoPA and DAG can activate secondary signal transduction pathways that could lead to activation of $cPLA_2$;[75-82] and (b) PC-PLD-dependent production of 1-acyl-2-arachidonyl PA from 1-acyl-2-arachidonyl PC followed by dephosphorylation to form 1-acyl-2-arachidonyl DAG. Free arachidonic acid can be subsequently liberated from PA by the action of PLA_2, and from DAG by the action of DAG lipase.

PC-PLD activity can be monitored on the basis of a transphosphatidylation reaction which is catalyzed by this enzyme (Fig. 10.4). When ethanol or propanol are included in the incubating medium of cells where PC-PLD is active, phosphatidylethanol or phosphatidylpropanol are synthesized respectively at the expense of phosphatidic acid. The liberation of free [^{3}H]-choline from [^{3}H]-choline labeled PC and the formation of [^{14}C]-phosphatidic acid from [^{14}C]-arachidonic acid labeled PC has also been used to monitor PLD activity. As choline can be liberated from PC by the action of PLC followed by dephosphorylation, and as PA can be formed by the action of PLC followed by DAG kinase (Fig. 10.3), on their own these parameters do not unequivocally demonstrate PC-PLD activation. Utilizing a combination of these approaches to monitor activity, rapid activation of PC-PLD following FcεRI aggregation has been observed in RBL cells[20,83-87] and peritoneal mast cells.[24,33,34,88-90] In RBL-2H3 cells, this activation maximizes within 10 minutes and is accompanied by an increase in PA and DAG production, and liberation of free arachidonic acid.[20] As mammalian PC-PLD has yet to be cloned and sequenced, it has been impossible to date to determine the precise role of PC-PLD in FcεRI dependent signal transduction pathways. Most studies have therefore relied upon the ability of high concentrations of the amphiphilic β-receptor antagonist, propranolol, to inhibit PAPase activity and/or the ability of ethanol to prevent PA and DAG production by shunting the transphosphatidylation pathway over to the production of phosphatidylethanol (Fig. 10.4). Due to the nonspecific nature of these approaches, however, the conclusions drawn from the studies described below should be viewed with a degree of caution.

The release of [^{14}C]-arachidonic acid from RBL-2H3 cells[20] and rat peritoneal mast cells[24] in response to FcεRI aggregation is preceded by an increase in label associated with PA then DAG. This time course suggests that the release of arachidonic acid requires the initial formation of PA and DAG and is consistent with PC-PLD activation being required for FcεRI mediated arachidonic acid mobilization.[20] Incubation of RBL-2H3 cells with ethanol (0.05% to 1%)[20] or propranolol (300 μM)[20,84,85] inhibits the increase in DAG production and the subsequent release of histamine and arachidonic acid metabolites normally observed following FcεRI aggregation. Incubation of peritoneal mast cells with ethanol[90] or propranolol also resulted in a 40% decrease in arachidonic acid release and a total inhibition of PGD_2 release in response to ionomycin.[88] When added in combination with parabromophenacyl bromide, however, ethanol, completely blocked arachidonic acid release[88] suggesting that both PLD and PLA_2 are responsible for FcεRI-dependent liberation of free arachidonic acid. The release of arachidonic acid from c-kit ligand receptor activated peritoneal mast cells was also inhibited by propranolol[89] suggesting that the release of arachidonic acid in response to stem cell factor is also dependent on PC-PLD activation. A role for DAG in

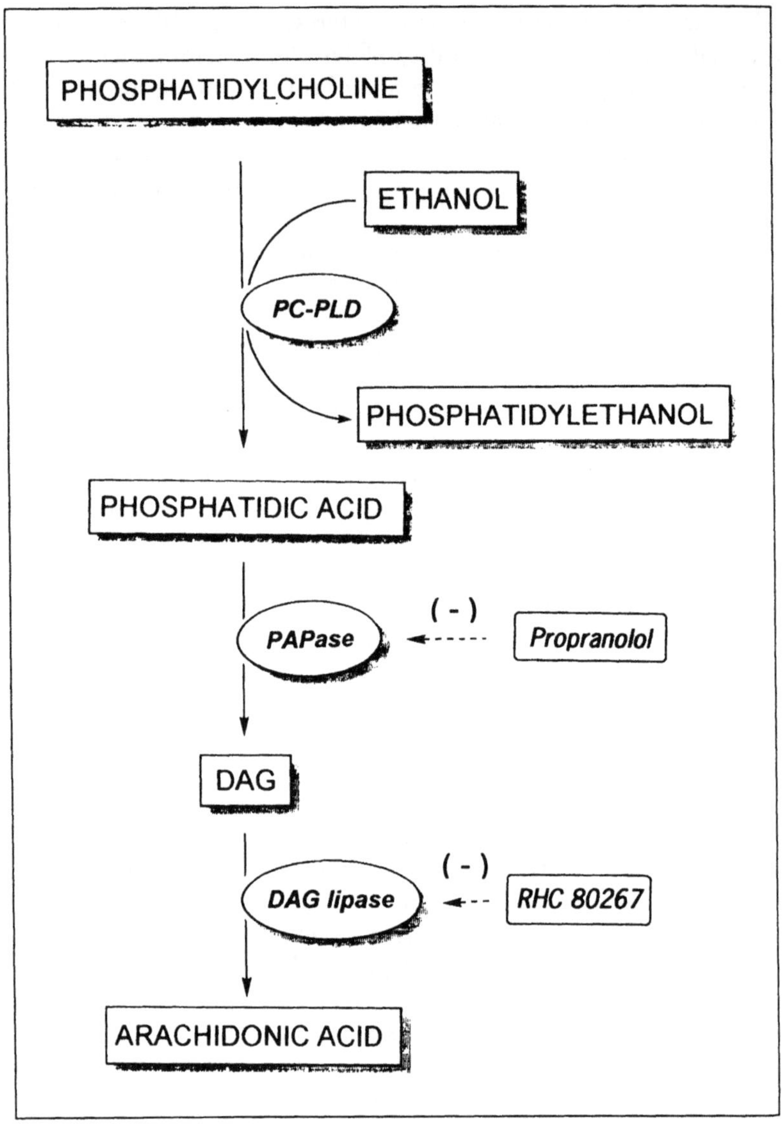

Fig. 10.4. PC-PLD catalyzed phosphatidylcholine hydolysis. The figure shows the pathway for the PC-PLD dependent liberation of arachidonic acid, the transphosphatidylation reaction for the formation of phosphatidylethanol from ethanol and the sites of action of propranolol and RHC 80267.

providing arachidonic acid substrate following FcεRI aggregation is supported by the observation that arachidonic acid metabolite release from rat peritoneal mast cells is blocked by incubating the cells with the DAG lipase inhibitor RHC 80267 (Fig. 10.4).[24]

Although the above data do support a role for the PLD/PAPase/DAG lipase pathway in the liberation of arachidonic acid following FcεRI aggregation, the precise role that this pathway plays in providing substrate and/or in the signal transduction cascade leading to $cPLA_2$ activation remains to be clearly defined. To elucidate the relative roles that these enzymes play will require more careful investigation perhaps with the use of knockout, antisense or dominant negative approaches.

5-LIPOXYGENASE

The first two steps of leukotriene synthesis from arachidonic acid are catalyzed by 5-lipoxygenase (5-LO) (Fig. 10.1). The ability of this enzyme to function in this capacity is dependent on translocation from the cytosol to the membrane. This is necessary to localize 5-LO to the arachidonic acid liberated by the action of phospholipases following cellular activation. Once translocated to the membrane, it is activated by the membrane associated protein, 5-LO activating protein (FLAP), which is also responsible for presenting free arachidonic acid to the catalytic domain of 5-LO.[91] Immunoblot and activity studies have demonstrated membrane translocation of 5-LO in both RBL-2H3 cells[92] and BMMCs[93] following FcεRI aggregation. Translocation is calcium dependent and both translocation and subsequent release of LTC_4 is blocked by incubating the cells with the FLAP inhibitor MK-886.[92] The translocation of 5-LO produced by antigen challenge is reversible and after 15 minutes the cytosolic pool returns to prestimulus levels.[93] In contrast, the translocation produced by A23187 is irreversible producing a significant depletion of cytosol-associated 5-LO.[93] This results in the A23187-treated, but not the antigen-treated cells, being refractory to subsequent challenge to A23187.[93]

Utilizing protocols to optimize the integrity of the subcellular fraction during cellular disruption, immunoblot analysis has revealed that 5-LO in quiescent RBL-2H3 cells is primarily located in the nuclear fraction with lesser amounts in the cytosolic and non-nuclear membrane fractions.[94] Within the nucleus, 5-LO appears to exist in two pools: a readily releasable form and one that is not removed by high salt concentrations. Brock et al[94] have hypothesized that, in addition to its role in arachidonic acid metabolism, the function of 5-LO associated with the nuclear fraction is to directly regulate gene transcription leading to cytokine release. Alternatively, 5-LO may act indirectly by catalyzing the synthesis of arachidonic acid metabolites that then act in the nucleus to modulate transcription. 5-LO contains a consensus guanine nucleotide binding site and a proline rich SH3 recognition site that binds to the SH3 domain of Grb2.[95] Grb2 is an adaptor protein that binds to the GTP/GDP exchanger Sos which in turn plays a role in gene transcription via the Ras/Raf pathway (see below). Peptides corresponding to the proline rich SH3 domain binding region of 5-LO were demonstrated to inhibit the translocation of 5-LO from the cytosol to the membrane in A23187 activated permeabilized neutrophils and the formation of the Grb2-5-LO complex in vitro[95] further supporting a possible role of 5-LO in gene transcription.

CYCLOOXYGENASE

The initial step in the production of prostaglandins from arachidonic acid is catalyzed by cyclooxygenase (alternatively termed prostaglandin endoperoxide synthase or prostaglandin H synthase; PGHS).[96] This enzyme results in the formation of the intermediate endoperoxides PGG_2 and PGH_2 which are then converted to prostaglandins by the actions of their respective synthases (alternatively termed synthetases or isomerases) (Fig. 10.1). PGHS is a 70 kDa protein that exists as two different isoforms: PGHS1 is constitutively expressed in mast cells and other cell types;[51,52,97-102] PGHS2 is not constitutively expressed in mast cells[51,52] but the expression of PGHS2 in mast cells and other cell types is induced in response to specific cytokines and c-kit ligand.[51,52,99] Two forms of PGD_2 synth(et)ase have also been described: a 26 kDa protein recovered from the spleen which requires glutathione for activity; and a 26 kDa protein recovered from the brain which is glutathione independent.[103,104] The glutathione-dependent form of PGD_2 synthase is expressed in mast cells where the expression levels of this protein is amongst the highest of all tissues examined.[105]

In BMMCs, the release of PGD_2 following FcεRI aggregation proceeds in the absence of expression of PGHS2.[51,99,106] The expression of PGHS2 in BMMCs can be induced by short term exposure to c-kit ligand (stem cell factor; scf) in association with IL-10 and IL-1β.[51,52,106] In response to scf, PGHS2 mRNA levels maximize within 2 hours, followed by an increase in PGHS2 protein which maximizes between 5-10 hours after exposure. There is a subsequent increase in cytokine-dependent PGD_2 generation.[99,106] The induction of

PGHS2 expression, however, does not potentiate PGD_2 release in response to FcεRI aggregation. Prolonged exposure of BMMCs with scf in the presence of IL-10 and IL-1β results in an increase in PGHS1 mRNA and protein levels after 24-48 hours.[51,98,99] These conditions also produce an increased expression of $cPLA_2$ and PGD_2 synthase.[51] In contrast to PGHS2, this increase in expression closely correlates with an increase in the ability of IgE to stimulate PGD_2 secretion but not LTC_4.[51] In addition, conditions that block cytokine induced PGHS2 induction failed to inhibit IgE-dependent PGD2 release.[52,99] The above data suggest that the major form of PGHS responsible for the generation of cyclooxygenase products following FcεRI aggregation is PGHS1.

PGHS has been reported to be present in purified secretory granules isolated from canine mastocytoma cells[97] and rat peritoneal mast cells.[107,108] Peritoneal mast cell secretory granules also appear to contain PLA_2[109] and high concentrations of phospholipids.[110] When incubated with exogenous arachidonic acid, granule components catalyzed the synthesis of PGD_2.[97,110] From these findings, it has been argued that the secretory granules represent the primary site for prostaglandin synthesis in mast cells.[107-109] It is difficult to envision how the generation of arachidonic acid metabolites could be so highly regulated under these circumstances. As discussed below, the signal transducing elements linking FcεRI to arachidonic acid metabolite generation are largely membrane- and cytosol-associated.

SIGNAL TRANSDUCTION PATHWAYS LINKING FcεRI AGGREGATION TO ARACHIDONIC ACID METABOLITE RELEASE

Compared to the volume of literature describing the signal transduction mechanisms that regulate FcεRI-dependent degranulation, until recently, little information existed about the signal transduction pathways that regulate the enzymes involved in the generation of arachidonic acid metabolites. Enhancement of arachidonic acid metabolite release following FcεRI is likely to be a consequence of upregulation of rate limiting signal transduction events which enhance the catalytic activities of phospholipases and translocation of the enzymes involved in arachidonic acid metabolism.

As discussed earlier, the major phospholipases involved in the liberation of arachidonic acid are likely to be $cPLA_2$ and perhaps PLD. Studies in other systems have determined that receptor-mediated increase in $cPLA_2$ catalytic activity is mediated by a calcium and tyrosine kinase dependent signal transduction pathway that leads to the MAP kinase catalyzed phosphorylation of $cPLA_2$ protein.[35-38,49,112] As discussed below, a similar pathway likely regulates $cPLA_2$ activity following FcεRI aggregation in mast cells. Less is currently known about the signal transduction pathways regulating PC-PLD activation.

THE ROLE OF CALCIUM AND G PROTEINS

Chelating extracellular calcium with 10 μM EGTA[55] or intracellular calcium with BAPTA (P Lin and AM Gilfillan, unpublished observations) inhibits arachidonic acid metabolite release from RBL-2H3 cells. Similar calcium dependency has been reported for the release of lipid derived mediators from peritoneal mast cells,[88] BMMCs[40,54] and human skin mast cells.[14] As described earlier, membrane translocation of $cPLA_2$ and 5-LO is calcium dependent. Mast cell $cPLA_2$ also has 0.1-1.0 μM calcium requirements for catalytic activity.[39] This is within the range of intracellular free Ca^{2+} concentrations ($[Ca^{2+}]_i$) observed in mast cells following FcεRI aggregation.[82,83,113-115] Although RBL-homogenate $cPLA_2$ activity is calcium dependent when assayed in vitro, Garcia-Gil and Siraganian[55] reported that $cPLA_2$ activation in vivo was still observed in the presence of 10 μM EGTA. These differences may be due to the presence of regulatory G proteins in the cell preparations. In this context, when calcium concentrations were buffered between 0.1 and 10 μM in streptolysin O permeabilized RBL-2H3 cells, GTPγS (40 μM) stimulated $cPLA_2$ activity by 5- to 6-fold.[117] At low concentrations of GTPγS (0.1 μM), antigen produced a synergystic increase in $cPLA_2$ activation.[117] In ATP permeabilized rat peritoneal mast cells, GTPγS (100 μM) stimulated [^{3}H]-arachidonic acid release in calcium-free buffer to an extent similar to that observed in the presence of calcium.[118] These data suggest that receptor mediated activation of G proteins can reduce the calcium dependency of $cPLA_2$.

It is unlikely that these responses are mediated by membrane-associated G proteins as pertussis toxin pretreatment failed to prevent arachidonic acid metabolite release in response to FcεRI aggregation.[119] In contrast, pertussis toxin treatment inhibits the arachidonic acid metabolite release in response to adenosine analogs,[119] compound 48/80[120] and C5a[5] suggesting that these stimuli initiate different signal transduction pathways than those initiated by aggregation of FcεRI. In RBL-2H3 cells transfected with the mi muscrarinic receptor, a role for small cytosolic G proteins in the IgE-dependent $cPLA_2$ activation, however, is supported by the ability of GDPβS to block the antigen, carbachol, ionomycin and thapsigargin induced increase in $cPLA_2$ catalytic activity.[121]

PC-PLD activity in RBL-2H3 cells[83] and peritoneal mast cells[88] is also enhanced by conditions which elevate $[Ca^{2+}]_I$. The increase in PC-PLD activity observed following FcεRI aggregation is inhibited by incubating the cells with EGTA or by loading the cells with BAPTA,[83] demonstrating an absolute requirement for calcium in FcεRI-dependent PC-PLD activation. The role of G proteins in mast cell PC-PLD activation has yet to be determined; however, studies in other cell systems have revealed that PC-PLD activity can be regulated by small cytosolic GTP binding proteins[122,123] including Rho[124] and ARF.[125-127] Therefore, it is possible that similar mechanisms exist to regulate PC-PLD activity in mast cells.

The Role of Protein Kinases and Protein Phosphorylation

FcεRI mediated mast cell degranulation is dependent on the activation of serine/threonine kinases such as protein kinase C (PKC)[128-134] and tyrosine kinases such as $p56^{lyn}$,[135-140] and $p72^{syk}$.[135-137,140,142,143] These enzymes catalyze the phosphorylation of specific proteins which are required for mast cell degranulation.[144-149] The activation of $cPLA_2$ and PC-PLD has been demonstrated to be dependent on protein kinases in a number of cell types[35-38,113,150-154] and recent evidence suggests that protein phosphorylation regulates the catalytic activity of these enzymes and subsequent arachidonic acid metabolite release following FcεRI aggregation.

Protein kinase C

There are conflicting data regarding the role of PKC in arachidonic acid metabolite generation in response to FcεRI aggregation. The PKC activating agent phorbol 12-myristate 13-acetate (PMA) potentiates the release of [^{14}C]-arachidonic acid metabolites from rat peritoneal mast cells in response to thapsigargin[156] and the release of LTC_4 from RBL-2H3 cells following FcεRI aggregation.[129] In contrast, PMA does not liberate LTC_4 from mature human basophils despite stimulating histamine release.[5] PKC inhibitors such as H7 and staurosporine block degranulation and the release of arachidonic acid metabolites from RBL-2H3 cells.[129] Staurosporine, however, failed to inhibit arachidonic acid and PGD_2 release from rat peritoneal mast cells following ionomycin stimulation.[89] Staurosporine and H7 have limited specificity toward PKC; therefore, the responses observed may be dependent on inhibition of other classes of kinases. Studies with more selective PKC inhibitors such as Ro31-7549 and Ro31-8425[21,128] have failed to detect inhibition of $cPLA_2$ or arachidonic acid metabolite release from RBL cells at concentrations required for PKC inhibition or prevention of IgE-dependent histamine release.[21] Only at high concentrations of Ro31-7549 and Ro31-8425 was any inhibition of $cPLA_2$ activity and LTC_4 and

PGD_2 release observed. These effects were likely to be independent of the ability of these compounds to inhibit PKC activity and may be dependent on inhibition of tyrosine kinases.[21] At lower concentrations of Ro31-7549 and Ro31-8425, IgE-dependent LTC_4 release was potentiated by up to 50% without a concommitant potentiation of other parameters.[21] In human basophils, the PKC inhibitors, Ro31-7549, ilmofosine, GF109203X and calphostin C, also potentiated IgE dependent LTC_4 production.[157] Furthermore, the priming effect of IL-3 on C5a induced LTC_4 release from mature human basophils was reversed by activating PKC with PMA.[5] These data suggest that PKC activation inhibits the release of LTC_4.

In contrast to PLA_2 activation, activation of PC-PLD does appear to be positively regulated by PKC. PMA added to RBL-2H3 cells in the absence of other stimuli activated PC-PLD.[83] This response was not dependent on the presence of extracellular calcium suggesting that the responses were due to the activation of calcium-independent isoforms of PKC. Prolonged exposure of RBL-2H3 cells to PMA results in the downregulation of PKC activity and an inability of FcεRI to elicit an increase in PC-PLD activity.[83] A role for PKC in the activation of PC-PLD following FcεRI aggregation has been further supported by the ability of PKC inhibitors such as staurosporine, Ro31-7549 and Ro31-8220 to prevent antigen dependent activation of PC-PLD.[83] Under calcium-free conditions, PMA did not liberate arachidonic acid metabolites suggesting that if PC-PLD does contribute to the release of arachidonic acid metabolites, other enzymes such as DAG lipase or PLA_2 need to be activated.[83]

Above data suggest that, although PKC activation may regulate PC-PLD activity and degranulation, the activation of $cPLA_2$ and other regulatory enzymes involved in arachidonic acid metabolite release are not positively regulated by PKC activation. If anything, the data suggest that the release of certain classes of arachidonic acid metabolites may be negatively regulated by PKC activation.

cAMP-dependent protein kinase

Elevation of intracellular cAMP results in the activation of cAMP-dependent protein kinase. Experimental conditions that increase intracellular cAMP concentrations in mast cells and basophils result in the inhibition of FcεRI-dependent arachidonic acid metabolite release from mast cells and basophils.[119,158-160] These effects are observed at concentrations of cAMP that have limited effect on histamine release. The mechanism of action of cAMP and the phosphorylation target(s) for cAMP-dependent protein kinase in mast cells are unknown; however, exogenously added dibutyryl cAMP results in the inhibition of FcεRI-dependent PC-PLD activation (P Lin and AM Gilfillan, unpublished observations). The target(s) for cAMP may, therefore, be one or more of the phospholipases responsible for the liberation of arachidonic acid following FcεRI aggregation.

Tyrosine kinases

Tyrosine kinase inhibitors such as genistein, levandustein, herbimycin A and tyrphostins inhibit arachidonic acid metabolite generation in RBL-2H3 cells,[21,161] MC/9 cells,[161] BMMCs[54] and human basophils.[5] In RBL-2H3 cells, the concentrations of tyrphostins required to inhibit arachidonic acid metabolite release are approximately 10 times lower than those required for the inhibition of degranulation (T Chen and AM Gilfillan, unpublished observations). These data suggest that the tyrosine kinase(s) required for arachidonic acid metabolite release may be different than the one required for degranulation.[21] Similarly, the kinase inhibitor quercitin was observed to block the release of arachidonic acid with a greater potency than that for the inhibition of degranulation.[121] Tyrosine kinase inhibitors have also been demonstrated to inhibit the normally observed activation of PC-PLD in RBL cells[86,87] and rat peritoneal mast cells[89] suggesting that PC-PLD activation may also be mediated by tyrosine kinase(s). This conclusion has been supported by studies demonstrating that PC-PLD activity could be recovered from activated RBL-2H3 cell membranes and cytosol following immunoprecipitation of the proteins utilizing an antiphosphotyrosine monoclonal antibody.[86]

The effects of tyrosine kinase inhibitors on PLA_2 and PC-PLD may be explained in terms of inhibition of tyrosine phosphorylation of the $cPLA_2$ and PC-PLD proteins or the inhibiton of tyrosine phosphorylation of other signaling proteins upstream of these enzymes. The increase in $cPLA_2$ activity observed in mi muscarinic receptor-transfected RBL-2H3 cells in response to carabachol, ionomycin, thapsigargin and FcεRI aggregation and in wild-type RBL-2H3 cells in response to FcεRI aggregation is associated with phosphorylation of $cPLA_2$ protein.[53,121] As phosphoamino acid analysis of the phosphorylated $cPLA_2$ protein was not conducted in these studies, it could not be determined if the phosphorus was on the tyrosine residues of $cPLA_2$. In BMMCs, a decrease in mobility of $cPLA_2$ on SDS PAGE gel attributable to phosphorylation of $cPLA_2$ protein was observed in extracts taken from BMMCs 30 seconds after FcεRI aggregation.[54] This gel shift could be reversed by treating the protein with potato acid phosphatase. Although tyrosine kinase inhibitors blocked $cPLA_2$ activation following FcεRI aggregation, antiphosphotyrosine antibodies revealed that the $cPLA_2$ protein was not tyrosine phosphorylated. The phosphorylated tyrosine residue(s) regulating $cPLA_2$ activity was likely present on a regulatory signal transducing protein upstream of $cPLA_2$.[54]

The MAP kinase pathway

The phosphorylation and activation of $cPLA_2$ in RBL cells is associated with phosphorylation of proteins of the Ras/Raf/MAP kinase pathway including Raf1, MEK1 and MAP kinase.[121] The phosphorylation of MAP kinase is in turn associated with an increase in MAP kinase activity[121] which closely correlates to the release of arachidonic acid.[121] It is likely that MAP kinase directly regulates cPLA2 activity by phosphorylation of serine residue 505 on the $cPLA_2$ protein.[112] Studies utilizing chimeric receptors stably transfected into RBL cells suggest that the MAP kinase pathway regulating the release of arachidonic acid metabolites is linked to the FcεRI-γ chain via the GDP/GTP exchange factor vav.[162] Stimulation of phosphorylation of the proteins in the MAP kinase pathway and $cPLA_2$ activation were enhanced in RBL cells by transient overexpression of wild-type form of $p72^{syk}$. Overexpression of a dominant negative form of $p72^{syk}$ blocked these responses suggesting that the MAP kinase pathway regulating $cPLA_2$ activation is also linked to the FcεRI-γ via $p72^{syk}$.[162]

CONCLUDING REMARKS

To summarize the studies described in this review, a model for the signal transduction cascade linking FcεRI to the liberation of arachidonic acid metbolites is presented below (see also Fig. 10.5).

Aggregation of FcεRI on the surface of mast cells and basophils initiates the activation of specific tyrosine kinases including $p56^{lyn}$ and $p72^{syk}$. $p56^{lyn}$ is constitutively bound to the FcεRI-β carboxy terminal and upon FcεRI aggregation, the ITAMs of the FcεRI-γ and FcεRI-β cytoplasmic domains are tyrosine phosphorylated[135,136] allowing binding of $p72^{syk}$ to the FcεRI-γ and increased binding of $p56^{lyn}$ to the FcεRI-β.[135,136,140] The increase in tyrosine kinase catalytic activity results in the tyrosine phosphorylation of a number of signal transducing proteins including vav. As vav has been demonstrated to bind to phosphorylated tyrosine residue(s) on the $p72^{syk}$-related tyrosine kinase $p70^{zap}$ by an SH2 domain-phosphotyrosine interaction,[163] vav may bind to $p72^{syk}$ in a similar manner. Vav binds to a number of other proteins including the adaptor protein Grb2.[164] This binding takes place by a dimeric crosslinking of the respective SH3 domains of Grb2 and vav.[164] The SH3 domains of Grb2 also bind to a proline rich region of Sos[165,166] and this complex results

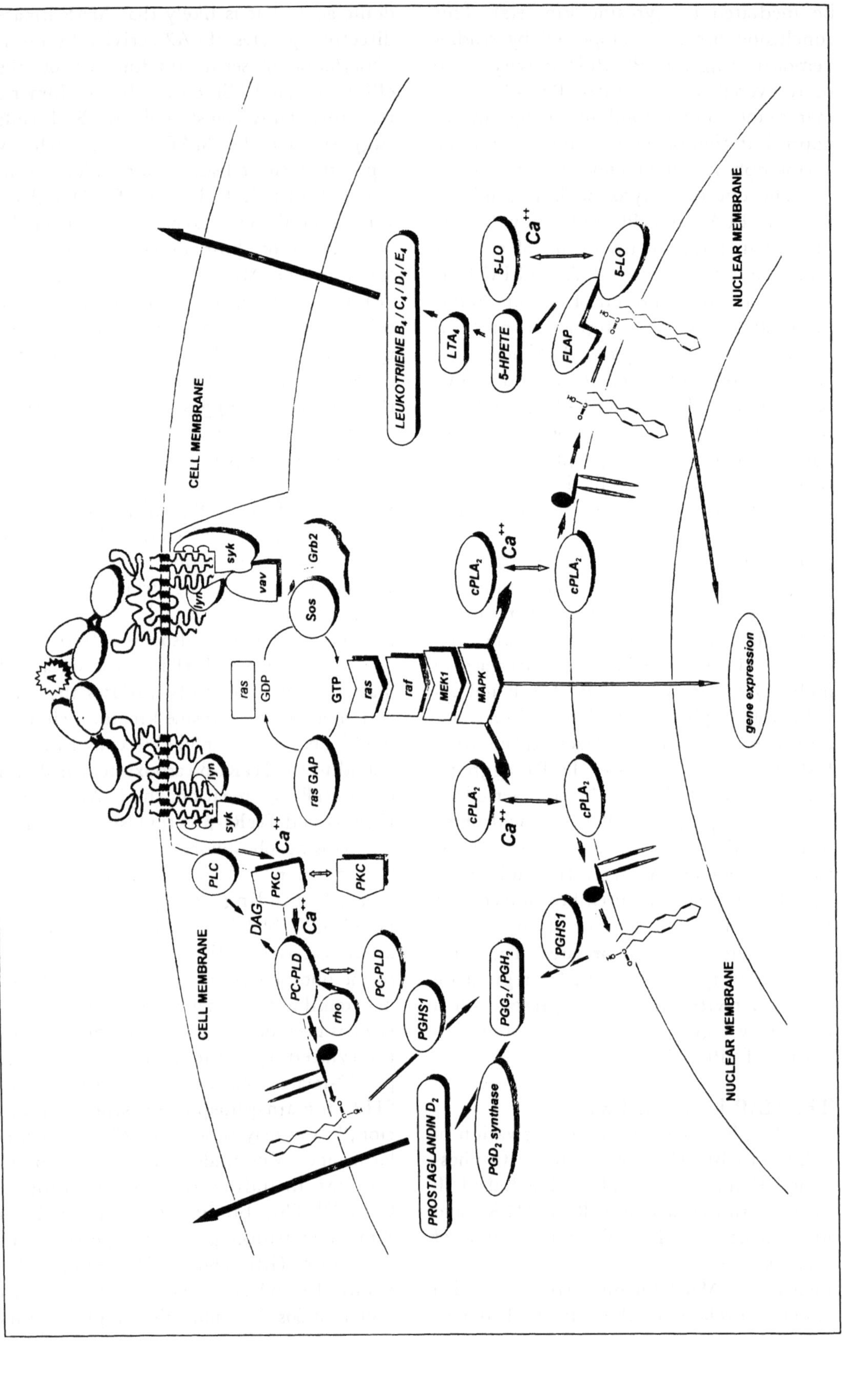
CELL MEMBRANE
CELL MEMBRANE
NUCLEAR MEMBRANE
NUCLEAR MEMBRANE
A
syk
lyn
syk
lyn
vav
Grb2
Sos
ras GAP
ras
GDP
GTP
ras
raf
MEK1
MAPK
gene expression
PLC
PKC
PKC
Ca++
DAG
PC-PLD
PC-PLD
rho
PGHS1
PGHS1
PGG2 / PGH2
PGD2 synthase
PROSTAGLANDIN D2
cPLA2
cPLA2
cPLA2
cPLA2
LEUKOTRIENE B4 / C4 / D4 / E4
LTA4
5-HPETE
5-LO
5-LO
FLAP

Fig. 10.5 (left). Possible signal transduction pathway linking the aggregation of FcεRI to the release of arachidonic acid metabolites. The intermediary steps between PC-PLD activation and arachidonic acid liberation have not been included in this figure as they have been represented in Figure 10.4. A: multivalent antigen. Other abbreviations are those used throughout the review. For full explanation of the figure, please refer to the text of the discussion section.

in activation of the cytosolic G protein Ras by Sos-mediated guanine nucleotide exchange of GDP with GTP.[167] Both Grb2 and Sos are expressed in RBL-2H3 cells (C Pullar, B Repetto and AM Gilfillan, unpublished observations), therefore, it is likely that these proteins also regulate Ras activity in mast cells. Ras initiates the MAP kinase pathway which involves sequential phosphorylation of Mek1 by Raf1 followed by the phosphorylation of MAP kinase by Mek1.[167]

Although the Ras/Raf/MAP kinase pathway is important for receptor mediated gene expression, the studies described in this chapter indicate that it also results in phosphorylation and activation of $cPLA_2$. The activated $cPLA_2$ is translocated to the nuclear membrane in a calcium-dependent manner where it catalyzes the liberation of arachidonic acid from PE and PC. The pool of free arachidonic acid is supplemented by arachidonic acid produced as a result of FcεRI-dependent activation of the PC-PLD/PAPase/DAG lipase pathway. The precise role of this pathway in either providing substrate or in regulating the catalytic activity of $cPLA_2$ remains to be established. The conversion of arachidonic acid to leukotrienes likely occurs at the nuclear membrane and this is a consequence of calcium-dependent translocation of 5-LO from the cytosol to the nuclear membrane. Activation of 5-LO and presentation of free arachidonic acid to the catalytic domain of 5-LO are regulated by the membrane-associated protein FLAP. The conversion of arachidonic acid to prostaglandins is catalyzed by PGHS1 and prostaglandin synthases, however: the sites of these reactions in mast cells are presently unknown. Once synthesized, leukotrienes and prostaglandins diffuse out of the cell into the extracellular spaces.

The signal transduction pathways leading to degranulation and cytokine gene expression are similar in the early stages to those described above for arachidonic acid metabolite release but these pathways diverge at later stages. The pathway for degranulation diverges at an earlier point (Fig. 10.6) than the pathway leading to cytokine gene expression suggesting that it may be easier to differentially regulate arachidonic acid metabolite production from degranulation than from cytokine production. The studies discussed in this review suggest that the signal transduction pathways leading to the release of the different classes of inflammatory mediators from mast cells may be influenced by both positively regulated and negatively regulated cross-talk mechanisms. For example, enzymes and products of the arachidonic acid metabolism pathway may influence gene transcription and, presumably, cytokine production, whereas lysophospholipids and 5-LO products may potentiate degranulation. Further studies will be required to fully elucidate these interactions.

Acknowledgments

Due to space limitations, all of the literature pertinent to the subject of this chapter could not be quoted. This does not imply that studies not quoted are of lesser importance.

I would like to thank Ms. Corinne Gilfillan for her help in preparing this manuscript, Drs. Barbara Repetto and Christine Pullar for their critical review of the manuscript and for their permission to discuss unpublished observations, Drs. Jarko Kochan and Peiyan Lin for permission to discuss unpublished observations and Ms. T. Chen for her help in preparing the figures.

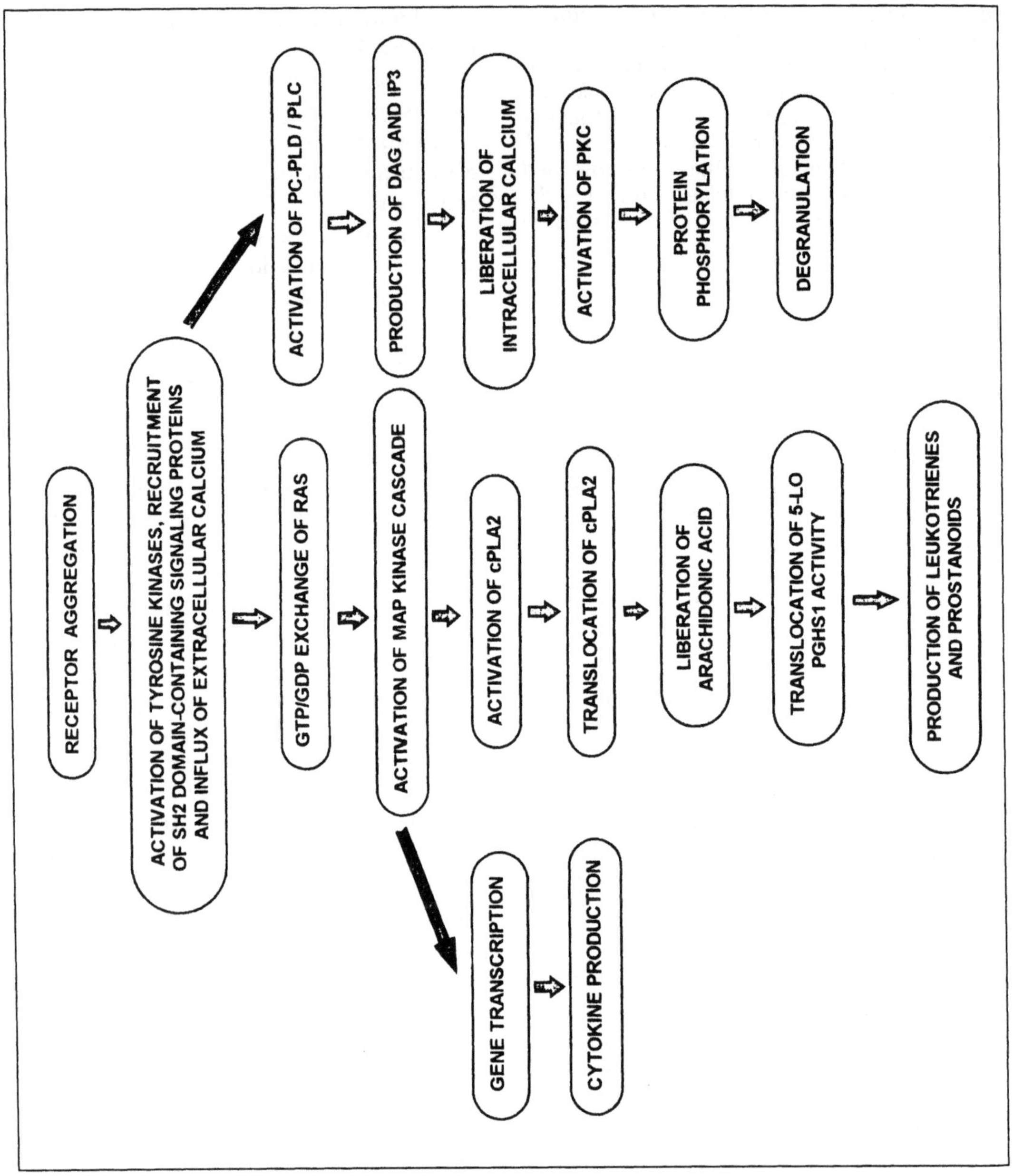

Fig. 10.6. Representation of the divergency of the signal transduction pathways linking aggregation of FcεRI to degranulation, cytokine gene expression and the release of arachidonic acid metabolites. For the sake of clarity, the possible role of the PC-PLD pathway in the production of arachidonic acid metabolites and "cross-talk" mechanisms between the pathways have not been included.

REFERENCES

1. Doran O, Stahl J, Cook E et al. Peptidoleukotriene (pLT) release from guinea pig lung mast cells. Inflammation 1994; 18:89-97.
2. Freeland HS, Schleimer RP, Schulman ES et al. Generation of leukotriene B_4 by human lung fragments and purified human lung mast cells. Am Rev Respir Dis 1988; 138:389-94.
3. Jakshick BA, Reugers T'A, Pinski JR. Nature of the mast cell macrophage interaction in immediate hypersensitivity. Adv Prostaglandin Thromboxane Leukotriene Res 1987; 17:180-85.
4. Jakschik BA, Wei Y, Owens LF. Interaction of macrophages and mast cells in the production of prostaglandins and leukotrienes. In: Nigan S, McBrien DCH, Slater TF, eds. Eicosanoids, Lipid Peroxidation and Cancer. New York: Springer Verlag, 1988:75-88.
5. Kreiger M, von Tscharner V, Dahinden CA. Signal transduction for interleukin-3 dependent leukotriene synthesis in normal human basophils: opposing role of tyrosine kinase and protein C. Eur J Immunol 1992; II:2907-13.
6. Lewis RA, Austen KF. Mediation of local homeostasis and inflammation by leukotrienes and other mast cell-dependent compounds. Nature 1981; 293:103-8.
7. Lewis RA, Soler NA, Diamond PT et al. Prostaglandin D_2 generation after activation of rat and human mast cells with anti-IgE. J Immunol 1982; 129:1627-31.
8. Levi-Schaffer F, Shalit M. Differential release of histamine and prostaglandin D_2 in rat peritoneal mast cells activated with peptides. Int Arch Allergy Appl Immunol 1989; 90:352-57.
9. Mencia-Huerta JM, Razin E, Ringel EW et al. Immunologic and ionophere-induced generation of leukotriene B_4 from mouse bone marrow derived mast cells. J Immunol 1983; 130:1885-90.
10. Musch MW, Siegel MI. Antigenic stimulated release of arachidonic acid, lipoxygenase activity and histamine release in a cloned murine mast cell MC9. Biochem Biophys Res Commun 1985; 126:517-25.
11. Neeleman P, Turk J, Jakschik BA et al. Arachidonic acid metabolism. Annu Rev Biochem 1986; 55:69-102.
12. Ramos BF, Zhang Y, Qureshi R et al. Mast cells are critical for the production of leukotrienes responsible for neutrophil recruitment in immune complex-induced peritonis in mice. J Immunol 1991; 147:1636-41.
13. Razin E, Mencia-Huerta JH, Stevens RL et al. IgE-mediated release of leukotriene C_4, chondroitin sulfate E proteoglycan, β-hexosaminidase and histamine from cultured bone marrow derived mast cells. J Exp Med 1983; 157:199-201.
14. Robinson C, Benyon RC, Holgate ST et al. The IgE- and calcium-dependent release of eicosanoids and histamine from human purified cutaneous mast cells. J Invest Dermatol 1989; 93:397-404.
15. Wells E, Harper ST, Jackson CG et al. Characterization of primate bronchoalveolar mast cells I. IgE-dependent release of histamine, leukotrienes and prostaglandins. J Immunol 1986; 137:3933-40.
16. Raible DG, Schulman ES, DiMuzio J et al. Mast cell mediators prostaglandin-D_2 and histamine activate human eosinophils. J Immunol 1992; 148:3536-42.
17. Peters SP, MacGlashan DW, Schulman ES et al. Arachidonic acid metabolism in purified human lung mast cells. J Immonol 1972; 132:1972-79.
18. Lewis RA, Austes KF, Sokerman RJ. Leukotrienes and other products of the 5-lipoxygenase pathway. Biochemistry and relation to pathobiology in human disease. N Engl J Med 1990; 323:645-55.
19. Marone G, Columbo M, Triggiani M. et al. Inhibition of IgE-mediated release of histamine and peptide leukotriene from human basophils and mast cells by forskolin. Biochem Pharmacol 1987; 36:13-20.
20. Lin P, Wiggan GA, Gilfillan AM. Activation of phospholipase D in rat mast (RBL-2H3) cell line. A possible unifying mechanism for IgE-dependent degranulation and arachidonic acid metabolite release. J Immunol 1991;146:1609-16.
21. Chen T, Heathers G, Hope W et al. Possible roles for protein kinase (PKC) and

tyrosine kinases (PTK) in the regulation of cytosolic phospholipase A ($cPLA_2$) activity in a rat mast (RBL) cell line. Mol Biol Cell 1994; 5:149a (abstract).
22. Fonteh AN, Chilton FH. Rapid remodeling of arachidonate from phosphatidylcholine to phosphatidylethanolanine pools during mast cell activation. J Immunol 1992; 148:1784-91.
23. Garcia-Gil M, Siraganian RP. Source of the arachidonic acid released on stimulation of rat basophilic leukemia cells. J Immunol 1986; 136:3825-28.
24. Yamada K, Okano Y, Miura K et al. A major role for phospholipase A_2 in antigen-induced arachidonic acid release in rat mast cells. Biochem J 1987; 247:95-99.
25. Fonteh AN, Bass DA, Marshall LA et al. Evidence that secretory phospholipase A_2 plays a role in arachidonic acid release and eicosanoid biosynthesis by mast cells. J Immunol 1994; 152:5438-46.
26. Colard O, Bidault J, Breton M et al. Biosynthesis of platelet-activating factor in cultured mast cells. Involvement of the CoA-independent transacylase demonstrated by analysis of the molecular species of platelet-activating factor. Eur J Biochem 1993; 216:835-40.
27. Winkler JD, Chilton FH. Inhibition of coenzyme A-independent transacylase may block PAF production and arachidonate release in inflammatory cells. CoA-IT's role in the production of PAF and the metabolism of arachidonate. DN&P 1993; 6: 133-38.
28. Fonteh AN, Chilton FH. Mobilization of different arachidonate pools and their roles in the generation of leukotrienes and free arachidonic acid during immnuologic activation of mast cells. J Immunol 1993; 150:563-70.
29. Majerus PW. Inositol Phosphate Biochemistry. Annu Rev Biochem 1992; 61:225-50.
30. Ali H, Cunha-Melo JR, Beaven MA. Receptor-mediated release of inositol 1,4,5-trisphosphate and inositol 1,4-bisphosphate in rat basophilic leukemia RBL-2H3 cells permeabilized with streptolysin O. Biochim Biophys Acta 1989; 1010:88-99.
31. Cunha-Melo JR, Dean NM, Moyer JD et al. The kinetics of phosphoinositide hydrolysis in rat basophilic leukemia (RBL-2H3) cells varies with the type of IgE receptor crosslinking agent used. J Biol Chem 1987; 262:11455-63.
32. Thomson FJ, Clark MA. Purification of a phosphatidic-acid-hydrolysing phospholipase A_2 from rat brain. Biochem J 1995; 306:305-309.
33. Kennerly DA. Phosphatidylcholine is a quantitatively more important source of increased 1,2-diacylglycerol than is phosphatidylinositol in mast cells. J Immunol 1990; 144:3912-19.
34. Gruchalla RS, Dinh TT, Kennerly DA. An indirect pathway of receptor-mediated 1,2-diacylglycerol formation in mast cells. I. IgE receptor-mediated activation of phospholipase D. J Immunol 1990; 144:2334-42.
35. Dennis EA. Diversity of group types, regulation and function of phospholipase A_2. J Biol Chem 1994; 269:13057-60.
36. Glaser KB, Mobilio D, Chang JY. Phospholipase A_2 enzymes: regulation and inhibition. TiPS 1993; 14:92-98.
37. Mayer RJ, Marshall LA. New insights on mammalian phospholipase A_2(s); comparison of arachidonoyl-selective and -nonselective enzymes. FASEB J 1993; 7:339-48.
38. Mukherjee AB, Miele L, Pattabiraman N. Phospholipase A_2 enzymes: regulation and physiological role. Biochem Pharmacol 1994; 48:1-10.
39. Nakatani Y, Hara S, Murakami M et al. Characterization of cytosolic phospholipase A_2 in rat mastocytoma RBL-2H3. Biol Pharm Bull 1994; 17:47-50.
40. Murakami M, Kudo I, Inoue K. Characteristics and possible functions of mast cell phospholipases A_2. In: Bazan NG, ed. Neurobiology of Essential Fatty Acids. New York: Plenum Press, 1992:27-34.
41. Murakami M, Kudo I, Umeda M et al. Detection of three distinct phospholipases A_2 in cultured mast cells. J Biochem 1992; 111:175-81.
42. Bellini F, Viola G, Menegus AM et al. Signalling mechanism in the lysophosphatidylserine-induced activation of mouse mast cells. Biochim Biophys Acta 1990; 1052: 216-20.

43. Horigone K. Tamori-Natori Y, Inoue K et al. Metabolism of lysophosphatidylserine, a potentiation of histamine release in rat mast cells. J Biochem 1986; 100:571-79.
44. Clark JD, Milona N, Knopf JL. Purification of a 110-kilodalton cytosolic phospholipase A_2 from the human monocytic cell line U937. Proc Natl Acad Sci 1990; 87:7708-12.
45. Lin L-L, Lin AY, Knopf JL. Cytosolic phospholipase A_2 is coupled to hormonally regulated release of arachidonic acid. Proc Natl Acad Sci 1992; 89:6147-51.
46. Lin L-L, Lin AY, DeWitt DL. Interleukin-1 α induces the accumulation of cytosolic phospholipase A_2 and the release of prostaglandin E_2 in human fibroblasts. J Biol Chem 1992; 267:23451-54.
47. Burghardt C, Gilfillan A, Chen T et al. Membrane translocation—a rate limiting step in FcεRI signalling. FASEB J 1995; 9:A781(abstract).
48. Clark JD, Lin L-L, Kriz RW et al. A novel arachidonic acid-selective cytosolic PLA_2 contains a Ca^{2+}-dependent translocation domain with homology to PKC and GAP. Cell 1991; 65:1043-51.
49. Kramer RM. Structure, function and regulation of mammalian phospholipases A_2. In: Brown BC, Dobson PRM, eds. Advances in Second Messenger and Phosphoprotein Research. Vol 28. New York: Raven Press Ltd., 1993:81-92.
50. Nalefski EA, Sultzman LA, Martin DM et al. Delineation of two functionally distinct domains of cytosolic phospholipase A_2, a regulatory Ca^{2+}-dependent lipid-binding domain and a Ca^{2+}-independent catalytic domain. J Biol Chem 1994; 269:18239-49.
51. Muramaki M, Matsumoto R, Urade Y et al. c-Kit ligand mediates increased expression of cytosolic phospholipase A_2, prostaglandin endoperoxide synthase-1 and hematopoietic prostaglandin D_2 synthase and increased IgE-dependent prostaglandin D_2 generation in immature mouse mast cells. J Biol Chem 1995; 270:3239-46.
52. Murakami M, Penrose JF, Urade Y et al. Interleukin 4 suppresses c-kit ligand-induced expression of cytosolic phospholipase A_2 and prostaglandin endoperoxide synthase 2 and their roles in separate pathways of eicosanoid synthesis in mouse bone marrow-derived mast cells. Proc Natl Acad Sci 1995; 92:6107-11.
53. Nakatani Y, Murakami M, Kudo I et al. Dual regulation of cytosolic phospholipase A_2 in mast cells after crosslinking of Fcε-receptor. J Immunol 1994; 153:796-803.
54. Currie S, Roberts EF, Spaethe SM et al. Phosphorylation and activation of Ca^{2+}-sensitive cytosolic phospholipase A_2 in MCII mast cells mediated by high-affinity Fc receptor for IgE. Biochem J 1994; 304: 923-28.
55. Garcia-Gil M, Siraganian RP. Phospholipase A_2 stimulation during cell secretion in rat basophilic leukemia cells. J Immunol 1986; 136:259-63.
56. Gawler DJ, Zhang L-J, Reedijk M et al. CaLB: a 43 amino acid calcium-dependent membrane/phospholipid binding domain in p120 Ras GTPase-activating protein. Oncogene 1995; 10:817-25.
57. Glover S, Bayburt T, Jonas M et al. Translocation of the 85 kDa phospholipase A_2 from cytosol to the nuclear envelope in rat basophilic leukemia cells stimulated with calcium ionophore or IgE/antigen. J Biol Chem 1995; 270:15359-67.
58. Butcher RD, Wojcik SJ, Lints T et al. Arachidonic acid, a growth signal in murine P815 mastocytoma cells. Cancer Res 1993; 53:3405-10.
59. Murakami M, Kudo I, Inoue K. Eicosanoid generation from antigen-primed mast cells by extracellular mammalian 14 kDa group II phospholipase A_2. FEBS Lett 1991; 294:247-51.
60. Murakami M, Kudo I, Suwa Y et al. Release of 14 kDa group-II phospholipase A_2 from activated mast cells and its possible involvement in the regulation of the degranulation process. Eur J Biochem 1992; 204:259-67.
61. Fonteh AN, Samet JM, Chilton FH. Regulation of arachidonic acid, eicosanoid and PLA_2 levels in murine mast cells by recombinant stem cell factor. J Clin Invest (in press).
62. Ishizaki J, Hanasaki K, Higashino K et al. Molecular cloning of pancreatic group I

phospholipase A_2 receptor. J Biol Chem 1994; 269:5897-904.
63. Lambeau G, Ancian P, Barhanin J et al. Cloning and expression of a membrane receptor for secretory phospholipases A_2. J Biol Chem 1994; 269:1575-78.
64. Ohara O, Ishizaki J, Arita H. Structure and function of phospholipase A_2 receptor. Prog Lipid Res 1995; 34:117-38.
65. Ishizaki J, Kishino J, Teraoka H et al. Receptor-binding capability of pancreatic phospholipase A_2 is separable from its enzymatic activity. FEBS Lett 1993; 324:349-52.
66. Murakami M, Hara N, Kudo I et al. Triggering of degranulation in mast cells by exogenous type II phospholipase A_2. J Immunol 1993; 151:5675-84.
67. Murakami M, Kudo I, Fujimori Y et al. Group II phospholipase A_2 inhibitors suppressed lysophosphatidylserine-dependent degranulation of rat peritoneal mast cells. Biochem Biophys Res Commun 1991; 181:714-21.
68. Chi E, Henderson WR, Klebanoff SJ. Phospholipase A_2-induced rat mast cell secretion. Lab Invest 1982; 47:579-85.
69. Peters SP, Schleimer RP, Marone G et al. Lipoxygenase products of arachidonic acid; role in modulation of IgE-induced histamine release. Leukotrines and other Lipoxygenase Products 1982.
70. Sullivan TJ, Parker CW. Possible role of arachidonic acid and its metabolites in mediator release from rat mast cells. J Immunol 1979; 122:431-36.
71. Bronner C, Cothenet V, Monte D et al. Role of phospholipase A_2 and G-proteins in the IgE-dependent activation of mast cells and macrophages. Agents and Actions 1990; 30:95-97.
72. Chand N, Pillar J, Diamantis W et al. Inhibition of histamine secretion from rat peritoneal mast cells and rabbit leukocytes by p-bromophenacyl bromide, a phospholipase A_2 inhibitor. Res Comm Chem Path and Pharm 1987; 55:17-24.
73. Marone G, Kagey-Sobotka A, Lichtenstein LM. Possible role of phospholipase A_2 in triggering histamine secretion from human basophils in vitro. Clin Immunol Immunopathol 1981; 20:231-39.
74. Morita Y, Aida N, Miyamoto T. Role of phospholipase A_2 activation in histamine release from human basophils. Allergy 1983; 38:413-18.
75. Chettibi S, Lawrence AJ, Stevenson RD et al. Effect of lysophosphatidic acid on motility, polarisation and metabolic burst of human neutrophils. FEMS Immunology and Medical Microbiology 1994; 8:271-82.
76. Durieux ME, Lynch KR. Signalling properties of lysophosphatidic acid. TiPS 1993; 14:249-54.
77. Fernandez B, Balboa MA, Solis-Herruzo JA et al. Phosphatidate-induced arachidonic acid mobilization in mouse peritoneal macrophages. J Biol Chem 1994; 269:26711-16.
78. Howe KR, Marshall CJ. Lysophosphatidic acid stimulates mitogen-activated protein kinase activation via a G-protein-coupled pathway requiring $p21^{ras}$ and $p74^{raf-1}$. J Biol Chem 1993; 268:20717-20.
79. Jones GA, Carpenter G. The regulation of phospholipase C-γ1 by phosphatidic acid. Assessment of kinetic parameters. J Biol Chem 1993; 268:20845-50.
80. Moolenaar WH. LPA: a novel lipid mediator with diverse biological actions. Trends In Cell Biol 1994; 4:213-19.
81. Pearce B, Jakobson K, Morrow C et al. Phosphatidic acid promotes phosphoinositide metabolism and DNA synthesis in cultured cortical astrocytes. Neurochem Int 1994; 24:165-71.
82. Seufferlein T, Rozengurt E. Lysophosphatidic acid stimulates tyrosine phosphorylation of focal adhesion kinase, paxillin and p130. J Biol Chem 1994; 269;9345-51.
83. Lin P, Gilfillan AM. The role of calcium and potein kinase C in the IgE-dependent activation of phosphatidylcholine-specific phospholipase D in a rat mast (RBL-2H3) cell line. Eur J Biochem 1992; 207:163-68.
84. Lin P, Wiggan GA, Welton AF et al. Differential effects of propranolol on the IgE-dependent, or calcium ionophore-stimulated, phosphoinositide hydrolysis and calcium mobilization in a mast (RBL-2H3) cell line. Biochem Pharmacol 1991; 41:1941-48.
85. Lin P, Fung WC, Gilfillan AM. Phosphatidylcholine-specific phospholipase D-derived 1,2-diacylglycerol does not initiate protein

kinase C activation in the RBL-2H3 mast-cell line. Biochem J 1992; 287:325-31.

86. Lin P, Fung WC, Li S et al. Temporal regulation of the IgE-dependent 1,2-diacylglycerol production by tyrosine kinase activation in a rat(RBL-2H3) mast-cell line. Biochem J 1994; 299:102-14.
87. Kumada T, Miyata H, Nozaway. Involvement of tyrosine phosphorylation in IgE receptor-mediated phospholipase D activation in rat basophilic leukemia (RBL-2H3) cells. Biochem Biophys Res Commun 1993; 191:1363-68.
88. Ishimoto T, Akiba S, Sato T et al. Contribution of phospholipases A_2 and D to arachidonic acid liberation and prostaglandin D_2 formation with increase in intracellular Ca^{2+} concentration in rat peritoneal mast cells. Eur J Biochem 1994; 219:401-406.
89. Koike T, Mizutani T, Hirai K et al. SCF/c-Kit receptor-mediated arachidonic acid liberation in rat mast cells. Biochem Biophys Res Commun 1993; 197:1570-77.
90. Sato T, Ishimoto T, Akiba S et al. Enhancement of phospholipase A_2 activation by phosphatidic acid endogenously formed through phospholipase D action in rat peritoneal mast cell. FEBS Lett 1993; 323: 23-26.
91. Dixon RAF, Diehl RE, Opas E et al. Requirement of a 5-lipoxygenase-activating protein for leukotriene synthesis. Nature 1990; 343:282-84.
92. Wong A, Cook MN, Hwang SM et al. Stimulation of leukotriene production and membrane translocation of 5-Lipoxygenase by crosslinking of the IgE receptors in RBL-2H3 cells. Biochemistry 1992; 31:4046-52.
93. Malaviya R, Malaviya R, Jakschik BA. Reversible translocation of 5-lipoxygenase in mast cells upon IgE/Antigen stimulation. J Biol Chem 1993; 268:4939-44.
94. Brock TG, Paine III R, Peters-Golden M. Localization of 5-Lipoxygenase to the nucleus of unstimulated rat basophilic leukemia cells. J Biol Chem 1994; 269: 22059-66.
95. Lepley RA, Fitzpatrick FA. 5-Lipoxygenase contains a functional Src homology 3-binding motif that interacts with the Src homology 3 domain of Grb2 and cytoskeletal proteins. J Biol Chem 1994; 269:24163-68.
96. Vane J. Towards a better aspirin. Nature 367:215-16.
97. Thomas PS, Wilson AN, Schreck RE et al. Prostaglandin D_2 production and identification of prostaglandin H synthase within canine mast cell granule. Am J Physiol 1992; 263:L168-L76.
98. Samet JM, Fasano MB, Fonteh AN et al. Selective induction of PGH synthase I by stem cell factor and dexamethasone in mast cells. J Biol Chem 1995; 270:8044-49.
99. Murakami M, Matsumoto R, Austen KF et al. Prostaglandin endoperoxide synthase-1 and -2 couple to different transmembrane stimuli to generate prostaglandin D_2 in mouse marrow-derived mast cells. J Biol Chem 1994; 269:22269-75.
100. Dewitt DL, Smith WL. Primary structure of prostaglandin G/H synthase for sheep vesicular gland determined from the complementary DNA sequence. Proc Natl Acad Sci 1988; 85:1412-16.
101. DeWitt DL, Harith EL, Kraemer EA et al. The aspirin and heme-binding sites or ovine and murine prostaglandin endoperoxide syntheses. J Biol Chem 1990; 265:5192-98.
102. Funk CD, Funk LB, Kennedy ME et al. Human platelet/erythroleukemic cell prostaglandin G/H synthase: cDNA cloning expression and gene chromosomal assignment. FASEB J 1991; 5:2304-12.
103. Urade Y, Fujimoto N, Hayaishi O. Purification and characterization of rat brain prostaglandin D synthetase. J Biol Chem 1985; 160:12410-15.
104. Urade Y, Fujimoto N, Ujihara M et al. Primary structure of rat brain prostaglandin D synthetase deduced from cDNA sequence. J Biol Chem 1989; 264:1041-45.
105. Urade Y, Ujihara M, Horiguchi Y et al. Mast cells contain spleen-type prostaglandin D synthetase. J Biol Chem 1990; 265:371-75.
106. Murakami M, Austen KF, Arm JF. The immediate phase of c-kit ligand stimulation of mouse bone marrow-derived mast cells elicits rapid leukotriene C_4 generation through posttranslational activation of cystolic phospholipase A_2 and 5-lipoxygenase. J Exp Med 1995; 182:197-206.
107. Chock SP, Schmauder-Chock EA. A new model for the mechanism of stimulus-secre-

tion coupling. BioFactors 1990; 2:133-46.

108. Schmauder-Chock EA, Chock SP. Localization of cyclo-oxygenase and prostaglandin E_2 in the secretory granule of the mast cell. J Histochem Cytochem 1989; 37:1319-28.

109. Chock SP, Schmauder-Chock EA, Cordella-Miele E et al. The localization of phospholipase A_2 in the secretory granule. Biochem J 1994; 300:619-22.

110. Chock SP, Schmauder-Chock EA. Phospholipid storage in the secretory granule of the mast cell. J Biol Chem 1989; 264:2862-68.

111. Chock SP, Schmauder-Chock EA. Synthesis of prostaglandins and eicosanoids by the mast cell secretory granule. Biochem Biophys Res Commun 1988; 156:1308-15.

112. Lin L-L, Wartmann M, Lin AY et al. $cPLA_2$ is phosphorylated and activated by MAP kinase. Cell 1993; 72:269-78.

113. Her E, Zor U. The role of external and internal free Ca^{2+} concentration on ionomycin induced leukotriene C_4 formation in rat basophilic leukemia cells. J Lipid Mediators 1991; 4:175-84.

114. Gilfillan AM, Kado-Fong H, Wiggan GA et al. Conservation of signal transduction mechanisms via the human FcεRI-α after transfection into a rat mast cell line, RBL-2H3. J Immunol 1992; 149:2445-51.

115. Ali H, Collado-Escobar DM, Beaven MA. The rise in concentration of free Ca^{2+} and of pH provides sequential synergistic signals for secretion in antigen-stimulated rat basophilic leukemia (RBL-2H3) cells. J Immunol 1989; 143:2626-33.

116. Hide M, Beaven MA. Calcium influx in a rat mast cell (RBL-2H3) line. Use of multivalent metal ions to define its characteristics and role in exocytosis. J Biol Chem 1991; 266:15221-29.

117. Narasimhan V, Holowka D, Baird B. A guanine nucleotide-binding protein participates in IgE receptor-mediated activation of endogenous and reconstituted phospholipase A_2 in a permeabilized cell system. J Biol Chem 1990; 265:1459-64.

118. Okano Y, Yamada K, Yano K et al. Guanosine 5'-(γ-thio) triphosphate stimulates arachidonic acid liberation in permeabilized rat peritoneal mast cells. Biochem Biophys Res Commun 1987; 145:1267-75.

119. Gilfillan AM, Wiggan GA, Welton AF. Pertussis toxin pretreatment reveals differential effects of adenosine analogs on IgE-dependent histamine and peptidoleukotriene release from RBL-2H3 cells. Biochimica et Biophysica Acta 1990; 1052:467-74.

120. Nakamura T, Ui M. Simultaneous inhibitions of inositol phospholipid breakdown, arachidonic acid release and histamine secretion in mast cells by islet-activating protein, pertussis toxin. J Biol Chem 1985; 260:3584-93.

121. Hiraswa N, Santini F, Beaven MA. Activation of the mitogen-activated protein kinase/cytosolic phospholipase A_2 pathway in a rat mast cell line. Indications of different pathways for release of arachidonic acid and secretory granules. J Immunol 1995; 154:5391-402.

122. Jiang H, Alexandropoulos K, Song J et al. Evidence that v-Src-induced phospholipase D activity is mediated by a G protein. Mol Cell Biol 1994; 14:3676-82.

123. Kusner DJ, Dubyak GR. Guanosine 5'[γ-thio]triphosphate induces membrane localization of cytosol-independent phospholipase D activity in a cell-free system from U937 promonocytic leucocytes. Biochem J 1994; 304:485-91.

124. Bowman EP, Uhlinger DJ, Lambeth JD. Neutrophil phospholipase D is activated by a membrane-associated Rho family small molecular weight GTPbinding protein. J Biol Chem 1993; 268:21509-12.

125. Brown HA, Gutowski S, Kahn RA et al. Partial purification and characterization of arf-sensitive phospholipase D from porcine brain. J Biol Chem 1995; 270:14935-43.

126. Brown HA, Gutowski S, Moomaw CR et al. ADP-ribosylation factor, a small GTP-dependent regulatory protein, stimulates phospholipase D activity. Cell 1993; 75:1137-44.

127. Cockcroft S, Thomas GMH, Fensome A et al. Phospholipase D: a downstream effector of ARF in granulocytes. Science 1994; 263:523-26.

128. Lambeth JD, Kwak J-Y, Bowman EP et al. ADP-ribosylation factor functions synergistically with a 50 kDa cytosolic factor in cell-free activation of human neutrophil phospholipase D. J Biol Chem 1995; 270: 2431-34.

129. Gilfillan AM, Wiggan GA, Welton AF. The effects of the protein kinase C inhibitors staurosporine and H7 on the IgE dependent mediator release from RBL-2H3 cells. Agents and Actions 1990; 30:418-25.
130. Morita Y, Takaishi T, Honda Z et al. Role of protein kinase C in histamine release from human basophils. Allergy 1988; 43:100-4.
131. Ozawa K, Szallasi Z, Kazanietz MG et al. Ca^{2+}-dependent and Ca^{2+}-independent isozymes of protein kinase C mediate exocytosis in antigen-stimulated rat basophilic RBL-2H3 cells. J Biol Chem 1993; 268:1749-56.
132. Ozawa K, Yamada K, Kazanietz MG et al. Different isozymes of protein kinase C mediate feedback inhibition of phospholipase C and stimulatory signals for exocytosis in rat RBL-2H3 cells. J Biol Chem 1993; 268:2280-83.
133. Yamada K, Jelsema CL, Beaven MA. Certain inhibitors of protein serine/threonine kinases also inhibit tyrosine phosphorylation of phospholipase Cγ1 and other proteins and reveal distinct roles for tyrosine kinase(s) and protein kinase C in stimulated, rat basophilic RBL-2H3 cells. J Immunol 1992; 149:1031-37.
134. Peleg I, Ludowyke RI, Beaven MA et al. The role of myosin phosphorylation in RBL-2H3 cell secretion. J Lab Clin Med 1992; 120:675-80.
135. Eiseman E, Bolen JB. Engagement of the high-affinity IgE receptor activates src protein-related tyrosine kinases. Nature 1992; 355:78-80.
136. Jouvin M-HE, Adamczewski M, Numerof R et al. Differential control of the tyrosine kinases lyn and syk by two signaling chains of the high affinity immunoglobin E receptor. J Biol Chem 1994; 269:5918-25.
137. Kihara H, Siraganian RP. Src homology 2 domains of syk and lyn bind to tyrosine-phosphorylated subunits of the high affinity IgE receptor. J Biol Chem 1994; 269:22427-32.
138. Minoguchi K, Swaim WD, Berenstein EH et al. Src family tyrosine kinase $p^{53/56lyn}$, a serine kinase and FcεRI associate with α-galactosyl derivatives of ganglioside GD_{1b} in rat basophilic leukemia RBL-2H3 cells. J Biol Chem 1994; 269:5249-54.
139. Wilson BS, Kapp N, Lee RJ et al. Distinct functions of the FcεRI γ and β submits in the control of FcεRI-mediated tyrosine kinase activation and signaling responses in RBL-2H3 mast cells. J Biol Chem 1995; 270:4013-22.
140. Yamashita T, Mao S-Y, Metzger H. Aggregation of the high-affinity IgE receptor and enhanced activity of $p53/56^{lyn}$ protein-tyrosine kinase. Proc Natl Acad Sci 1994; 91:11251-55.
141. Benhamou M, Ryba NJP, Kihara H et al. Protein-tyrosine kinase $p72^{syk}$ in high affinity IgE receptor signaling. J Biol Chem 1993; 268:23318-24.
142. Minoguchi K, Benhamou M, Swaim WD et al. Activation of protein tyrosine kinase $p72^{syk}$ by FcεRI aggregation in rat basophilic leukemia cells. J Biol Chem 1994; 269:16902-8.
143. Stephan V, Benhamou M, Gutkind JS et al. FcεRI-induced protein tyrosine phosphorylation of pp72 in rat basophilic leukemia cells (RBL-2H3). J Biol Chem 1992; 267:5434-41.
144. Benhamou M, Gutkind JS, Robbins KC et al. Tyrosine phosphorylation coupled to IgE receptor-mediated signal transduction and histamine release. Proc Natl Acad Sci 1990; 87:5327-30.
145. Hamawy MM, Swaim WD, Minoguchi K et al. The aggregation of the high affinity IgE receptor induces tyrosine phosphorylation of paxillin, a focal adhesion protein. J Immunol 1994; 153:4655-62.
146. Kawakami T, Inagaki N, Takei M et al. Tyrosine phosphorylation is required for mast cell activation by FcεRI crosslinking. J Immunol 1992; 148;3513-19.
147. Park DJ, Min HK, Rhee SG. IgE-induced tyrosine phosphorylation of phospholipase C-γ1 in rat basophilic leukemia cells. J Biol Chem 1991; 266:24237-40.
148. Pribluda VS, Pribluda C, Metzger H. Transphosphorylation as the mechanism by which the high-affinity receptor for IgE is phosphorylated upon aggregation. Proc Natl Acad Sci 1994; 91:11246-50.
149. Santini F, Beaven MA. Tyrosine phosphorylation of a mitogen-activated protein kinase-like protein occurs at a late step in exocytosis. J Biol Chem 1993; 268:22716-22.

150. Yu K-T, Lyall R, Jariwala N et al. Antigen- and ionophore-induced signal transduction in rat basophilic leukemia cells involves protein tyrosine phosphorylation. J Biol Chem 1991; 266:22564-68.
151. Balboa MA, Firestein BL, Godson C et al. Protein kinase Cα mediates phospholipase D activation by nucleotides and phorbol ester in madin-darby canine kidney cells. J Biol Chem 1994; 269:10511-16.
152. Gustavsson L, Moehren G, Torres-Marquez ME et al. The role of cytosolic Ca^{2+}, protein kinase C and protein kinase A in hormonal stimulation of phospholipase D in rat hepatocytes. J Biol Chem 1994; 269:849-59.
153. Hoek JB. Tyrosine kinase activation and signal transduction mediated by phospholipase D. Lab Invest 1993; 69:1-4.
154. Song J, Foster DA. v-Src activates a unique phospholipase D activity that can be distinguished from the phospholipase D activity activated by phorbol esters. Biochem J 1993; 294:711-17.
155. Uings IJ, Thompson NT, Randall RW et al. Tyrosine phosphorylation is involved in receptor coupling to phospholipase D but not phospholipase C in the human neutrophil. Biochem J 1992; 281:597-600.
156. Jacobsen S, Hansen HS, Jensen B. Synergism beween thapsigargin and the phorbol ester 12-0-tetradecancyl phorbol -13-acetate on the release of [^{14}C] arachidonic acid and histamine from rat peritoneal mast cells. Biochem Pharmacol 1987; 36:621-26.
157. Amon U, von Stebut E, von Gizycki et al. Histamine release and mast cells. Control function of protein kinase C isozymes on leukotriene generation from human basophils. Agents Actions 1994; 41:C9-C10.
158. Undem BJ, Torphy TJ, Goldman D et al. Inhibition by adenosine 3':5'-monophosphate of eicosanoid and platelet-activating factor biosynthesis in the mouse PT-18 mast cell. J Biol Chem 1990; 265:6750-58.
159. Peachell PT, MacGlashan DW Jr, Lichtenstein LM et al. Isoproterenol induced inhibition of immunoglogbin E mediated release of histamine and arachidonic acid metablilites from the human lung mast cell. J Immunol 1988; 140:571-79.
160. Undem BJ, Peachell PT, Lichtenstein LM. Regulation of human basophil and lung mast cell function by cyclic adenosine monophosphate. J Pharmacol Exp Ther 1988; 247:209-17.
161. Hagmann W. Cell proliferation status, cytokine action and protein tyrosine phosphorylation modulate leukotriene biosynthesis in a basophil leukaemia and a mastocytoma cell line. Biochem J 1994; 299:467-72.
162. Hirasawa N, Scharenberg A, Yamamura H et al. A requirement for Syk in the activation of the microtubule-associated protein kinase/phospholipase A_2 pathway by FcεR1 is not shared by a G protein-coupled receptor. J Biol Chem 1995; 270:10960-67.
163. Katzav S, Sutherland M, Packham G et al. The protein tyrosine kinase ZAP-70 can associate with the SH2 domain of proto-vav. J Biol Chem 1994; 269:32579-85.
164. Ye Z-S, Baltimore D. Binding of Vav to Grb2 through dimerization of Src homology 3 domains. Proc Natl Acad Sci 1994; 91:12629-33.
165. Margolis B. The GRB family of SH2 domain proteins. Prog Biophys molec Biol 1994; 62:223-24.
166. Buday L, Egan SE, Viciana PR et al. A complex of Grb2 adaptor protein, Sos exchange factor and a 36 kDa membrane-bound tyrosine phosphoprotein is implicated in Ras activation in T cells. J Biol Chem 1994; 269:9019-23.
167. Seger R, Krebs EG. The MAPK signaling cascade. FASEB J 1995; 9:726-35.

CHAPTER 11

FcεRI-induced Cytokine Production and Gene Expression

John R. Gordon

INTRODUCTION

The concept that cells which express high affinity IgE receptors (FcεRI) can produce cytokines is a relatively new one, for it was less than 10 years ago that is was first shown that some transformed mast cell lines could secrete these proteins. Since then we have come to understand that normal mast cells and basophils, the cells that we most commonly associate with the FcεRI and IgE-dependent responses, may also secrete a wide array of cytokines. We have known for some time of course that these cells are important to allergic responses, and we have had a good appreciation of some of the underlying mechanisms, but for many the discovery of "mast cell cytokines" and the important roles that at least some of them play in allergic diseases has provided some rather straightforward explanations for what had seemed to be very complex questions about the molecular pathophysiology of allergic responses. These comments should not be taken as a minimization of the roles of non-cytokine mast cell mediators in the pathogenesis of atopy, for many of these are also critical components of the process. Nevertheless, the concept of cytokine production by mast cells is truly an exciting one that has enormous potential ramifications not only for atopic diseases, but perhaps also for the fundamental regulation of other immune responses.

Mast cells can produce numerous cytokines but, just as important, they also employ these (and other mediators) to recruit additional sources of many of these same cytokines, as well as others that they do not produce. This mast cell-initiated cytokine cascade is important to the pathogenesis of atopic diseases. In this chapter we will survey the available

IgE Receptor (FcεRI) Function in Mast Cells and Basophils, evidence about FcεRI-induced cytokine secretion by mast cells and basophils, including what is known of the regulation of this process, and then we will discuss what is known of their impact on the pathophysiology of the host. While numerous other cells also express FcεRI (i.e., Langerhan's cells,[1,2] dermal dendritic cells,[3] monocytes,[4] eosinophils[5] and perhaps T cells[6]), these cells will not be discussed since, with the exception of antigen presentation by Langerhan's cells, little is known of the roles of the FcεRI receptors in these cells.

THE NATURAL HISTORY OF MAST CELLS AND BASOPHILS

Mast cells are large resident tissue cells that migrate via the circulation only as physiologically immature precursors, while basophils are peripheral blood leukocytes that normally mature in the bone marrow and circulate as fully functional, highly granulated cells. Both populations contain histamine and large metachromatic cytoplasmic granules comprising a sulfated proteoglycan core to which numerous mediators are electrostatically bound (reviewed in ref. 7). Both also express receptors for multiple cytokines and growth factors, as well as high affinity IgE (FcεRI) and low affinity IgG receptors.[8,9] A number of the distinguishing phenotypic characteristics of these cells are outlined in Table 11.1.

Basophils are considered to represent a relatively homogenous population, although substantial maturational variation can occur, while mast cells comprise at least two functionally distinct phenotypes. Mast cell precursors that seed mucosal sites in rodents can proliferate upon exposure to T cell-derived cytokines such as interleukin-3 (IL-3), as would occur during some helminth infections, while those that differentiate in the connective tissues mature under the influence of tissue fibroblasts and proliferate when exposed to high concentrations of SCF[10] or during chronic inflammatory conditions.[11] Thus, the primary distinguishing phenotypic features of mast cells have historically fallen along anatomic lines, with "mucosal" mast cells being T cell-dependent (i.e., IL-3-dependent) and "connective tissue" mast cells T cell-independent (reviewed in ref. 7). While these two populations are often thought of as "terminally" differentiated in situ, in fact each retains the ability to "transdifferentiate"[12] and proliferate, such that if transferred into a genetically mast cell-deficient WBB6F1-*W*/*W^v* (*W*/*W^v*) mouse, one cell of either phenotype is sufficient for complete reconstitution of the mouse with mast cells of both types.[13] More recently it has been appreciated that the constituent granule proteases of mast cells can be used to distinguish their phenotype. In humans, mast cells located at the airway and gut epithelial surfaces contain relatively high levels of mast cell-specific tryptase (and little chymase) and are designated MC_T, while those in the dermis and subepithelial sites (e.g., airway and gastrointestinal submucosa) typically contain both tryptase and chymase activities and are referred to as MC_{TC}; admixtures of these two phenotypes can occur in one compartment, such as the nasal mucosa (reviewed in ref. 14). The protease expression-phenotyping of mast cells in mice is more complex, since numerous mouse mast cell-specific proteases have been identified (Table 11.1).[15]

Many of our insights into the basic biology of mast cells have arisen through the discovery that virtually unlimited numbers and pure populations of these cells could be generated in vitro simply by culturing mouse bone marrow cells with IL-3.[7] Large numbers of mast cells can alternately be obtained by using spontaneously or virally transformed in vitro-derived mast cells or mastocytoma cell lines. Indeed the first reports of mast cell cytokine production were from laboratories seeking the molecular basis for the autonomous growth of such cells.[16-18] More recently it has been determined that the FcεRI-positive cells that normally reside in the bone marrow of mice are largely immature basophils or basophilic myelocytes, although when placed in long-term culture with appropriate growth factors these basophils are rapidly

Table 11.1. Phenotypic characteristics of mast cells and basophils

	Basophils	Mast Cells (in vivo)		Mast Cells (culture-derived)	
Phenotype		"mucosal "	"connective tissue"	IL-3- dependent	SCF- dependent
Origin of precursor cells[7]	bone marrow	bone marrow	bone marrow	mouse bone marrow	**mouse:** bone marrow **human:** chord blood, fetal liver
Site of maturation	bone marrow	intestinal mucosa, lung alveolar wall[14]	normal skin, intestinal submucosa[14]	in vitro	in vitro
Site of residency:[7] normal- pathologic-	-circulation -tissues	-tissues -tissues	-tissues -tissues	-	-
Primary growth factor	IL-3 (in vitro[182])	**mouse**: IL-3[7]	**mouse:** SCF[10]	IL-3[183]	SCF[184]
Nuclear morphology[7]	segmented	nonsegmented	nonsegmented	nonsegmented	nonsegmented
Major granule proteoglycan[185]	chrondroitin sulfate	chrondroitin sulfate	heparin	chrondroitin sulfate	heparin
Phenotypic granule proteases[14]		**mouse:** MMCP-1, -2 **human:** tryptase	**mouse:** MMCP-3, -4, -5, & -6 **human:** tryptase & chymase	**mouse:** MMCP-5 & -7	**mouse:** MMCP-4, -5, & -6 **human:** chord blood > tryptase; fetal liver > tryptase & chymase
Histamine content	low	low	high	low	high
c-kit expression[176]	low	high	high	high	high

overgrown by mast cells,[19] so that they are not routinely available in large numbers. Since the resident FcεRI-positive bone marrow or spleen cells were originally isolated by removing all cells that expressed T cell or B cell markers (e.g., Thy1, CD3, CD4, CD8, B220 and class II MHC), they are most often referred to not as basophils but as FcεRI$^+$ non-B, non-T cells.[20] Human FcεRI$^+$ non-B, non-T cells can also be purified from bone marrow,[21] and human mast cell populations can be generated, albeit in relatively low numbers, by culturing fetal liver or chord blood cells in SCF.[22] In mice, IL-3-dependent bone marrow-derived cultured mast cells (BMMC or BMCMC) are phenotypically quite immature, but nevertheless express a functional FcεRI and share a number of characteristics with mucosal compartment mast cells (e.g., low histamine content, chondroitin sulfate proteoglycan). However, when maintained either in the presence of fibroblasts or SCF, BMMC tend to acquire some phenotypic characteristics associated with connective tissue-type mouse mast cells (e.g., increased histamine levels, heparin sulfate proteoglycan). Cultured human fetal liver-derived mast cells grown in SCF do not possess a functional FcεRI (the α chain of the α, β, γ complex is not expressed; ref. 22), but are reactive to other traditional mast cell secretagogues (e.g., calcium ionophores). Even though BMMC from mice have the ability to mature with time into fully differentiated mast cells of either phenotype after transplantation into mice,[23] the precise physiological relationship between BMMC and the mast cells found in vivo is not completely understood.

CYTOKINE PRODUCTION BY FcεRI-POSITIVE CELLS

Mast Cells as a Source of Cytokines[1]

As alluded to above, the first observations of cytokine production by FcεRI-positive (FcεRI$^+$) cells arose from studies with AbMuLV-transformed mouse mast cells—it was soon determined that such mast cell lines variably secrete several hemopoietic cytokines. Some constitutively synthesize and secrete granulocyte-monocyte colony stimulating factor (GM-CSF), but not IL-3,[16] while others produce IL-3, IL-6 and GM-CSF.[17] Brown and co-workers reported constitutive IL-3, IL-4 and GM-CSF production by such cells, as well as by spontaneously transformed and nontransformed mouse mast cell lines.[18] The first cytokine activity to be associated with normal mast cells was that of a natural cytotoxic (as opposed to natural killer) cell activity,[24-26] but it was not until 1986 that immunologic evidence strongly suggested that this activity was TNF-α;[27-29] in 1990 it was confirmed by molecular analysis that the mast cell TNF-α-like product was indeed a product of the TNF-α locus.[30] We also reported at that time that very substantial amounts of TNF-α are present in a preformed state in freshly purified mouse serosal mast cells.[30] Since then it has been determined by immunohistochemistry that mast cells can also contain several other preformed cytokines in vivo, including IL-4, IL-5, IL-6 and TNF-α, among others (see Table 11.2).

FcεRI Signaling and Mast Cell Cytokine Secretion

In 1989 four laboratories independently reported that in vitro-derived mouse mast cell lines or clones contain specific mRNA for and/or secrete a wide array of cytokines following cellular stimulation. FcεRI crosslinking on these cells was shown to lead to the expression of mRNA for IL-1,[31] IL-3,[31-33] IL-4, IL-5,[31,32] IL-6,[31,32,34] GM-CSF,[31,33] interferon-gamma (IFN-γ), monocyte chemoattractant protein-1 (MCP-1 or JE), macrophage inflammatory protein-1α (MIP-1α), MIP-1β and TCA3.[31] In addition it was confirmed that these cells secrete significant amounts of IL-1,[31] IL-3,[32,33] IL-4, IL-6,[31,32,34] and GM-CSF.[31-33] Since that time we have determined that activated mouse mast cells may also pro-

[1] *The array of cytokines for which we have evidence of production or potential production by mast cells is presented in Table 9.2.*

duce or at least contain transcripts for numerous other cytokines, including IL-12,[35] IL-13,[36] leukemia inhibitory factor (LIF),[37] TNF-α,[30,38] TGF-β,[39,40] endothelin-1 (ET-1),[41] and the chemokine MARC[42] (the mouse homologue of human MCP-3[43]). Purified mouse serosal mast cells have been shown to produce IL-3,[32] IL-6, TGF-β1,[39,40] and TNF-α,[30,38] while similar cells from rats secrete IL-6[44] and TNF-α.[45] Canine mastocytoma cells also produce TGF-β1.[46] Finally, a small body of evidence suggests that mast cells can secrete at least some cytokines (e.g., TNF-α) in response to crosslinking of their Fcγ receptors.[47]

Cytokine Production by Human Mast Cells

The evidence for cytokine production by human mast cells is not as broad as that for rodent mast cells but there is no doubt that human mast cells represent a considerable source of some of these mediators. TNF-α was first detected (by in situ hybridization) within cultured human cells with a "mast cell/basophil-like" phenotype.[48] This observation was quickly followed by immunohistochemical and other evidence of TNF-α in human skin and lung mast cells[49-53] and confirmation of the FcεRI-dependent release of TNF-α from these cells.[53-55] Ex vivo studies with freshly purified human tissue mast cells and/or mast cell lines have confirmed that human mast cells can secrete IL-3,[56] IL-4,[53,57] IL-5,[53,58] IL-6,[53] and IL-8,[53,59] and further immunohistochemical evidence indicates that bronchial or nasal mucosal mast cells contain preformed IL-4,[51,60-63] IL-5,[51,60,63] IL-6,[51,63] and GM-CSF.[56,60]

Studies with the human mast cell leukemia line HMC-1, which does not express a functional FcεRI, have shown that these cells constitutively contain transcripts for I-309,[64] IL-1β,[65] MCP-1,[64] M-CSF, TGF-β[66] and TNF-α.[65,66] Following cellular activation with phorbol esters and/or calcium ionophores, HMC-1 cells undergo specific mRNA induction and cytokine secretion responses for each of the chemokines I-309, IL-8, MCP-1, MIP-1α, MIP-1β and RANTES,[64] as well as for IL-1β, IL-3, IL-6, GM-CSF, TNF-β and PDGF-A.[66] Recent studies in our laboratory indicate that SCF-dependent human fetal liver-derived cell populations which comprise mostly immature mast cells secrete a number of cytokines following cellular activation with PMA and calcium ionophore A23187, including IL-6,[67] IL-4, IL-5, GM-CSF, TNF-α and TGF-β (JR Gordon, C Rouse, JC Kendall et al, manuscript in preparation).

Cytokine Production by Basophils

The seminal observations on basophil cytokine secretion arose through experiments in Paul's laboratory, where they demonstrated that basophils (or more correctly, FcεRI$^+$ non-B, non-T cells) from fresh bone marrow or spleen could produce IL-4 when cultured on plates coated with IgE antibodies.[20] The levels of non-B, non-T cell IL-4 production are increased substantially by adding IL-3 to the cultures.[20] This inductive effect of IL-3 (and/or other T cell-derived cytokines) is apparently reproduced in vivo in mice with *Nippostrongylus brasiliensis* infections, since increased numbers of non-B, non-T cells can be recovered from these animals during their IL-3-dependent parasite rejection responses.[68] Non-B, non-T cells have also been shown to be a major source of IL-6 in immunized mice.[69] Unlike the growth factor-dependent BMMC which arise from extended culture of bone marrow cells in IL-3, FcεRI-positive non-B, non-T cells do not produce IL-3 following FcεRI crosslinking.[9] Of great interest, too, is the fact that coating the culture flasks with IgG1, IgG2 or IgG3 instead of IgE also subserves IL-4 production by these cells.[9] Non-B, non-T cells purified from human bone marrow produce IL-4 and IL-5 in response to FcεRI crosslinking, but apparently do not produce IL-2, IFN-γ or IL-6.[21]

Freshly purified peripheral blood basophils from humans also secrete IL-4[70] (and perhaps also IL-1[71]) following FcεRI signaling and, as observed with non-B, non-T cells, the addition of IL-3 to the

Table 11.2. List of cytokines known to be induced within or secreted by mast cells (MC) or mast cell lines from mice and humans

Cytokine	Cell Type	Protein[1]	mRNA[1]	Cytokine Effects (targets)	Ref.
endothelin-1	**mouse:** MC lines	inducible: FcεRI	inducible: FcεRI	long-term microenvironmental effects	41
FKBP12[2]	**human:** HMC-1 cells	preformed: FcεRI	?	activates neutrophils	186
GM-CSF	**mouse:** MC lines **human:** HMC-1 cells, lung mast cells	inducible: FcεRI	inducible: FcεRI	multi-potent colony stimulating factor	16-18, 31-33
IFN-γ	**mouse:** MC lines	FcεRI	inducible: some cells, via FcεRI	immunoregulatory cytokine	31
I-309	**human:** HMC-1 cells	PMA	PMA	CC chemokine (monocytes)	64
IL-1	**mouse:** MC lines **human:** HMC-1 cells	inducible: FcεRI	inducible: FcεRI	inflammatory effects	31,65
IL-2	**mouse:** some MC lines	?	inducible in some cells	enhances T cell responses	96,97
IL-3	**mouse:** MC lines, serosal MC	preformed: FcεRI	inducible: FcεRI	MC growth factor	17,18, 31-33
IL-4	**mouse:** MC lines, serosal MC **human:** lung & other MC	preformed: FcεRI	inducible: FcεRI	MC growth/maturation factor, B cell IgG>IgE switch factor	31,32,51, 57,60,62, 63,138, 162
IL-5	**mouse:** MC lines, serosal MC **human:** lung & other MC	preformed: FcεRI	inducible: FcεRI	eosinophil growth and activating factor	31,32,51, 56,60,63, 138,162
IL-6	**mouse:** MC lines **rat:** serosal MC **human:** lung MC	preformed: FcεRI	inducible: FcεRI	augment antibody production, acute phase response...	17,31,32, 34,44,51, 60,63,138
IL-8	**human:** HMC-1 cells, skin MC	PMA/A23187	inducible: FcεRI	CXC chemokine (granulocytes,T cells, endothelial cells, mast cells)	59,64

Table 11.2. Continued...

Cytokine	Cell Type	Protein	mRNA	Cytokine Effects (targets)	Ref.
IL-12	**mouse:** MC lines	?	inducible: SCF	immunoregulatory	35
IL-13	**mouse:** MC line	inducible: FcεRI	inducible: FcεRI	IL-4-like activity	36
MARC (MCP-3)	**mouse:** MC lines, serosal MC	inducible: FcεRI	inducible: FcεRI	CC chemokine (monocytes, eosinophils(?))	42
MCP-1 (JE)	**mouse:** MC lines **human:** HMC-1 cells	inducible: various	inducible: FcεRI	CC chemokine (monocytes, basophils)	31,64
MIP-1α	**mouse:** MC lines **human:** HMC-1 cell line	inducible: various	inducible: FcεRI	CC chemokine (monocytes, T & B cells, granulocytes)	31,64
MIP-1β	**mouse:** MC lines **human:** HMC-1 cells	inducible: various	inducible: FcεRI	chemokine (monocytes, T cells)	31,64
PDGF-A	**human:** HMC-1 cells	inducible: PMA	inducible: PMA	fibroblast mitogen	66
RANTES	**human:** HMC-1 cells	PMA	PMA?	CC chemokine (monocytes, T cells, granulocytes)	64
TCA3	**mouse:** MC lines	?	inducible: FcεRI	inflammatory chemo-attractant (neutrophils?)	31,108, 187
TGFβ	**mouse:** MC lines, serosal MC **dog:** mastocytoma cells.	preformed: FcεRI	preformed	immunoregulatory cytokine, activates fibroblasts	39,40,46
TNFα	**rodent:** MC lines, serosal MC, **human:** HMC-1 cells, skin, lung & other MC	preformed & inducible: FcεRI	inducible: FcεRI	activate endothelial cells, mediate granulocyte & fibroblast recruitment	30,38,40, 45,50,51, 54-56,60, 139
TNFβ	**human:** HMC-1 cells	inducible: PMA	inducible: PMA	T cell cytotoxin	66

[1]Protein or mRNA for some cytokines is constitutively present in some mast cell polulations, while in some its expression can also or must be induced by agents such as PMA or A23187, or by stimulation via the FcεRI.
[2] While the immunophilin FKBP12 is not a cytokine per se, it is secreted by activated mast cells and in turn can activate neutrophils.[185]

cultures increases this IL-4 response.[70] Immunolocalization studies with purified basophils indicate that ≈6% of unstimulated cells contain immunoreactive IL-4, while following in vitro priming with IL-3 and stimulation with anti-IgE or C5a the numbers of IL-4-positive cells climb to ≈19% of the basophil population.[72] The levels of IL-4 production by basophils of different individuals may vary from 40-630 pg of IL-4 per million cells,[73] although this in vitro IL-4 production by cells from low IL-4-producer subjects can be increased substantially by increasing the levels of calcium in the system.[73] The signal transduction mechanisms for IL-4 production versus cellular degranulation (i.e., histamine release) in basophils may be different—the optimal levels of FcεRI signaling for these two functions are substantially different, for some signals that induce high histamine release from these cells (e.g., C5a, fMLP) induce only limited IL-4 production.[73] Finally, a large number of cytokines and chemokines can affect histamine release from basophils,[74-80] but their effects on cytokine secretion by these cells have not been reported.

Cytokines Comprise a Unique Class of Mast Cell Mediators

Mast cell mediators have traditionally been assigned to one of two categories—the preformed, granule-associated components (e.g., proteases, histamine, proteoglycans) that are released in an anaphylactic fashion immediately upon cellular activation or, alternately, the arachidonic acid metabolites (e.g., PGD_2, LTC_4) synthesized for a short period of time following mast cell degranulation, in association with cell membrane-bound enzymes (reviewed in ref. 7). But a review of the evidence demonstrates conclusively that mast cell cytokine secretion patterns do not fit neatly into either of these two categories. Some cytokines are released from IgE and antigen-activated mast cells immediately from a preformed pool and after a brief lag period from a de novo-synthesized pool, while others may be derived only from de novo-synthesized product.

TNF-α might be considered prototypical of those mast cell cytokines released from both preformed and de novo-synthesized pools.[38] The precise levels of preformed TNF-α in mouse mast cells vary roughly in tandem with the granulation status of the cells—some in vitro-derived rodent mast cell lines have poorly developed cytoplasmic granules and contain very little TNF-α, while freshly purified serosal mast cells have large cytoplasmic granules and contain abundant cytokine activity (for example, see ref. 30). Several studies have confirmed the association of TNF-α with the cytoplasmic granules of mast cells.[28,48]

We have shown experimentally that when mouse mast cells (cell lines or freshly purified serosal cells) are activated via the FcεRI, they degranulate as expected, releasing their contents of preformed TNF-α within 30 minutes. In addition, within 15-30 minutes significant levels of TNF-α mRNA begin to accumulate in the cytoplasm (Fig. 11.1A) and, shortly thereafter, the cells begin a second wave of TNF-α secretion that lasts for another 3-4 hours, until after the TNF-α transcription response wanes.[38] When this type of experiment is repeated, except in the presence of the transcription inhibitor actinomycin D, the mast cells release their preformed pool of TNF-α as above (i.e., within 30 minutes), but then TNF-α secretion ceases[38] (Fig. 11.1B). These early studies demonstrated conclusively that mast cell cytokines may be secreted from both preformed and de novo-synthesized pools and predicted that the actions of mast cell cytokines would extend well beyond the immediate degranulation response.[38] Evidence indicates that a similar type of activation-induced secretion of de novo-synthesized TNF-α occurs with freshly purified human skin[54] and lung[55] mast cells challenged ex vivo via the FcεRI.

The intracellular mechanisms by which mast cells direct TNF-α secretion following FcεRI-dependent challenge have re-

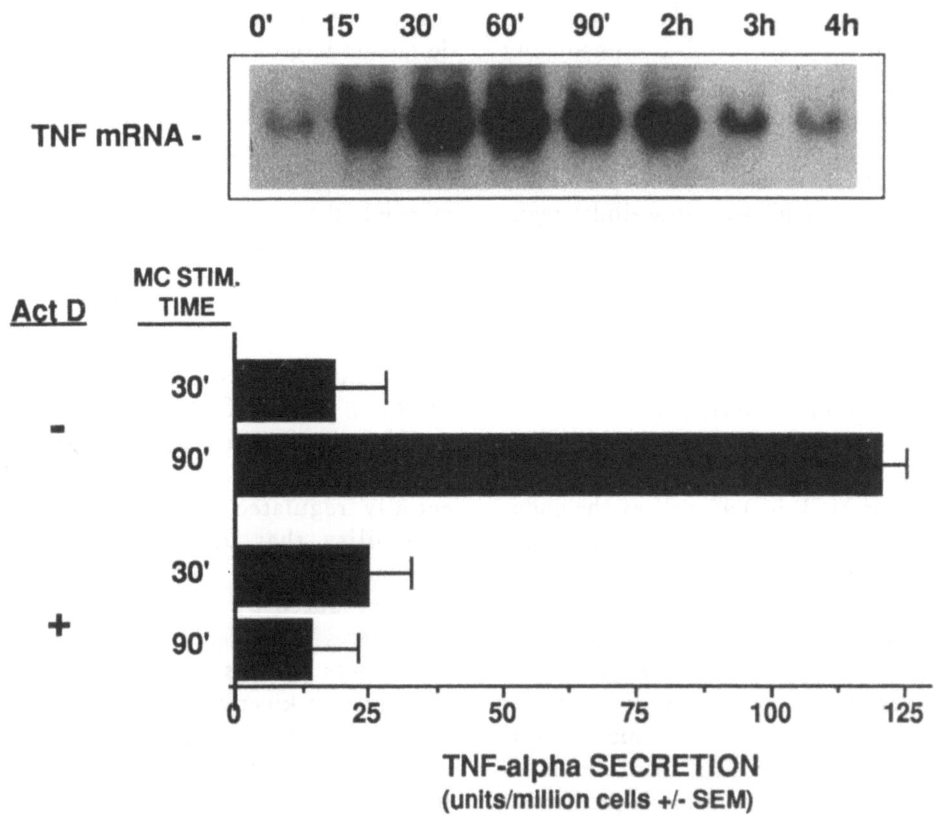

Fig. 11.1. Evidence that the TNF-α secreted by mouse mast cells is derived from both preformed and de novo-synthesized pools of cytokine. Mouse mast cells (Cl.MC/C57.1 cells, upper panel; freshly purified serosal mast cells, lower panel) were sensitized with IgE antibodies and challenged with specific allergen. (Upper panel) *At various times thereafter the mRNA was extracted from the Cl.MC/C57.1 cells and processed for Northern analysis of TNF-α mRNA expression, using a* ^{32}*P-labeled cDNA probe.* (Lower panel) *The serosal mast cells were challenged with antigen in the absence (Act D-) or presence (Act D+) of the transcription inhibitor actinomycin D and, either 30 or 90 minutes later, the cell supernatants were harvested and assayed for TNF-α bioactivity using an L929 cell cytotoxicity assay.*

The results confirm that the Cl.MC/C57.1 cells undergo a strong TNF-α transcription response following FcεRI signaling (upper panel), but when these cells were challenged in the presence of actinomycin D, TNF-α mRNA expression remained at background levels (not shown). Similarly, in freshly isolated serosal mast cells, TNF-α is secreted from an actinomycin D-insensitive (i.e., preformed) pool of cytokine that is released within 30 minutes and from an actinomycin D-sensitive (i.e., transcription-dependent) pool that is secreted later on. Actinomycin D did not affect the secretion of serotonin from these cells (not shown).

cently been addressed experimentally using RBL cells, which do not contain preformed TNF.[81] FcεRI challenge of RBL cells induces immediate degranulation (i.e., hexosaminidase release), followed some 30-60 minutes later by the appearance of TNF-α in the culture medium. Brefeldin A treatment of the cells either before or 30 minutes after activation completely blocks the TNF-α *secretion* and thereby results in a buildup of cytokine within the cells.[81] (Brefeldin A selectively disrupts Golgi membranes and vesicular trafficking, but doesn't affect degranulation responses.) Thus, these data indicate that RBL cells process their TNF-α through the Golgi apparatus and ultimately secrete it through a vesicular transport mechanism, without a need for routing through the intracellular cytoplasmic granule system.[81] However, for other mast cells that contain preformed TNF-α within their cytoplamic granules and therefore secrete TNF-α immediately from these cytoplasmic granules as well as for more prolonged periods via vescicular transport (e.g., mouse serosal or human tissue mast cells), it would seem that there must be as yet undefined intracellular regulatory mechanisms that govern the routing of newly synthesized cytokine into either the vesicular transport system (for activated cells secreting cytokine) or nascent cytoplasmic granules (for cells recovering from prior activation[82]).

Other patterns of cytokine secretion also exist in mast cells. The chemokine MARC[42] (or MCP-3[43]) as well as the endothelin ET-1[41] are both found at significant levels in unstimulated mouse mast cells. However, while FcεRI signaling induces their release, this is a greatly delayed process that does not occur at the same time as cellular degranulation, so that neither cytokine is detectable in the supernatants of the activated cells until many hours after cellular challenge. A similar production pattern has been reported for IL-5 in freshly purified human lung mast cells.[51,58,60] For each of these cytokines, FcεRI signaling induces specific mRNA expression in the cells.[41,42,58] Quite distinct from the pattern for TNF-α or IL-5 is that of IL-8, which is not detectable in quiescent human skin mast cells but can be detected within these cells many hours after activation via the FcεRI.[59] Consistent with this observation is the fact that IL-8 secretion by HMC-1 cells does not begin until at least 4 hours following challenge with PMA and A23187, and is accompanied by a strong IL-8 transcriptional response, with peak mRNA levels being achieved at ≈8 hours and maximal cytokine secretion occurring over 24 hours.[59]

Finally, classical degranulation responses are not necessarily a prerequisite for cytokine secretion by mast cells. While IL-6 is secreted from rodent mast cells following activation via the FcεRI,[31,32,34,44] it has recently been shown that freshly purified rat serosal mast cells also become robust secretors of IL-6 following exposure to either PGE_1 or PGE_2, even though they do not undergo degranulation or otherwise release histamine in response to these agonists.[83] Thus, while the signaling pathway(s) that leads to each of the observed modes of cytokine secretion has not been mapped, the observations do clearly indicate that cytokine secretion in mast cells must be regulated by multiple distinct mechanisms.

Can Mast Cells Differentially Secrete Their Cytokines?

The suggestion that the cytokines expressed by mouse mast cells are differentially regulated also provides for the possibility that under the appropriate circumstances they could also be differentially produced by these cells. One of the first pieces of evidence in support of this notion was the observation that the overall induction kinetics of mast cell cytokines very roughly fall into two overlapping categories, with some undergoing more rapid induction and others being induced with a slightly delayed time course[84,85] (e.g., see Fig. 11.2). Following FcεRI-dependent challenge of either growth factor-dependent or -independent mouse mast cell lines or primary cultures of BMMC, mRNA for the more "proinflammatory" mast cell cytokines (i.e., IL-6,[84,85] MARC,[42] MCP-1, MIP-1α, MIP-1β, TCA3[84,85] and TNF-α[38]) can be detected in the cellular cytoplasm within 15-30 minutes. The levels of specific cytokine mRNA progressively increase in the cells to a peak at about 60-120 minutes and then wane relatively rapidly thereafter.[84,85] As might be expected, however, even within this group substantial differences in induction kinetics may be observed (e.g., MIP-1α versus MCP-1; Fig. 11.2). Somewhat distinct from this can be the mRNA induction kinetics for the more "growth factor-like" cytokines (e.g., IL-3, IL-4, IL-5, GM-CSF) which reportedly are first detectable in the cytoplasm of IgE and antigen-stimulated BMMC at 60-90 minutes post-challenge, reach peak levels at 2-4 hours, and then decline there-

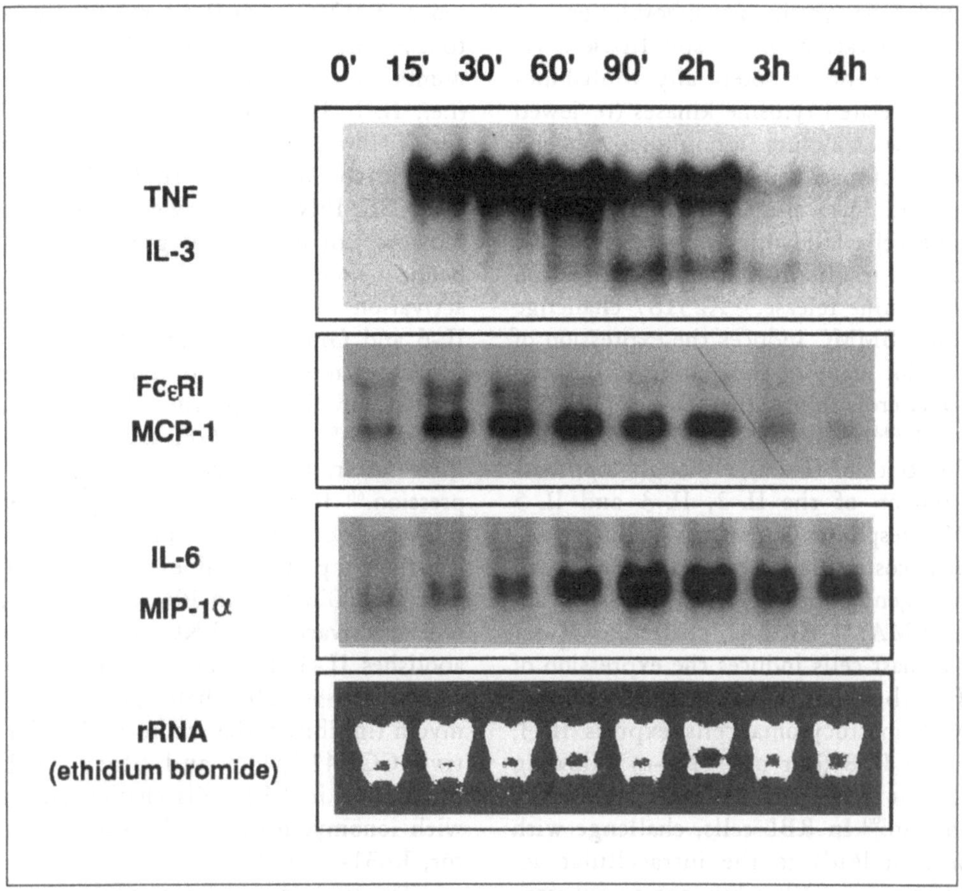

Fig. 11.2. Differential cytokine mRNA levels and expression kinetics in a mast cell population stimulated via the FcεRI. Cl.MC/C57.1 cells were sensitized with IgE antibodies and challenged with specific antigen and, at varying times thereafter, the mRNA was extracted from the cells and processed for Northern analysis (5 μg of total cellular RNA/lane). The blots were sequentially probed with pairs of ^{32}P-labeled cytokine cDNAs, stripped by stringent washing and then reprobed, for a total of three depicted probing cycles. The cDNA pairs used were TNF-α/IL-3, FcεRI-β chain/MCP-1, and IL-6/MIP-1α. In each case, the probes correctly identified mRNA species of the correct size (not shown). Autoradiograph exposure times: TNF-α/IL-3, ≈36 hours; FcεRI-β/MCP-1, ≈18 hours; IL-6/MIP-1α, ≈18 hours.

It can be seen that the TNF-α mRNA signal is substantially stronger than that for IL-3 and that the TNF-α mRNA response becomes detectable at ≈15', peaks at ≈60' and then wanes thereafter, while the IL-3 response begins at ≈60', peaks at ≈90'-2 hours, and then wanes. Similarly, increases in the accumulation of mRNA for MCP-1, IL-6, and MIP-1α are also detectable at 15' postchallenge. While the kinetics for MCP-1 induction largely mirror those for TNF-α, the IL-6 and MIP-1α kinetics is slightly delayed in these cells, and the IL-6 mRNA levels are low relative to MCP-1 and MIP-1α. The FcεRI-β mRNA expression is moderately downregulated by FcεRI signaling.

after[85] (see also Fig. 11.2). In general the peak levels of "proinflammatory" cytokine mRNA which accumulate in mouse mast cell lines or primary cultures are greater than those of the "growth factor-like" cytokines[85] (Fig. 11.2).

The hypothesized differences between these two groups may be stretched beyond their kinetics and levels of induction, to their differential utilization of the available intracellular signaling pathways and their variable sensitivity to the transcription inhibitor cycloheximide. FcεRI triggering of mast cells activates multiple signal transduction mechanisms, including the phosphoinositide pathway, with its mobilization of intracellular calcium ions and activation of protein kinase C (PKC). These events are initiated upon tyrosine phosphorylation and thereby activation of phospholipase C-γ1, via recruitment of potentially multifunctional *src*-related tyrosine kinases (reviewed in ref. 86). Calcium ionophore A23187 treatment of mast cells mimics many of the changes in intracellular calcium observed in these cells following activation via the FcεRI, and results in cellular degranulation and histamine release.[87] A23187 challenge of mouse BMMC induces the expression of mRNA for many of the cytokines induced by FcεRI crosslinking (i.e., IL-3,[88,89] IL-5,[89] GM-CSF,[88,89] MIP-1α, MIP-1β, TCA3 and TNF-α, but not IL-1[89]), although "optimal" stimulation of the IL-2, IL-3 and IL-5 mRNA responses apparently requires additional costimulation with the PKC-activating agent phorbol-12, 13-myristate acetate (PMA).[84] A23187 challenge of rat serosal mast cells induces the expression of TNF-α (but not IL-6),[44] while human HMC-1 mastocytoma cells express IL-3, IL-4 and IL-8 (but not TNF-α) following treatment with the calcium ionophore ionomycin.[90] In RBL cells, challenge with ionomycin leads to the intracellular accumulation of TNF-α but not to substantial cytokine release.[81] However, even when these cells are stimulated with both ionomycin and PMA, their TNF secretion is still only a fraction of that observed following challenge via the FcεRI.[81] And challenge of mouse growth factor-independent cells with ionomycin reportedly leads to a differential induction of IL-3 and IL-4 mRNA responses—IL-3 production progresses for some 24 hours after cellular stimulation but IL-4 secretion wanes within 4 hours,[91] as it does following FcεRI-crosslinking or challenge with A23187.[91] Taken together, these data indicate that, while these responses can vary substantially with different mast cell populations and ionophores, calcium mobilization nevertheless leads to a relatively broad-spectrum cytokine induction response.

Selective activation of the PKC pathway in mast cells may have a different effect on the cytokine response. Challenging mouse BMMC with PMA reportedly leads to a preferential induction of mRNA for a number of the "proinflammatory" cytokines (i.e., IL-1, IL-6, MIP-1α, MIP-1β, TCA3 and TNF-α), but not the "growth factor-like" cytokines (i.e., IL-3, IL-4, IL-5, GM-CSF, IFN-γ)[84,85] (reviewed in ref. 89). However, in at least one growth factor-dependent mouse mast cell line, PB-3c, PKC activation induces the expression of both IL-6 and GM-CSF,[92] and in HMC-1 cells PMA treatments lead to dramatic increases in IL-1 and IL-3 expression, lesser increases in GM-CSF and PDGF, and still smaller increases in IL-6, TNF-α and M-CSF expression.[66] It has also been reported that selective activation of protein kinase A leads to a preferential expression of IL-3 over IL-6/GM-CSF in PB-3c cells.[92] Selective inactivation of PKC enzyme activity abolishes IL-3, IL-6 and GM-CSF expression in PB-3c cells challenged with ionomycin (inhibitor, the staurosporine derivative CGP41251)[92] and blocks TNF-α production in RBL cells challenged either with ionomycin or via the FcεRI (inhibitor, Ro31-7549).[81]

The protein synthesis inhibitor cycloheximide has little if any detrimental effect on the induction of mRNA for MIP-1α, MIP-1β, MCP-1, TCA3,[31,89] or TNF-α,[30] but the FcεRI-induced "growth factor-like" cytokine mRNA responses (i.e., IL-1, IL-3, IL-4 and GM-CSF) are reportedly inhibited in BMMC by cycloheximide,[89] indicating that protein synthesis is required for accumulation of mRNA for these cytokines in mouse BMMC. On the other hand, the addition of cycloheximide to PB-3c mouse mast cells several hours after cellular challenge weakly protects existing stocks of the IL-3 and GM-CSF mRNA from rapid degradation,[88] indicating that the IL-3 and GM-CSF mRNA degradative processes require the input of de novo-synthesized proteins (i.e., either ribonucleases or ribonuclease-regulating proteins).

This arguable division of the mast cell cytokines into two groups (i.e., "pro-inflammatory" versus "growth factor-like") based on the signaling pathways by which they may be preferentially induced in mouse BMMC suggests the possibility that as yet undefined pathogenic stimuli could exist which lead mast cells to secrete inflammatory mediators such as IL-6, MARC, MCP-1, MIP-1α, MIP-1β, TCA3 and TNF-α, but not, or to a lesser extent, hemopoietic cytokines such as IL-3, IL-4, IL-5, GM-CSF and IFN-γ. Such putative differential cytokine production could bias the role(s) the mast cell plays in the balance between pathology and normal physiology. An important point to remember here however is that the relationship between cultured BMMC and mast cells in mice or humans is not well understood, so that hypothetical extrapolations from work with growth factor-dependent mast cells to the pathophysiology of mast cell-dependent or other responses in vivo must be undertaken with great care. Indeed the few studies beyond the preliminary reports on differential cytokine expression in mouse BMMC suggest that this differential induction (i.e., inflammatory versus growth factor-like cytokines) may not be a hard and fast rule for all mast cell populations or stimuli. A case in point is that of freshly purified rat serosal mast cells, which produce IL-6, but not TNF-α following stimulation with PGE_1 or PGE_2.[83] Thus, a thorough examination of these issues using freshly purified mast cells may be quite instructive in this regard.

An alternate and intriguing consideration related to a potential differential secretion of cytokines by mast cells is the question of whether the phenotype of the mast cell could affect its cytokine secretion pattern. Mouse BMMC that are cultured in IL-3 or SCF preferentially express phenotypic markers of "mucosal" or "connective tissue" mast cells respectively;[15] they also reportedly express preferentially mRNA for IL-4 or IL-12, as determined by RT-PCR, such that BMMC that are cultured in IL-3-containing medium express mRNA for IL-4, but not IL-12, and cells grown in SCF express mRNA for IL-12, but not IL-4.[35] Cells cultured in both SCF and IL-3 express a mucosal mast cell phenotype,[93] and have low levels of IL-4 transcripts (relative to cells grown in IL-3 alone) and no IL-12 mRNA.[35]

Human mast cells may also express a phenotypic heterogeneity with respect to cytokine production. A recent study of allergic rhinitis patients suggests that of their bronchial and nasal mucosal mast cells which contain immunoreactive IL-4, 85% are MC_{TC} and 15% are MC_T, but virtually all of the IL-5- and IL-6-positive cells are MC_T. The same report indicates that approximately 90% of purified human foreskin mast cells, comprising almost entirely MC_{TC}, are positive for IL-4, and only a very small number express weak immunoreactivity for IL-5 or IL-6.[63] These observations are consistent with a suggestion that MC_{TC} produce IL-4 and very little IL-5 or IL-6,[63] but it must be remembered that this study addressed the presence of <u>preformed</u> cytokine in "unstimulated" lung and skin mast cells.[63] It is noteworthy then that freshly purified human lung mast cells release little if any IL-5 over the first 4 hours following challenge via the FcεRI, but thereafter they secrete abundant IL-5 activity (i.e., $\approx$1500 pg/10^6 cells over 48 hours).[58] Thus, a question arises as to whether the observed heterogeneity of IL-5 expression by MC_T and MC_{TC} could feasibly be associated with a differential storage of preformed cytokine by the two populations and not necessarily with the potential to secrete cytokine(s) following cellular activation.

MOLECULAR REGULATION OF CYTOKINE SECRETION IN MAST CELLS

The observation that calcium ionophores can induce the production of many of the cytokines associated with FcεRI signaling in mast cells indicates that the intracellular calcium levels are an important part of cytokine signaling in these cells. Indeed, the addition of the calcium chelating agents EDTA[94] or EGTA[81] to mast cells undergoing FcεRI-dependent responses blocks subsequent cytokine secretion, and the use of calcium-deficient medium during IgE and antigen challenge of mouse BMMC substantially reduces the IL-4 mRNA response.[95] That calcium is important to FcεRI-dependent mast cell cytokine production may also be indicated by the sensitivity of this process to cyclosporin A (CSA).[31,84,96-99] CSA is a clinically important immunosuppressive agent and has been used extensively in the experimental analysis of cytokine expression in T cells.[100] Like the macrolide antibiotic FK506, CSA acts by forming a complex with specific intracellular immunophilins, which then bind to and inactivate calcineurin, a calcium- and calmodulin-dependent phosphatase that is critically involved in numerous cellular activation processes, including the generation of some transcription activating factors.[100] The CSA-specific immunophilin is called cyclophilin, while the FK506-specific immunophilin is FKBP12.[100] CSA is able to block GM-CSF,[84,98] IL-2, IL-3, IL-4,[96] IL-6,[84,98] MCP-1, MIP-1α, MIP-1β, TCA3[84] and TNF-α[98,99] induction mediated by the FcεRI,[84,98,99] A23187[84] or by combined PMA/A23187[96] challenge of mouse mast cell lines. In addition, it has been demonstrated that CSA can block the mast cell-dependent induction of TNF-α in vivo during cutaneous allergic responses in mice.[99] Comparisons of the doses of CSA required to inhibit calcineurin activity (IC_{50} = 8 nM) versus those required to inhibit IL-1 (IC_{50} = 4 nM), TNF-α (IC_{50} = 65 nM) or IL-6 (IC_{50} = 130 nM) in mouse growth factor-dependent mast cells suggest that IL-1 expression may be largely dependent on or related to calcineurin expression, while TNF-α and IL-6 expression are much less dependent on this enzyme.[98] And we have observed that MCP-1, MIP-1α, MIP-1β and TCA3 responses are also blocked at a dose of CSA that is largely ineffective in blocking IL-6 induction in the same cells.[31] While FcεRI- and A23187-mediated mast cell cytokine responses are sensitive to CSA, at least some PMA-induced cytokine responses (i.e., MIP-1α, MIP-1β and TCA3) are insensitive to its effects.[84] On the other hand, since BMMC do not express FKBP12, these mast cells are resistant to the effects of FK506,[98] although they can be made sensitive by molecular transfection with FKBP12-expressing plasmids.[101] However, some cloned mast cell lines are naturally sensitive to the effects of FK506, while others may be only partially susceptible—FK506 completely blocks IL-2, IL-3, IL-4 and GM-CSF production in MC/9 mast cells, but while it blocks the IL-3 and GM-CSF responses in MCIII mast cells, it has no effect on IL-2 and IL-4 in these cells.[96,97] The related macrolide antibiotic rapamycin, which works in a subtly different manner from CSA or FK506,[100] has no inhibitory effect on mast cell cytokine production.[96] Indeed, rapamycin apparently reverses the inhibitory effects of FK506 on MC/9 and MCIII mast cells, but not those of CSA.[96] The general effects of CSA, FK506 and rapamycin on mast cell biology have recently been reviewed in detail.[102]

Regulation of Cytokine Gene Expression

Analysis of cytokine expression in mast cells has also shown that the regulatory processes for these mediators variably occur at the transcriptional, post-transcriptional or translational levels, or combinations thereof. IL-3 and GM-CSF are constitutively transcribed in growth factor-dependent mouse mast cells, although specific mRNA cannot be detected in unstimulated cells by Northern blotting.[33,88] (IL-3 mRNA can be detected by RT-PCR technology in some, but not all unstimulated mouse serosal mast cells.)[103] Although a low level upregulation of GM-CSF transcription takes place following A23187 challenge, much of the GM-SCF and all of the IL-3 expression in these cells arises through a calcium-dependent stabilization of the constitutively transcribed mRNA. Thus IL-3 expression in these mast cells is regulated at the post-transcriptional level, while GM-CSF is subject to both transcriptional and post-transcriptional regulation.[88] TGF-β mRNA is present at very high levels in resting mast cells, and cellular activation does not result in a substantial increase in specific mRNA levels in these cells,[39] although it can induce TGF-β secretion in some populations.[40,46] Thus this cytokine can be regulated at the translational and/or post-translational levels in these cells.

A number of laboratories have begun the process of mapping the genomic regulatory domains for mast cell cytokine expression. At this point in time data are available only for IL-4,[104,105] IL-5,[106] MARC (MCP-3)[107] and TCA3[108]—the evidence suggests that in each case transcriptional enhancement is an important feature of their expression. Although IL-4 is constitutively transcribed in some transformed mast cell lines,[18] this does not occur in non-transformed cells.[104] Multiple transcription enhancer elements reside within the promotor upstream of the transcription initiation site (TIS) for the IL-4 gene in non-transformed growth factor-dependent mast cells. In addition, two enhancer elements with binding specificities for the nuclear transcriptional activating factors GATA-1, GATA-2 and Pu.1 are located downstream of the TIS, within intron 2 of the gene.[104,105] This DNAse I-hypersensitive intronic enhancer element is a mast cell-specific feature of the IL-4 gene.[104,105] Thus, the gene for at least one cytokine that is produced by both mast cells and T cells is decidedly differentially regulated. The human IL-5 promotor contains a large number of potential (i.e., consensus) transcription factor binding sites, but in mouse mast cells only two enhancer elements have been identified.[106] One is a constitutive, weakly active GATA binding site (located at position -80 to -70) and the other is a strong, FcεRI-dependent inducible enhancer element with binding specificity for an NF-AT-like factor. This mast cell NF-AT-like factor is related antigenically to and shares at least partial sequence identity with the NF-AT of T cells.[106] Although AP-1 binding activity (or at least *c-jun* and *c-fos*) is induced in mast cells stimulated via the FcεRI,[106,109] no AP-1 binding is observed within the IL-5 promotor. Thus, FcεRI-dependent IL-5 gene transcription is driven in these cells by an NF-AT-like enhancer supported by a relatively weak, constitutively active GATA-binding enhancer element.[106] As noted above, the GM-CSF gene is constitutively transcribed in mast cells and this expression is substantially enhanced by cellular activation,[88] although the nuclear regulatory mechanisms in mast cells have not been examined. The gene product for the chemokine MARC (MCP-3), a member of the RANTES/sis gene family, has to date only been found in mast cells.[107] It is relatively strongly regulated at the transcriptional level, but some post-transcriptional regulation also occurs,[107] probably in part through the AUUUA mRNA "destabilization" motif expressed in the 3' untranslated region of the mRNA for IL-5,[107] as well as numerous other cytokines.[110] Promotor expression analysis indicates that the promotor sequence spanning positions -113 to -72 relative to the TIS are needed for full FcεRI-dependent MARC gene activation,[42] and an AP-3 or AP-3-like transcriptional factor binding site seems to be largely responsible for driving MARC gene expression.[107] AP-3 is known to enhance IL-2 and IL-2R gene transcription in T cells.[111] Finally, evidence indicates that the TCA3 gene is also transcriptionally regulated in mast cells.[108] As mapped to date, the TCA3 promotor comprises a relatively weak enhancer element between positions -136 to -82, a putative NF-κB binding site between positions -324 to -136, a strong FcεRI-dependent enhancer element between positions -1324 to -324 and a putative inhibitory element between positions -2000 to -1324.[108] Taken together, these data clearly indicate that there is at least one difference between mast cells and T cells in their regulation of cytokine gene expression.

Effects of Exogenous Cytokines on Mast Cell Cytokine Production

The activation of mast cells in vivo often takes place in the context of inflammatory diseases in which these cells are constantly exposed to an ongoing symphony of the host's immunomodulatory mediators. Mast cells express receptors for and are potentially reactive to a large number of ligands including, for example, cytokines, neurotransmitters, extracellular matrix molecules and many others. Cytokines themselves have a broad spectrum of effects on the mast cell, ranging from overt activation to modulation of the cell's responses to third party stimuli. For example, injection of SCF,[74] MCP-1 (10^{-7} M) or MIP-1α (10^{-7} M)[112] into the skin of humans or mice causes substantial local mast cell degranulation. In vitro, high concentrations of SCF also directly induce histamine secretion, but not leukotriene production by purified human lung mast cells; lower concentrations of this cytokine do not directly activate these cells in this fashion, but they do potentiate the histamine release induced by FcεRI crosslinking.[113,114] A number of other mediators, many of which are potent histamine-releasing factors (HRF) for human basophils,[115] do not release histamine from human lung, skin or uterine mast cells—the list of inactive cytokines includes GM-CSF,[114] GRO,[116] IL-1β,[114] IL-3, IL-4,[113,114] IL-5,[114] IL-8,[116] IL-9,[113] IP-10, MCP-1, MIP-1α, MIP-1β,[116] NGF[117] and RANTES.[114,116] Nor do chemokines from this list induce the expression of other mast cell 'activation' markers (i.e., CD11b/C3biR, CD45/IL-2R, CD63, IL-3Rα or CD117/c-kit) when added to HMC-1 cells.[116] The effects of SCF by itself on the induction of cytokine production by mast cells have not been addressed, but it has been reported that purified human lung mast cells do not produce IL-4 after FcεRI crosslinking unless the cells are maintained in SCF-containing medium.[58] Interestingly, IL-5 production by these cells has no such requirement for SCF.[58] The neurotropic cytokine nerve growth factor (NGF) is the only cytokine of which we are aware that has been reported to directly effect cytokine induction responses in mast cells—it transiently induces the appearance of MIP-1α mRNA in growth factor-dependent mouse BMMC.[118]

A number of other cytokines also do not have the capacity to directly induce mast cell degranulation but nevertheless do potentiate their responses to FcεRI crosslinking. For example, long-term incubation (i.e., 48 hours) of mouse serosal mast cells with either IL-3 or IL-4 significantly augments the IgE-dependent release of serotonin from these cells.[119,120] Although calcium ionophores can to some extent mimic the effects of FcεRI challenge in mast cells, the potentiating effects of IL-3 versus IL-4 on mast cells challenged via the two routes are quite different—IL-4 augments calcium ionophore-induced serotonin secretion from mouse serosal mast cells, but IL-3 has no such effect (and SCF actually dampens this response).[120] Both IL-3 and IL-4 also affect cytokine production by mast cells. Preculture of growth factor-dependent mouse mast cells with IL-3, as opposed to IL-4, leads to a 10- to 100-fold increase in cytokine production in response to FcεRI crosslinking,[121] mediated perhaps in part through IL-3 receptor occupancy altering the spectrum of proteins normally tyrosine phosphorylated upon FcεRI-dependent activation of these cells[121] and thereby intensifying the activation cascade. When IL-4 is added to cultures of resting human HMC-1 mastocytoma cells the levels of IL-3 and IL-8 mRNA in the cells increase very slightly,[90] but their baseline levels of IL-4[90] and TNF-α[65,90] mRNA remain unaffected, and their constitutive expression of IL-1β mRNA is dampened.[65] IL-4 treatment of these cells for 24 hours augments their expression of IL-3 and IL-8 (but not IL-4 or TNF-α) mRNA in response to ionomycin challenge.[90] IL-4 receptor occupancy in HMC-1 mastocytoma cells induces the cellular expression of *c-fos* and CD54 (the intracellular adhesion molecule ICAM-1) on the cell surface, and reduces the expression of *c-kit*.[122] FcεRI-dependent activation of mast cells also has the effect of down-regulating their expression of *c-kit*[123] and at least the β chain of the FcεRI complex (see Fig. 11.2).

IFN can dampen some mast cell responses, including the direct TNF-α-dependent killing of WEHI-164 tumor cells by rat peritoneal mast cells.[45] In this system, mast cell activation is not required for target cell killing, for the mast cells constitutively express their TNF-α. Both IFN-α/β and IFN-γ suppress this cytotoxic activity, and their suppressive effects are additive.[45] IFN-γ can also affect mast cells in other ways, including the downregulation of MHC class II molecule expression.[124] Preliminary evidence from Marshall's laboratory indicates that IL-10 can inhibit the production of IL-6 by rat peritoneal mast cells challenged via the FcεRI, or with LPS or PGE_2. When IL-10 is added to the cultures at 5 ng/ml, the IL-6 production by these mast cells is reduced by ≈75%, although at substantially lower concentrations, IL-10 can still significantly suppress IL-6 production (J Marshall, 1995; personal communication).

Effects of Non-Cytokine Mediators on Mast Cell Cytokine Production

There is an increasing body of evidence that neurotransmitters can modulate or initiate mast cell responses. A large number of these mediators, including substance P, calcitonin gene-related product (i.e., CGRP), neurokinin A and others can induce wheal and flare responses upon injection into humans (reviewed in ref. 125), and substance P has been shown to directly induce mast cell degranulation upon intradermal injection into mice (reviewed in ref. 126). At least one neurotransmitter can affect cytokine expression in primary cultures of mouse BMMC—the addition of substance P (10^{-4}M) to these cells results in a moderately low level induction of MIP1α mRNA, which peaks at 2 hours post-challenge and is back to baseline by 4 hours.[118] These effects occur in the absence of detectable histamine release.[118] And substance P has been shown also to induce the secretion of TNF-α from isolated mouse serosal mast cells and growth factor-dependent mouse mast cell lines, but it has no effect on the expression of IL-1, IL-3, IL-4, IL-6 or GM-CSF in these cells.[127]

As noted earlier, prostanoids can also affect mast cell cytokine production. The addition of PGE_1 or PGE_2 (but not PGD_2) to cultures of rat peritoneal mast cells directly induces IL-6 secretion[83] and, like IFNα/β and -γ,[45] dampens the constitutive release of TNF-α by these cells.[83] Both PGE_1 and PGE_2 additively affect FcεRI-dependent IL-6 production and synergistically increase endotoxin-induced IL-6 secretion.[44,83] Thrombin, another mediator produced within inflammatory reactions, also can affect cytokine expression in mast cells. Like substance P and NGF, the addition of thrombin (2 μg/ml) to mouse BMMC reportedly induces a transient expression of MIP-1α mRNA within the cells.[118]

Mast cells also express receptors for extracellular matrix molecules (i.e., β integrins), including those for laminin, fibronectin and vitronectin,[128-134] and some of these can also influence mast cell cytokine production. For example, TNF-α release from IL-3-dependent PT-18 mouse mast cells is significantly enhanced when the cells are challenged via the FcεRI in the presence of fibronectin.[135] Since fibronectin receptor occupancy does not affect the levels of TNF-α mRNA in these cells,[135] but does markedly enhance FcεRI-mediated mast cell secretory responses,[128] it might be assumed that it also leads to a more robust release of *preformed* cytokine. However, fibronectin signaling does substantially increase the levels of IL-3 mRNA constitutively expressed in PT-18 cells.[135] The observed effects of fibronectin on TNF-α release can be blocked by the CS-1 fibronectin receptor antagonist peptide, suggesting that VLA-5 is the dominant fibronectin receptor for this response.[135]

Inhibition of Mast Cell Cytokine Production

The appreciation that mast cells secrete multiple cytokines which are important to the pathogenesis of allergic diseases has also rekindled questions about the effector mechanisms of some of the agents used to treat atopic diseases. For example, corticosteroids are a mainstay in the treatment of chronic allergic diseases, and while they inhibit multiple functions of other cells involved in allergic reactions (e.g., lymphocytes, eosinophils), they do not interfere with mast cell degranulation responses (reviewed in ref. 136). Observations such as this may contribute to a conceptual minimization of the overall roles of mast cell secretion products in the pathogenesis of atopic diseases. Recent evidence indicates however that mast cell cytokines do play a highly significant role in allergic pathology (below) and that steroids and other immunosuppressive agents directly affect the production of at least some mast cell cytokines. Thus, a significant proportion of the effects of these agents on the pathogenesis of atopic responses could potentially be related directly to their effects on *mast cell* cytokines.

We have already discussed the broad scale inhibition of mast cell cytokine secretion by agents like CSA—it can block the expression of essentially all cytokines in mouse mast cells stimulated via the FcεRI, although some like IL-6 are much less sensitive to its effects than others.[31,98] The corticosteroid dexamethasone also affects cytokine secretion by these cells. When added to cultures of mouse growth factor-dependent mast cells or purified rat serosal mast cells, it blocks TNF-α[99,137] and IL-6 [83] production respectively in response to FcεRI crosslinking. The effects of glucocorticoids on mast cell cytokine production may be quite selective. Thus, while methyl-prednisolone blocks MCP-1 expression in human HMC-1 mastocytoma cells challenged with PMA, it does not affect the expression of several other chemokines, including I-309, IL-8, MIP-1β, MCP-1 or RANTES.[64] A differential suppression of mast cell cytokine expression may also occur in vivo. It has been reported that IL-4 and IL-6 are expressed primarily by mast cells (i.e., not by lymphocytes or other cells) in the nasal mucosa and submucosa of patients suffering from allergic rhinitis.[138] Fluticasone therapy substantially reduces the seasonal hypertrophy of IL-4-positive mast cells in these patients, but does not affect the numbers of IL-6-positive mast cells,[138] so it has been hypothesized that this effect may reflect a steroid suppression of IL-4, but not IL-6, expression in these mast cells.[138] Dexamethasone has been shown to block the FcεRI- and mast cell-dependent expression of TNF-α in the skin of allergen-challenged mice and also the associated mast cell TNF-α-dependent late phase cellular influx into the lesions.[99,137] However, dexamethasone blocks not only mast cell TNF-α expression in these lesions, but also the abilities of normally TNF-α-responsive cells in the tissue to respond to this cytokine.[99] Unfortunately, these broad-spectrum effects of corticosteroids on the full array of tissue cells will reduce their usefulness as tools for dissecting the roles of *mast cell* cytokines in disease processes.

MAST CELL CYTOKINE EXPRESSION IN VIVO

Animal Model Systems

We have discussed the ex vivo production of cytokines by mast cells isolated from the serosal cavities of rodents—these cells have been shown to produce IL-3, IL-6, TGF-β and TNF-α under the influence of the FcεRI (above). The analysis of cytokine production by mast cells in vivo is of course substantially more problematic but, nevertheless, the few studies that have been done have been very fruitful in terms of providing insights into the extent to which these cells regulate atopic responses.

The first analyses in animals were aimed at testing the hypothesis that TNF-α of mast cell origin was important to the induction of allergic late phase responses (LPR). They were partly fueled by the knowledge that anti-TNF-α antibodies could block endothelial leukocyte adhesion molecule (ELAM-1) expression in human skin challenged ex vivo with mast cell secretagogues.[49] It was known at that time that: (a) mast cells freshly isolated from mice produce abundant TNF-α in response to FcεRI crosslinking;[30,38] (b) TNF-α is a strong inducer of ELAM-1 expression by vascular endothelial cells; and (c) ELAM-1 expression is important for the recruitment of neutrophils to sites of inflammation.[139] We found that specific allergen challenge of IgE-sensitized BALB/c and WBB6F1-+/+ (+/+) mice results in the expression of TNF-α mRNA within the reaction sites. However, TNF-α induction does not occur in tissues of genetically mast cell-deficient W/W^v mice that are treated in an identical fashion, unless the animals are first reconstituted with mast cells from the +/+ mice.[38] At about the same time we reported that the intradermal injection of recombinant TNF-α induces delayed neutrophil infiltration responses in the skin of BALB/c mice, and also that W/W^v mice do not develop a neutrophilic LPR 4-8 hours after specific allergen challenge.[140] This pathogenic deficiency in the mast cell-deficient mice could be reconstituted by transplantation of mast cells from littermate +/+ mice and, conversely, it was also shown that the late phase neutrophil influx could be significantly reduced by injection of anti-TNF-α antibodies at the time of allergen challenge.[140] Dexamethasone or CSA treatments of these mice block both the IgE- and mast cell-dependent TNF-α mRNA induction in the skin, as well as the delayed neutrophil influx.[99,137] These results confirmed the importance of mast cells for both IgE and antigen-dependent TNF-α induction and the delayed neutrophil recruitment in the skin of mice. They also documented that TNF-α largely mediates this neutrophilic component of the LPR. They do not rule out roles for other mediators—indeed it seems evident that other mediators would be involved in the neutrophil response, for TNF-α antibody alone does not completely abolish the late phase neutrophil influx (reviewed in ref. 141).

TNF-α is a pleiotropic cytokine with a capacity to activate numerous cells within the mast cell's microenvironment. Many of these could potentially contribute to the progression of the atopic responses through elaboration of their own cytokines. During an experimental survey in our laboratory of the effects of mast cell secretion products on the various types of cells normally resident in the skin of mice, we observed that supernatants of activated, but not quiescent mouse mast cells induced strong proliferative responses in primary cultures of skin fibroblasts.[142] Given the array of cytokines potentially produced by activated fibroblasts (e.g., MCP-1, SCF), we then sought direct evidence of an association between mast cell-mediated responses and fibroblast activation in vivo, using the expression of type 1 collagen as a marker of successful fibroblast activation. We found that within 1-2 hours of FcεRI-dependent mast cell challenge, before circulating inflammatory cells begin to infiltrate the reaction sites, the local fibroblasts begin to express type I collagen mRNA and they continue to do so increasingly for about 16-20 hours[40] (Fig. 11.3A). This fibroblast activation does not occur in the skin of identically treated W/W^v mice, although it too can be reconstituted by prior transplantation of mast cells from +/+ mice.[40] Finally, using a cytokine subtraction approach with freshly purified mouse serosal mast cells activated via the FcεRI and primary cultures of mouse skin fibroblasts, we determined that mast cell TNF-α and TGF-β together are critical to the mast cell-dependent fibroblast activation—selective absorption of these two cytokines from the serosal mast cell secretion products reduces fibroblast type I collagen expression to near baseline levels.[40]

A further screening of cytokine gene expression in the mast cell-activated fibroblasts indicated that they contain abundant mRNA for the C-C chemokine MCP-1,[143] a monocyte chemoattractant that induces a delayed monocyte infiltration response following intradermal injection in mice.[75,112] Our data indicates that MCP-1 is highly expressed in the skin of BALB/c and WBB6F1-+/+ mice undergoing allergic responses (Fig. 11.3B), but not in identically-treated W/W^v mice unless they are first reconstituted with mast cells.[143] In situ hybridization analysis indicate that the cells expressing MCP-1 during these cutaneous allergic responses are fibroblast-like in morphology and distribution. Furthermore, we also found that anti-MCP-1, but not normal control IgG antibodies reduce monocyte recruitment into the reaction sites by about 65% (Fig. 11.3C), indicating that this mediator is important to the pathogenesis of these responses.[143]

Studies from Metcalfe's laboratory indicate that IL-6 is also highly expressed in the skin of mice undergoing passive cutaneous anaphylaxis responses. IL-6 can be produced by numerous types of cells, including mast cells and fibroblasts, and the in situ hybridization data confirm that multiple cell types within the mouse lesions are strongly positive for IL-6 mRNA.[144] While its role in atopic responses has not been determined as yet, IL-6 has the potential to impact strongly on multiple phases in the pathogenesis of allergic diseases, including T cell activation and IgE and other antibody production responses (reviewed in ref. 145), as well as the induction of the acute phase response.[146]

Taken together, this animal model data documents important roles for mast cell TNF-α and TGF-β in the mast cell-initiated cytokine cascades that drive allergic pathology.[140,143] There is no doubt however that other mast cell mediators, both cytokine and non-cytokine in nature (e.g., IL-1, tryptase), could also contribute importantly to these responses. The data on the mast cell-induced expression of MCP-1 in fibroblasts[143] and broad spectrum IL-6 expression[144] also highlight the importance of considering alternate sources of cytokines that are recruited into mast cell-initiated responses. It is also important to keep in mind that these responses are IgE-mediated and that other cells that express the FcεRI are or can be present in these compartments (e.g., Langerhans cells, eosinophils). Indeed a recent report which addressed the induction of cutaneous IgE-mediated reactions in mice indicates that a form of eosinophil infiltration and skin thickening response can occur some 24 hours after challenge with IgE and antigen, even in mice that are mast cell-deficient.[147] And it has been shown that genetically mast cell-deficient *W/W^v* mice can suffer fatal IgE-dependent pulmonary anaphylaxis responses.[148]

Mast Cell Cytokine Production in Human Tissues

Mast cells purified from enzymatically dispersed human tissues have been shown to produce IL-4,[57] IL-5[149] and TNF-α[54,55] following signaling via the FcεRI. Just as in animal models, the first experiments which addressed the production of cytokines by mast cells in human tissues were directed towards the demonstration that these cells produce TNF-α. It was thus confirmed in an ex vivo tissue system that human cutaneous mast cells contain TNF-α,[50] that their degranulation leads to the induction of ELAM-1 on the local vascular endothelial cells, and that this ELAM-1 expression can be inhibited with anti-TNF-α antibodies.[49,50] However, both anti-TNF and anti-IL-1 antibodies may be required for complete inhibition of ELAM-1 expression in allergen-challenged tissues.[150] Like TNF-α, IL-1 is produced by activated mouse[31] and human[66] mast cells; it is also produced during the initiation phases of allergic LPR in human skin.[151]

Other proinflammatory cytokines which are induced during cutaneous allergic responses in human tissues include IL-6,[152,153] MCP-1,[153] MCP-3 and RANTES.[154] Human mast cells can produce IL-6,[51,60,138] and it is found at very high levels within allergen-challenged tissues in vivo,[152] but the cellular sources of IL-6 shift substantially during the development of the LPR in human skin.[153] It is expressed in these lesions with variable but often biphasic kinetics, such that the IL-6 activity peaks at very high levels at ≈3 hours post-challenge and then again at lower levels at ≈10 hours.[152] In our experiments, we found that early on (i.e., ≈2-4 hours) after cutaneous anti-IgE challenge there was very intense expression of IL-6 within interstitial cells (probably mast cells and, to a greater extent, fibroblast) and endothelial cells, while at later time intervals (i.e., 16 hours) the endothelial cell IL-6 expression ceases and the interstitial cell IL-6 expression modulates substantially.[153] The expression of MCP-1 within these tissues is also associated with the interstitial cells and perhaps also the basal epidermal cells.[153] MCP-3 (or MARC) is produced by mouse serosal mast cells following FcεRI cross-linking, and RANTES may be produced by

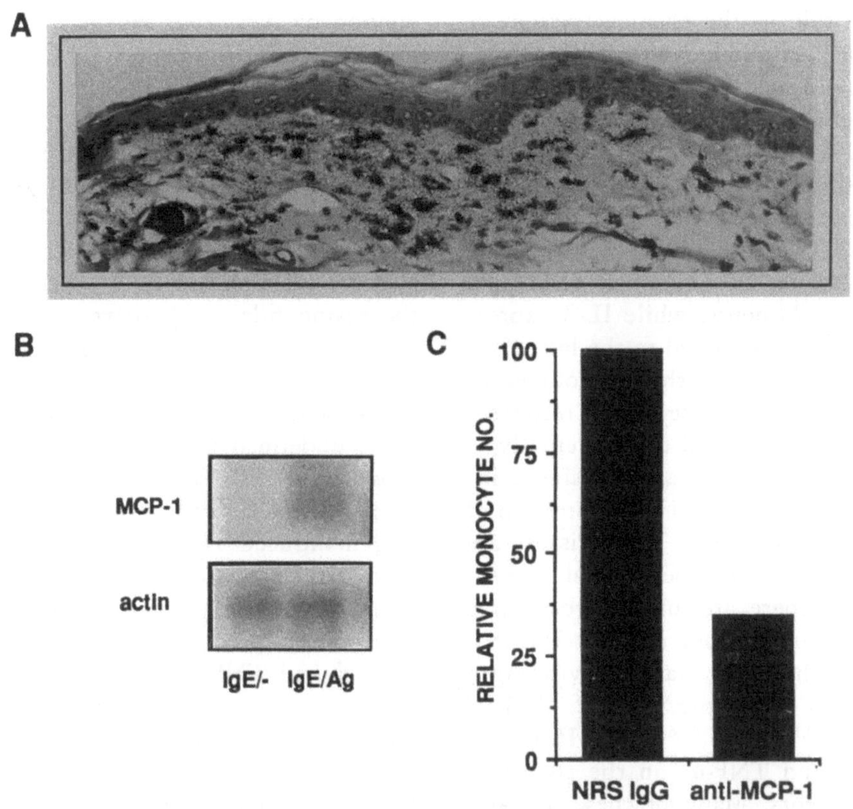

Fig. 11.3 Mast cell-dependent fibroblast activation and MCP-1 expression are important to monocyte recruitment into allergic responses. The ears of BALB/c mice were injected with IgE antibodies and then 24 hours later the animals were challenged intravenously with specific allergen. (A) At 16 hours after allergen challenge the tissues were assessed for levels of mast cell-dependent fibroblast activation by in situ hybridization with ^{35}S-labeled antisense cRNA probes for type α–1 (I) collagen. It can be seen that many cells within the dermis are very strongly positive for this activation marker. Tissues that were probed with sense probes were uniformly negative, as were antisense-probed control tissues not challenged with antigen. (B) Northern blot of mRNA taken at ≈6 hours from ears of mice that were either challenged with allergen (IgE/Ag) or injected with medium alone (IgE/-). The blot was probed with ^{32}P-labeled MCP-1 and γactin cDNA probes and processed for autoradiography. Allergen challenge quite clearly induces the high level expression of MCP-1 mRNA, while no such induction occurs in the control tissues. (C) At the time of allergen challenge, the ears of the mice were injected with 25 μl of a 1:100 dilution (in saline) of either normal rabbit serum (NRS) or rabbit anti-MCP-1 antisera. Sixteen hours later the numbers of monocytes within the reaction sites were quantified by direct counting of the numbers of monoclonal antibody F4/80-reactive cells. NRS had no affect on the monocyte responses, while the anti-MCP-1 antibody treatments reduced monocyte recruitment by ≈65% ($p \leq 0.001$). Monoclonal antibody F4/80 specifically recognizes mouse monocytes/macrophages.

HMC-1 cells,[64] but the sources of these chemokines in allergen-challenged human tissues have not been documented.[154] Irrespective of their sources, the correlations between the kinetics of MCP-3 and RANTES expression in these lesions and the timing of the eosinophil and T cell infiltration responses respectively are consistent with these chemokines playing a significant role in the chemoattraction of these cells.[154] A number of growth factor-like cytokines are also expressed in allergen-challenged human skin, including IL-3,[155] IL-4, IL-5[155,156] and GM-CSF,[155,157] but the cellular sources of these cytokines also have not been determined. IL-4, IL-5[156] and GM-CSF[157] are found in the lesions as early as 6-7 hours post-challenge, as well as at 24 hours, while IL-3 expression is apparently confined to the later time points.[155] The roles of these cytokines in the pathogenesis of these atopic responses have not been addressed experimentally.

The nasal mucosa is another site of intensive investigation into the expression of cytokines by mast cells. These tissues are readily available for experimental manipulation and there are ample prospective subjects suffering from allergic rhinitis. Taken together, the data convincingly demonstrate the expression of IL-3,[158] IL-4,[60,62,63,158,159] IL-5,[60,63,158,160] IL-6,[51,138] GM-CSF[158] and TNF-α[52] in the tissues of rhinitic patients. Many of these are also expressed in the tissues of normal subjects. TNF-α reportedly localizes almost exclusively to nasal epithelial mast cells, and not to the eosinophils or T cells,[52] despite the fact that eosinophils can be a rich source of TNF-α.[160] Following local allergen challenge, high levels of TNF-α can be detected in nasal secretions within 2 minutes,[52] substantiating that it is secreted from an immediately available pool of cytokine. There is agreement on the expression of IL-4 and IL-5 in the nasal mucosa of atopic subjects, with both T cells and mast cells producing both cytokines (IL-5 is also produced by eosinophils is these tissues[138]), but the estimates of the relative contributions of each cell to the total cytokine pool vary substantially (e.g., refs. 60, 63 see also refs. 62, 160-162)—these discrepancies could well be related to technical differences between the subjects and methods employed. Nevertheless, there is little doubt that mast cells in these tissues can produce both IL-4 and IL-5, as well as IL-6.[52]

An increasingly large body of data is becoming available regarding cytokine production by mast cells within the lungs of humans suffering from chronic allergic asthma and similar diseases. The spectrum of cytokines for which there is evidence of mast cell production include IL-3,[56] IL-4, IL-5,[51,162] IL-6,[51] GM-CSF and TNF-α.[51,56,163] In addition, the infiltrating monocytes/macrophages in allergic asthma patients can produce IL-1[56] and TNF-α,[163] the eosinophils are positive for IL-5 (and could potentially also produce IL-1, IL-3, IL-6, IL-8, GM-CSF, MIP-1α, TGF-α, TGF-β and TNF-α; reviewed in refs. 164, 165), undefined airway mononuclear cells produce IL-13[166] and T cells may release IL-3 [56] IL-4, IL-5[56,162] and GM-CSF.[56] Allergen-induced IL-8 expression has also been detected in alveolar lavage fluids of asthmatic patients,[167] and we have reported a substantial production of MCP-1 (as well as IL-6 and TNF-α) in lung tissues challenged ex vivo with anti-IgE antibodies.[153] Similar to what we had observed in human skin,[153] IL-6 is expressed at very high levels in anti-IgE-challenged lung tissues and its cellular source changes with time, from an early combined production by endothelial cells and fibroblast-like cells to a later waning of the endothelial cell contribution to this IL-6 pool.[153] Here too, MCP-1 expression in response to mast cell degranulation is associated with the interstitial fibroblast-like cells.[153] Given the role of MCP-1 in the recruitment of monocytes into allergen-induced lesions in the skin of mice,[143] and the allergen-dependent activation of monocytes and alveolar macrophages in allergic asthma patients (reviewed in ref. 168), it is feasible that the MCP-1 expressed in this compartment under the influence of FcεRI signaling could be an integral part of this monocyte recruitment.

Monocytes in turn can produce mediators that exacerbate eosinophil responses.[164,168]

Taken together, these experiments indicate that human mast cells can produce IL-3, IL-4, IL-5, IL-6, GM-CSF and TNF-α in vivo. It is noteworthy that these cytokines are detectable within the mast cells of non-atopic individuals, although the numbers of cytokine-positive mast cells are lower in these subjects than in atopic subjects not under continuing allergen challenge, and lower still than in atopics undergoing chronic (e.g., seasonal) exposure to their allergens. Thus, it seems that ongoing pathogenic FcεRI or other mast cell ligand signaling is not requisite for mast cell cytokine expression in vivo, but the evidence clearly indicates that such pathogenic stimuli do induce increases in cytokine expression within mast cells, or at least they induce the accumulation of increased numbers of cytokine-expressing mast cells. The evidence also clearly demonstrates that the concept of the mast cell-initiated cytokine cascades that we have shown to occur in animal model systems is also operative in human allergic responses. However, to date TNF-α remains the only mast cell cytokine with an assigned pathogenic role in human tissues undergoing such responses.

POTENTIAL ROLES OF CYTOKINES PRODUCED BY FcεRI-POSITIVE CELLS

The evidence that mast cells from humans or animals produce multiple cytokines and that at least some of them have substantial, perhaps even pivotal, roles in the pathogenesis of atopic responses is quite compelling. A case in point is that of TNF-α, which largely affects the recruitment of neutrophils in cutaneous LPR.[50,140] The evidence is less weighty, however, for most other cytokines. Conceptually, "mast cell" cytokines could have one of several different effects in vivo. The first is that, despite the abilities of cloned mastocytoma cells to produce a cytokine or transcribe its mRNA following stimulation with PMA or any other stimulus, mast cells as they are found in vivo for whatever reason, may not actually express (i.e., ultimately secrete) that cytokine. The cytokine could still be expressed by other cells within atopic responses, and its expression could be entirely IgE- and mast cell-dependent if the actual source were recruited by other mast cell mediators. For example, mast cell TNF-α and TGF-β1 recruit fibroblasts to produce MCP-1,[153] but despite the fact that some mastocytoma cell populations can produce MCP-1 in vitro,[31,64] at this point in time there is no truly clear evidence that mast cells do so in vivo. At the opposite end of the spectrum of possibilities for mast cell cytokine production is that the mast cell is the only source of the cytokine, although its effects may range from being so exquisitely subtle as to be functionally indiscernible to being quite obviously the overwhelming pathogenic influence. In both of these cases (and the next) it is quite likely that the expression and/or effect of the cytokine could be modulated by the environment of the mast cell at the time of cellular activation. There is abundant evidence that numerous exogenous ligands (e.g., cytokines, extracellular matrix molecules) can dramatically affect cytokine production by mast cells, so that in a chronically inflamed tissue environment the mast cell could feasibly produce more or less of any one cytokine than it would in a physiologically "quiescent" tissue. In addition, in inflamed tissues there may be cellular targets available to the mast cell which might not otherwise be present. Finally, and much more likely for most mast cell cytokines is that the mast cell secretes the cytokine, but that alternate sources of the same cytokine are recruited into the reactions.[141,165] TNF-α, for example, can be produced by many of the cells we associate with atopic diseases, including mast cells,[30,52] neutrophils,[169] eosinophils[170] and monocytes/macrophages.[163] While TNF-α may be detectable primarily within the resting mast cells of allergic rhinitis patients,[60] and its early (e.g., <4 hours) release from these cells may largely mediate the granulocyte recruitment responses which accompany allergen challenges,[50,140] the tissue and airway macrophages of allergic asthma patients may also produce TNF-α during the later stages of allergic reactions.[163,171] So, too, might the granulocytes.[169,170] The production of the same cytokine by multiple sources during a response would not necessarily represent a redundant process—each of the sources could be independently important to the pathogenesis of the response if indeed each released their product at a distinct phase of the reaction. On the other hand, in a case wherein two of more productive sources of a cytokine were simultaneously present, their relative contributions to the responses would depend on factors such as their relative numbers and activation status.

One of the more important roles for mast cell/basophil cytokines may not be related to their abilities to initiate pathogenic processes during allergic responses. Conceptually more important perhaps could be the potential roles of some of these cytokines in maintaining or ensuring the maintenance of atopy, or even in the initiation or establishment of the allergen-specific atopic state to novel allergens. Mast cells/basophils potentially have the resources to affect each of these processes through their elaboration of cytokines such as IL-4, IL-6 and TNF-α, to name a few. With these cytokines, mast cells can induce B cells to switch from IgG antibody production to high level IgE antibody secretion, without a requirement for input from other cells.[172,173] IL-4 is a critical IgG to IgE switch factor for B cells, but mast cells also express the CD40 ligand, another important signal for this switch,[172] as well as TNF-α and IL-6, both of which may significantly augment B cell antibody production.[173] Furthermore, mast cell IL-4 could both dampen potential Th1 responses (reviewed in ref. 174) and reinforce the expansion of already differentiated allergen-specific effector Th2 (*e*Th2) populations and perhaps reactivate memory Th2 (*m*Th2) cells,[174] and thereby ensure a continued biasing of the response towards the atopic phenotype[175] (see also ref. 176).

Mast cells/basophils could also impact on much earlier stages in the ontogeny of the immune response however,—perhaps at the level of priming naive or precursor Th (*p*Th) cells to differentiate towards the *e*Th2 phenotype at the time of first exposure to the new allergen. Ostensibly, the presence of IL-4 by itself is sufficient to drive a novel allergen-specific *p*Th cell response towards the *e*Th2 phenotype,[175] and, as reviewed above, mast cells/basophils represent potentially important preformed sources of this cytokine. One theoretical problem faced by proponents of a mast cell/basophil IL-4-driven *p*Th-to-*e*Th2 cell differentiation is how the mast cells/basophils might secrete IL-4 in the absence of FcεRI signaling. IgE would be required for such signaling, and IgE production is considered to be largely driven by allergen-specific T cell responses,[173] which of course would not be in place at this time in the genesis of the allergen-specific response. However, mast cells can be induced to release their mediators in response to multiple IgE-independent ligands (e.g., C5a, SCF) that could accompany some antigenic challenges. In addition, there is also evidence that at least some highly allergenic and enzymatically active proteins (e.g., bee venom phospholipase A_2) can directly induce IL-4 secretion by mast cells in the absence of IgE signaling.[177] If a particular antigen could nonspecifically induce mast cell/basophil IL-4 secretion in an otherwise immunologically uncommitted microenvironment, in some settings (e.g., in individuals genetically prone to atopy) this might be adequate to shift the balance of the initial *p*Th cell differentiation process for that antigen towards the Th2 phenotype. If indeed mast cell IL-4 did have a role in this process, then therapies which could specifically target mast cell, as opposed to T cell, IL-4 production could be of benefit. The recent discovery in M. Brown's laboratory of mast cell-specific IL-4 genomic enhancer elements might be just the sort of target required for such hypothetical interventions.[104,105]

Mast cells are apparently "nonspecifically" activated during a variety of chronic inflammatory diseases (e.g., chronic graft-versus-host disease; GVHD[178]), probably through the actions of mediators such as SCF[74] or other mast cell activators produced within the inflamed tissues. Tissue fibrosis also accompanies many of these chronic inflammatory processes, such that it has been speculated that the mast cell itself may be responsible in part for this fibrosis.[178] Indeed, we and others have shown that multiple mast cell cytokines (as well as non-cytokine mediators such as tryptase) can induce these cells to proliferate in vitro,[40,142,179,180] and we have demonstrated that mast cell TNF-α and TGF-β1 induce high level expression of type 1 collagen within allergic reaction sites in mice[40] (Fig. 11.3A). Not surprisingly though, this acute mast cell-dependent fibroblast activation does not lead to a significant deposition of collagen within these tissues,[40] probably in part because of the simultaneous induction of collagenases and other metalloproteinases designed to dampen inappropriate responses. But it has been demonstrated using genetically mast cell-deficient as well as normal mice that, in one form or another, mast cells are important contributors to tissue fibrosis in animals undergoing chronic inflammatory challenges.[181] In the context of these diseases in which the mast cells are also chronically activated, the balance between the fibrogenic and anti-fibrogenic processes is obviously tipped in favor of tissue fibrosis,[178] but we know little of the roles of mast cell cytokines in the regulation of these processes. Thus, much remains to be accomplished in terms of analyzing this system.

In summary, we continue to accumulate data indicating that mast cells have a capacity to produce cytokines—they contain preformed cytokines in vivo or mastocytoma cells may be induced to produce these mediators—but few resources are being allocated for an experimental determination of the effector functions of these *mast cell* cytokines in vivo. Until we come to an understanding as to whether the elaboration of each cytokine by mast cells represents an important mechanism for atopic and other pathologies or simply a vestigial or redundant oddity of the immune system, we will not be able to effectively direct our energies in the management of atopic and selected other diseases.

References

1. Wang B, Rieger A, Kilgus O et al. Epidermal Langerhans cells from normal human skin bind monomeric IgE via the FcεRI. J Exp Med 1992; 175:1353.
2. Bieber H, DelaSalle H, Wollenburg A et al. Human epidermal Langerhans cells express the high affinity receptor for immunoglobulin E (FcεRI). J Exp Med 1992; 175:1285.
3. Osterhoff B, Rappersberger C, Wang B et al. Immunomorphological characterization of FcεRI-bearing cells within the human dermis. J Invest Dermatol 1994; 102: 315-23.
4. Maurer D, Fiebiger E, Reininger B et al. Expression of functional high affinity IgE receptors (FcεRI) on monocytes of atopic individuals. J Exp Med 1994; 179:745.
5. Gounni AS, Lamkhioued B, Delaporte E et al. The high-affinity IgE receptor on eosinophils: from allergy to parasites or from parasites to allergy? J Allergy Clin Immunol 1994; 94:1214-6.
6. Louahed J, Kermouni A, Van Snick J et al. IL-9 induces expression of granzymes and high-affinity IgE receptor in murine T helper clones. J Immunol 1995; 154: 5061-70.
7. Galli SJ, Lichtenstein LM. Biology of mast cells and basophils. In: Middleton E, Reed CE, Ellis EF et al, eds. Allergy. Principles and Practice. St. Louis: CV Mosby Co., 1988:106-34.
8. Alber G, Kent U, Metzger H. Functional comparison of Fc epsilon RI, Fc gamma RII, and Fc gamma RIII in mast cells. J Immunol 1992; 149:2428-36.
9. Ben Sasson SZ, Le Gros G, Conrad DH et al. Crosslinking Fc receptors stimulate splenic non-B, non-T cells to secrete interleukin 4 and other lymphokines. Proc Natl Acad Sci (USA) 1990; 87:1421-5.

10. Tsai M, Shih L, Newlands GFK et al. The rat *c-kit* ligand, stem cell factor, induces the development of connective tissue-type and mucosal mast cells in vivo. Analysis by anatomical distribution, histochemistry and protease phenotype. J Exp Med 1991; 174:125-31.
11. Gordon JR, Galli SJ. Phorbol 12-myristate-13-acetate-induced development of functionally active mast cells in W/W^v but not Sl/Sl^d genetically mast cell-deficient mice. Blood 1990; 75:1637-45.
12. Kitamura Y, Nakano T, Sonoda T et al. Probable transdifferentiation between connective tissue and mucosal mast cells. In: Befus A, Bienenstock J, Denburg J, eds. Mast Cell Differentiation and Heterogeneity. New York: Raven Press, 1986:135-40.
13. Kanakura Y, Thompson H, Nakano T et al. Multiple bidirectional alterations of phenotype and changes in proliferative potential during the in vitro and in vivo passage of clonal mast cell populations derived from mouse peritoneal mast cells. Blood 1988; 72:877-85.
14. Schwartz L. Mast cells: function and contents. Curr Opin Immunol 1994; 6:91-7.
15. Gurish M, Ghildyal N, McNeil M et al. Differential expression of secretory granule proteases in mouse mast cells exposed to interleukin 3 and *c-kit* ligand. J Exp Med 1992; 175:1003-12.
16. Chung SW, Wong PMC, Shen-Ong G et al. Production of granulocyte-macrophage colony stimulating factor by Abelson virus-induced tumorigenic mast cell lines. Blood 1986; 68:1074-81.
17. Humphries K, Abraham S, Krystal G et al. Activation of multiple hempoietic growth factor genes in Abelson virus-transformed myeloid cells. Exp Hematol 1988; 16: 774-81.
18. Brown MA, Pierce JH, Watson CJ et al. B cell stimulatory factor-1/ interleukin-4 mRNA is expressed by normal and transformed mast cells. Cell 1987; Cell 50:809-18.
19. Seder RA, Paul WE, Dvorak AM et al. Mouse splenic and bone marrow cell populations that express high-affinity Fc epsilon receptors and produce interleukin-4 are highly enriched in basophils. Proc Natl Acad Sci (USA) 1991; 88:2835-9.
20. Le Gros G, Ben Sasson SZ, Conrad DH et al. IL-3 promotes production of IL-4 by splenic non-B, non-T cells in response to Fc receptor crosslinkage. J Immunol 1990; 145:2500-6.
21. Piccinni MP, Macchia D, Parronchi P et al. Human bone marrow non-B, non-T cells produce interleukin-4 in response to crosslinkage of Fc epsilon and Fc gamma receptors. Proc Natl Acad Sci (USA) 1991; 88:8656-60.
22. Irani AM, Nilsson G, Miettenen U et al. Recombinant human stem cell factor stimulates differentiation of mast cells from dispersed human fetal liver cells. Blood 1992; 80:3009-21.
23. Galli SJ, Tsai M, Gordon JR et al. Analyzing mast cell development and function using mice carrying mutations at *W/c-kit* or *Sl/MGF* (SCF) loci. Ann N Y Acad Sci 1992; 664:69-88.
24. Farram E, Nelson DS. Mouse mast cells as anti-tumor effector cells. Cell Immunol 1980; 55:294.
25. Ghiara P, Boraschi D, Villa L et al. In vitro-generated mast cells express natural cytotoxicity against tumor cells. Immunol 1985; 55:317-23.
26. Tharp M, Kasper C, Thiele D et al. Studies of connective tissue mast cell-mediated cytotoxicity. J Invest Dermatol 1989; 93: 423-28.
27. Jadus M, Schmunk G, Djeu J et al. Morphology and lytic mechanisms of interleukin-3-dependent natural cytotoxic cells: tumor necrosis factor as a possible mediator. J Immunol 1986; 137:2774-83.
28. Young JD-E, Liu C-C, Butler G et al. Identification, purification and characterization of a mast cell-associated cytolytic factor related to tumor necrosis factor. Proc Natl Acad Sci (USA) 1987; 84:9175-79.
29. Richards A, Okuno T, Takagaki Y et al. Natural cytotoxic cell-specific cytotoxic factor produced by IL-3-dependent basophilic mast cells. J Immunol 1988; 141:3061-66.
30. Gordon JR, Galli SJ. Mast cells as a source of both preformed and immunologically-inducible tumor necrosis factor-alpha. Nature 1990; 346:274-76.

31. Burd PR, Rogers HW, Gordon JR et al. Interleukin-3-dependent and -independent mast cells stimulated with IgE and antigen express multiple cytokines. J Exp Med 1989; 170:245-57.
32. Plaut M, Pierce JH, Watson CJ et al. Mast cell lines produce lymphokines in response to crosslinkage of FcεRI or to calcium ionophores. Nature 1989; 339:64-7.
33. Wodnar-Filipowicz A, Heusser C, Moroni C. Production of the hemopoietic growth factors GM-CSF and interleukin 3 in response to IgE receptor-mediated activation. Nature 1989; 339:150-52.
34. Hulter L, Szots H, Welle M et al. Mouse bone marrow-derived interleukin 3-dependent mast cells and an autonomous subline produce interleukin-6. Immunol 1989; 67:408-13.
35. Smith TJ, Ducharme LA, Weis JH. Preferential expression of IL-12 or IL-4 by murine bone marrow mast cells derived in mast cell growth factor or IL-3. Eur J Immunol 1994; 24:822-26.
36. Burd PR, Thompson WC, Max EE et al. Activated mast cells produce interleukin 13. J Exp Med 1995; 181:1373-80.
37. Marshall J, Gauldie J, Nielsen L et al. Leukemia inhibitory factor production by rat mast cells. Eur J Immunol 1993; 23: 2116-20.
38. Gordon JR, Galli SJ. Release of both preformed and newly synthesized TNF-α/cachectin by mouse mast cells stimulated via the FcεRI. A mechanism for the sustained action of mast cell-derived TNF-α during IgE-dependent biological responses. J Exp Med 1991; 174:103-7.
39. Tsai M, Gordon JR, Galli SJ. Mast cells constitutively express transforming growth factor β mRNA. FASEB J 1990; 4: A1944 [abstr].
40. Gordon JR, Galli SJ. Promotion of mouse fibroblast collagen gene expression by mast cells stimulated via the $Fc_{\epsilon}RI$. Role for mast cell-derived transforming growth factor-β and tumor necrosis factor-α. J Exp Med 1994; 180:2027-37.
41. Ehrenreich H, Burd P, Tottem M et al. Endothelins belong to the assortment of mast cell-derived and mast cell-bound cytokines. New Biol 1992; 4:147-56.
42. Kulmberg P, Huber N, Scheer B et al. Immunoglobulin E plus antigen challenge induces a novel intercrine/chemokine in mouse mast cells. J Exp Med 1992; 176:1773-8.
43. Minty A, Chalon P, Guillemot J et al. Molecular cloning of the MCP-3 chemokine gene and regulation of its expression. Eur Cytokine Netw 1993; 4:99-110.
44. Leal-Berumen I, Conlon P, Marshall JS. IL-6 production by rat peritoneal mast cells is not necessarily preceded by histamine release and can be induced by bacterial lipopolysaccharide. J Immunol 1994; 152: 5468-76.
45. Bissonnette E, Befus AD. Interferon-gamma can suppress IgE-mediated TNF release from rat peritoneal mast cells. J Immunol 1990; 145:3385-93.
46. Pennington DW, Lopez AR, Thomas PS et al. Dog mastocytoma cells produce transforming growth factor-β1. J Clin Invest 1992; 90:35-41.
47. Latour L, Bonnerot C, Fridman W et al. Induction of tumor necrosis factor-alpha production by mast cells via Fc gamma R. Role of the Fc gamma RIII gamma subunit. J Immunol 1992; 149:2155-62.
48. Steffen M, Abboud M, Potter G et al. Presence of tumor necrosis factor or a related factor in human basophils/mast cells. Immunol 1989; 66:445-51.
49. Klein LM, Lavker RM, Matis WL et al. Degranulation of human mast cells induces an endothelial antigen central to leukocyte adhesion. Proc Natl Acad Sci (USA) 1989; 86:8972-80.
50. Walsh LJ, Trinchieri G, Waldorf HA et al. Human dermal mast cells contain and release tumor necrosis factor alpha which induces endothelial leukocyte adhesion molecule-1. Proc Nat Acad Sci (USA) 1991; 88:4220-4.
51. Bradding P, Roberts JA, Britten KM et al. Interleukin-4, interleukin-5, and interleukin-6 and tumor necrosis factor-alpha in normal and asthmatic airways-evidence for the human mast cell as a source of these cytokines. Am J Resp Cell Mol Biol 1994; 10:471-80.

52. Bradding P, Mediwake R, Feather IH et al. TNF-alpha is localized to nasal mucosal mast cells and is released in acute allergic rhinitis. Clin Exp Immunol 1995; 25:406-15.
53. Okayama Y, Semper A, Holgate ST et al. Multiple cytokine mRNA expression in human mast cells stimulated via Fc-epsilon-R1. Internat Arch All Immunol 1995; 107:158-9.
54. Benyon RC, Bissonnette EY, Befus AD. Tumor necrosis factor-α dependent cytotoxicity of human skin mast cells is enhanced by anti-IgE antibodies. J Immunol 1991; 147:2253-62.
55. Gordon JR, Post T, Schulman ES et al. Characterization of mouse mast cell TNFα induction in vitro and in vivo, and demonstration that purified human lung mast cells contain TNFα. FASEB J 1991; 5: A1009.
56. Ackerman V, Marini M, Vittori E et al. Detection of cytokines and their cell sources in bronchial biopsy specimens from asthmatic patients. Relationship to atopic status, symptoms, and level of airway hyperresponsiveness. Chest 1994; 105:687-96.
57. Bradding P, Feather IH, Howarth PH et al. Interleukin 4 is immunolocalized to and released by human mast cells. J Exp Med 1992; 176:1381-92.
58. Okayama Y, Petit-Frere C, Kassel O et al. IgE-dependent expression of mRNA for IL-4 and IL-5 in human lung mast cells. J Immunol 1995; 155:1796-808.
59. Moller A, Lippert U, Lessmann D et al. Human mast cells produce IL-8. J Immunol 1993; 151:3261-66.
60. Bradding P, Feather IH, Wilson S et al. Immunolocalization of cytokines in the nasal mucosa of normal and perennial rhinitic subjects. The mast cell as a source of IL-4, IL-5, and IL-6 in human allergic mucosal inflammation. J Immunol 1993; 151: 3853-65.
61. Horsmanheimo L, Harvima I, Jarvikalllia A et al. Mast cells are one major source of interleukin-4 in atopic dermatitis. Br J Dermatol 1994; 131:348-54.
62. Ying S, Durham S, Jacobson M et al. T lymphocytes and mast cells express messenger RNA for interleukin-4 in the nasal mucosa in allergen-induced rhinitis. Immunol 1994; 82:200-6.
63. Bradding P, Okayama Y, Howarth PH et al. Heterogeneity of human mast cells based on cytokine content. J Immunol 1995; 155:297-307.
64. Selvan R, Butterfield J, Krangel M. Expression of multiple chemokine genes by human mast cell leukemia. J Biol Chem 1994; 269:13893-8.
65. Sillaber C, Bevec D, Butterfield J et al. Tumor necrosis factor alpha and interleukin-1 beta mRNA expression in HMC-1 cells: differential regulation of gene product expression by recombinant interleukin-4. Exp Hematol 1993; 21:1271-5.
66. Nilsson G, Svensson V, Nilsson K. Constitutive and inducible cytokine mRNA expression in the human mast cell line HMC-1. Scand J Immunol 1995; 42:76-81.
67. Gordon JR, Rouse C, Irani A-MA et al. Induction of IL-6 expression in and by human mast cells following cellular activation. Cytokine 1994; 6:546A.
68. Conrad DH, Ben Sasson SZ, Le Gros G et al. Infection with *Nippostrongylus brasiliensis* or injection of anti-IgD antibodies markedly enhances Fc-receptor-mediated interleukin 4 production by non-B, non-T cells. J Exp Med 1990; 171:1497-508.
69. Aoki I, Kinzer C, Shirai A et al. IgE receptor-positive non-B/non-T cells dominate the production of interleukin 4 and interleukin 6 in immunized mice. Prod Natl Acad Sci (USA) 1995; 92:2534-8.
70. Brunner T, Heusser CH, Dahinden CA. Human peripheral blood basophils primed by IL-3 produce IL-4 in response to immunoglobulin E receptor stimulation. J Exp Med 1993; 177:605-14.
71. Schroeder JT, Kagey-Sobotka A, Macglashan DW et al. The interaction of cytokines with human basophils and mast cells. Internat Arch All Immunol 1995; 107:79-81.
72. Mueller R, Heusser CH, Rihs S et al. Immunolocalization of intracellular interleukin-4 in normal human peripheral blood basophils. Eur J Immunol 1994; 24: 2935-40.
73. Schroeder JT, Macglashan DW, Kagey-Sobotka A et al. IgE-dependent IL-4 secretion by human basophils-the relationship

between cytokine production and histamine release in mixed leukocyte cultures. J Immunol 1994; 153:1808-17.
74. Columbo M, Horowitz EM, Botana LM et al. The human recombinant c-kit receptor ligand, rhSCF, induces mediator release from human cutaneous mast cells and enhances IgE-dependent mediator release from both skin mast cells and peripheral blood basophils. J Immunol 1992; 149:599-608.
75. Alam R, Forsythe P, Stafford S et al. Macrophage inflammatory protein-1 alpha activates basophils and mast cells. J Exp Med 1992; 176:781-6.
76. Kuna P, Reffiga SR, Schall TJ et al. Characterization of the human basophil response to cytokines, growth factors and histamine releasing factors of the intercrine/ chemokine family. J Immunol 1993; 150:1932-43.
77. Alam R, Forsythe P, Stafford S et al. Monocyte chemotactic protein-2, monocyte chemotactic protein-3, and fibroblast-induced cytokine-three new chemokines induce chemotaxis and activation of basophils. J Immunol 1994; 153:3155-9.
78. Kuna P, Reddigari SR, Rucinski D et al. Chemokines of the alpha, beta-subclass inhibit human basophils responsiveness to monocyte chemotactic and activating factor monocyte chemoattractant protein-1. J All Clin Immunol 1995; 95:574-86.
79. Radermecker M, Louis R, Leclercq M et al. Cytokine modulation of basophil histamine release in wasp-venom allergy. Allergy 1994; 49:641-4.
80. Columbo M, Horowitz EW, Kagey-Sobotka A et al. Histamine release from human basophils induced by platelet activating factor-the role of extracellular calcium, interleukin-3, and granulocyte-macrophage colony-stimulating factor. J All Clin Immunol 1995; 95:565-73.
81. Baumgartner RA, Yamada K, Deramo VA et al. Secretion of TNF from a rat mast cell line is a brefeldin A-sensitive and a calcium/protein kinase C-regulated process. J Immunol 1994; 153:2609-17.
82. Dvorak A. Human mast cells: ultrastructural observations of in situ, ex vivo, and in vitro sites, sources and systems. In: Kaliner M, Metcalfe D, eds. The Role of the Mast Cell in Health and Disease. New York: Marcel Dekker, 1993:1-90.
83. Leal-Berumen I, O'Byrne P, Gupta A et al. Prostanoid enhancement of interleukin-6 production by rat peritoneal mast cells. J Immunol 1995; 154:4759-67.
84. Burd PR, Costa JJ, Metcalfe DD et al. Cytokine gene expression occurs via multiple pathways in primary murine bone marrow mast cells (BMCMC). FASEB J 1990; 4:A1705 [abstr].
85. Costa JJ, Burd PR, Siebenlist U et al. Primary mouse bone marrow mast cell cultures transcribe multiple cytokines in response to $Fc_εRI$ crosslinking. FASEB J 1990; 4:A1943.
86. Benhamou M, Siraganian R. Protein-tyrosine phosphorylation: an essential early component of FcεRI signalling. Immunol Today 1992; 13:195-201.
87. Ishizaka T. Mechanisms of IgE-mediated hypersensitivity. In: Middleton E, Reed CE, Ellis EF et al, eds. Allergy. Principles and Practice. CV Mosby Co: St. Louis, 1988: 71-93.
88. Wodnar-Filipowicz A, Moroni C. Regulation of interleukin 3 mRNA expression in mast cells occurs at the posttranscriptional level and is mediated by calcium ions. Proc Natl Acad Sci (USA) 1990; 87:777-81.
89. Costa J, Burd P, Metcalfe D. Mast cell cytokines. In: Kaliner M, Metcalfe D, eds. The Mast Cell in Health and Disease. New York: Marcel Dekker, Inc., 1993:443-66.
90. Buckley MG, Williams CMM, Thompson J et al. IL-4 enhances IL-3 and IL-8 gene expression in a human leukemic mast cell line. Immunol 1995; 84:410-5.
91. Wershil BK, Levine AD, Nauert JP et al. Ionomycin stimulation of a growth factor-independent mouse mast cell line induces differential IL-3 and IL-4 gene expression and release of product. FASEB J 1990; 4: A1943.
92. Hahn S, Moroni C. Modulation of cytokine expression in PB-3c mastocytes by IBMX and PMA. Lymphokine Cytokine Res 1994; 13:247-52.
93. Galli SJ, Zsebo KM, Geissler EN. The kit ligand, stem cell factor. Adv Immunol 1994; 55:1-96.

94. Plaut M, Kagey-Sobotka A, Niv Y et al. Regulation of mast cell cytokine secretion. FASEB J 1990; 4:A1705.
95. Nakamura M, MacGlashan D. Desensitization of IL-4 secretion from mouse bone marrow-derived mast cells. Immunol Lett 1994; 41:129-33.
96. Hatfield S, Mynderse J, Roehm N. Rapamycin and FK506 differentially inhibit mast cell cytokine production and cytokine-induced proliferation and act as reciprocal antagonists. J Pharmacol Exp Therap 1992; 261:970-6.
97. Hatfield S, Roehm N. Cyclosporine and FK506 inhibition of murine mast cell cytokine production. J Pharmacol Exp Therap 1992; 260:680-8.
98. Kaye R, Fruman D, Bierer B et al. Effects of cyclosporin A and FK506 on Fc epsilon receptor type I-initiated increases in cytokine mRNA in mouse bone marrow-derived progenitor mast cells: resistance to FK506 is associated with a deficiency in FK506-binding protein FKBP12. Proc Natl Acad Sci (USA) 1992; 89:8542-6.
99. Wershil BK, Furuta GT, Lavigne JA et al. Dexamethasone or cyclosporin A suppress mast cell-leukocyte cytokine cascades-multiple mechanisms of inhibition of IgE- and mast cell-dependent cutaneous inflammation in the mouse. J Immunol 1995; 154:1391-8.
100. Fruman DA, Burakoff SJ, Bierer BE. Immunophilins in protein folding and immunosuppression. FASEB J 1994; 8: 391-400.
101. Fruman D, Bierer B, Benes J et al. The complex of FK506-binding protein 12 and FK506 inhibits calcineurin phosphatase activity and IgE activation-induced cytokine transcripts, but not exocytosis in mouse mast cells. J Immunol 1995; 154:1846-51.
102. Hohman R, Hultsch T. Modulation of mediator release from mast cells. In: Kaliner MA, Metcalfe DD, eds. The Role of the Mast Cell in Health and Disease. New York: Marcel Dekker, 1993:269-300.
103. Razin E, Leslie K, Schrader J. Connective tissue mast cells in contact with fibroblasts express IL-3 mRNA. Analysis of single cells by polymerase chain reaction. J Immunol 1991; 146:1479-86.
104. Henkel G, Weiss DL, McCoy R et al. A DNAse I-hypersensitive site in the second intron of the murine IL-4 gene defines a mast cell-specific enhancer. J Immunol 1992; 149:3239-46.
105. Henkel G, Brown MA. Pu.1 and GATA-components of a mast cell-specific interleukin 4 intronic enhancer. Proc Natl Acad Sci (USA) 1994; 91:7737-41.
106. Prieschl EE, Gouilleuxgruart V, Walker C et al. A nuclear factor of activated T cell-like transcription factor in mast cells is involved in IL-5 gene regulation after IgE plus antigen stimulation. J Immunol 1995; 154:6112-9.
107. Jarmin DI, Kulmburg PA, Huber NE et al. A transcription factor with AP3-like binding specificity mediates gene regulation after an allergic triggering with IgE and Ag in mouse mast cells. J Immunol 1994; 153:5720-9.
108. Oh C, Metcalfe D. Transcriptional activation of the TCA3 gene in mast cells after Fc epsilon RI crosslinking. J Immunol 1994; 153:325-32.
109. Chaikin E, Hakeem I, Razin E. Enhancement of interleukin-3-dependent mast cell proliferation by suppression of c-jun expression. J Biol Chem 1994; 269:8498-503.
110. Malter JS. Identification of an AUUUA-specific messenger RNA binding protein. Science 1989; 246:664-6.
111. Granelli-Piperno A, Nolan P. A nuclear transcription factor that binds to elements of the IL-2 promotor. J Immunol 1991; 147:2734-41.
112. Alam R, Kuman D, Anderson-Walters D et al. Macrophage inflammatory protein-1α and monocyte chemoattractant peptide-1 elicit immediate and late cutaneous reactions and activate murine mast cells in vivo. J Immunol 1994; 152:1298-303.
113. Bischoff SC, Dahinden CA. c-kit ligand: a unique potentiator of mediator release by human lung mast cells. J Exp Med 1992; 175:237-44.
114. Takaishi T, Morita Y, Hirai K et al. Effect of cytokines on mediator release from human dispersed lung mast cells. Allergy 1994; 49:837-42.
115. Kaplan A, Reddigari S, Baeza M et al.

Histamine releasing factors and cytokine-dependent activation of basophils and mast cells. Adv Immunol 1991; 50:237-60.
116. Fureder W, Agis H, Semper H et al. Differential response of human basophils and mast cells to recombinant chemokines. Annal Hematol 1995; 70:251-8.
117. Bischoff S, Dahinden C. Effect of nerve growth factor on the release of inflammatory mediators by mature basophils. Blood 1992; 79:2662-9.
118. Costa J, Burd P, Metcalfe D. Cytokine mRNA induction by non-FcεRI-mediated stimuli in mouse mast cells. FASEB J 1992; 5: A1064.
119. Coleman JW, Holliday M, Kimber I et al. Regulation of mouse peritoneal mast cell secretory function by stem cell factor, IL-3 or IL-4. J Immunol 1993; 150:556-62.
120. Holliday M, Banks E, Dearman R et al. Interactions of IFN-gamma with IL-3 and IL-4 in the regulation of serotonin and arachidonate release from mouse peritoneal mast cells. Immunol 1994; 82:70-4.
121. Keegan A, Pierce J, Artrip J et al. Ligand stimulation of transfected and endogenous growth factor receptors enhances cytokine production by mast cells. EMBO J 1991; 10:3675-82.
122. Nilsson G, Nilsson K. Effects of interleukin (IL)-13 on immediate-early response gene expression, phenotype and differentiation of human mast cells-comparison with IL-4. Eur J Immunol 1995; 25:870-3.
123. Geissler EN, Gordon JR, Tsai M et al. Activation of mast cells down-regulates mRNA for c-kit tyrosine kinase receptor. FASEB J 1990; 4: A1704.
124. Frandji P, Tkaczyk C, Oskeritzian C et al. Presentation of soluble antigens by mast cells-upregulation by IL-4 and GM-CSF and downregulation by IFN-gamma. Cell Immunol 1995; 163:37-46.
125. White M. Mast cell secretagogues. Histamine releasing factors and neuropeptides. In: The Role of the Mast Cell in Health and Disease. Kaliner M, Metcalfe D, eds. NY: Marcell Dekker 1993:109-28.
126. Galli SJ, Wershil BK, Yano H et al. Analysis of the roles of phenotypically distinct mast cell populations in nonimmunological responses. In: Galli SJ, Austen KF, eds. Mast Cell and Basophil Differentiation and Function in Health and Disease. New York: Raven Press, 1989:255-73.
127. Ansel JC, Brown J, Payan D et al. Substance P selectively activates TNF-alpha gene expression in murine mast cells. J Immunol 1993; 150:4478-85.
128. Hamawy MM, Oliver C, Mergenhagen SE et al. Adherence of rat basophilic leukemia (RBL-2H3) cells to fibronectin-coated surfaces enhances secretion. J Immunol 1992; 149:615-21.
129. Hamawy MM, Mergenhagen SE, Siraganian RP. Cell adherence to fibronectin and the aggregation of the high affinity immunoglobulin E receptor synergistically regulate tyrosine phosphorylation of 105-115 kDa proteins. J Biol Chem 1993; 268:5227-33.
130. Sperr W, Agis H, Czerwenka K et al. Differential expression of cell surface integrins on human mast cells and human basophils. Ann Hematol 1992; 65:10-8.
131. Taub D, Dastych J, Inamura N et al. Bone marrow-derived murine mast cells migrate, but do not degranulate in response to chemokines. J Immunol 1995; 154: 2393-402.
132. Nilsson G, Butterfield JH, Nilsson K et al. Stem cell factor is a chemotactic factor for human mast cells. J Immunol 1994; 153:3717-23.
133. Columbo M, Bochner B, Marone G. Human skin mast cells express functional β1 integrins that mediate adhesion to extracellular matrix proteins. J Immunol 1995; 154:6058-64.
134. Hamawy MM, Mergenhagen SE, Siraganian RP. Adhesion molecules as regulators of mast cell and basophil function. Immunol Today 1994; 15:62-6.
135. Ra CS, Yasuda M, Yagita H et al. Fibronectin receptor integrins are involved in mast cell activation. J All Clin Immunol 1994; 94:625-8.
136. Schleimer R. Glucocorticosteroids. In: Middleton E, Reed CE, Ellis EF, eds. Allergy. Principles and Practice. St. Louis: CV Mosby Co., 1988:739-65.
137. Wershil BK, Gordon JR, Wang Z-S et al. Dexamethasone (DEX) inhibits IgE-depen-

dent TNF-α production by mouse mast cells (MC's) in vitro and suppresses IgE-, mast cell- and TNF-α-dependent inflammation in vivo. FASEB J 1992; 6:A1716.
138. Bradding P, Feather IH, Wilson S et al. Cytokine immunoreactivity in seasonal rhinitis-regulation by a topical corticosteroid. Am J Resp Crit Care Med 1995; 151:1900-6.
139. Bevilacqua MP, Wheeler ME, Pober JS et al. Endothelial-dependent mechanisms of leukocyte adhesion: regulation by interleukin 1 and tumor necrosis factor. In: Movat HZ, ed. Leukocyte Emigration and Its Sequellae. New York: S Karger, 1987:79-93.
140. Wershil BK, Wang Z-S, Gordon JR et al. The recruitment of neutrophils during IgE-dependent cutaneous late phase reactions in the mouse is mast cell-dependent: partial inhibition of the reaction with antiserum against tumor necrosis factor-alpha. J Clin Invest 1991; 87:446-53.
141. Gordon JR, Burd PR, Galli SJ. Mast cells as a source of multifunctional cytokines. Immunol Today 1990; 11:458-64.
142. Kendall JC, Li X-H, Galli SJ, Gordon JR. Promotion of mouse fibroblast proliferation by IgE-dependent activation of mast cells: Role for tumour necosis factor-α and transforming growth factor-β. J Allerg Clin Immunol 1996; (in press).
143. Gordon JR. Mast cell-dependent induction of monocyte responses. Role for monocyte chemoattractant protein-1 (MCP-1) expression during cutaneous allergic responses. In: Metzger H, Pecht I, Razin E, eds. Proceedings of the Signal Transduction in Mast Cell Activation and Development Workshop. Bethesda, 1995.
144. Mican JA, Arora N, Burd PR et al. Passive cutaneous anaphylaxis in mouse skin is associated with local accumulation of interleukin-6 mRNA and immunoreactive interleukin-6 protein. J All Clin Immunol 1992; 90:815-24.
145. Van Snick J. Interleukin-6: An Overview. Ann Rev Immunol 1990; 8:253-78.
146. Gauldie J, Richards C, Harnish D et al. Interferon-β2/BSF-2 shares identity with monocyte-derived hepatocyte stimulating factor (HSF) and regulates the major acute phase protein response in liver cells. Proc Natl Acad Sci (USA) 1987; 84:7251-55.
147. Nagai H, Sakurai T, Inagaki N et al. An immunopharmacological study of the biphasic allergic skin reaction in mice. Biol Pharm Bull 1995; 18:239-45.
148. Martin TR, Galli SJ, Katona IM et al. Role of mast cells in anaphylaxis. Evidence for the importance of mast cells in pulmonary alterations and death induced by anti-IgE. J Clin Invest 1989; 83:1375-83.
149. Okayama Y, Quint D, Hunt T et al. Expression of IL-4 mRNA in human dermal mast cells in response to Fc receptor crosslinkage in the presence of SCF. J All Clin Immunol 1993; 91:256 (abstract).
150. Leung DYM, Pober JS, Cotran RS. Expression of endothelial-leukocyte adhesion molecule-1 in elicited late phase allergic reactions. J Clin Invest 1991; 87:1805-13.
151. Bochner BS, Charlesworth EN, Lichtenstein LM et al. Interleukin-1 is released at sites of cutaneous allergic reactions. J Allergy Clin Immunol 1990; 86:830-38.
152. Lee CE, Neuland ME, Teaford HG et al. Interleukin-6 is released in the cutaneous response to allergen challenge in atopic individuals. J Allergy Clin Immunol 1992; 89:1010-20.
153. Gordon JR. Allergic late phase response cytokine production in human skin and lung tissues challenged in vitro via the FcεRI. Proceedings of the 9th International Congress of Immunology. San Francisco, July 1995.
154. Ying S, Tabordabarata L, Meng Q et al. The kinetics of allergen-induced transcription of messenger RNA for monocyte chemotactic protein-3 and RANTES in the skin of human atopic subjects—relationship to eosinophil, T cell, and macrophage recruitment. J Exp Med 1995; 181:2153-9.
155. Kay A, Ying S, Varney V et al. Messenger RNA expression of the cytokine gene cluster, IL-3, IL-4, IL-5, and GM-CSF in allergen-induced late phase cutaneous reactions in atopic subjects. J Exp Med 1991; 173:775-84.
156. Vowels BR, Rook AH, Cassin M et al. Expression of interleukin-4 and interleukin-

5 mRNA in developing cutaneous late-phase reactions. J All Clin Immunol 1995; 96:92-6.
157. Massey W, Friedman B, Kato M et al. Appearance of granulocyte-macrophage colony-stimulating factor activity at allergen-challenged cutaneous late-phase reaction sites. J Immunol 1993; 150:1084-92.
158. Durham S, Ying S, Varney V et al. Cytokine mRNA expression for IL-3, IL-4, IL-5, and GM-CSF in the nasal mucosa after allergen provocation: relationship to tissue eosinophilia. J Immunol 1992; 148:2390-99.
159. Karlsson MG, Davidsson A, Viale G et al. Nasal messenger RNA expression of interleukins 2, 4, and 5 in patients with allergic rhinitis. Diag Mol Pathol 1995; 4:85-92.
160. Ying S, Durham S, Barkans J et al. T cells are the major source of IL-5 mRNA in allergen-induced rhinitis. Am J Resp Cell Mol Biol 1993; 9:356-63.
161. Kay AB, Ying S, Durham SR. Phenotype of cells positive for interleukin-4 and interleukin-5 mRNA in allergic tissue reactions. Int Arch All Immunol 1995; 107:208-10.
162. Ying S, Durham SR, Corrigan CJ et al. Phenotype of cells expressing mRNA for Th2-type (interleukin 4 and interleukin 5) and Th1-type (interleukin 2 and interferon gamma) cytokines in bronchoalveolar lavage and bronchial biopsies from atopic asthmatic and normal control subjects. Am J Resp Crit Care Med 1995; 12:477-87.
163. Ohkawara Y, Yamauchi K, Tanno Y et al. Human lung mast cells and pulmonary macrophages produce tumor necrosis factor-alpha in sensitized lung tissue after IgE receptor triggering. Am J Respir Cell Mol Biol 1992; 7:385-92.
164. Gosset P, Tonnel AB, Joseph M et al. Secretion of a chemotactic factor for neutrophils and eosinophls by alveolar macrophages from asthmatic patients. J All Clin Immunol 1984; 74:827-34.
165. Galli SJ, Gordon JR, Wershil BK et al. Mast cell and eosinophil cytokines in allergy and inflammation. In: Gleich GJ, Kay AB, eds. Eosinophils: Immunological and Clinical Aspects. New York: Marcell Decker, 1994:255-80.
166. Huang SK, Xiao HQ, Kleinetebbe J et al. IL-13 expression at the sites of allergen challenge in patients with asthma. J Immunol 1995; 155:2688-94.
167. Virchow JC, Walker C, Hafner D et al. T cells and cytokines in bronchoalveolar lavage fluid after segmental allergen provocation in atopic asthma. Am J Resp Crit Care Med 1995; 151:960-8.
168. Fowler AA, Schwartz LB. Macrophage-derived chemotactic factors and allergic asthma. J All Clin Immunol 1984; 74: 777-80.
169. Djeu JY, Serbousek D, Blanchar DK. Release of tumor necrosis factor by human polymorphonuclear leukocytes. Blood 1990; 76:1405-11.
170. Costa JJ, Matossian K, Resnick MB et al. Human eosinophils can express the cytokines tumor necrosis factor-alpha and macrophage inflammatory protein-1 alpha. J Clin Invest 1993; 91:2673-84.
171. Lassalle P, Gosset P, Delneste Y et al. Modulation of adhesion molecule expression on endothelial cells during the late asthmatic reaction: role for macrophage-derived tumour necrosis factor-alpha. Clin Exp Immunol 1993; 94:105-10.
172. Gauchat JF, Henchoz S, Mazzei G et al. Induction of human IgE synthesis in B cells by mast cells and basophils. Nature 1993; 365:340-43.
173. Leung DY. Mechanisms controlling the human immunoglobulin E response: new directions in the therapy of asthma. Pediatr Res 1993; 33:S56-62.
174. Seder RA, Paul WE. Acquisition of lymphokine-producing phenotype by CD4+ T cells. Ann Rev Immunol 1994; 12:635-73.
175. Seder RA, Paul WE, Davis MM et al. The presence of IL-4 during in vitro priming determines the lymphokine-producing potential of CD4+ T cells from T cell receptor transgenic mice. J Exp Med 1992; 176: 1091-98.
176. Paul WE, Seder RA, Plaut M. Lymphokine and cytokine production by Fc epsilon RI+ cells. Adv Immunol 1993; 53:1-29.
177. Dudler T, Machado DC, Kolbe L et al. A link between catalytic activity, IgE-inde-

pendent mast cell activation, and allergenicity of bee venom phospholipase A(2). J Immunol 1995; 155:2605-13.
178. Claman HN. Mast cells, T cells and abnormal fibrosis. Immunol Today 1985; 6:192-4.
179. Ruoss SJ, Hartmann T, Caughey GH. Mast cell tryptase is a mitogen for cultured fibroblasts. J Clin Invest 1991; 88:493-99.
180. Monroe JG, Haldar S, Prystowsky MB et al. Lymphokine regulation of inflammatory processes: interleukin-4 stimulates fibroblast proliferation. Clin Immunol Immunopathol 1988; 292-98.
181. Everett ET, Pablos JL, Harley RA et al. The role of mast cells in the development of skin fibrosis in tight-skin mutant mice. Comp Biochem Physiol A-Comp Physiol 1995; 110:159-65.
182. Ishizaka T, Conrad D, Huff T et al. Unique features of human basophilic granulocytes developed in in vitro culture. Int Arch All App Immunol 1985; 77:137-43.
183. Galli SJ. New concepts about the mast cell. New Engl J Med 1993; 328:257-65.
184. Tsai M, Takeishi T, Thompson H et al. Induction of mast cell proliferation, maturation and heparin synthesis by the rat *c-kit* ligand, stem cell factor. Proc Natl Acad Sci (USA) 1991; 88:6382-86.
185. Holgate ST, Robinson C, Church MK. Mediators of immediate hypersensitivity. In: Middleton E, Ellis CE, Adkinson NF, eds. Allergy: Principles and Practice. St. Louis: Mosby, 1988:135-63.
186. Bang H, Muller W, Hans M et al. Activation of Ca^{2+} signaling in neutrophils by the mast cell-released immunophilin FKBP12. Proc Natl Acad Sci (USA) 1995; 92:3435-8.
187. Wilson SD, Kuchroo VK, Israel DI et al. Expression and characterization of TCA3: a murine inflammatory protein. J Immunol 1990; 145:2745-50.

CHAPTER 12

Regulation of FcεRI Function by Integrins and Other Molecules on Mast Cells and Basophils

Majed M. Hamawy

INTRODUCTION

Preceding chapters in this book have described and discussed in detail the structure of FcεRI as well as the intracellular signals transduced following FcεRI aggregation. This chapter is concerned with reviewing and discussing the regulation of FcεRI-mediated responses by signals initiated by molecules expressed on mast cells and basophils. There are numerous examples in the literature describing the ability of membrane molecules to modulate the signal transduction pathways and the function of different surface receptors. Examples are the following: (a) integrins, receptors for the extracellular matrix, modulate CD3-mediated T cell proliferation; (b) the accessory molecules CD4 and CD8 are required for optimal T cell receptor activation; (c) gp130 is a surface molecule that is essential for interleukin-6 receptor function; (d) gangliosides modulate epidermal growth factor-induced signal transduction; (e) CD45, a tyrosine phosphatase, modulates T cell activation. Indeed, recent data implicate several surface molecules in regulating FcεRI signaling and function in mast cells and basophils. Here, some of these surface molecules are discussed in detail.

INTEGRINS

Integrins are surface receptors composed of noncovalently linked heterodimers of α and β polypeptides.[1-3] There is no homology between

IgE Receptor (FcεRI) Function in Mast Cells and Basophils,
edited by Majed M. Hamawy.

the integrin α and β chains. The extracellular domain of the α subunit has three or four tandem repeats of a putative Ca^{2+}-binding motif. There are at least 8 different β and 14 α subunits. The integrins are divided into groups based on the β subunit, therefore there are at least eight different integrin subfamilies. Many α chains associate with only a single β chain, while other α chains associate with several different β chains. Integrins are expressed on many cells, and most cells express several integrins. A further level of complexity is that there is alternative splicing of the cytoplasmic domain of several of the β and α subunits.[4] This variation may play a role in the different interactions of the cytoplasmic domain of integrins with the actin-based cytoskeleton.[5-7]

Integrins mediate the binding of cells to other cells and to the extracellular matrix (ECM). The ECM consists of glycoproteins and proteoglycans that are secreted by cells and assembled locally into an organized network.[8,9] The components of the ECM vary between tissues, but generally include one or more types of collagen and proteoglycans. Other molecules frequently incorporated into the ECM are fibronectin, fibrin(ogen), entactin, tenascin, thrombospondin, von Willebrand factor, vitronectin and laminin. Members of the β_1, β_3, β_5, β_6 and β_7 subfamilies mediate cell binding to the ECM, whereas the β_2 subfamily mediates cell-cell interaction. Some β_1 integrins ($\alpha4\beta_1$), β_3 integrins ($\alpha_v\beta_3$) and β_7 ($\alpha_4\beta_7$) can also bind adhesion molecules on cells and, therefore, mediate cell-cell interactions.

Some integrins bind to several ECM proteins and there is more than one integrin that binds to the same ECM glycoprotein. Short peptide sequences in the ECM constitute some of the ligands for integrin binding. Members of the β_1, β_3, β_5 and β_6 subfamilies recognize, in addition to other sequences, the Arg-Gly-Asp (RGD) sequence that is present on many ECM glycoproteins.[10,11] The $\alpha IIb\beta_3$ also recognizes the sequence KQAGDV in fibrinogen. Unlike other β_1 integrins, the $\alpha_4\beta_1$ (also called VLA-4) binds the Leu-Asp-Val (EILDV)-sequence in the alternatively spliced CS1 region of fibronectin.[11,12]

The specificity and affinity of the integrins appear to be regulated by cell activation.[13-22] Thus, although cells can adhere spontaneously to other cells and to the ECM glycoproteins, cell activation usually enhances such binding. The enhanced binding of the cells may be a consequence of an increase in the binding affinity, an increase in the number of receptors or due to the expression of a new type of receptor. Some of the changes in adherence are regulated by the cytoplasmic domain of the integrins, e.g., the tyrosine phosphorylation of the cytoplasmic domain of β_1 results in inactivation of adherence to fibronectin.[15,23]

By mediating cell-cell and cell-matrix interactions, integrins affect many cellular processes including cell migration, anchorage, morphology, differentiation and proliferation.[24-26] This, in turn, influences several physiological processes including embryological development, hemostasis, thrombosis, wound healing, immune response, tumor metastasis and oncogenic transformation.

Integrins not only establish a physical link between cells and the ECM but they also function as true receptors. Thus, upon aggregation integrins transduce complex signals into the cell that include protein tyrosine phosphorylation, phosphatidylinositol hydrolysis, changes in intracellular pH and calcium concentration.[27-37] Integrins also induce the expression of several genes.[38-42] These integrin-induced signals are independent from those induced by the changes in the cell shape and in the organization of the cytoskeleton that result from cell adhesion to the ECM. Integrins not only transduce intracellular signals but also modulate intracellular signals induced by other surface receptors.[43-47]

Mast cells and basophils have several integrins that mediate binding to other cells and to the ECM proteins.[19,21,48-60] Table 12.1 summarizes the expression of integrins on mast cells and basophils and the ligand that they can bind. Recent data indicate that adhesion by integrins causes

Table 12.1. Integrins on mast cells and basophils

Integrin	CD	Ligand	Human Cells						Rodent Cells			
			Bas	Ku-812	L-,U-MC	S-MC	HMC-1	FL-MC	RBL-2H3	PT-18	BMMC	PRMC
$\alpha_1\beta_1$	49a/29	Lm, Co	?	?	?	?	?	?	?	?	+	?
$\alpha_2\beta_1$	49b/29	Lm, Co	?	+	–	–	–	–	?	?	+	?
$\alpha_4\beta_1$	49d/29	Fn,VACM-1	+	+	+	?	+	+	+	+	+	+
$\alpha_5\beta_1$	49e/29	Fn	+	+	+	?	+	+	+	+	+	+
$\alpha_6\beta_1$	49f/29	Lm	–	–	–	+	–	–	?	?	+	?
$\alpha_L\beta_2$	11a/18	ICAM-1/2	+	+	–	–	–	?	?	?	?	?
$\alpha_M\beta_2$	11b/18	C3bi, Fb	+	+	–	–	–	–	?	?	?	?
$\alpha_X\beta_2$	11c/18	Fb	+	+	–	–	–	–	?	?	?	?
$\alpha_V\beta_3$	51/61	Vn, Fb	?	+	+	+	+	+	+	+	+	+
$\alpha_4\beta_7$	49d/–	Fn, VCAM-1	+	?	?	?	?	?	?	?	+	?

Abbreviation: Bas, peripheral-blood basophils; BMMC, cultured mouse bone-marrow derived mast cells; Co, collagen; Fb, fibrinogen; Fn, fibronectin; FL-MC, cultured human fetal-liver mast cells; HMC-1, human leukemia mast-cell line; ICAM, intercellular adhesion molecule; KU-812, human basophilic leukemia cell line; Lm, laminin; L-MC, human lung mast cells, PRMC, Peritoneal rat mast cells; PT-18, mouse mast cell line; RBL-2H3, rat basophilic leukemia cell line. S-MC, human skin mast cells; U-MC, human uterine mast cells; VCAM, vascular cell adhesion molecule; VN, vitronectin; +, molecules are present on most cells; -, molecules absent or found on few cells; ?, unknown.

important regulatory changes in FcεRI signaling and function in mast cells and basophils.

Modulation of FcεRI-Signaling

Protein tyrosine phosphorylation is vital for FcεRI signaling.[61,62] In the absence of FcεRI aggregation, pp105-115, focal adhesion kinase pp125^{FAK} and paxillin are poorly phosphorylated on tyrosine in fibronectin-adherent RBL-2H3 cells.[49,63,64] None of these proteins are phosphorylated on tyrosine in nonadherent cells. Aggregating FcεRI in nonadherent cells induces low level, if any, tyrosine phosphorylation of pp105-115, pp125^{FAK} and paxillin, whereas it causes strong tyrosine phosphorylation of 38, 72, 80, 97 and 140 kDa proteins. Crosslinking the FcεRI in adherent cells induces a slight increase in the tyrosine phosphorylation of the 38, 72, 80, 97 and 140 kDa proteins when compared with that induced by FcεRI in nonadherent cells. More striking, however, FcεRI aggregation in adherent cells induces strong tyrosine phosphorylation of the pp105-115, pp125^{FAK} and paxillin. Hence, adherence to fibronectin by integrins regulates FcεRI-induced tyrosine phosphorylation of pp105-115, pp125^{FAK} and paxillin. The mechanism(s) by which cell adhesion controls FcεRI-mediated tyrosine phosphorylation may involve any of the following: (a) induction of massive reorganization of the cytoskeleton which, in turn, leads to the redistribution of cellular proteins and to the accumulation of several tyrosine kinases and their substrates at the sites of cell-substratum contact. Thus, in fibronectin-adherent cells, redistributed tyrosine kinases and/or substrates could be more accessible to FcεRI signal transduction pathways; (b) induction of tyrosine phosphorylation of pp105-115, pp125^{FAK} and its substrate paxillin. Adhesion of RBL-2H3 cells to fibronectin induces tyrosine phosphorylation of these proteins. Thus, the phosphorylated tyrosine residues on pp105-115, pp125^{FAK} and paxillin may act as docking sites for Src homology 2, SH2, domains of other proteins including kinases activated by FcεRI. The recognition of the tyrosine phosphorylated proteins by FcεRI-activated kinases may cause further phosphorylation.

The exaggerated protein tyrosine phosphorylation of pp105-115, pp125^{FAK} and paxillin in adherent cells may affect several aspects of FcεRI functions. For example, protein tyrosine phosphorylation is important for FcεRI-mediated secretion. Indeed, FcεRI aggregation releases significantly more histamine from adherent than from nonadherent cells (see below). Also, protein tyrosine phosphorylation is important for cell growth and differentiation. As FcεRI aggregation induces cell proliferation,[65] the enhanced protein tyrosine phosphorylation may modulate FcεRI-mediated cell proliferation. Figure 12.1 depicts some of the intracellular signals induced by FcεRI and integrins in mast cells.

Modulation of FcεRI-Mediated Secretion

Adherence of RBL-2H3 cells to fibronectin does not induce secretion.[22] Yet, it markedly potentiates the release of mediators induced by aggregating FcεRI.[22] Accordingly, FcεRI aggregation releases significantly more histamine from fibronectin-adherent RBL-2H3 cells compared to that released from nonadherent cells. The enhancement is significant at all antigen concentrations tested. This adhesion-induced enhancement of histamine release was not due to the change in the rate of FcεRI-mediated secretion.[22] Thus, adhesion leads to a fundamental change in the amounts of histamine that can be released by FcεRI aggregation. Since both adherent and nonadherent cells contain similar amount of histamine, the enhanced release must be due to a change in the responsiveness of the cells to activation by FcεRI aggregation. Importantly, the enhancement lasts for at least 6 hours, indicating that this is not a transient phenomenon.[22] Recently, Ra et al[66] have reported that adhesion of the mouse mast cell line PT-18 by integrins to fibronectin enhances FcεRI-mediated cytokine release and cytokine gene expression. The role of integrins in

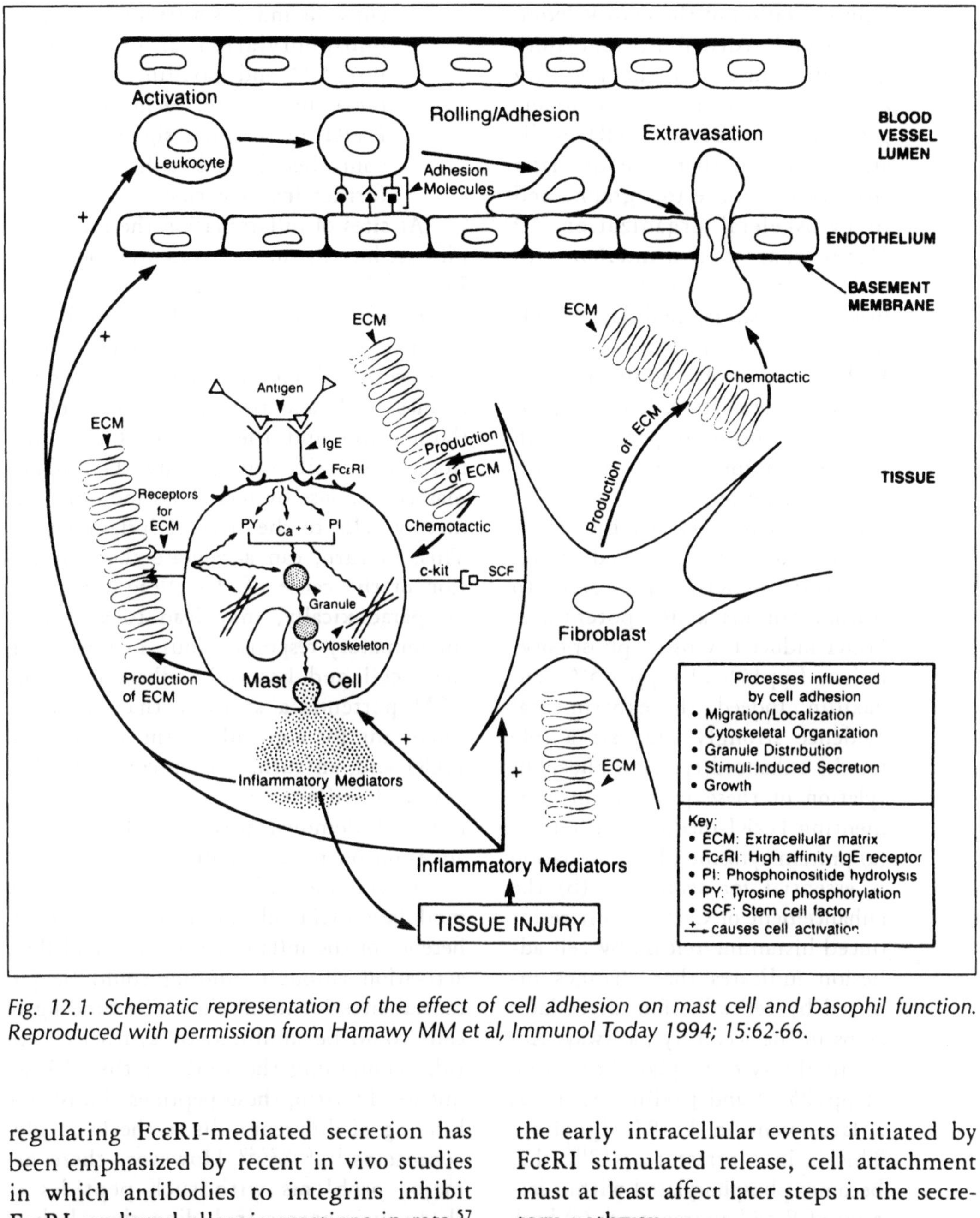

Fig. 12.1. Schematic representation of the effect of cell adhesion on mast cell and basophil function. Reproduced with permission from Hamawy MM et al, Immunol Today 1994; 15:62-66.

regulating FcεRI-mediated secretion has been emphasized by recent in vivo studies in which antibodies to integrins inhibit FcεRI-mediated allergic reactions in rats.[57]

Mast cells and basophils can also be stimulated by Ca^{2+} ionophores to secrete by a receptor-independent pathway.[67] Interestingly, this pathway is also regulated by cell adhesion. Accordingly, RBL-2H3 cells plated on fibronectin release significantly more histamine when activated with the Ca^{2+} ionophore A23187 than do nonadherent cells.[22] In as much as ionophore-induced release bypasses some of the early intracellular events initiated by FcεRI stimulated release, cell attachment must at least affect later steps in the secretory pathway.

Adherence-induced enhancement of secretion could be the consequence of any of the following.

1. Modification of cell shape. This could activate stretch-sensitive ion channels and thus alter the level of intracellular ions.[68] Secretion from mast cells and basophils is sensitive to changes in intracellular Ca^{2+} and Na^{+}.[69]

2. Reorganization of the cytoskeleton. This could redistribute regulatory molecules and alter the mobility and function of integral membrane proteins.[70,71] During FcεRI-mediated secretion, there are dramatic changes in surface topography and in cytoskeletal organization.[72,73] Moreover, agents that interfere with actin polymerization, such as dihydrocytochalasin B, modulate FcεRI-induced secretion.[74]
3. Redistribution of granules to areas closer to the surface. This may facilitate the fusion of granules with the plasma membrane, a critical step in secretion.[75]
4. Changes in intracellular biochemical reactions. As discussed above, adhesion of RBL-2H3 cells to fibronectin markedly potentiates FcεRI-induced tyrosine phosphorylation of pp105-115, $pp125^{FAK}$ and paxillin. Indeed, the experimental evidence strongly suggests a role for the enhanced tyrosine phosphorylation of these proteins in augmenting FcεRI-mediated secretion: (a) protein tyrosine phosphorylation is essential for secretion; (b) the enhancement of Ca^{2+} ionophore-induced histamine release by cell adhesion indicates that changes induced by cell adhesion affect later steps in the secretory pathway. Accordingly, tyrosine phosphorylation of $pp125^{FAK}$ and paxillin occurs as a late event in FcεRI signaling. This is in contrast to $p72^{syk}$ that becomes phosphorylated within seconds of FcεRI aggregation; (c) both $pp125^{FAK}$ and paxillin can be tyrosine phosphorylated by the direct increase in intracellular Ca^{2+} and the direct activation of protein kinase C (PKC). In FcεRI signaling, both Ca^{2+} influx and PKC activation are late events and are required for secretion; (d) when compared with that in the nonadherent cells, FcεRI aggregation in adherent cells induces a stronger tyrosine phosphorylation of pp105-115, $pp125^{FAK}$ and paxillin and releases more histamine. Whether other FcεRI-intracellular signals are also controlled by cell adhesion awaits further investigation.

At sites of inflammation there is a significant increase in the concentration of the ECM.[76-79] In vitro these molecules are chemotactic for mast cells[80,81] and they enhance FcεRI-mediated secretion from these cells.[22,57] The increase in the number of mast cells and basophils at sites of inflammation and the enhanced secretion from these cells can amplify the inflammatory response by recruiting more inflammatory cells to the sites of inflammation. This, in turn, can accelerate the elimination of the cause of the allergic response or, paradoxically, may lead to a chronic inflammatory response. Thus, manipulating mast cell and basophil adhesion to the ECM portends a strategy that could be important for controlling the function of FcεRI and, in turn, the outcome of allergic and inflammatory responses. Figure 12.2 demonstrates the effect of cell adhesion on mast cell function.

Several methods may be employed to modulate ECM-cell interactions. The adherence of the inflammatory cells and their activation through binding could be potential sites for inhibition. The binding of cells could be inhibited by synthetic peptides containing the RGD or the CS1 sequence. In vitro, these peptides abolish the binding of different cells to the ECM glycoproteins.[12,16-19,22,49] However, there are major problems with such peptides as therapeutic agents including their lack of specificity, short half-life in vivo and problems of administration.[10] Short synthetic peptides containing the RGD sequence can be designed to exhibit varying integrin specificities by restricting the conformation of the peptide. Other approaches to regulate cell-ECM interaction will be the design of drugs that can specifically suppress or downregulate integrins on the surface of inflammatory cells or to inhibit the

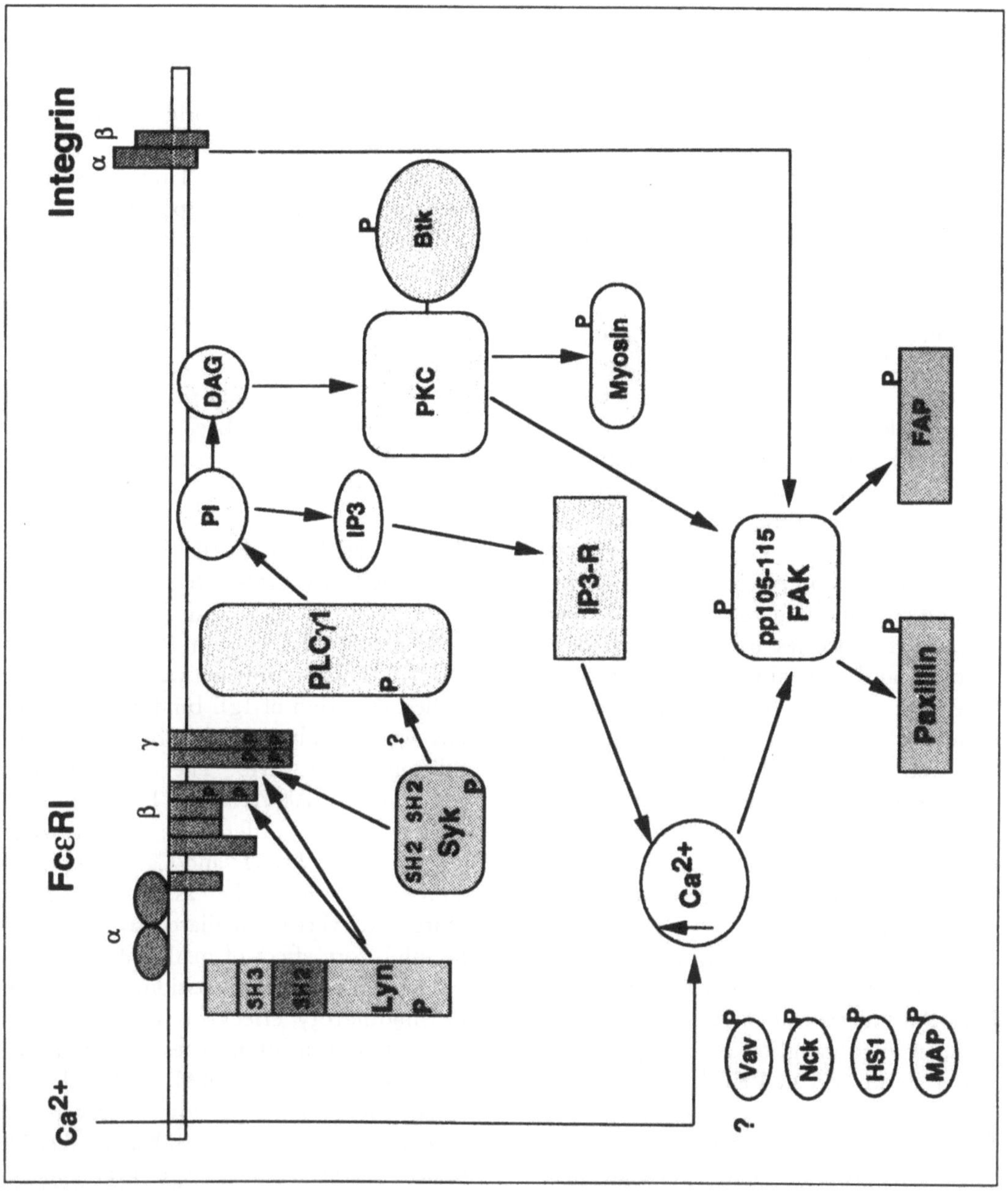

Fig. 12.2. Intracellular signals induced in mast cells by FcεRI aggregation. Reproduced by permission of the publisher from Hamawy MM et al. Protein Tyrosine Phosphorylation as a Mechanism of Signaling in Mast Cells and Basophils. Cell Signal 1995; 7:535-44. Copyright 1995 by Elsevier Science, Inc.

activation sequence that is induced by the adherence of the cells to the ECM glycoproteins.

GANGLIOSIDES

Gangliosides are sialic acid-containing glycosphingiolipids found on the outer leaflet of the plasma membrane of all vertebrate tissues and species.[82] Although predominantly located in plasma membranes, they also present at lower concentration in the endoplasmic reticulum, the golgi apparatus and secondary lysosomes. Because gangliosides are primarily localized in the outer leaflet of the plasma membrane, they are ideally positioned to interact with other cells, the extracellular matrix and other components of the extracellular vicinity. Recent studies implicate gangliosides as playing an important role in several aspects of cellular processes including cell development and adhesion.[82-86] Gangliosides have been shown to have an important biological role both in the recognition and in the modulation of the function of membrane-associated proteins.[83,84] For example, gangliosides inhibit epidermal growth factor signal transduction, probably by inhibiting the autophosphorylation of the receptor. Interestingly, the ganglioside modulatory effect is not limited to membrane associated proteins but it also regulates the function of intracellular proteins; thus, in vitro gangliosides are specific and potent inhibitors of PKC.[87]

GANGLIOSIDE G_{D1b} IN MAST CELLS

Our knowledge of the regulatory role of gangliosides in mast cells came primarily from studies with two mAb produced in different laboratories.[88,89] Although both mAb AA4 (IgG) and B17 (IgM) bind ganglioside G_{D1b}, mAb AA4 binds also to novel α-derivatives of the ganglioside G_{D1b}.[88,90] Binding studies have shown that a maximum of 2 X 10^5 mAb B17 molecules and 38 X 10^5 mAb AA4 molecules can bind per RBL-2H3 cells. The reason for this discrepancy in the number of binding sites for the mAb is not known, but may be due to the binding of the mAb to different components of G_{D1b}. With the exception of bone marrow, mAb AA4 bound only to mast cells in rat tissues. In bone marrow, large, poorly differentiated cells, most likely mast cell precursors, also stained with mAb AA4. Thus, these G_{D1b} derivatives are mast cell specific.

The spatial proximity of the G_{D1b} to FcεRI is supported by several observations: (a) Fab, $F(ab)_2$, or intact mAb inhibited IgE binding to FcεRI; (b) mAb AA4 coprecipitated FcεRI subunits (see below). The proximity of the binding site for mAb AA4 to the FcεRI has also been documented by fluorescence resonance energy transfer experiments.[91]

MODULATION OF FcεRI-MEDIATED SECRETION BY GANGLIOSIDES

Both mAb B17 and AA4 inhibited FcεRI-mediated secretion from RBL-2H3 cells.[88,89] This inhibition was independent of the inhibition of IgE binding; thus, the cells were incubated with IgE before the addition of the mAb. The modulation of secretion by mAb AA4 and B17, however, exhibited several differences: (a) mAb AA4 inhibited both FcεRI- and Ca^{2+} ionophore-induced secretion, whereas, mAb B17 inhibited only FcεRI-mediated secretion; (b) the inhibitory effect of mAb AA4 required at least 6 hours preincubation. In contrast, the modulatory effect of mAb B17 was apparent within 30 minutes of incubation; (c) unlike mAb AA4, mAb B17 had a bimodal effect. Thus, at low concentration of mAb B17 there was enhanced FcεRI-induced secretion, whereas, at high concentration of B17 secretion was inhibited. The reasons for the differences in their inhibition of secretion is not known but may be due to their binding to different components of G_{D1b} or could be related to the different glycolipid aggregation patterns produced by virtue of their different binding sites (IgG vs. IgM).

The mechanisms by which gangiliosides modulate FcεRI function are not known. Importantly, gangiliosides initiate morphological changes that are almost identical to those initiated in response to

FcεRI aggregation,[91] associate with intracellular proteins involved in FcεRI signaling,[92] and initiate intracellular signals similar to those induced by FcεRI aggregation.[91,93]

Ganglioside G_{D1b}-induced morphological changes

Binding of mAb AA4 to RBL-2H3 cells induces dramatic morphological changes.[91,93] As a result, the cells lose their normal spindle shaped appearance, increase their ruffling, and spread over the surface of the culture dish (Fig. 12.3). These changes are accompanied by a redistribution of the cytoskeletal elements, actin, tubulin and vimentin, but only the actin was associated with the membrane ruffles. These morphological changes did not affect the surface distribution of mAb AA4 and did not induce the internalization of the mAb. The morphological changes induced by mAb AA4 were identical to those induced by crosslinking FcεRI. In contrast, mAb AA4-induced redistribution of tyrosine phosphorylated proteins was different from that induced by FcεRI aggregation.[93] Thus, the tyrosine phosphorylated proteins after mAb AA4 binding were localized in patches at the cell membrane and in areas of cell-cell contact, whereas after FcεRI aggregation, there was a reticular cytoplasmic pattern. Although Fab fragments of mAb AA4 bind to the cells, no morphological changes were observed.

Ganglioside G_{D1b}-mediated signal transduction

Treatment of RBL-2H3 cells with mAb AA4 results in a significant hydrolysis and release of inositol phospholipid, although this was considerably less than that seen with FcεRI aggregation.[91]

Although mAb B17 did not by itself induce an increase in intracellular Ca^{2+} ions, it did enhance or inhibit FcεRI-induced changes in Ca^{2+} level in a manner that correlated with its effect on secretion.[88] In contrast, mAb AA4 binding induced a moderate, but significant, rise in intracellular Ca^{2+} levels as measured by Quin-2 fluorescence.[91] Unlike mAb B17, mAb AA4 had little effect on the cells' capacity to respond to crosslinking FcεRI, and the final intracellular Ca^{2+} level was not significantly different from that induced by aggregating FcεRI alone.

mAb AA4 binding also affected PKC.[91] Changes in PKC induced by mAb AA4 were measured by staining with NBD-dodencanoyl-phorbol acetate. Changes in fluorescence intensity were observed as early as 5 minutes in treated cells and reached a maximum at 30 minutes. More evidence for the effect of mAb AA4 on PKC comes from the fact that mAb AA4-induced morphologic changes can be inhibited by staurosporine, an inhibitor of PKC.

Binding of mAb AA4 also results in the tyrosine phosphorylation of several proteins involved in FcεRI-signaling including FcεRI-β, FcεRI-γ, $p72^{syk}$ and phospholipase C.[92,93] There was also activation of the $p53/56^{lyn}$ and $p72^{syk}$ protein tyrosine kinases. However, unlike FcεRI aggregation, mAb AA4 failed to induce tyrosine phosphorylation of pp105-115 and $pp125^{FAK}$. Furthermore, the time course of tyrosine phosphorylation induced by mAb AA4 was slower than that induced by FcεRI aggregation.

Association of gangliosides G_{D1b} with cellular proteins

The restriction of ganglioside to the outer leaflet necessitates an association of the ganglioside with other surface proteins. Thus, to exert their biological effects, gangliosides must interact and rely on membrane proteins for transducing intracellular signals. Immunoprecipitation studies suggest the association of gangliosides G_{D1b} with FcεRI in mast cells.[92,93] Both the β and the γ subunits of FcεRI were detected in mAb AA4 precipitates. Yet, only 2% of cellular FcεRI was detected in the mAb AA4 immunoprecipitate. This suggests that there is only a weak association of the ganglioside with FcεRI. Strikingly, mAb AA4 failed to induce protein tyrosine phosphorylation either when FcεRI

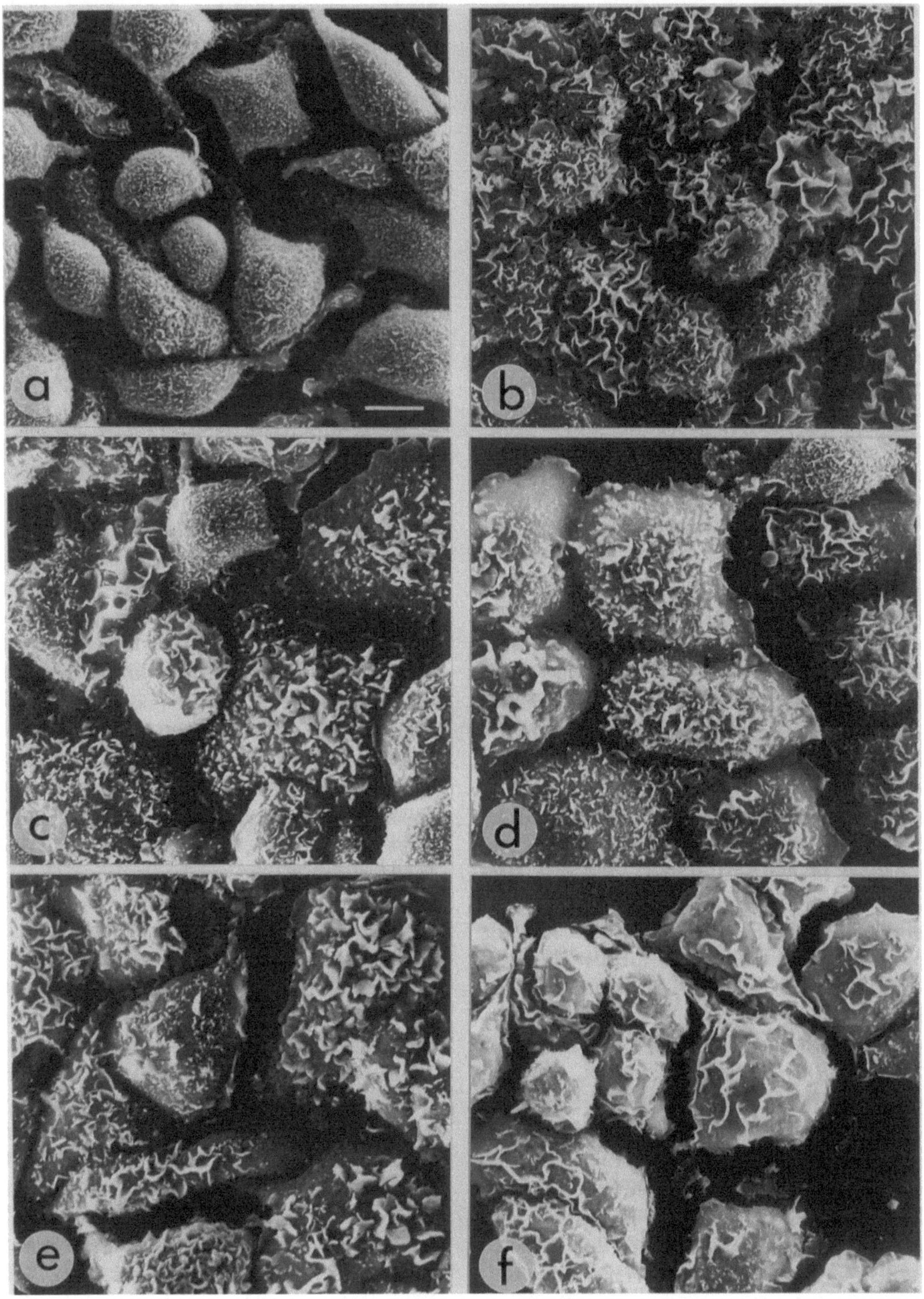

Fig. 12.3. Morphological changes induced in RBL-2H3 cells by anti-ganglioside mAb AA4. Cells were unstimulated (a) or incubated at 37°C with mAb AA4 for 5 minutes (b); 10 minutes (c); 1 minutes (d); 20 minutes (e); 30 minutes (f). Scale bar = 10 μm. (Courtesy W. D. Swaim, NIDR, NIH).

was occupied by IgE or when $F(ab)_2$ fragments of mAb AA4 were used, although these fragments still induced morphological changes.[93] These data suggest that ganglioside G_{D1b} requires FcεRI for inducing protein tyrosine phosphorylation. Thus, a model can be envisioned in which the $F(ab)_2$ sites of mAb AA4 bind directly to the ganglioside, whereas the Fc portion of the antibody binds FcεRI. In addition to FcεRI, ganglioside G_{D1b} appears to associate with several other membrane proteins. For example, mAb AA4 immunoprecipitated proteins of 50-60, 120 and 135 kDa from I^{125}-surface labeled RBL-2H3 cells.[94] The association of these proteins with the ganglioside was dependent on the detergent used for cell solubilization; thus, these proteins were detected when the cells were solubilized in CHAPS but not with Triton X-100.

When mAb AA4 precipitates were subjected to in vitro kinase assays, there were phosphorylated proteins of 71-80, 53/56 and 41/42 kDa.[92] The 53/56 kDa proteins were phosphorylated only on tyrosine, whereas, the 71-80 and the 41/42 kDa proteins were phosphorylated both on tyrosine and on serine. These proteins were observed when several different detergents were used for solubilizing the cells. The 53/56 kDa proteins were identified as the Src-related tyrosine kinase $p53/56^{lyn}$. At least 10% of the cellular Lyn was present in mAb AA4 precipitates. Although both Lyn and gangliosides associate with FcεRI, the association of Lyn with the gangliosides was not secondary to the association of the kinase with FcεRI. The protein(s) mediating the association between Lyn and the gangliosides is not known. The association of the gangliosides with protein kinases and FcεRI may explain the modulatory effect excreted by gangliosides on FcεRI function.

PROTEIN TYROSINE PHOSPHATASE CD45

CD45, also known as the leukocyte common antigen (LCA), consists of a group of high molecular weight glycoproteins which are expressed on the surface of all leukocytes and hematopoietic progenitor cells.[95,96] The cytoplasmic domain is highly conserved between species, while the extracellular domain is much less conserved. The cytoplasmic domain of CD45 consists of two tandemly repeated protein tyrosine phosphatase (PTP) domains of about 240 residues separated by a 56 residue spacer region. Furthermore, the cytoplasmic domain consists of a membrane proximal region of 77 residues and a carboxy-terminal tail of 78 residues. The second PTP domain includes a unique acidic region of 21 residues with several potential sites for serine phosphorylation. The mechanism(s) by which the PTP activity of CD45 is regulated is not known. Although CD45 has several Ser/Thr phosphorylation sites, the PTP activity of purified CD45 is not influenced by phosphorylation.[97] The expression of CD45 on the cell surface suggests that the PTP may be a receptor whose enzymatic activity would be regulated by an external ligand. However, such a ligand has not yet been found. Furthermore, deletion of the external domain of CD45 does not seem to affect its PTP activity.

The potential role for CD45 in regulating FcεRI signal transduction came from studies with human basophils.[98] In those studies, mAb to CD45 inhibited FcεRI-mediated histamine release from human basophils. Notably, the mAb required several hours of incubation to induce maximum inhibition. Maximum inhibition by the mAb also required incubation at 25°C or 37°C, suggesting a requirement for actively metabolizing basophils. The mAb did not inhibit histamine release induced by formylmethionine tripeptide, Ca^{2+} ionophore or phorbol esters. Thus, CD45 changes are most likely affecting early events in FcεRI-signal transduction pathway.

Berger et al[99] have shown similar inhibition by mAb to CD45 of FcεRI-mediated degranulation in bone marrow derived mast cells. A more direct evidence for the role of CD45 in FcεRI-signal transduction came from studies in which mast cells from CD45-deficient mice were used.[99] Cross-linking FcεRI on mast cells deficient in CD45 did not induce degranulation. Yet, these cells degranulated in response to stimulation by Ca^{2+} ionophores. These data support the notion that CD45 regulates early events in the FcεRI-signal transduction pathway.

Although tyrosine phosphatases are important for FcεRI signal transduction in RBL-2H3 cells, the role of CD45 in regulating FcεRI-signaling in these cells is not clear (for details see chapter 7). For example, the secretory response, as well as the inositol 1,4,5-triphosphate (InsP3) formation to FcεRI, was not attenuated in CD45-deficient RBL-2H3 cells.[100]

The mechanism by which CD45 regulates FcεRI-signal transduction is not known. Data from CD45-deficient RBL-2H3 cells suggest that CD45 might directly or indirectly regulate the kinase activity of PKC.[100] Recently, Adamczewski et al[101] transfected FcεRI in Jurkat cell line and in a CD45-deficient variant of Jurkat. Although FcεRI aggregation induced Ca^{2+} influx in CD45-positive Jurkat cells, it failed to do so in CD45-deficient cells. Reintroducing CD45 restored, at least in part, FcεRI-mediated increase in intracellular Ca^{2+}. As in RBL-2H3 cells, aggregating FcεRI in Jurkat cells led to an increase in the phosphorylation of the β and γ subunits. Surprisingly, aggregating FcεRI in CD45-deficient Jurkat cells led only to a slight increase in the phosphorylation of FcεRI subunits. Reintroducing CD45 into the cells restored the phosphorylation of the subunits to levels similar to that induced by FcεRI aggregation in Jurkat cells. These data suggest a role for CD45 in facilitating FcεRI-induced tyrosine phosphorylation. In T cells, CD45 mediates its effect, at least in part, by regulating the tyrosine phosphorylation of Src-family of protein kinases. The data obtained by Adamczewski et al[101] suggest a similar mechanism for CD45 in regulating FcεRI-signal transduction. Thus, the basal activity of the Src-related tyrosine kinase, Lck, is increased in CD45-deficient Jurkat cells when compared with that in CD45-positive cells. When FcεRI was aggregated in the CD45 deficient cells it failed to increase the kinase activity of Lck. Reintroducing CD45 decreased the basal activity of Lck, and as in the CD45-positive cells, there was an FcεRI-induced increase in the basal level of Lck.

CYTOKINE RECEPTORS

The cytokines are soluble factors that include several subfamilies including the hematopoietins, the interferons, tumor necrosis factor (TNF) and related molecules, the immunoglobulin superfamily and the chemokines. These molecules exert their biological functions by binding to high affinity receptors expressed on the surface of different types of cells and by inducing biochemical signals that profoundly affect the functions of the cell. Thus, cytokines play important regulatory roles in the immune, hematopoietic, endocrine, hepatic and neural systems.

There are numerous examples in the literature of the regulation of FcεRI responses by different cytokines (see also chapters 11 and 13). Here, the effects of the interaction of interleukins and stem cell factor (SCF) with their respective receptors on FcεRI responses are discussed.

INTERLEUKIN RECEPTORS

These receptors are type I membrane glycoproteins, with their N-termini outside the cell and with a single hydrophobic transmembrane domain. The extracellular part of the receptor is the most conserved between receptors. The highly conserved parts are four cystine residues located in the N-terminal half of the region, and a WSXWS sequence found just outside the transmembrane domain. Binding of leukotrienes to their respective receptors initiates intracellular signals that include the activation of tyrosine kinases and tyrosine phosphorylation of several proteins including the receptor itself.[102]

IL-1, -3, -5 and -6 markedly enhance FcεRI-dependent secretion from human basophils, but IL-3 is the most potent.[103-111] In contrast, IL-2 and -4 are ineffective in modulating FcεRI-mediated secretion.[107] IL-3 enhances FcεRI-induced histamine release from both normal and atopic basophils. By contrast, IL-3 does not influence FcεRI responses on mast cells, presumably due to the lack of an IL-3 receptor on these cells.[112] IL-3-mediated enhancement of FcεRI-induced secretion is temperature dependent, rapid and reaches a plateau in a few minutes. Furthermore, histamine is released more rapidly in response to FcεRI aggregation after IL-3 priming. The enhancement of histamine release by IL-3 did not result from de novo syntheses of histidine decarboxylase; thus, IL-3 primed and unprimed cells contained similar amounts of histamine. Unlike other ILs, IL-3 releases histamine from basophils, mainly from allergic donors.[104,105,107,113,114] Elution of basophil-bound IgE by acid treatment renders the cells unresponsive to IL-3, suggesting that surface IgE may be necessary for the histamine releasing activity of IL-3.

The mechanisms by which IL modulates FcεRI-mediated secretion are not known. In the absence of IL-3, the PKC inhibitor staurosporine inhibited FcεRI-mediated histamine release.[104] Yet, in the presence of IL-3, saturasoprine appeared ineffective. These data suggest that IL-3 may modulate FcεRI-responses by activating staurosporine-insensitive PKC. Pertussis toxin did not inhibit FcεRI-mediated secretion, or the IL-3 priming effect, suggesting that pertussis toxin-sensitive G proteins are not involved in the initiation of this process.[104] Beauvais et al[115] reported that the enhancing effect of IL-3 may involve Na^+/Ca^{2+} exchanges.[115] In their studies, the enhancement by IL-3 was dependent on the presence of extracellular Na^+ and was reduced by benzamil, an inhibitor of Na^+/Ca^{2+} exchanges.

Kit

The proto-oncogene c-*kit* encodes the Kit protein, a 145-160 kDa cell surface receptor with tyrosine kinase activity.[116] This protein is a member of the receptor tyrosine kinase family that includes the platelet-derived growth factor receptor and the receptor for colony stimulating factor-1 on macrophages. Kit is a transmembrane protein that consists of three domains: an N-terminal, extracellular, ligand-binding domain with five immunoglobulin-like loops, a small transmembrane segment, and a C-terminal intracellular tyrosine kinase region. This receptor is expressed on different cells including mast cells; yet, it is sparsely expressed on basophils. The ligand for Kit is the growth factor SCF, which is a gene product of the Steel locus in the mouse. SCF has been referred to by many names, including mast cell growth factor, steel factor and Kit ligand. It exists as a soluble molecule as well as a cell-associated protein. SCF-Kit interaction influences several cellular processes including anchorage, migration, proliferation, differentiation and secretion.

In human lung mast cells, a mucosal type mast cell, recombinant SCF (rhSCF) at concentrations 10-100 times less than that required to promote cell proliferation, enhances FcεRI-mediated histamine and leukotrine release.[117] By contrast, other cytokines including IL-1, IL-3, IL-4, IL-9 and GM-CSF failed to modulate FcεRI-mediated release from these cells. SCF by itself induced minimal histamine release in some experiments and failed to cause leukotrine release. Although the presence of Kit on human lung mast cells has not been demonstrated, SCF-mediated effects are presumed to occur as a result of activating Kit on the surface of these cells. SCF did not alter the time course of FcεRI-induced mediator release from mast cells. This is in contrast to the effect of IL-3 on FcεRI-mediated release from basophils (see above). However, SCF did lower the threshold of FcεRI-mediated activation of mast cells. Maximal potentiation of FcεRI-mediated histamine release occurred when Kit and FcεRI were activated simultaneously. In contrast, Kit had to be preactivated for at lest 10 minutes in order to enhance FcεRI-mediated leukotrine release.

Incubation of rhSCF with human skin mast cells, a connective tissue type mast cell, induces a rapid increase in intracellular Ca^{2+} levels that is associated with histamine release.[118] Only a few minutes of incubation of these cells with rhSCF significantly augments FcεRI-mediated histamine release. In contrast, a 24 hour incubation of mast cells with rhSCF did not affect FcεRI-induced secretion, probably due to Kit desensitization. The potentiating activity of rhSCF appears to be specific for FcεRI since it did not enhance secretion in response to other stimuli. Studies with mouse connective tissue mast cells showed that SCF also enhances FcεRI-mediated secretion in these cells.[112] In basophils, rhSCF does not induce an increase in intercellular Ca^{2+} or secretion.[118] Nonetheless, it potentiates FcεRI-mediated secretion. However, rhSCF was far less effective in potentiating FcεRI-mediated secretion from basophils than from mast cells.

The mechanism(s) by which Kit modulates FcεRI-mediated secretion is not known. The binding of SCF to Kit results in Kit autophosphorylation.[119] As a result, several molecules involved in FcεRI signaling associate with tyrosine phosphorylated Kit. These include SH2-containing protein tyrosine phosphatases;[120,121] Grb2, a small adaptor protein with one SH2 and two SH3 domains that may be involved in regulating the activity of Ras;[121] and Shc which has two SH2 domains and one SH3 domain.[122] As protein tyrosine phosphorylation is essential for FcεRI-mediated signaling, it is feasible that the modulatory effect of Kit involves the modulation of FcεRI-mediated protein tyrosine phosphorylation. Kit-signaling also involves the increase in intracellular Ca^{2+} concentrations[123,124] and the activation of several enzymes that are involved in FcεRI signaling including phospholipase C-γ,[119,125] PKC[123] and mitogen activated protein kinase.[126]

It is important when analyzing data generated with human mast cells and basophils to bear in mind the role of contaminating cells, i.e., cells that copurify with mast cells and basophils. Thus, contaminating cells may respond to the stimuli by releasing factors that could influence mast cell and basophil responses.

LOW AFFINITY IgG RECEPTOR (FcγRII)

Human FcγRII (CD32) is a heavily glycosylated 40 kDa transmembrane glycoprotein.[127] Human FcγRII exists as two types at least: FcγRIIA/C and FcγRIIB. Mouse FcγRII is a 40-60 kDa protein and resembles closely human FcγRIIB. Human and mouse Fc bind monomeric IgG with low affinity K_a $10^7 M^{-1}$. Mouse FcγRII can also bind IgE with low affinity. In the mouse and human, two isoforms generated by alternative splicing of sequences encoded by the single FcγRIIB gene have been identified as FCγRIIB1 and FCγRIIB2. The mouse isoforms are identical to each other with the exception of a 47 amino acid deletion in the cytoplasmic domain of FCγRIIB2. The human isoforms differ by a 19 amino acid deletion in the cytoplasmic domain of FcγRIIB2. Both mouse mast cells and human basophils express FcγRII.

Recent studies by Daeron et al[128] suggest that FcγRII may regulate FcεRI function. These researchers stably transfected RBL-2H3 cells with cDNAs encoding FcγRII and other Fcγ receptors. In the transfected cells, FcγRII was inactive, that is, it failed to trigger the cells for mediator release. Importantly, secretion induced by FcεRI was inhibited when FcεRI was crosslinked to FcγRII. Inhibition affected only the crosslinked FcεRI and was reversible if the crosslinked receptors were disengaged from crosslinking. Both FcγRIIB1 and FCγRIIB2 were equally effective in inhibiting FcεRI-mediated secretion. Notably, deleting the cytoplasmic domain of FcγRII abolished its inhibitory effect.

The inhibitory effect of FcγRII might be due to the presence of an immunoreceptor tyrosine-based inhibitory motif (ITIM) in its cytoplasmic region. ITIM contains a single YXXL sequence which is a potential ligand for SH2 domains of tyrosine kinases/phosphatases.[129] This is in contrast to the cytoplasmic region of FcεRI that contains the immunoreceptor tyrosine-based activation motif (ITAM), a twice repeated YxxL sequence separated by seven amino acids (see chapter 3). Recently, D'Ambrosio et al[130] demonstrated that the protein tyrosine phosphatase PTP1C (also known as SHP) binds to the ITIM within the cytoplasmic domain of FCγRII in activated B cells. Importantly, the binding of the phosphatase to the ITIM appears to increase its phosphatase activity. It has been proposed that the SH2 of the phosphatase may be involved in regulating the phosphatase activity by an allosteric mechanism. Thus, the binding of the SH2 domains to ITIM may induce conformational changes and/or may render the PCP1C susceptible to modifications that upregulate the phosphatase activity. The association of PCP1C and FcγRII in B cell is believed to be important for the regulation of the activation of the B cell receptor. Thus, the tyrosine phosphorylation of FcγRII recruits PCP1C to the membrane. This in turn leads to an increase in the phosphatase activity of PCP1C. By virtue of proximity, the FcγRII-associated phosphatase may then terminate B cell receptor signal transduction by dephosphorylating appropriate substrates. Whether a similar model exists in mast cells and basophils awaits further studies.

CONCLUDING REMARKS

FcεRI function is markedly influenced by other molecules on mast cells and basophils. In vitro studies on FcεRI function often neglect the effects of membrane molecules other than FcεRI subunits on the signals generated by aggregating the receptor. As discussed above, cell adhesion induces intracellular signals and modifies those initiated by FcεRI signaling. Hence, drugs that inhibit or modulate FcεRI signaling in nonadherent cells may not exert the same effects on FcεRI function in adherent cells. Another important issue to consider when interpreting in vitro data is that the intracellular signals induced by FcεRI aggregation on cell lines grown in a selective medium or in a medium containing growth factors may not mimic (in magnitude and/or in nature) those signals initiated by FcεRI aggregation in vivo. This is particularly important to bear in mind when designing drugs to abort FcεRI signaling. Finally, surface molecules and their ligands which modulate FcεRI function should be also considered as a target when treating allergic diseases.

REFERENCES

1. Hynes RO. Integrins: versatility, modulation and signaling in cell adhesion. Cell 1992; 69:11-25.
2. Albelda SM, Buck CA. Integrins and other cell adhesion molecules. FASEB J 1990; 4:2868-80.
3. Ruoslahti E. Integrins. J Clin Invest 1991; 87:1-5.
4. Balzac F, Belkin AM, Koteliansky VE et al. Expression and functional analysis of a cytoplasmic domain variant of the beta 1 integrin subunit. J Cell Biol 1993; 121: 171-78.
5. Burn P, Kupfer A, Singer SJ. Dynamic membrane-cytoskeletal interactions: specific association of integrin and talin arises in vivo after phorbol ester treatment of peripheral blood lymphocytes. Proc Natl Acad Sci USA 1988; 85:497-501.
6. Burridge K, Fath K, Kelly T et al. Focal adhesions: transmembrane junctions between the extracellular matrix and the cytoskeleton. Annu Rev Cell Biol 1988; 4:487-25.
7. Otey CA, Pavalko FM, Burridge K. An interaction between alpha-actinin and the beta 1 integrin subunit in vitro. J Cell Biol 1990; 111:721-29.
8. Hay ED. Cell Biology of Extracellular Matrix. New York: Plenum Press, 1992.
9. Thompson HL, Burbelo PD, Gabriel G et al. Murine mast cells synthesize basement membrane components. A potential role in early fibrosis. J Clin Invest 1991; 87:619-23.
10. Ruoslahti E, Pierschbacher MD. New perspectives in cell adhesion: RGD and integrins. Science 1987; 238:491-97.
11. Yamada KM. Adhesive recognition sequences. J Biol Chem 1991; 266:12809-12.

12. Komoriya A, Green LJ, Mervic M et al. The minimal essential sequence for a major cell type-specific adhesion site (CS1) within the alternatively spliced type III connecting segment domain of fibronectin is leucine-aspartic acid-valine. J Biol Chem 1991; 266:15075-79.
13. Yoon PS, Boxer LA, Mayo LA et al. Human neutrophil laminin receptors: activation-dependent receptor expression. J Immunol 1987; 138:259-65.
14. Shaw LM, Mercurio AM. Interferon gamma and lipopolysaccharide promote macrophage adherence to basement membrane glycoproteins. J Exp Med 1989; 169:303-08.
15. Shaw LM, Messier JM, Mercurio AM. The activation dependent adhesion of macrophages to laminin involves cytoskeletal anchoring and phosphorylation of the alpha 6 beta 1 integrin. J Cell Biol 1990; 110:2167-74.
16. Owen CA, Campbell EJ, Hill SL et al. Increased adherence of monocytes to fibronectin in bronchiectasis. Regulatory effects of bacterial lipopolysaccharide and role of CD11/CD18 integrins. Am Rev Respir Dis 1992; 145:626-31.
17. Roth P, Polin RA. Lipopolysaccharide enhances monocyte adherence to matrix-bound fibronectin. Clin Immunol Immunopathol 1990; 57:363-73.
18. Bianchine PJ, Burd PR, Metcalfe DD. IL-3-dependent mast cells attach to plate-bound vitronectin. Demonstration of augmented proliferation in response to signals transduced via cell surface vitronectin receptors. J Immunol 1992; 149:3665-71.
19. Dastych J, Costa JJ, Thompson HL et al. Mast cell adhesion to fibronectin. Immunology 1991; 73:478-84.
20. Thompson HL, Burbelo PD, Metcalfe DD. Regulation of adhesion of mouse bone marrow-derived mast cells to laminin. J Immunol 1990; 145:3425-31.
21. Thompson HL, Burbelo PD, Segui-Real B et al. Laminin promotes mast cell attachment. J Immunol 1989; 143:2323-27.
22. Hamawy MM, Oliver C, Mergenhagen SE et al. Adherence of rat basophilic leukemia (RBL-2H3) cells to fibronectin-coated surfaces enhances secretion. J Immunol 1992; 149:615-21.
23. Tapley P, Horwitz A, Buck C et al. Integrins isolated from Rous sarcoma virus-transformed chicken embryo fibroblasts. Oncogene 1989; 4:325-33.
24. Shimizu Y, Shaw S. Lymphocyte interactions with extracellular matrix. FASEB J 1991; 5:2292-99.
25. Springer TA. Adhesion receptors of the immune system. Nature 1990; 346:425-34.
26. Hamawy MM, Siraganian RP, Mergenhagen SE. The role of the extracellular matrix in inflammation. In: Genco R, Hamada S, Lehner T, eds. Molecular Pathogenesis of Periodontal Disease. Washington, D.C.: American Society of Microbiology, 1994: 235-46.
27. Morino N, Mimura T, Hamasaki K et al. Matrix/integrin interaction activates the mitogen-activated protein kinase, p44erk-1 and p42erk-2. J Biol Chem 1995; 270: 269-73.
28. Schlaepfer DD, Hanks SK, Hunter T et al. Integrin-mediated signal transduction linked to Ras pathway by GRB2 binding to focal adhesion kinase. Nature 1994; 372:786-91.
29. Guan JL, Trevithick JE, Hynes RO. Fibronectin/integrin interaction induces tyrosine phosphorylation of a 120 kDa protein. Cell Regul 1991; 2:951-64.
30. Kornberg LJ, Earp HS, Turner CE et al. Signal transduction by integrins: increased protein tyrosine phosphorylation caused by clustering of beta 1 integrins. Proc Natl Acad Sci USA 1991; 88:8392-96.
31. Nojima Y, Rothstein DM, Sugita K et al. Ligation of VLA-4 on T cells stimulates tyrosine phosphorylation of a 105 kD protein. J Exp Med 1992; 175:1045-53.
32. Schaller MD, Borgman CA, Cobb BS et al. pp125 FAK, a structurally distinctive protein-tyrosine kinase associated with focal adhesions. Proc Natl Acad Sci USA 1992; 89:5192-96.
33. Jaconi ME, Theler JM, Schlegel W et al. Multiple elevations of cytosolic-free Ca^{2+} in human neutrophils: initiation by adherence receptors of the integrin family. J Cell Biol 1991; 112:1249-57.
34. Richter J, Ng-Sikorski J, Olsson I et al. Tumor necrosis factor-induced degranula-

tion in adherent human neutrophils is dependent on CD11b/CD18-integrin-triggered oscillations of cytosolic free Ca^{2+}. Proc Natl Acad Sci USA 1990; 87:9472-76.

35. Werb Z, Tremble PM, Behrendtsen O et al. Signal transduction through the fibronectin receptor induces collagenase and stromelysin gene expression. J Cell Biol 1989; 109:877-89.
36. Groux H, Huet S, Valentin H et al. Suppressor effects and cyclic AMP accumulation by the CD29 molecule of CD4+ lymphocytes. Nature 1989; 339:152-54.
37. Nathan C, Sanchez E. Tumor necrosis factor and CD11/CD18 (beta 2) integrins act synergistically to lower cAMP in human neutrophils. J Cell Biol 1990; 111:2171-81.
38. Sporn SA, Eierman DF, Johnson CE et al. Monocyte adherence results in selective induction of novel genes sharing homology with mediators of inflammation and tissue repair. J Immunol 1990; 144:4434-41.
39. Haskill S, Johnson C, Eierman D et al. Adherence induces selective mRNA expression of monocyte mediators and protooncogenes. J Immunol 1988; 140:1690-94.
40. Eierman DF, Johnson CE, Haskill JS. Human monocyte inflammatory mediator gene expression is selectively regulated by adherence substrates. J Immunol 1989; 142: 1970-76.
41. Yurochko AD, Liu DY, Eierman D et al. Integrins as a primary signal transduction molecule regulating monocyte immediate-early gene induction. Proc Natl Acad Sci USA 1992; 89:9034-38.
42. Lin TH, Yurochko A, Kornberg L et al. The role of protein tyrosine phosphorylation in integrin-mediated gene induction in monocytes. J Cell Biol 1994; 126: 1585-93.
43. Matsuyama T, Yamada A, Kay J et al. Activation of CD4 cells by fibronectin and anti-CD3 antibody. A synergistic effect mediated by the VLA-5 fibronectin receptor complex. J Exp Med 1989; 170:1133-48.
44. Yamada A, Nikaido T, Nojima Y et al. Activation of human CD4 T lymphocytes. Interaction of fibronectin with VLA-5 receptor on CD4 cells induces the AP-1 transcription factor. J Immunol 1991; 146: 53-56.
45. Davis LS, Oppenheimer-Marks N, Bednarczyk JL et al. Fibronectin promotes proliferation of naive and memory T cells by signaling through both the VLA-4 and VLA-5 integrin molecules. J Immunol 1990; 145:785-93.
46. Shimizu Y, Van Seventer GA, Horgan KJ et al. Costimulation of proliferative responses of resting CD4+ T cells by the interaction of VLA-4 and VLA-5 with fibronectin or VLA-6 with laminin. J Immunol 1990; 145:59-67.
47. Nojima Y, Humphries MJ, Mould AP et al. VLA-4 mediates CD3-dependent CD4+ T cell activation via the CS1 alternatively spliced domain of fibronectin. J Exp Med 1990; 172:1185-92.
48. Hamawy MM, Mergenhagen SE, Siraganian RP. Adhesion molecules as regulators of mast cell and basophil function. Immunol Today 1994; 15:62-66.
49. Hamawy MM, Mergenhagen SE, Siraganian RP. Cell adherence to fibronectin and the aggregation of the high affinity IgE receptor synergistically regulate tyrosine phosphorylation of 105-115 kDa proteins. J Biol Chem 1993; 268:5227-33.
50. Ducharme LA, Weis JH. Modulation of integrin expression during mast cell differentiation. Eur J Immunol 1992; 22:2603-7.
51. Gurish MF, Bell AF, Smith TJ et al. Expression of murine beta 7, alpha 4 and beta 1 integrin genes by rodent mast cells. J Immunol 1992; 149:1964-72.
52. Shimizu Y, Irani AM, Brown EJ et al. Human mast cells derived from fetal liver cells cultured with stem cell factor express a functional CD51/CD61 (alpha v beta 3) integrin. Blood 1995; 86:930-39.
53. Thompson HL, Burbelo PD, Yamada Y et al. Identification of an amino acid sequence in the laminin A chain mediating mast cell attachment and spreading. Immunology 1991; 72:144-49.
54. Bochner BS, Luscinskas FW, Gimbrone MA, Jr. et al. Adhesion of human basophils, eosinophils and neutrophils to interleukin 1-activated human vascular endothelial cells: contributions of endothelial cell adhesion molecules. J Exp Med 1991; 173:1553-57.
55. Walsh LJ, Kaminer MS, Lazarus GS et al.

Role of laminin in localization of human dermal mast cells. Lab Invest 1991; 65:433-40.
56. Bochner BS, McKelvey AA, Sterbinsky SA et al. IL-3 augments adhesiveness for endothelium and CD11b expression in human basophils but not neutrophils. J Immunol 1990; 145:1832-37.
57. Yasuda M, Hasunuma Y, Adachi H et al. Expression and function of fibronectin binding integrins on rat mast cells. Int Immunol 1995; 7:251-58.
58. Bochner BS, Sterbinsky SA. Altered surface expression of CD11 and Leu 8 during human basophil degranulation. J Immunol 1991; 146:2367-73.
59. Bochner BS, MacGlashan Jr DW, Marcotte GV et al. IgE-dependent regulation of human basophil adherence to vascular endothelium. J Immunol 1989; 142:3180-86.
60. Bochner BS, Sterbinsky SA, Knol EF et al. Function and expression of adhesion molecules on human basophils. J Allergy Clin Immunol 1994; 94 Suppl:1157-62.
61. Benhamou M, Siraganian RP. Protein-tyrosine phosphorylation: an essential component of Fc epsilon RI signaling. Immunol Today 1992; 13:195-97.
62. Hamawy MM, Mergenhagen SE, Siraganian RP. Protein tyrosine phosphorylation as a mechanism of signaling in mast cells and basophils. Cell Signal 1995; 7:535-44.
63. Hamawy MM, Mergenhagen S, Siraganian RP. Tyrosine phosphorylation of pp125FAK by the aggregation of high affinity immunoglobulin E receptors requires cell adherence. J Biol Chem 1993; 268:6851-54.
64. Hamawy MM, Swaim WD, Minoguchi K et al. The aggregation of the high affinity IgE receptor induces tyrosine phosphorylation of Paxillin, a focal adhesion protein. J Immunol 1994; 153:4655-62.
65. Takagi M, Nakahata T, Koike K et al. Stimulation of connective tissue-type mast cell proliferation by crosslinking of cell-bound IgE. J Exp Med 1989; 170:233-44.
66. Ra C, Yasuda M, Yagita H et al. Fibronectin receptor integrins are involved in mast cell activation. J Allergy Clin Immunol 1994; 94 Suppl:625-28.
67. Beaven MA, Moore JP, Smith GA et al. The calcium signal and phosphatidylinositol breakdown in 2H3 cells. J Biol Chem 1984; 259:7137-42.
68. Lansman JB, Hallam TJ, Rink TJ. Single stretch-activated ion channels in vascular endothelial cells as mechanotransducers? Nature 1987; 325:811-13.
69. Stump RF, Oliver JM, Cragoe Jr EJ et al. The control of mediator release from RBL-2H3 cells: roles for Ca^{2+}, Na+ and protein kinase C1. J Immunol 1987; 139:881-86.
70. Cervera M, Dreyfuss G, Penman S. Messenger RNA is translated when associated with the cytoskeletal framework in normal and VSV-infected HeLa cells. Cell 1981; 23:113-20.
71. Yahara I, Edelman GM. Modulation of lymphocyte receptor mobility by locally bound concanavalin A. Proc Natl Acad Sci USA 1975; 72:1579-83.
72. Pfeiffer JR, Seagrave JC, Davis BH et al. Membrane and cytoskeletal changes associated with IgE-mediated serotonin release from rat basophilic leukemia cells. J Cell Biol 1985; 101:2145-55.
73. Sahara N, Siraganian RP, Oliver C. Morphological changes induced by the calcium ionophore A23187 in rat basophilic leukemia (2H3) cells. J Histochem Cytochem 1990; 38:975-83.
74. Oliver JM, Seagrave J, Stump RF et al. Signal transduction and cellular response in RBL-2H3 mast cells. Prog Allergy 1988; 42:185-45.
75. Bonifacino JS, Perez P, Klausner RD et al. Study of the transit of an integral membrane protein from secretory granules through the plasma membrane of secreting rat basophilic leukemia cells using a specific monoclonal antibody. J Cell Biol 1986; 102:516-22.
76. Akiyama S, Yamada KM. Fibronectin in disease. In: Wagner BM, Fleischmajer R, Kaufman N, eds. Connective Tissue Diseases. Baltimore: Williams and Wilkins, 1983:55-96.
77. Clark RA, Horsburgh CR, Hoffman AA et al. Fibronectin deposition in delayed-type hypersensitivity. Reactions of normals and a patient with afibrinogenemia. J Clin In-

vest 1984; 74:1011-16.
78. Mosher DF. Fibronectin. San Diego: Academic Press, Inc., 1989.
79. Roche WR, Beasley R, Williams JH et al. Subepithelial fibrosis in the bronchi of asthmatics. Lancet 1989; 1:520-24.
80. Thompson HL, Burbelo PD, Yamada Y et al. Mast cells chemotax to laminin with enhancement after IgE-mediated activation. J Immunol 1989; 143:4188-92.
81. Thompson HL, Thomas L, Metcalfe DD. Murine mast cells attach to and migrate on laminin-, fibronectin-, and matrigel-coated surfaces in response to FcEpsilonRI-mediated signals. Clin Exp Allergy 1993; 23:270-75.
82. Zeller CB, Marchase RB. Gangliosides as modulators of cell function. Am J Physiol 1992; 262:C1341-C55.
83. Igarashi Y, Nojiri H, Hanai N et al. Gangliosides that modulate membrane protein function. Methods Enzymol 1989; 179: 521-41.
84. Bremer EG, Hakomori S, Bowen-Pope DF et al. Ganglioside-mediated modulation of cell growth, growth factor binding and receptor phosphorylation. J Biol Chem 1984; 259:6818-25.
85. Bremer EG, Schlessinger J, Hakomori S. Ganglioside-mediated modulation of cell growth. Specific effects of GM3 on tyrosine phosphorylation of the epidermal growth factor receptor. J Biol Chem 1986; 261: 2434-40.
86. Hakomori S, Igarashi Y. Gangliosides and glycosphingolipids as modulators of cell growth, adhesion and transmembrane signaling. Adv Lipid Res 1993; 25:147-62.
87. Kreutter D, Kim JY, Goldenring JR et al. Regulation of protein kinase C activity by gangliosides. J Biol Chem 1987; 262: 1633-37.
88. Ortega E, Licht A, Biener Y et al. A glycolipid-specific monoclonal antibody modulates Fc epsilon receptor stimulation of mast cells. Mol Immunol 1990; 27:1269-77.
89. Basciano LK, Berenstein EH, Kmak L et al. Monoclonal antibodies that inhibit IgE binding. J Biol Chem 1986; 261:11823-31.
90. Guo NH, Her GR, Reinhold VN et al. Monoclonal antibody AA4, which inhibits binding of IgE to high affinity receptors on rat basophilic leukemia cells, binds to novel alpha-galactosyl derivatives of ganglioside GD1b. J Biol Chem 1989; 264:13267-72.
91. Oliver C, Sahara N, Kitani S et al. Binding of monoclonal antibody AA4 to gangliosides on rat basophilic leukemia cells produces changes similar to those seen with Fc epsilon receptor activation. J Cell Biol 1992; 116:635-46.
92. Minoguchi K, Swaim WD, Berenstein EH et al. Src family tyrosine kinase p53/56lyn, a serine kinase and Fc epsilon RI associate with alpha-galactosyl derivatives of ganglioside GD1b in rat basophilic leukemia RBL-2H3 cells. J Biol Chem 1994; 269:5249-54.
93. Swaim WD, Minoguchi K, Oliver C et al. The anti-ganglioside monoclonal antibody AA4 induces protein tyrosine phosphorylations, but not degranulation, in rat basophilic leukemia cell. J Biol Chem 1994; 269:19466-73.
94. Stephan V, Guo N, Ginsburg V et al. Immunoprecipitation of membrane proteins from rat basophilic leukemia cells by the antiganglioside monoclonal antibody AA4. J Immunol 1991; 146:4271-77.
95. Trowbridge IS, Thomas ML. CD45: An emerging role as a protein tyrosine phosphatase required for lymphocyte activation and development. Annu Rev Immunol 1994; 12:85-16.
96. Tonks NK, Diltz CD, Fischer EH. CD45, an integral membrane protein tyrosine phosphatase. Characterization of enzyme activity. J Biol Chem 1990; 265:10674-80.
97. Streuli M, Krueger NX, Tsai AY et al. A family of receptor-linked protein tyrosine phosphatases in humans and Drosophila. Proc Natl Acad Sci USA 1989; 86: 8698-702.
98. Hook WA, Berenstein EH, Zinsser FU et al. Monoclonal antibodies to the leukocyte common antigen (CD45) inhibit IgE-mediated histamine release from human basophils. J Immunol 1991; 147:2670-76.
99. Berger SA, Mak TW, Paige CJ. Leukocyte common antigen (CD45) is required for immunoglobulin E-mediated degranulation of mast cells. J Exp Med 1994; 180:471-76.

100. Schneider H, Korn M, Haustein D. CD45-deficient RBL-2H3 cells: Cellular response to FCEpsilonR-and ionophore-induced stimulation. Immunol Invest 1993; 22: 503-15.
101. Adamczewski M, Numerof RP, Koretzky GA et al. Regulation by CD45 of the tyrosine phosphorylation of high affinity IgE receptor beta- and gamma-chains. J Immunol 1995; 154:3047-55.
102. Kishimoto T, Taga T, Akira S. Cytokine signal transduction. Cell 1994; 76:253-62.
103. Kurimoto Y, De Weck AL, Dahinden CA. The effect of interleukin 3 upon IgE-dependent and IgE-independent basophil degranulation and leukotriene generation. Eur J Immunol 1991; 21:361-68.
104. Okayama Y, Begishvili TB, Church MK. Comparison of mechanisms of IL-3 induced histamine release and IL-3 priming effect on human basophils. Clin Exp Allergy 1993; 23:901-10.
105. Dah inden CA, Bischoff SC, Brunner T et al. Regulation of mediator release by human basophils: importance of the sequence and time of addition in the combined action of different agonists. Int Arch Allergy Appl Immunol 1991; 94:161-64.
106. Bischoff SC, Brunner T, De Weck AL et al. Interleukin 5 modifies histamine release and leukotriene generation by human basophils in response to diverse agonists. J Exp Med 1990; 172:1577-82.
107. Miadonna A, Roncarolo MG, Lorini M et al. Inducing and enhancing effects of IL-3, -5 and -6 and GM-CSF on histamine release from human basophils. Clin Immunol Immunopathol 1993; 67:210-15.
108. Massey WA, Randall TC, Kagey-Sobotka A et al. Recombinant human IL-1 alpha and -1 beta potentiate IgE-mediated histamine release from human basophils. J Immunol 1989; 143:1875-80.
109. Brunner T, Heusser CH, Dahinden CA. Human peripheral blood basophils primed by interleukin 3 (IL-3) produce IL-4 in response to immunoglobulin E receptor stimulation. J Exp Med 1993; 177:605-11.
110. Hirai K, Morita Y, Misaki Y et al. Modulation of human basophil histamine release by hemopoietic growth factors. J Immunol 1988; 141:3958-64.
111. Schleimer RP, Derse CP, Friedman B et al. Regulation of human basophil mediator release by cytokines. I Interaction with antiinflammatory steroids. J Immunol 1989; 143:1310-17.
112. Coleman JW, Holliday MR, Kimber I et al. Regulation of mouse peritoneal mast cell secretory function by stem cell factor, IL-3 or IL-4. J Immunol 1993; 150:556-62.
113. Alam R, Welter JB, Forsythe PA et al. Comparative effect of recombinant IL-1, -2, -3, -4 and -6, IFN-gamma, granulocyte-macrophage-colony-stimulating factor, tumor necrosis factor-alpha and histamine-releasing factors on the secretion of histamine from basophils. J Immunol 1989; 142:3431-35.
114. MacDonald SM, Schleimer RP, Kagey-Sobotka A et al. Recombinant IL-3 induces histamine release from human basophils. J Immunol 1989; 142:3527-32.
115. Beauvais F, Echasserieau K, Burtin C et al. Regulation of human basophil activation; The role of Na+ and Ca^{2+} in IL-3-induced potentiation of IgE-mediated histamine release from human basophils. Clin Exp Immunol 1994; 95:191-94.
116. Galli SJ, Zsebo KM, Geissler EN. The kit ligand, stem cell factor. Adv Immunol 1994; 55:1-96.
117. Bischoff SC, Dahinden CA. c-kit ligand: a unique potentiator of mediator release by human lung mast cells. J Exp Med 1992; 175:237-44.
118. Columbo M, Horowitz EM, Botana LM et al. The human recombinant c-kit receptor ligand, rhSCF, induces mediator release from human cutaneous mast cells and enhances IgE-dependent mediator release from both skin mast cells and peripheral blood basophils. J Immunol 1992; 149:599-608.
119. Rottapel R, Reedijk M, Williams DE et al. The Steel/W transduction pathway: kit autophosphorylation and its association with a unique subset of cytoplasmic signaling proteins is induced by the Steel factor. Mol Cell Biol 1991; 11:3043-51.
120. Yi T, Ihle JN. Association of hematopoietic cell phosphatase with c-Kit after stimulation with c-Kit ligand. Mol Cell Biol 1993;

13:3350-58.
121. Tauchi T, Feng GS, Marshall MS et al. The ubiquitously expressed Syp phosphatase interacts with c-kit and Grb2 in hematopoietic cells. J Biol Chem 1994; 269:25206-11.
122. Liu L, Damen JE, Cutler RL et al. Multiple cytokines stimulate the binding of a common 145-kilodalton protein to Shc at the Grb2 recognition site of Shc. Mol Cell Biol 1994; 14:6926-35.
123. Yee NS, Langen H, Besmer P. Mechanism of kit ligand, phorbol ester, and calcium-induced downregulation of c-kit receptors in mast cells. J Biol Chem 1993; 268: 14189-201.
124. Asano Y, Brach MA, Ahlers A et al. Phorbol ester 12-O-tetradecanoylphorbol-13-acetate downregulates expression of the c-kit protooncogene product. J Immunol 1993; 151:2345-54.
125. Gold MR, Duronio V, Saxena SP et al. Multiple cytokines activate phosphatidylinositol 3-kinase in hemopoietic cells. Association of the enzyme with various tyrosine-phosphorylated proteins. J Biol Chem 1994; 269:5403-12.
126. Tsai M, Chen RH, Tam SY et al. Activation of MAP kinases, pp90rsk and pp70-S6 kinases in mouse mast cells by signaling through the c-kit receptor tyrosine kinase or FcEpsilonRI: Rapamycin inhibits activation of pp70-S6 kinase and proliferation in mouse mast cells. Eur J Immunol 1993; 23:3286-91.
127. Hulett MD, Hogarth PM. Molecular basis of Fc receptor function. Adv Immunol 1994; 57:1-127.
128. Daeron M, Malbec O, Latour S et al. Regulation of high-affinity IgE receptor-mediated mast cell activation by murine low-affinity IgG receptors. J Clin Invest 1995; 95:577-85.
129. Muta T, Kurosaki T, Misulovin Z et al. A 13-amino-acid motif in the cytoplasmic domain of FcgammaRIIB modulates B cell receptor signalling. Nature 1994; 368: 70-73.
130. D'Ambrosio D, Hippen KL, Minskoff SA et al. Recruitment and activation of PTP1C in negative regulation of antigen receptor signaling by FcgammaRIIB1. Science 1995; 268:293-97.

CHAPTER 13

Signal Transduction Issues in Studies of Human Basophils and Mast Cells

Donald MacGlashan, Jr.

INTRODUCTION

In humans, there are now known to be several types of cells which bear FcεRI. The list currently includes mast cells, basophils, monocytes,[1] Langerhans cells,[2] and eosinophils[3] (unpublished reports of epithileal cells also). Of these, expression on eosinophils appears to be controversial with only one laboratory currently supporting this possibility; however, the functional consequences of FcεRI aggregation on monocytes and Langerhans cells is also not yet entirely clear. In addition, it appears that FcεRI on monocytes and Langerhans cells may be composed only of the α and γ subunits, the significance of which is also not clear. On the other hand, the mast cell and basophil are representative of cells which secrete in response to FcεRI aggregation and these cells clearly play a number of roles in various diseases. Understanding the nature of the signal transduction apparatus that exists in mast cells and basophils will have ready application in developing new therapies for the treatment of diseases like atopy and asthma. Mast cells and basophils, however, are not readily obtainable in numbers or purities that allow for the easy discovery of the underlying signal transduction mechanisms. This type of discovery work is best accomplished in either human or animal cell lines. Nevertheless, it is also quite clear that the development of some therapies will only succeed when the function of the normal mast cell and basophil is understood. A good example of how animal cells are not representative of their human counterpart can be found in the effort in the 1970s to develop potent analogs of cromolyn. An enormous effort by the pharmaceutical industry found that drugs which inhibited rat mast cell

IgE Receptor (FcεRI) Function in Mast Cells and Basophils,
edited by Majed M. Hamawy.

FcεRI-mediated release with high potency and efficacy were ineffective in humans. While cromolyn is used today by some physicians for some allergic conditions, the mechanism of its actions don't clearly revolve around mast cell inhibition, but remain unknown and have weak to moderate effects at best. The lesson seems to be that drug development should be well integrated with studies on normal human mast cells or basophils. Apparently, the precise details of the signal transduction mechanisms vary enough among cells bearing FcεRI that not all approaches will translate well among the various cell types. These issues will be discussed against a backdrop of information derived from extensive studies in both RBL-2H3 cells, rat peritoneal mast cells or other cell lines and studies of human basophils and mast cells.

Human mast cells and basophils both possess FcεRI and granules containing histamine. Both cell types appear to be involved in allergic diseases although the precise involvement of the basophil is not fully understood nor universally accepted. In experimental antigen challenge models of upper airways, lower airways or skin, investigators have become interested in the biphasic nature of the tissue response that occurs in some individuals. The first phase of this response is characterized by the release of histamine and leukotrienes from mast cells activated by exposure to antigens for which they have antigen-specific IgE bound to their FcεRI.[4] As is now apparent and will be discussed further below, there may also be a number of cytokines secreted both early or late in the mast cell reaction. The secretion of histamine or other mediators from the mast cell activate nearby endothelial cells to recruit circulating leukocytes. This recruitment is somewhat specific; apparently the mediators initiate the expression of specific adhesion molecules on the endothelium such that basophils,[4] eosinophils[5] and lymphocytes[6] are the predominant leukocytes that migrate into the tissue. Investigators know that the specific expression of VCAM-1 on the endothelium and the VLA-4 ligand that is found only on basophils, eosinophils and lymphocytes provides one possible explanation for this specific recruitment[7,8] but a full understanding and/or experimental demonstration of this particular model has not yet been made. The migration and accumulation of these leukocytes into the tissue constitute the second phase of the whole response. Once in the tissue, these cells may be activated and contribute further mediators which aggrevate the reaction. Many atopic patients or patients with asthma show the histological signs of a chronic inflammatory reaction so it is felt that the experimental challenge model provides a reasonable approximation to the events that occur in vivo in such patients.

As a participant in these inflammatory reactions, the basophil is able to respond in the same manner as the mast cell, i.e., antigens induce the crosslinking of IgE bound to FcεRI and cause the basophil to secrete a large number of mediators (although eosinophils have been reported to have FcεRI, the studies from the one group that makes this observation have noted that its expression is restricted to a rare condition, hyper-eosinophilia, so that for the moment, it is not apparent that the eosinophil would contribute to inflammatory reaction in typical atopic patients by virtue of antigen interacting with IgE bound to the infiltrating eosinophils). In experimental challenge models, the second phase of the reaction is characterized by the release of mediators that are more typical of the basophil.[9-11] Most notably, there is histamine release in the absence of observable mast cell tryptase. It has been clearly demonstrated that the majority of the histamine containing cells in bronchoalveolar washings are basophils. This is based on cell surface phenotype, the type of stimuli which induce secretion and the profile of mediators released. Thus, it appears that the way the basophil may contribute to the chronic atopic condition is much the way the mast cell initiates the entire cascade of events.

It is possible to view the basophil as a surrogate for mast cells. The basophil is generally more available for study and it would be quite useful if we were able to use it both as a surrogate for the mast cell as well as use it to study basophil-specific behavior. However, this desire may not fully be realized since, as noted above, the precise details of the signal transduction apparatus might be different and these differences may be critical in the development of some new therapies. On the other hand, the basophil is one step closer to the human mast cell than animal cell models and because basophils appear to participate in allergic reactions, it would be useful to understand how they function as well. This chapter will sometimes interchange data derived from mast cell and basophil studies in order to develop a line of thought but it is acknowledged that this must be done carefully. Consequently, there will be some direct comparisons made between human mast cells and basophils just as some comparisons need to be made between human cells and the RBL-2H3 cell which is the dominant model for FcεRI-mediated signal transduction.

FcεRI AGGREGATION IN BASOPHILS AND MAST CELLS

An early hallmark of FcεRI-mediated activation was the requirement for aggregation of the receptor in order to initiate and maintain secretion from mast cells or basophils. Only in recent years have the reasons for this requirement for aggregation begun to emerge and the explanation has naturally awaited a clear picture of the molecules involved in FcεRI-mediated signal transduction. But it was clear shortly after the first demonstration of a need for aggregation that very few aggregates were necessary to initiate a half-maximal response in RBL-2H3 cells. Early studies by Fewtrell and Metzger noted that only hundreds of trimeric aggregates were sufficient to trigger RBL-2H3 cells.[12] It is useful to keep in mind that RBL-2H3 cells typically express hundreds of thousands of FcεRI, suggesting a vast receptor excess. Not only did RBL-2H3 cells show this high sensitivity but in some still unexplained manner, the cell was almost unresponsive to dimeric aggregates. This was unlike the rat peritoneal mast cell where dimeric aggregates were sufficient.[13] The story in the RBL-2H3 cell is recapitulated in human basophils. Here, dimeric and trimeric IgE (covalently crosslinked as in the RBL-2H3 cell studies) produced qualitatively different responses.[14] In addition, studies of cellular sensitivity noted that only hundreds of receptor aggregates were sufficient to initiate histamine release in some donors' basophils.[15] Human basophils have median FcεRI densities of ≈100,000, ranging from densities as low as 5,000 to densities as high as 1,000,000 per cell.[16] This indicates a vast receptor excess. It could be hypothesized that this may be necessary since the cells must bind a large variety of antigen-specific IgE antibodies. Since any one antigen specificity might represent but a small fraction of the total circulating IgE, lower receptor densities might reduce the responsiveness of the basophil below the range required for activation. However, the median densities of FcεRI on human tissue lung mast cells appear to be in the range of 5,000-20,000[17] although densities similar to basophils have been observed. Since these cells often respond quite well to the same antigens as the basophil, the need for higher densities on the basophil is made less clear.

It has also been noted that the density of the FcεRI on basophils is positively correlated with the serum total IgE titer.[18] Surprisingly, the density of unoccupied FcεRI is relatively constant across the entire range of receptor densities (or IgE titer).[16] One possible explanation for these observations is that bound IgE induces the expression of additional FcεRI. The mature basophil in the bone marrow may begin its "life" with few receptors, but throughout the lifespan of the circulating basophil, more receptors are steadily accumulated under the drive of serum IgE. Low IgE titers induce a slow accumulation such that by the end of the basophil's normal lifespan (2 days to 2 weeks depending on the study) only a few receptors have accumulated. At the high end of IgE titers, many more receptors would have accumulated under this scheme. In a sense, it is as if the cell strives to maintain a certain level of unoccupied FcεRI. In recent pilot studies of anti-IgE antibody therapy (using a monoclonal anti-IgE specific for an epitope in the region of IgE normally bound to FcεRI, so that it does not bind to cells or induce aggregation) we have found basophil FcεRI densities decrease dramatically. Based on previous studies with this antibody, this occurs at a time when IgE titers are expected to be low although these data are not yet available for this new study. Such data lend support to the hypothesis that the FcεRI density is driven by serum IgE titer. In unpublished studies, we have found that alcian blue positive cells derived from IL-3-treated 3-week old mouse bone marrow cells appear to upregulate their FcεRI density in response to a 48 hour incubation with IgE. This upregulation appears specific for the FcεRI and 3- to 10-fold increases in the density of the receptor can be observed. We have not been able to replicate these observations using cultured human basophils although it is not entirely clear which cytokines need to be present to maintain a normally active basophil in culture or whether the effect on basophil FcεRI is mediated through the basophil itself.

KNOWN SIGNAL TRANSDUCTION EVENTS IN HUMAN CELLS

INTRODUCTION

There are an enormous number of signal transduction biochemistries known to operate in a wide variety of cell types. While a good number of these have been studied in RBL-2H3 cells the list is far shorter for human basophils or mast cells. However, because studies of histamine release are relatively straightforward, there have been a large number of studies that examine how pharmacological agents modify mast cell or basophil histamine release. Since these same drugs have often been examined in the context of signal transduction biochemistries in RBL-2H3 cells, some extrapolation to the human cells is possible.

Throughout the 1980s, two signal transduction events received a great deal of attention, elevations in free cytosolic calcium concentrations ($[Ca^{2+}]_I$) and activation of protein kinase C (PKC). It is well accepted that these two events are central to many types of receptor signal transduction. Not surprisingly, elevations in $[Ca^{2+}]_I$ and activation of PKC follow FcεRI-mediated stimulation of human basophils.[19,20] Calcium studies in mast cells have been performed[21] while PKC studies have not. Given the ubiquitous nature of this form of signal transduction, it is not the fact of their occurrence that is interesting today but the various conditions of their occurrence that begins to shed some light on how the whole activation cascade functions. In this section, I will discuss five different aspects of basophil and mast cell biology that have been studied in the context of $[Ca^{2+}]_I$ signaling or PKC activation: requirements for mediator release, single cell vs. whole population behavior, desensitization, cytokine regulation and releasability. The first three focus on a more basic aspect of basophil or mast cell biology, while the later two address issues that may have some clinical relevance.

REQUIREMENTS FOR MEDIATOR RELEASE

Basophils and mast cells secrete a wide variety of chemicals in response to receptor stimulation. They can be grouped in several ways but a common grouping focuses on the temporal aspects of their secretion. The granules contain a number of preformed mediators of which histamine is but one example, albeit the one most often studied. This group of mediators is the first observed following any receptor-mediated stimulation of human basophils or mast cells. Newly synthesized lipid mediators are released shortly after degranulation begins. The time lag between the release of histamine and lipid mediators is short: for example, in basophils, histamine release following FMLP stimulation begins within seconds while LTC4 release isn't apparent for 30-60 seconds.[22] This time lag is not a consequence of synthesis without secretion since Bligh-Dyer extraction of basophils during the reaction also shows the same kinetics.[15] The profile of lipid mediators secreted differs among basophils and mast cells from different tissues. The most clear differentiation is the complete absence of any cyclooxygenase product during basophil secretion.[22] Leukotriene release in the basophil is restricted to LTC_4 and a 5HETE[22] with no measurable LTB4.[23] Mast cells from any tissue secrete both PGD_2 and LTC_4 (with some 5HETE as well).[24-26] However, the relative amounts of these two lipid products vary among tissue mast cells, being secreted in nearly equimolar quantities from human lung mast cells while PGD_2 release dominates the lipids secreted from skin mast cells (LTC_4 release is almost unmeasurable in many cases).[27,28]

The last mediator grouping involves the secretion of cytokines from basophils and mast cells. This is a relatively new field of study for these cell types and certain issues remain somewhat controversial. For human basophils, IL-4 secretion has received the lion share of attention. Like LTC4 release, this appears to be a newly synthesized mediator although it is secreted on a time scale of hours, not minutes and it requires protein synthesis while histamine and LTC4 secretion do not.[29,30] In other words, cycloheximide inhibits IL-4 release while not inhibiting histamine or LTC4 secretion. Transcription is also apparent; IL-4 mRNA levels increase at least 10-fold following FcεRI-mediated secretion and nearly 100-fold during ionomycin stimulation.[30] There is some controversy as to whether some basophils have preformed stores of IL-4, although in our laboratories, we have not been able to measure IL-4 from lysed resting basophils. Other cytokine or protein synthesis dependent secretion from basophils may occur but hasn't yet received too much attention in the published literature although Dahinden's group suggests that other TH1- or TH2-like cytokines are not secreted.[31] Human mast cells have been reported to secrete a variety of cytokines although it has been difficult to obtain a consensus about any selected cytokine. For example, IL-4 secretion was first reported by Bradding et al[32] but this observation is not universal[33] and awaits verification by other laboratories. Other cytokines have been reported, including IL-5[33] and TNF-α, with TNF-α being reported by several laboratories to be both preformed and newly synthesized.[34,35] Consequently, this cytokine could be secreted both minutes and hours after stimulation. Although the details remain unclear, it seems apparent, certainly from murine mast cell studies,[36] that mast cells also respond to FcεRI aggregation by initiating transcriptional signal transduction events which cause a later secretion of proteins.

We have been examining how elevations in free cytosolic calcium and the activation of PKC relate to the secretion of each of these three classes of mediators. The results are not yet conclusive, but there is evidence that the requirements for the three types of mediator secretion differ. Beginning with degranulation, i.e., histamine release, we find a complicated picture. Human basophils are somewhat unique in being able to release histamine in response to stimulation with phorbol esters like PMA[37] human mast cells do not, although PMA can synergize with calcium ionophores to secrete histamine in these cells. Although, calcium ionophores and PMA synergize in basophils, the remarkable feature is that in the absence of extracellular calcium and with basophils heavily loaded with the calcium chelator BAPTA, PMA continues to cause histamine release, albeit somewhat slower (unpublished studies). The BAPTA loading is critical for interpreting these experiments; the resting level of $[Ca^{2+}]_i$ is below normal resting levels as assessed by labeling the cells with FURA-2. Indeed, in the presence of BAPTA, the FURA-2 ratio is as low as is ever measured, defining the low end of the FURA-2 calibration curve. Furthermore, FMLP stimulation (see below) of loaded basophils results in no measurable changes in $[Ca^{2+}]_I$ and even stimulation with 1-2 μM ionomycin results in no increases in $[Ca^{2+}]_I$ for several minutes. Although $[Ca^{2+}]_I$ elevations do eventually occur under these conditions, ionomycin normally causes a rise in $[Ca^{2+}]_I$ that reaches more than 1 μM within seconds. Thus, the suppression of this normal increase by loading with BAPTA indicates that it is effective in chelating or buffering any reasonable $[Ca^{2+}]_I$ changes. The fact that PMA induces histamine release under these conditions suggests that the granule fusion apparatus does not require even resting levels of $[Ca^{2+}]_I$.

We have also attempted to determine the requirements for $[Ca^{2+}]_I$ elevations during receptor-mediated histamine release. Our initial studies of FMLP-induced histamine release noted a temporal association between the kinetics of histamine release and the initial rise in $[Ca^{2+}]_I$ generally associated with the release of calcium from internal stores. The addition of EGTA to the extracellular buffer at the same time FMLP was added to the reaction would ablate the second phase of the $[Ca^{2+}]_I$ response, that associated with calcium influx, but would have only a small effect on histamine release (≈20% inhibition of the response that occurs in the presence of extracellular calcium).[38] However, in similarly designed experiments using either anti-IgE antibody or platelet activating factor (PAF), this initial $[Ca^{2+}]_I$ transient was not accompanied by histamine release.[39] These data indicated that such a $[Ca^{2+}]_I$ rise was insufficient for histamine release. Parenthetically, it is interesting to note that an examination of the FMLP dose response curve, using histamine release and the peak of the initial $[Ca^{2+}]_I$ response or the average of the second phase $[Ca^{2+}]_I$ response as measures of the cell's response, shows that only the second phase of the $[Ca^{2+}]_I$ response precisely follows the dose response curve as measured by histamine release. The initial phase of the response doesn't begin decreasing until very low concentrations of FMLP are used, at a point when histamine release is nearly absent. Since the second phase of the response can be eliminated with EGTA and not affect histamine release, we conclude that the magnitude of this second phase $[Ca^{2+}]_I$ response mirrors some other aspect of activation that is independent of the release of calcium from internal stores but is critical for histamine release.

While these studies addressed the 'sufficiency condition', the 'necessary condition' is murkier. A long-term treatment of basophils with EGTA appears to slowly deplete the cell's internal stores; 30 minutes of incubation with EGTA reduces the initial $[Ca^{2+}]_I$ response to FMLP by half. Likewise, histamine release is also reduced by half (release tested in the presence of EGTA). This proportional loss in itself is puzzling given the FMLP dose response data mentioned above. However, the results suggest the dependence of histamine release on some $[Ca^{2+}]_I$ response. On the other hand, if the basophil is first treated with thapsagargin (an inhibitor of a Ca^{2+} ATPase associated with returning calcium to the internal store compartment) in the presence of extracellular EGTA, a subsequent stimulation with FMLP shows an initial $[Ca^{2+}]_I$ response which is blunted by ≈ 90%. A similar blunting by the long-term incubation procedure above would result in complete inhibition of histamine release but in these thapsagargin experiments, histamine release was unaffected.[39] Such an experiment suggests that the $[Ca^{2+}]_I$ elevation is not necessary, although one could argue that the small remaining response in these experiments was all that was necessary to drive a normal response.

Two other experiments support the view that the $[Ca^{2+}]_I$ response is not necessary. If basophils are challenged in the presence of a buffer containing 140 mM saline and 140 mM choline (a hypertonic solution), there is a small rise in the resting $[Ca^{2+}]_I$ level. However, FMLP stimulation causes nor further elevation of $[Ca^{2+}]_I$, neither initial or second phase. Histamine release under these conditions is approximately one-half control release (under normal conditions of stimulation).[40] This is an accidental discovery and is odd in design; however, it does also suggest that the further elevation of $[Ca^{2+}]_I$ beyond that which occurs in the resting cells is not necessary for histamine release.

Lastly, we have developed image processing algorithms[41] which can measure the extent of basophil degranulation by measuring the size and number of degranulation sacs[42,43] that form after histamine release has occurred during observations by optical microscopy. Although the precise relationship between these degranulation sacs and histamine release is not known, these sacs appear only in degranulating cells and their numbers (or cytoplasmic "holes") measured in 50-100 cells (those under observation in the microscope chamber used for these studies) correlate with the extent of histamine release measured in the supernatant of the same challenged cells (which is sampled from the entire microscope chamber containing 30,000-100,000 cells). Using this measure of single cell degranulation, we find that there is no correlation between either the first or second phase of the $[Ca^{2+}]_I$ response and the number and size of degranulation sacs in individual cells when FMLP is the stimulus. With anti-IgE antibody as a stimulus, the relationship appears somewhat stronger but is still statistically uncorrelated.

Each of these experimental designs has some flaws: long-term preincubation with EGTA could be having other effects on cell function; pretreatment with thapsagargin could "prime" the cell to be more responsive to FMLP; the hyperosmolar choline experiments stress the cell in unknown ways; and the degranulation sac measurements are not fully proven to directly reflect histamine release on a cell to cell basis. Therefore, there is no definitive experiment. Instead the data more strongly support the possibility that if a $[Ca^{2+}]_I$ elevation is required for histamine release, such an elevation need only be a small fraction of the $[Ca^{2+}]_I$ response that the cell is capable of producing. These conclusions may also vary according to the stimulus. It is well known that FcεRI-mediated release requires extracellular calcium. Whether this need reflects a need for elevated $[Ca^{2+}]_I$ levels or indicates that some external plasma membrane processes require this calcium isn't entirely clear. Both calcium ionophores and PMA alone do cause histamine release, but quite poorly relative to the degree of change in either $[Ca^{2+}]_I$ or activation of PKC, respectively, and the secretion process is oddly slow relative to receptor-mediated secretion. Together, these two artificial stimuli induce a rapid secretion. As noted by Beaven et al,[44] the implication is that secretion is a synergistic application of several signal transduction components. In this light, the variable results relating $[Ca^{2+}]_I$ changes to degranulation might not be surprising.

In contrast with the variable relationship between $[Ca^{2+}]_I$ elevations and histamine release, all experimental maneuvers that inhibit the second phase of the $[Ca^{2+}]_I$ response, inhibit both leukotriene C4 and IL-4 release from basophils. As noted above, LTC_4 release lags behind histamine release although the release of both mediators ends about the same time. The initial phase of the $[Ca^{2+}]_I$ response, following either FMLP or anti-IgE antibody, precedes LTC_4 release and is usually complete by the time LTC_4 release begins. Consequently, if EGTA is included with the stimulus, the initial $[Ca^{2+}]_I$ transient remains intact, the influx phase is eliminated and so is LTC_4 release.[38] For most donors, the anaphylatoxin C5a causes histamine release in the absence of LTC_4 release and appears to be a natural analog of the addition of FMLP with EGTA, i.e., there is often little or no second phase to the $[Ca^{2+}]_I$ response.[38] In addition to the experimental maneuvers noted above, it is also possible to cause FMLP to show only a transient $[Ca^{2+}]_I$

response by the inclusion of 5 μM lanthanum or by switching the buffer to 140 mM K^+. In these later experiments, the initial transient $[Ca^{2+}]_I$ response remains intact; histamine release remains intact but LTC4 release is abolished.[39,40] This dependence is not particularly surprising since 5-lipoxygenase is a calcium sensitive enzyme.[45] The rate limiting step for FMLP-induced LTC4 release appears to be the generation of arachidonic acid which seems to explain its dependence on the second phase of the $[Ca^{2+}]_I$ response.[46] This AA generation can be accelerated to occur within the time frame of the first phase of the $[Ca^{2+}]_I$ response by preincubating basophils with IL-3 (see below).[46] Under these circumstances LTC4 release occurs even in the presence of EGTA and abruptly stops when the initial $[Ca^{2+}]_I$ transient is complete.[46]

We have less experience with IL-4 release, but it appears that like LTC4 release, it is dependent on a prolonged $[Ca^{2+}]_I$ elevation. EGTA added after histamine release is complete but before IL-4 release ablates its secretion.[47] Ionomycin is a better stimulus for IL-4 release than anti-IgE antibody which may be related to its ability to maintain an elevated $[Ca^{2+}]_I$. In a similar context, FMLP is generally a poor stimulus for IL-4 release and shows a half-time for the return of $[Ca^{2+}]_I$ to resting levels (in the continued presence of stimulus) 10 times faster than anti-IgE antibody. Although, the initial $[Ca^{2+}]_I$ response following FMLP is often two times greater than that observed following anti-IgE antibody, the rate of return of the $[Ca^{2+}]_I$ signal may obviate later IL-4 synthesis.

The activation of PKC may be another major determinant of mediator release. Clearly, stimulation with PMA is sufficient to induce degranulation. During FcεRI-mediated release, PKC activity rises and returns to basal levels on a time frame consistent with the cessation of histamine release.[19] There exist several PKC inhibitors that are currently felt to be specific for PKC but their ability to inhibit histamine release is varied. For example, calphostin C inhibits both FMLP and FcεRI-mediated histamine release while bisindoylmaleimide II has a weak effect on FMLP-induced histamine release and actually enhances FcεRI-mediated release. The α/β specific PKC isozyme inhibitor Go-6976 very potently inhibits FcεRI-mediated release (a somewhat higher potency than its inhibition of PMA-induced release) but we have found that like staurosporin, it very effectively inhibits the $[Ca^{2+}]_I$ response. This suggests that it is not a PKC specific inhibitor. These results with PKC inhibitors are intriguing because most do not inhibit release, with the exception of calphostin C (which we haven't yet tested for its effects on the $[Ca^{2+}]_I$ response), only inhibitors which also affect the $[Ca^{2+}]_I$ response, inhibit release. Others have no effect.

With some selectivity, isozymes of PKC are translocated during FcεRI-mediated stimulation of RBL-2H3 cells and both Ca^{2+}-dependent and Ca^{2+}-independent forms appear to participate in active secretion[48] (it is not always clear whether translocation is really the equivalent of activation, so the reconstitution of specific PKC isozymes in permeabilized cells in these studies was an important observation as well). Long-term PMA treatment markedly downregulated the Ca^{2+}-dependent forms without effecting δ and ε isozymes and the cells still degranulated.[48] We have preliminary studies of PKC isozyme profiles in human basophils (purified to near 100% and also compared to the contaminating cells that normally accompany less purified basophil preparations, a mix of lymphocytes and monocytes) and like RBL-2H3 cells, basophils don't contain η or γ isozymes. However, while RBL-2H3 cells have α:β or α:δ ratios of 2:1 and ≈1:1, respectively, thus far (n = 6), α is nearly undetectable in human basophils giving α:β and α:δ ratios of 0.05:1 and .02:1, respectively (α ratios in the contaminating cells were more like RBL-2H3 cells). The significance of this observation is not yet known. Long-term PMA treatments haven't be done in basophils because they degranulate fully and adhere to culture wells. However, these studies are possible in skin and lung mast cells and like the RBL-2H3 cell, histamine release still occurred in skin mast cells but at 20% of the release from untreated cells.[49] Surprisingly, these long-term incubations with PMA had only modest effects on lung mast cell histamine release (10-40% inhibition). This will be discussed further below under desensitization studies.

PMA does not stimulate LTC_4 release but this is not surprising since it completely inhibits an FcεRI-mediated $[Ca^{2+}]_I$ response and markedly suppresses FMLP induced $[Ca^{2+}]_I$ responses. In support of this interpretation, it has minor variable effects on ionomycin-induced LTC_4 release. In contrast to this later observation, PMA effectively suppresses IL-4 release in ionomycin stimulated basophils and PMA alone also does not cause IL-4 release. This is in contrast to IL-4 release from lymphocytes where ionomycin and PMA markedly synergize to cause IL-4 release. This suppression of IL-4 release appears to occur at the level of mRNA transcription since IL-4 mRNA increases are also inhibited. A recent study by Casolaro and Ono in this division has demonstrated the existence of a IL-4 promoter element that may normally be induced by NFATp ($[Ca^{2+}]_I$ sensitive) but the binding of NFkB (induced by PKC) to a nearby promotor element competitively reduces NFATp binding, thereby shutting off the IL-4 promotor.[50]

Single Cell Issues

A long standing question regarding cell function is the issue of whether studies of whole cell populations accurately reflect the behavior of single cells. It has become clear, especially for the case of $[Ca^{2+}]_I$ responses, that whole cell population measurements misrepresent the single cell response. In a wide variety of cells, it has now been observed that the single cell $[Ca^{2+}]_I$ response is often composed of a train of $[Ca^{2+}]_I$ spikes, much the way a nerve cell may respond with a train of transmembrane potential reversal spikes. Studies in rat peritoneal mast cells first identified $[Ca^{2+}]_I$ transients[51] but extensive studies of FcεRI-mediated $[Ca^{2+}]_I$ oscillations were then carried out in RBL-2H3 cells.[52,53] Studies in human basophils and mast cells revealed another aspect to this behavior. In human lung mast cells, spiking behavior was a relatively infrequent event, more often single cells would demonstrate a time lag before a rapid transition to a new level of $[Ca^{2+}]_I$. There were many variations on this basic theme but spiking was only apparent, and in a limited way, with suboptimal conditions of stimulation.[21,54] In contrast, $[Ca^{2+}]_I$ oscillations or spiking was the most common characteristic of FcεRI-mediated stimulation in human basophils.[54] As with mast cells, we found that there was a latent period after the addition of anti-IgE antibody or antigen which lasted 15 seconds to 4 minutes depending on the concentration of the stimulus or more appropriately, the strength of the stimulation (i.e., there can be long time lags even for a stimulus known to establish IgE crosslinks within seconds but which generally induces a "weak" signal, bivalent penicillin is one example). Following this latent period there is usually an abrupt change in $[Ca^{2+}]_I$, the transition occurring in 1-5 seconds. For basophils, the first change is actually a $[Ca^{2+}]_I$ spike while for mast cells a new level of $[Ca^{2+}]_I$ is maintained. Spiking generally has an irregular periodicity although periodic oscillations are occasionally observed. For mast cells, the existence of a latency period followed by the abrupt change to a new level gives the appearance of an all-or-nothing transition; there is even a 'pacemaker'-like subtle rise in $[Ca^{2+}]_I$ just prior to the transition. However, the magnitude of the change to a higher level of $[Ca^{2+}]_I$ is a variable as well, decreasing as the strength of stimulus decreases. The final result is a kinetic pattern that appears all-or-nothing but the magnitude of the change is graded according to the stimulus strength. The same conclusions can be drawn for basophils. We had predicted at the time of these studies that there was a continuum of behavior among mast cells and basophils. The generation of a $[Ca^{2+}]_i$ signal is quite nonlinear relative to the stimulus strength, with feedforward and feedback loops and an IP3 receptor that has positive and negative cooperativity built in. One would expect a variety of behaviors and a recent model of early $[Ca^{2+}]_i$ signaling events developed by Keizer's group[55] does indeed predict that some cells will exist in a domain of steady state behavior while others may be oscillating. We have subsequently found that skin mast cells often show $[Ca^{2+}]_i$ spiking and thus resemble the basophil.[56] At a minimum, these kinds of experiments provide a proper perspective on the nature of calcium signaling and raise the possibility that other signal transduction events are not properly represented in average population studies. It is not at all clear whether there is a functional significance to this behavior. Lung mast cells secrete histamine and leukotrienes as readily as basophils, so in this instance the presence or absence of oscillations has no apparent functional consequence.

The single cell $[Ca^{2+}]_i$ studies also indicated that single cells have graded responses to stimulation. We have extended this particular observation to include a number of measures of basophil activation, including F-actin polymerization, CD11b upregulation, spreading and flattening on a solid surface, and two indirect measures of degranulation—forward scatter changes and loss of acridine orange labeling (which appears to metachromatically label basophil granules).[41,57] In each case, there are unimodal shifts in the single cell response distributions that are a function of stimulation strength. It should be noted, however, that previous studies have come to conclusion that basophil degranulation is an all-or-nothing event.[58,59] Our data do not directly question whether there are subpopulations of circulating basophils but they do make it clear that individual basophils show graded responses rather than the all-or-nothing behavior proposed in some studies.

Desensitization

Basophils from most donors do not usually secrete all their available histamine even under conditions of optimal FcεRI-mediated stimulation. An examination of the kinetics of histamine release shows that, within 10-30 minutes, release reaches a plateau which is well under 100%. Adding other FcεRI-independent stimuli induces further release at this plateau, or combinations of ionomycin and PMA induce 100% release, so it is clear that the cell is capable of releasing all of its histamine. Therefore, the fact that a plateau of release is reached suggests that there is a process which shuts off secretion. Such a process can be operationally demonstrated in basophils by incubating the cells with antigen or anti-IgE antibody in the absence of extracellular calcium. At some point after the addition of stimulus, calcium is added back to the buffer allowing the cell to secrete. The longer one waits before adding back the calcium, the poorer the cell responds relative to controls which aren't treated with antigen during the calcium-free phase of the experiment.[60,61] It appears that this desensitization process is an active process; low concentrations of DFP included in the calcium-free phase of the experiments inhibit desensitization.[62] If the same concentrations of DFP are included during active secretion, release is enhanced and has a more linear appearance, partially eliminating the plateau of release normally observed (DFP at somewhat higher concentrations can inhibit release so these experiments appear balanced between inhibition of desensitization and inhibition of activation). We have shown that desensitization does not result from downregulation of IgE and its receptor.[63] A fully desensitized cell expresses the same density of IgE (antigen-specific or total) and can bind similar quantities of applied antigen.

The process of desensitization can be divided into three categories, specific, nonspecific and large aggregate-dependent. Specific desensitization leads only to desensitization of the cell to the antigen used in the calcium-free phase. Nonspecific desensitization refers to the observation that the response of the cell to a noncross-reacting antigen is also inhibited by the desensitization phase of the experiment. For example, in a donor with ragweed and rye grass specific IgE, complete desensitization with ragweed antigen in the calcium-free phase of the experiment would not only ablate the response of the cell to ragweed antigen when calcium was returned to the buffer, but also partially or completely ablate the response to rye grass antigen. The response of the cell to other non-IgE-dependent stimuli is unaffected. If IgE aggregates are dissociated and antigen washed away, the cell appears to be able to recover, over 24 hours, from nonspecific desensitization but does not recover from specific desensitization.[63] In addition, the amount of nonspecific desensitization appears to depend on the number of crosslinks established by the desensitizing antigen, suggesting that there is downregulation of a shared intermediate in the activation cascade.[64] These observations have implications for future modeling of these signal transduction events (see below). The final form of desensitization is not necessarily the consequence of an active process regulating secretion but results from receptor aggregates that are too large. It appears that a very large FcεRI aggregate is no better or worse at inducing signals than smaller aggregates.[63,65] Under these conditions, the slow accumulation of large aggregates would result in a kind of desensitization. It is this behavior of large aggregates that may explain different optimums of histamine and IL-4 release in basophils challenged with polyclonal anti-IgE antibody.

As will be discussed further below, among the general population, maximal IgE-mediated histamine release varies from 0-100%. This ability to respond to IgE-dependent stimulation has been called releasability and is a term generally restricted to IgE-mediated histamine release. There are many possible explanations for this phenomenon; one possibility is that cells which release poorly have overactive desensitization processes. The *rate* that a donor's basophils desensitize using polyclonal anti-IgE antibody has been shown to be inversely correlated with the maximal histamine release. Since the initial rate of histamine release does not differ among the donor's basophils, the implication is that a rapid desensitization event would stop secretion prematurely while a slow desensitization process would allow release to progress to nearly 100%. Since the ability of the basophil or mast cell to secrete mediators probably plays some role in the severity of an individuals allergic disease, this regulation by desensitization may have some clinical relevance.

With the recognition that basophils secrete IL-4 (and possibly other cytokines) long after histamine and leukotriene release are complete, it became necessary to reevaluate the nature of desensitization. When only histamine release was considered, the operative hypothesis was that desensitization events shut off the activation cascade. In light of today's knowledge that nonreceptor tyrosine kinases like SYK mediate the very early signals emanating from the aggregated FcεRI, one might postulate that desensitization processes downregulated these early tyrosine kinases (e.g., CSK might induce phosphorylation of *lyn*, see below). However, the release of IL-4 is also dependent on FcεRI aggregation and it has been observed in both murine cell lines and human basophils that these aggregates must be maintained throughout the period of cytokine release. If crosslinks are dissociated after histamine release is complete but before the onset of or during cytokine release (e.g., at the 30 minute time point in basophils), IL-4 release is stopped. The implication is that some signaling from the FcεRI is required throughout IL-4 secretion and that whatever desensitization events stopping histamine and leukotriene release don't affect IL-4 release. This viewpoint does not mesh well with desensitization modifying early signal transduction biochemistries and yet it is difficult to visualize specific desensitization not affecting some event proximal to the FcεRI receptor. One possibility is that there are multiple branch points in signal transduction that occur at the level of the receptor. For example, there are multiple

ARAMs on both the β and γ FcεRI subunits they may mediate the activation of several different associated tyrosine kinases (or other functional molecules),[66] and desensitization events associated with the cessation of histamine release only modify a subset of these interacting molecules, not the receptor itself. This may represent a form of nonspecific desensitization.

PMA has been known for some time to actively suppress the $[Ca^{2+}]_I$ response in mast cells of various types.[67] This also occurs in human basophils[20] and to a lesser extent in human lung mast cells. As will be discussed in the next paragraph, desensitization could result from a downregulation of the $[Ca^{2+}]_I$ response, so one working hypothesis has been that PKC mediates desensitization. In RBL-2H3 cell studies, the Ca^{2+}-independent isozymes of PKC are translocated during stimulation in the absence of extracellular calcium.[48] Furthermore, α and ε isozymes appear capable of downregulating the calcium response by inhibiting the activation PLC-γ.[68] In a calcium-free medium this would leave the ε isozyme able to modify PLC-γ activation. In human mast cells, a short preincubation with PMA will partially inhibit (30-50%) the FcεRI-mediated $[Ca^{2+}]_I$ response. Long-term treatment with PMA produces a cell that is still capable of generating a $[Ca^{2+}]_I$ response and degranulating after stimulation with anti-IgE antibody but a further short term treatment with PMA results in no further inhibition of the $[Ca^{2+}]_I$ response. Using the RBL-2H3 data as a guide, we might speculate that α isozyme activation is responsible for the short-term PMA effect in untreated cells and that the long-term treatment downregulates the α isozyme so that this effect is not observable. It is difficult to reconcile this view with our observations that basophils have little α isozyme because in basophils, in contrast to mast cells, acute PMA treatment essentially eliminates the $[Ca^{2+}]_I$ response. This suggests that other PKC isozymes are important to this PMA effect or that the small amount of α present is sufficient. However, an interesting aspect to these long-term PMA treatments is that the decay of $[Ca^{2+}]_I$ to resting levels after FcεRI-mediated stimulation is the same in treated and untreated cells. In other words, the putative presence or absence of α isozymes is not relevant to the normal process of $[Ca^{2+}]_I$ regulation during secretion.

The differential desensitization of IL-4 and histamine release carries over to the $[Ca^{2+}]_I$ response. It was noted several ago that $[Ca^{2+}]_I$ elevations returned to resting levels far slower than expected, if downregulation of $[Ca^{2+}]_I$ response was the explanation for the cessation of histamine release.[69] In mast cells, histamine release is typically complete within 5 minutes while the $[Ca^{2+}]_I$ response has only reached its peak by this point. The decay to resting levels has a $T_{1/2}$ of approximately 11 minutes. The relationships in basophils are similar, histamine release (at optimal anti-IgE antibody concentrations) is typically complete within 10 minutes while the $T_{1/2}$ for return of $[Ca^{2+}]_I$ to resting levels is greater than 2 hours (and indeterminate is some cases because $[Ca^{2+}]_I$ can be studied for limited periods under the microscope and the calcium curve shows no signs of decreasing within a typical 30 minute observation). Under appropriate conditions, adding monovalent hapten to such a reaction, after histamine release is complete, still causes the $[Ca^{2+}]_I$ to immediately return to resting levels. The conclusion has been that histamine release is not downregulated by a downregulation of the $[Ca^{2+}]_I$ signal. A similar conclusion is reached when the data are looked in another way. The rate in which the $[Ca^{2+}]_I$ response is desensitized (in a desensitization-style experiment) is uncorrelated with the rate of desensitization measured by the extent of histamine release and is frequently much slower. For example, following a 5 minute incubation of mast cells with anti-IgE antibody in the absence of extracellular calcium, histamine release is 1/2 the control response (after the addition of calcium) while the $[Ca^{2+}]_I$ response can be the same as nondesensitized cells.[69] These data fit several other observations. The first is that

IL-4 release appears to require a sustained $[Ca^{2+}]_I$ signal and the observed $[Ca^{2+}]_I$ signal may serve to allow cytokine release despite the fact that the histamine release is complete. Second, desensitization measured by the inhibition of histamine release is not the result of receptor loss because these receptors still signal both $[Ca^{2+}]_I$ elevations and ultimately IL-4 release. The conclusions drawn from these experiments is that desensitization is a complex set of events, probably involving both homeostatic control processes as well as specific alterations in signal transduction cascades that differ for each of the types of secreted mediators.

Cytokine Modulation of Release

There have been several important insights concerning immediate hypersensitivity diseases in the last decade but one which appears to be critically important is the modulation of infiltrating leukocytes by secreted cytokines. Generally speaking, eosinophils and basophils share similar sensitivities to cytokines such as GM-CSF, IL-5 or IL-3 and these cytokines have significant effects on cell function and longevity.[70-74] Of the list of cytokines which alter basophil function, IL-3 appears to be the most efficacious upregulator of function. Both maximal histamine release and cellular sensitivity (number of crosslinks for a half-maximal response) are increased 2- to 10-fold. Histamine, LTC4 and IL-4 release are all increased 2- to 10-fold. Upregulation is not specific for FcεRI-mediated release, the response to all known stimuli is enhanced as is the release of all known mediators enhanced. This should be an important clue as to how IL-3 induces its changes. The signal transduction cascades differ for each of the stimuli, as evidenced in both pharmacological and biochemical studies, and the signal transduction cascades differ for each of the mediators, so that IL-3 must induce a change(s) that is fundamental to all secretory function. Another clue is that upregulation appears biphasic; a short preincubation of 3-5 minutes induces some changes while an 18-24 hour incubation induces additional changes. However, we have not discovered a signal transduction modification which explains these observations. There is no apparent change in ATP levels for example. IL-3 receptors are now thought to function through associated tyrosine kinases of the Jak family and to regulate DNA transcription through Stat proteins.[75,76] Surprisingly, the early upregulation (5-15 minutes treatment) induced by IL-3 is not sensitive to staurosporin, which is generally a potent and nonspecific tyrosine kinase inhibitor. Herbimycin is also ineffectual during the early upregulation.

A $[Ca^{2+}]_I$ response, although varied in character among the stimuli which induce mediator release in basophils, is a common theme of activation. Following a short preincubation with IL-3, there is no upregulation of the stimulated $[Ca^{2+}]_I$ elevation for either FMLP, anti-IgE antibody or C5a stimulation.[46] Despite an absence of any enhancement of the $[Ca^{2+}]_I$ response, histamine and LTC_4 release were enhanced. Following a 24 hour incubation with IL-3 there is a marked enhancement of the $[Ca^{2+}]_I$ response following any of these three stimuli. This is one of several lines of study that suggest a biphasic nature to IL-3 effect. In the previous study of $[Ca^{2+}]_I$ responses following short incubations we also found that IL-3 substantially upregulated the stimulated generation of arachidonic acid. We now believe that AA generation results from more than one enzymatic process. Current evidence indicates that AA generated through the action of a 14 kDa PLA_2 (so-called secretory PLA_2) is not generally observable but is efficiently coupled to 5-lipoxygenase for the generation of LTC_4.[77] The AA that is normally observed during stimulation comes from some other source, possibly a cytosolic PLA_2 or diglyceride lipase. Since IL-3 upregulates both LTC_4 release and observable free AA, we would have to conclude that the modification also jointly affects both lipase pathways. Finally, IL-3 also upregulates PMA induced release. This stimulus induces neither $[Ca^{2+}]_I$ responses nor AA generation, so the enhancement by IL-3 indicates once again that the change involves an alteration that doesn't necessarily involve the $[Ca^{2+}]_i$ response.

Releasability

One intriguing aspect of studying human basophil histamine release is the observation that there is wide variability in the responsiveness of basophils to FcεRI-mediated stimulation. A given donor's basophil response is relatively stable over a significant period of time. For example, if the maximum FcεRI-mediated histamine release that can be evoked is 50%, then this value will only vary ± 20% over several months or even years of testing. The underlying mechanisms that result in a particular level of response are not clear.[78] There may be technical or cell handling artifacts that generate this observation but tests using cells stimulated in whole blood indicate that the releasability of given donor's basophils is similar in and out of whole blood. As noted above, the rate that a cell desensitizes is inversely related to the maximum histamine release obtained but this analysis is not ideal because the only way to assess desensitization is to measure the amount of release after the desensitization phase of the experiment. Therefore the need to include a release phase may confound a clear interpretation of what is happening in the desensitization phase of the experiment.

Releasability can also be defined by the sensitivity of the basophil to stimulation, i.e., how many crosslinks are required to produce 50% of the donors' maximum response. We have found that there is a reasonable range of sensitivity among donors, with some requiring only 300-500 antigen specific IgE molecules for 50% of their maximum response and others requiring 20,000.[15] Maximal histamine release and basophil sensitivity do not correlate and appear to be independently regulated.[16] There are several other ways of viewing "releasability" and there are a multitude of possible explanations for the phenomenon and it may ultimately require several to explain all the observations. In our studies of this problem, we have chosen to focus attention on a previously unstudied subset of donors, the so-called nonreleasers.[79] A small group of donors, 15-20% of the nonatopic population, have basophils that don't respond to FcεRI-mediated stimulation. We have been strict in this definition: the cells must release less than two standard deviations of the noise in the replicates of spontaneous release (the amount of histamine release from unstimulated cells). Spontaneous release is typically 1-5% of the total histamine content and noise is typically ± 1%, so a nonreleaser shows less than 2% histamine release. This must happen in multiple tests over a several month period. The reason for these criteria is to eliminate individuals who release poorly or cycle through low levels of release. Enhancing agents (deuterium oxide, cytochalasin B, Sr^{++} or IL-3) markedly enhance poor releasers but have no apparent effect on nonreleasers. However, it should be pointed out that nonreleaser's basophils do respond to non-FcεRI mediated stimuli, such as FMLP or C5a. For FcεRI-dependent stimuli, nonreleasers don't release histamine, LTC_4 or IL-4. These donors do have typical densities of endogenous IgE,[79] typical FcεRI densities, and express mRNA for the α and β subunits of FcεRI.[80] Indeed, it is possible to observe normal binding of anti-IgE antibody and a typical effect of anti-IgE antibody on the basophil's responsiveness to the phorbol ester, PMA.[79] It was demonstrated in the early 1980s that preincubation of basophils of releasers with anti-IgE antibody in the absence of extracellular calcium would increase the sensitivity of the cells to PMA (≈5-fold).[81] Since PMA could cause release without extracellular calcium, this effect could be ascertained at nearly any time after addition of the anti-IgE antibody[81,82] without concern for IgE-mediated release. It was later shown that the same effect occurred in nonreleasing basophils.[79] It was also found that stimulating nonreleaser cells with anti-IgE inhibited their subsequent response to FMLP.[83] Interestingly, while this was not normally observed in releaser basophils, a pretreatment with wortmannin, which can completely inhibit FcεRI-mediated release, would reveal this IgE-dependent inhibition of FMLP in releasing basophils. It is now believed that wortmannin specifically inhibits PI3-kinase. If the PI3 kinase pathway is a critical step in FcεRI-mediated histamine release, its selective inhibition may leave intact a pathway that both inhibits FMLP and enhances PMA-induced release. Whatever the mechanistic explanation of these joint observations, the implication is that some aspect of the FcεRI-mediated signal transduction cascade is still activated in nonreleaser's basophils. If so, then the deficiency of these cells may be a critical component of secretion (or have an overactive inhibitory pathway) and such a deficiency may result from the absence of a component or its poor function.

There are several other observations relevant to this nonreleaser problem. As in T cells, it appears that CD45 is important to FcεRI-mediated release (see chapter 7).[84,85] This phosphatase appears to regulate the phosphorylation state of receptor associated src kinases. The inhibition of CD45 function would allow phosphorylation of src kinases such as lck (T cells) or lyn (mast cells) which inhibits their function. Hook and Siraganian developed anti-CD45 antibodies that inhibit FcεRI-mediated release from human basophils.[86] In a collaborative study we found that these same monoclonals inhibit FcεRI-mediated $[Ca^{2+}]_i$ changes but not FMLP or its associated $[Ca^{2+}]_i$ changes. We subsequently examined CD45 expression in releasing and nonreleasing basophils and found no differences. Unless there is an alteration in the function of CD45 in nonreleaser's basophils, this does not appear to be an explanation.

Several studies have noted that FcεRI-dependent stimulation of nonreleaser, basophils doesn't lead to meaningful $[Ca^{2+}]_i$ responses. Our own studies do indicate that a smaller than usual initial $[Ca^{2+}]_i$ transient is observable in nonreleasing cells but that there is essentially no second phase. The exact implications of this observation are not entirely clear. The initial phase of the $[Ca^{2+}]_i$ response is responsive to IP_3 but our studies indicate that this $[Ca^{2+}]_i$ transient is relatively insensitive to decreases in the strength of the stimulus (presumably because IP3 levels normally exceed the EC_{50} of the IP3 receptor). The time lag (see above) in individual cells is a function of the strength of the stimulus. This is also true when crosslinks are established rapidly but the stimulus is weak, as occurs with bivalent hapten stimulation. In nonreleasers, the time lag following anti-IgE addition is three times longer than is typical for releasers. These observations together suggest that there are weak activation events in nonreleaser basophils. Whether a weak initial activation fails to initiate meaningful calcium influx or whether there is another deficiency that explains this particularly weak response is also not clear but we have found that the PKC activation kinetics in nonreleasers also suggests a deficiency in calcium influx. In releasers, in the presence of extracellular calcium, FcεRI-mediated membrane PKC activity is transient, returning to resting levels within 30 minutes. However, in the absence of extracellular calcium, peak PKC activity is somewhat less than in calcium containing buffers but remains at a plateau for at least 30 minutes. In nonreleasers, PKC activity (which is much weaker than a releaser's response) also remains at a plateau for at least 30 minutes, suggesting that there is no calcium influx.

These data suggest that deficiency of the nonreleasers response is some early signal transduction activity. Some effort has also been applied to finding a reason for this phenotype. It could represent an phenotype intrinsic to the basophil itself or it could result from extrinsic factors such as the level of cytokine expression. In this context, we have examined whether IL-3 changes the nonreleaser phenotype. As noted above, a 24 hour incubation with IL-3 has little effect on mediator release in nonreleasers. This is in contrast to poor releasers where 10-fold changes in histamine and LTC_4 release are typically observed. We have also noted that IL-3 does upregulate the nonreleasers basophil response to FMLP, so these basophils do express functional IL-3 receptors.[79] In a recent meeting report, long-term treatment with IL-3 was found to convert nonreleasers to releasers.[87] In this study, 24 hours was also not sufficient to observe a change but 4-7 days of culture induced a progressive increase in responsiveness. Thus, the endogenous expression of IL-3 in nonreleaser donors might contribute to this particular basophil phenotype or IL-3 is simply able to overcome the absence of some other intrinsic or extrinsic factor necessary for the basophil to fully express an FcεRI-mediated response.

EARLY SIGNAL TRANSDUCTION EVENTS

The last 4 years have seen an explosion of information concerning the immediate post-receptor aggregation events occurring during FcεRI-dependent activation. Nearly all these studies have been made using RBL-2H3 cells and a translation to human basophils and mast cells has awaited adequate tools and a refinement of methods. A description of these events is provided in other chapters, so it is useful to speculate on how this new information may aid in our understanding of human basophil or mast cell function. It will also be useful to provide some pharmacological data that bear on these issues.

The role of nonreceptor tyrosine kinases in FcεRI signal transduction is now well established[66,88-95] and preliminary studies from Warner's laboratory indicate that human basophils and mast cells express both *lyn* and Syk, as determined by Western blots.[96] We have also found moderate levels of mRNA expression for these two proteins. The involvement of tyrosine kinases early in the signal transduction cascade helped to explain an early pharmacological experiment. Several years ago, staurosporin was used as a PKC specific inhibitor. However, in human basophils, staurosporin inhibited the FcεRI-mediated (but not FMLP) $[Ca^{2+}]_i$ response with the same IC_{50} as its inhibition of histamine release. The acknowledgment that staurosporin is also an equally potent inhibitor of many tyrosine kinases and the need for *lyn* or SYK activation in FcεRI-dependent release provided a re-interpretation of the $[Ca^{2+}]_i$ data. We have recently found a similar problem with the second generation PKC inhibitor, Go-6976, which is supposed to be specific for α and β isozymes. This inhibitor effectively inhibits the FcεRI-mediated $[Ca^{2+}]_i$ response and histamine release at concentrations lower than required to inhibit PMA-induced release. Therefore, we would guess that this inhibitor inhibits tyrosine kinases. A number of tyrosine kinase inhibitors have been examined for their effects on human basophil or mast cell release and some interesting contrasts have appeared. For example, methyl 2,5 dihydroxy cinnimate inhibits mast cell release while having no effect on basophil release while the opposite situation occurs for genistein.[97] Staurosporin only inhibits FcεRI-mediated human lung mast cell histamine release at concentrations (>100 μM) where it is reported to lose its specificity for PKC or tyrosine kinases.[49] Surprisingly, skin mast cells have a sensitivity to staurosporin that is similar to basophils (it is not known whether there is differential metabolism of staurosporin in these different cell preparations).[49] These unexpected differences between basophils and mast cells may indicate that there are some subtle differences in the tyrosine kinases involved in FcεRI-mediated signaling or they may result from effects of these agents on other aspects of the cell's signaling. The pharmacological data are not in themselves useful for reconciling these different interpretations.

Interestingly, in basophils, genistein significantly lengthens the time lag (the interval between the addition of stimulus and the first $[Ca^{2+}]_i$ upswing measured in individual cells, see above) while staurosporin does so weakly at best. If Syk activity were required for PLC-γ activation, we might expect a lengthening of the time lag, which is presumably dependent on the rate of IP_3 generation. The discrepancy between staurosporin and genistein in affecting this parameter is therefore unexpected. Lavendustin A does not inhibit the basophil response and a variety of tyrphostins have been examined but only B46 appears to cause inhibition of FcεRI-mediated release and only after long incubations (18+ hours). Herbimycin A has the same characteristic. It has been noted in other cells that herbimycin A may actually lead to the downregulation of src kinases, which may explain this time dependency of its effects on basophils.

Other kinases and phosphatases have been recently implicated in the functioning or modulation of aggregation-dependent receptors. For example, the C-terminal src kinase (Csk) has been demonstrated to phosphorylate src kinases in T cells, downregulating lck function and therefore downregulating the T cell response.[98,99] It is conceivable that Csk provides a potential means of downregulating *lyn* kinase in FcεRI-dependent activation (we have initial studies that indicate that Csk is operative in RBL-2H3 cells). The phosphatase PTP1c has also been shown to mediate the sensitivity of B cells to antigen activation presumably by dephosphorylating ARAMs on the B cell IgM receptor.[100,101] Preliminary studies indicate that human basophils have high levels of PTP1c expression (compared to the lymphocytes/monocytes that typically accompany these preparations). Using these various pieces of information, it is possible to provide some new hypotheses regarding either the releasability problem or desensitization. For example, releasability may be a function of *lyn*, Syk, CSK or PTP1c expression. Nonreleasers may express low levels of *lyn* or Syk although very preliminary studies of mRNA expression for these two kinases don't support this view. But some early modification in the function of nonreceptor tyrosine kinases by CSK or PTP1c might explain the $[Ca^{2+}]_i$ observations in nonreleaser basophils. Some form of desensitization might also be explained by activity of CSK or PTP1c. These are areas of future study.

References

1. Maurer D, Fiebiger E, Reininger B et al. Expression of functional high affinity immunoglobulin E receptors (FcεRI) on monocytes of atopici individuals. J Exp Med 1994; 179:745-750.
2. Wang B, Rieger A, Kilgus O et al. Epidermal Langerhans cells from normal human skin bind monomeric IgE via Fc epsilon RI. J Exp Med 1992; 175:1353-65.
3. Gounni AS, Lamkhioued B, Ochiai et al. High-affinity IgE receptor on eosinophils is involved in defence against parasites. Nature 1994; 367:183-6.
4. Liu MC, Hubbard WC, Proud D et al. Immediate and late inflammatory responses to ragweed antigen challenge of the peripheral airways in allergic asthmatics. Cellular, mediator and permeability changes. Am Rev Respir Dis 1991; 144:51-8.
5. Wardlaw AJ, Dunette S, Gleich GJ et al. Eosinophils and mast cells in broncheoalveilar lavage in mild asthma: relationship to bronchial hyperreactivity. Am Rev Resp Dis 1988; 137:62-69.
6. Azzawi M, Bradley B, Jeffery PK et al. Identification of activated T-lymphocytes and eosinophils in bronchial biopsies in stable atopic asthmatics. Am Rev Resp Dis 1990; 142:1407-1413.
7. Bochner BS, Schleimer RP. The role of adhesion molecules in human eosinophil and basophil recruitment. J Allergy Clin Immunol 1994; 94:427-438.
8. Iademarco MF, Barks JL, Dean DC. Regulation of vascular cell adhesion molecule-1 expresssion by IL-4 and TNF-alpha in cultures of endothelial cells. J Clin Invest 1995; 95:264-271.
9. Bascom R, Wachs M, Naclerio RM et al. Basophil influx occurs after nasal antigen challenge: effects of topical corticosteroid pretreatment. J Allergy Clin Immunol 1988; 81:580-9.
10. Iliopoulos O, Baroody F, Naclerio RM et al. Histamine containing cells obtained from the nose hours after antigen challenge have

functions and phenotypic characteristics of basophils. J Immunol 1992; 148: 2223-2228.
11. Guo CB, Liu MC, Galli SJ et al. Identification of IgE bearing cells in the late response to antigen in the lung as basophils. Am J Resp Cell Mol Biol 1993; 10:384-390.
12. Fewtrell C, Metzger H. Larger oligomers of IgE are more effective than dimers in stimulating rat basophilic leukemia cells. J Immunol 1980; 125:701-10.
13. Segal DM, Taurog JD, Metzger H. Dimeric immunoglobulin E serves as a unit signal for mast cell degranulation. Proc Natl Acad Sci USA 1977; 74:2993-7.
14. MacGlashan DW, Schleimer RP, Lichtenstein LM. Qualitative differences between dimeric and trimeric stimulation of human basophils. J Immunol 1983; 130:4-6.
15. MacGlashan DW, Peters SP, Warner J et al. Characteristics of human basophil sulfidopeptide leukotriene release: releasability defined as the ability of the basophil to respond to dimeric crosslinks. J Immunol 1986; 136:2231-9.
16. MacGlashan DW. Releasability of human basophils: Cellular sensitivity and maximal histamine release are independent variables. J Allergy Clin Immunol 1993; 91:605-615.
17. Kagey SA, MacGlashan DW, Lichtenstein LM. Role of receptor aggregation in triggering IgE-mediated reactions. Fed Proc 1982; 41:12-6.
18. Malveaux FJ, Conroy MC, Adkinson NFJ et al. IgE receptors on human basophils. Relationship to serum IgE concentration. J Clin Invest 1978; 62:176-81.
19. Warner JA, MacGlashan DW. Protein kinase C (PKC) changes in human basophils. IgE-mediated activation is accompanied by an increase in total PKC activity. J Immunol 1989; 142:1669-77.
20. Warner JA, MacGlashan DW. Signal transduction events in human basophils—A comparative study of the role of protein kinase-C in basophils activated by anti-IgE antibody and formyl-methionyl-leucyl-phenylalanine. J Immunol 1990; 145: 1897-1905.
21. MacGlashan Jr DW. Single-cell analysis of Ca^{2+} changes in human lung mast cells: graded vs. all-or-nothing elevations after IgE-mediated stimulation. J Cell Biol 1989; 109:123-34.
22. Warner JA, Peters SP, Lichtenstein LM et al. Differential release of mediators from human basophils: differences in arachidonic acid metabolism following activation by unrelated stimuli. J Leukoc Biol 1989; 45:558-71.
23. Warner JA, Freeland HS, MacGlashan Jr DW et al. Purified human basophils do not generate LTB4. Biochem Pharmacol 1987; 36:3195-9.
24. MacGlashan Jr DW, Schleimer RP, Peters SP et al. Generation of leukotrienes by purified human lung mast cells. J Clin Invest 1982; 70:747-51.
25. Peters SP, MacGlashan Jr DW, Schulman ES et al. Arachidonic acid metabolism in purified human lung mast cells. J Immunol 1984; 132:1972-9.
26. Lewis RA, Holgate ST, Roberts LJD et al. Preferential generation of prostaglandin D2 by rat and human mast cells. Kroc Found Ser 1981; 14:239-54.
27. Lawrence ID, Warner JA, Cohan VL et al. Purification and characterization of human skin mast cells. Evidence for human mast cell heterogeneity. J Immunol 1987; 139:3062-9.
28. Benyon RC, Robinson C, Church MK. Differential release of histamine and eicosanoids from human skin mast cells activated by IgE-dependent and nonimmunological stimuli. Br J Pharmacol 1989; 97:898-904.
29. Brunner T, Heusser CH, Dahinden CA. Human peripheral blood basophils primed by interleukin-3 (IL-3) produce IL-4 in response to immunoglobulin E receptor stimulation. J Exp Med 1993; 177:605-611.
30. MacGlashan Jr DW, White JM, Huang SK et al. Secretion of interleukin-4 from human basophils: The relationship between IL-4 mRNA and protein in resting and stimulated basophils. J Immunol 1994; 152:3006-3016.
31. Ochensberger B, Rihs S, Brunner T et al. IgE-independent interleukin-4 expression and induction of a late phase of leukotriene

C4 formation in human blood basophils. Blood 1995; 86:4039-4049.

32. Bradding P, Feather IH, Howarth PH et al. Interleukin 4 is localized to and released by human mast cells. J Exp Med 1992; 176:1381-1386.
33. Jaffe JS, Glaum MC, Raible DG et al. Human lung mast cell IL-4 gene and protein expression: Temporal analysis of upregulation following IgE-mediated activation. Am J Respir Cell Mol Biol 1995; 13:665-675.
34. Okayama Y, Semper A, Holgate ST et al. Multiple cytokine mRNA expression in human mast cells stimulated via Fc epsilon RI. International Archives of Allergy & Immunology 1995; 107:158-9.
35. Walsh LJ, Trinchieri G, Waldorf HA et al. Human dermal mast cells contain and release tumor necrosis factor alpha, which induces endothelial leukocyte adhesion molecule 1. Proc Natl Acad Sci USA 1991; 88:4220-4.
36. Paul WE, Seder RA, Plaut M. Lymphokine and cytokine production by FceRI positive cells. Adv in Immunol 1993; 53:1-29.
37. Schleimer RP, Gillespie E, Lichtenstein LM. Release of histamine from human leukocytes stimulated with the tumor promoting phorbol esters. I. Characterization of the response. J Immunol 1981; 126:570-574.
38. MacGlashan Jr DW, Warner JA. Stimulus-dependent leukotriene release from human basophils: A comparative study of C5a and Fmet-leu-phe. J Leukoc Biol 1991; 49:29-40.
39. MacGlashan Jr DW, Botana L. Biphasic Ca^{2+} responses in human basophils: Evidence that the initial transient elevation associated with mobilization of intracellular calcium is an insufficient signal for degranulation. J Immunol 1993; 150:980-991.
40. Botana LM, MacGlashan DW. Fmet-leu-phe induced degranulation of human basophils can occur in the absence of increases in cytosolic free calcium. J Immunol 1993; 150:221A
41. MacGlashan Jr DW, Bochner B, Warner JA. Graded changes in the response of individual human basophils to stimulation: Distributional behavior of early activation events. J Leukoc Biol 1994; 55:13-23.
42. Dvorak AM, Warner JA, Morgan E et al. An ultrastructural analysis of tumor-promoting phorbol diester-induced degranulation in human basophils. Am J Pathology 1992; 141:1309-1322.
43. Dvorak AM, Warner JA, Kissell S et al. F-met peptide-induced degranulation of human basophils. Lab Invest 1991; 64: 234-253.
44. Beaven MA, Guthrie DF, Moore JP et al. Synergistic signals in the mechanism of antigen-induced exocytosis in 2H3 cells: evidence for an unidentified signal required for histamine release. J Cell Biol 1987; 105:1129-36.
45. Rouzer CA, Samuelsson B. Demonstration of multiple regulatory factors for purified human leukocyte 5-lipoxygenase. Adv Enzyme Regul 1987; 26:133-44.
46. MacGlashan Jr DW, Hubbard WC. Interleukin-3 alters free arachidonic acid generation in C5a-stimulated human basophils. J Immunol 1993; 151:6358-6369.
47. Schroeder JT, MacGlashan Jr DW, Kagey-Sobotka A et al. The IgE-dependent IL-4 secretion by human basophils: The relationship between cytokine production and histamine release in mixed leukocyte cultures. J Immunol 1994; 153:1808-1817.
48. Ozawa K, Szallasi Z, Kanzanietz MG et al. Ca^{2+}-dependent and Ca^{2+}-independent isozymes of protein kinase C mediate exocytosis in antigen-stimulated rat basophilic RBL-2H3 cells: Reconstitution of secretory responses with Ca^{2+} and purified siozymes in washed permeabilized cells. J Biol Chem 1993; 268:1749-1756.
49. Massey WA, Cohen VL, MacGlashan Jr DW et al. Protein Kinase C modulates IgE-mediated activation of human mast cells from lung and skin. I. Pharmacologic inhibition. J Pharm Exp Ther 1991; 258:824-829.
50. Casolaro V, Georas S, Song Z et al. Inhibition of NFATp-dependent transcription by NFkB. Implications for differential gene expression in T cell subsets. Proc Natl Acad Sci USA 1995; 92:11623-11626.
51. Neher E, Almers W. Fast calcium transients in rat peritoneal mast cells are not sufficient to trigger exocytosis. Embo J 1986; 5:51-3.

52. Millard PJ, Gross D, Webb WW et al. Imaging asynchronous changes in intracellular Ca^{2+} in individual stimulated tumor mast cells. Proc Natl Acad Sci USA 1988; 85:1854-8.
53. Millard PJ, Ryan TA, Webb WW et al. Immunoglobulin E receptor crosslinking induces oscillations in intracellular free ionized calcium in individual tumor mast cells. J Biol Chem 1989; 264:19730-9.
54. MacGlashan Jr DW, Guo CB. Oscillations in free cytosolic calcium during IgE-mediated stimulation distinguish human basophils from human mast cells. J Immunol 1991; 147:2259-2269.
55. De Young GW, Keizer J. A single-pool inositol 1,4,5-trisphosphate-receptor-based model for agonist-stimulated oscillations in Ca^{2+} concentration. Proc Natl Acad Sci USA 1992; 89:9895-9.
56. Columbo M, Botana LM, Horowitz EM et al. Studies of the intracellular Ca^{2+} levels in human adult skin mast cells activated by the ligand for human c-kit receptor, rhSCF and anti-IgE antibody. Biochem Pharm 1994; 47:2137-2145.
57. MacGlashan Jr DW. Graded changes in the response of individual human basophils to stimulation: Distributional behavior of events temporally coincident with degranulation. J Leukoc Biol 1995; 58: 177-188.
58. Pruzansky JJ, Zeiss CR, Patterson R. A linear correlation between histamine release and degranulation of human basophils by specific antigen or the ionophore A23187. Immunology 1980; 40:411-416.
59. Knol EF, Mul FPJ, Jansen H et al. Monitoring human basophil activation via CD63 monoclonal antibody 435. J Allergy Clin Immunol 1991; 88:328-338.
60. Lichtenstein LM, De BR. IgE mediated histamine release: in vitro separation into two phases. Int Arch Allergy Appl Immunol 1971; 41:56-71.
61. MacGlashan Jr DW, Lichtenstein LM. Basic characteristics of human lung mast cell desensitization. J Immunol 1987; 139: 501-5.
62. Kazimierczak W, Meier HL, MacGlashan Jr DW et al. An antigen-activated DFP-inhibitable enzyme controls basophil desensitization. J Immunol 1984; 132:399-405.
63. MacGlashan Jr DW, Mogowski M, Lichtenstein LM. Studies of antigen binding on human basophils. II. Continued expression of antigen-specific IgE during antigen-induced desensitization. J Immunol 1983; 130:2337-42.
64. MacGlashan Jr DW, Lichtenstein LM. The transition from specific to nonspecific desensitization in human basophils. J Immunol 1981; 127:2410-4.
65. Ishizaka T, Ishizaka K. Cell surface IgE on human basophil granulocytes. Ann NY Acad Sci 1975; 254:462-75.
66. Jouvin MH, Adamczewski M, Numerof R et al. Differential control of the tyrosine kinase lyn and syk by the two signaling chains of the high affinity immunoglobulin E receptor. J Biol Chem 1994; 269: 5918-5925.
67. Sagi ER, Lieman H, Pecht I. Protein kinase C regulation of the receptor-coupled calcium signal in histamine-secreting rat basophilic leukaemia cells. Nature 1985; 313:59-60.
68. Ozawa K, Yamada K, Kanzanietz MG et al. Different isozymes of protein kinase C mediate feedback inhibition of phospholipase C and stimulatory signals for exocytosis in rat RBL-2H3 cells. J Biol Chem 1993; 268:2280-2283.
69. MacGlashan Jr DW. Single cell analysis of free cytosolic calcium changes in human lung mast cells: II. The relationship between desensitization and the cellular regulation of calcium changes. Mol Immunol 1991; 28:585-597.
70. Schleimer RP, Derse CP, Friedman B et al. Regulation of human basophil mediator release by cytokines. I. Interaction with antiinflammatory steroids. J Immunol 1989; 143:1310-7.
71. Hirai K, Morita Y, Misaki Y et al. Modulation of human basophil histamine release by hematopoetic growth factors. J Immunol 1988; 141:3957-3961.
72. Bischoff SC, Brunner T, De WAL et al. Interleukin 5 modifies histamine release and leukotriene generation by human basophils in response to diverse agonists. J Exp Med 1990; 172:1577-82.

73. Bischoff SC, de Weck AL, Dahinden CA. Interleukin 3 and granulocyte/macrophage-colony-stimulating factor render human basophils responsive to low concentrations of complement component C3a. Proc Natl Acad Sci USA 1990; 87:6813-7.
74. Lamas AM, Marcotte GV, Schleimer RP. Human endothelial cells prolong eosinophil survival: Regulation by cytokines and glucocorticoids. J Immunol 1989; 142: 3978-3984.
75. Pazdrak K, Stafford S, Alam R. The activation of the Jak-STAT 1 signaling pathway by IL-5 in eosinophils. J Immunol 1995; 155:397-402.
76. Ihle JN, Kerr IM. Jaks and Stats in signaling by the cytokine receptor superfamily. Review. Trends in Genetics 1995; 11:69-74.
77. Hundley TR, Marshall L, Hubbard WC et al. Arachidonic acid release for leukotriene C4 synthesis in human basophils is mediated by a secretory phospholipase A2. submitted to J Immunol 1995;
78. Lichtenstein LM, MacGlashan Jr DW. The concept of basophil releasability. J Allergy Clin Immunol 1986; 77:291-4.
79. Nguyen KL, Gillis S, MacGlashan Jr DW. A comparative study of releasing and nonreleasing human basophils: nonreleasing basophils lack an early component of the signal transduction pathway that follows IgE crosslinking. J Allergy Clin Immunol 1990; 85:1020-9.
80. Kuster H, Zhang L, Brini AT et al. The gene and cDNA for the human high affinity immunoglobulin E receptor beta chain and expression of the complete human receptor. J Biol Chem 1992; 267: 12782-12787.
81. Schleimer RP, Gillespie E, Daiuta R et al. Release of histamine from human leukocytes stimulated with the tumor-promoting phorbol diesters. II. Interaction with other stimuli. J Immunol 1982; 128:136-40.
82. Warner J, MacGlashan DW. Persistence of early crosslink-dependent signal transduction events in human basophils after desensitization. Immunol Lett 1988; 18:129-37.
83. Knol EF, Mul FP, Kuijpers TW et al. Intracellular events in anti-IgE nonreleasing human basophils. J Allergy Clin Immunol 1992; 90:92-103.
84. Berger SA, Mak TW, Paige CJ. Leukocyte common antigen (CD45) is required for immunoglobulin E-mediated degranulation of mast cells. J Exp Med 1994; 180: 471-476.
85. Adamczewski M, Numerof RP, Koretzky GA et al. Regulation by CD45 of the tyrosine phosphorylation of high affinity IgE receptor beta- and gamma-chains. J Immunol 1995; 154:3047-55.
86. Hook WA, Berenstein EH, Zinsser FU et al. Monoclonal antibodies to the leukocyte common antigen (CD45) inhibit IgE-mediated histamine release from human basophils. J Immunol 1991; 147:2670-2676.
87. Yamaguchi M, Hirai K, Kitani S et al. Culturing in the presence of IL-3 converts anti-IgE nonresponding basophils into responding basophils. J Allergy Clin Immunol 1994; 93:193.
88. Benhamou M, Gutkind JS, Robbins KC et al. Tyrosine phosphorylation coupled to IgE receptor-mediated signal transduction and histamine release. Proc Natl Acad Sci USA 1990; 87:5327-30.
89. Stephan V, Benhamou M, Gutkind SJ et al. IgE-mediated tyrosine phosphorylation in RBL-2H3 cells: Relationship to GTP-binding protein activation and phosphoinositol (PI) hydrolysis. FASEB J 1991; 5:A1676.
90. Stephan V, Benhamou M, Gutkind JS et al. Fc epsilon RI-induced protein tyrosine phosphorylation of pp72 in rat basophilic leukemia cells (RBL-2H3). Evidence for a novel signal transduction pathway unrelated to G protein activation and phosphatidylinositol hydrolysis. J Biol Chem 1992; 267:5434-41.
91. Hutchcroft JE, Geahlen RL, Deanin GG et al. Fc epsilon RI-mediated tyrosine phosphorylation and activation of the 72 kDa protein-tyrosine kinase, PTK72, in RBL-2H3 rat tumor mast cells. Proc Natl Acad Sci USA 1992; 89:9107-11.
92. Li W, Deanin GG, Margolis B et al. FceRI-mediated tyrosine phosphorylation of multiple proteins including phospholipase C gamma 1 and the receptor beta/gamma2 complex in RBL-2H3 cells. Mol Cell Biol 1992; 12:3176-3182.
93. Eiseman E, Bolen JB. Engagement of the

high-affinity IgE receptor activates src protein-related tyrosine kinases. Nature 1992; 355:78-80.

94. Scharenberg AM, Lin S, Cuenod B et al. Reconstitution of interactions between tyrosine kinases and the high affinity IgE receptor which are controlled by receptor clustering. EMBO J 1995; 14: 3385-94.
95. Hirasawa N, Scharenberg A, Yamamura H et al. A requirement for Syk in the activation of the microtubule-associated protein kinase/phospholipase A2 pathway by Fc epsilon R1 is not shared by a G protein-coupled receptor. J Biol Chem 1995; 270:10960-7.
96. Lavens SE, Warner JA. The role of the tyrosine kinase, SYK, in IgE-mediated signal transduction in human lung mast cells and basophils. FASEB J 1995; 9:A781.
97. Lavens SE, Peachell PT, Warner JA. Role of tyrosine kinases in IgE-mediated signal transduction in human lung mast cells and basophils. Am J Resp Cell Mol Biol 1992; 7:637-44.
98. Chow LM,. Fournel M, Davidson D et al. Negative regulation of T cell receptor signalling by tyrosine protein kinase p50csk. Nature 1993; 365:156-60.
99. Chow LM, Jarvis C, Hu Q et al. Ntk: a Csk-related protein-tyrosine kinase expressed in brain and T lymphocytes. Proc Natl Acad Sci USA 1994; 91:4975-9.
100. Pani G, Kozlowski M, Cambier JC et al. Identification of the tyrosine phosphatase PTP1C as a B cell antigen receptor-associated protein involved in the regulation of B cell signaling see comments. J Exp Med 1995; 181:2077-84.
101. Cyster JG, Goodnow CC. Protein tyrosine phosphatase 1C negatively regulates antigen receptor signaling in B lymphocytes and determines thresholds for negative selection. Immunity 1995; 2:13-24.

INDEX

Items in italics denote figures (f) or tables (t).

A

B

C

D

I

J

K

L

M

N

O

P

Q

R

V

W

Z

MIX
Papier aus verantwortungsvollen Quellen
Paper from responsible sources
FSC® C105338

If you have any concerns about our products,
you can contact us on
ProductSafety@springernature.com

In case Publisher is established outside the EU,
the EU authorized representative is:
Springer Nature Customer Service Center GmbH
Europaplatz 3, 69115 Heidelberg, Germany

Printed by Libri Plureos GmbH
in Hamburg, Germany